Remote Sensing for Water Resources and Environmental Management

Remote Sensing for Water Resources and Environmental Management

Guest Editors

Timothy Dube
Munyaradzi Davis Shekede
Christian Massari

MDPI

Basel • Beijing • Wuhan • Barcelona • Belgrade • Novi Sad • Cluj • Manchester

Guest Editors
Timothy Dube
Earth Sciences
University of the Western Cape
Cape Town
South Africa

Munyaradzi Davis Shekede
Department of Geography
Geospatial Sciences and Earth
Observation
University of Zimbabwe
Harare
Zimbabwe

Christian Massari
National Research Council
Perugia
Italy

Editorial Office
MDPI AG
Grosspeteranlage 5
4052 Basel, Switzerland

This is a reprint of the Special Issue, published open access by the journal *Remote Sensing* (ISSN 2072-4292), freely accessible at: https://www.mdpi.com/journal/remotesensing/special_issues/RS_WREM.

For citation purposes, cite each article independently as indicated on the article page online and as indicated below:

Lastname, A.A.; Lastname, B.B. Article Title. *Journal Name* **Year**, *Volume Number*, Page Range.

ISBN 978-3-7258-5329-8 (Hbk)
ISBN 978-3-7258-5330-4 (PDF)
https://doi.org/10.3390/books978-3-7258-5330-4

Contents

About the Editors

Timothy Dube

Timothy Dube is an academic Professor in the field of Geoinformation Science and Earth Observation, based in the Department of Earth Sciences at the University of the Western Cape, South Africa, where he is also Director for the Institute for Water Studies. He is additionally affiliated with the United Nations University (UNU-FLORES) as an Adjunct Professor and Data Scientist. His research applies satellite remote sensing, GIS, and data-driven approaches to better understand water resources, land-use change, and environmental variability. He has contributed to several regional and international initiatives on water and climate, and he continues to work on projects that link Earth observation technologies with practical water management solutions. He has published in peer-reviewed journals, supervised postgraduate students, and collaborates widely on capacity-building efforts on Earth observation in Africa.

Munyaradzi Davis Shekede

Munyaradzi Davis Shekede is a spatial ecologist and a Senior Lecturer with experience in Geographic Information Science and Earth observation applications in natural resource management. Dr Shekede's research interests are in the applications of earth observation and geoinformation science to understand earth system dynamics, focusing on climate change, socio-ecological systems, and water resources. He has expertise in GIS, earth observation, environmental modeling, natural hazard modeling and assessment, environmental impact assessments, and spatial ecology. He has more than 25 publications in peer-reviewed international journals and three book chapters.

Christian Massari

Christian Massari has extensive experience in hydraulic and hydrological research, focusing on water resource management, flood forecasting, and drought monitoring using land surface models and remote sensing data. He has contributed over 80 peer-reviewed publications. His expertise includes hydrological and land surface water balance modeling, development of land data assimilation systems for integrating remote sensing retrievals, microwave and radar remote sensing of soil moisture and precipitation, drought monitoring and its impacts on food production, hydrological drought development, supply pipe system diagnostics to reduce energy and water waste, flood and drought hazard assessment through remote sensing, and ecohydrology.

Preface

This Reprint focuses on innovative remote sensing and geospatial modeling techniques that enhance monitoring, assessment, and sustainable management of freshwater resources. It supports SDG 6 by addressing water scarcity, quality, and environmental challenges across diverse spatial scales worldwide.

Timothy Dube, Munyaradzi Davis Shekede, and Christian Massari
Guest Editors

remote sensing

MDPI

Editorial

Remote Sensing for Water Resources and Environmental Management

Timothy Dube [1,*], Munyaradzi D. Shekede [2] and Christian Massari [3]

1 Institute for Water Studies, Department of Earth Science, University of the Western Cape, Private Bag X17, Bellville, Cape Town 7535, South Africa
2 Department of Geography Geospatial Sciences and Earth Observation (GGEO), University of Zimbabwe, P. Bag MP 167, Mt Pleasant, Harare, Zimbabwe
3 National Research Council, Via Madonna Alta 126, 06128 Perugia, Italy
* Correspondence: tidube@uwc.ac.za ; Tel.: +27-21-959-4130

Abstract: In line with the United Nations Sustainable Development Goal (SDG) 6, the main goal of the Special Issue on "Remote sensing for water resources and environmental management" was to solicit papers from a diverse range of scientists around the world on the use of cutting-edge remote sensing technologies to assess and monitor freshwater quality, quantity, availability, and management to ensure water security. Special consideration was given to scientific manuscripts that covered, but were not limited to, the development of geospatial techniques and remote sensing applications for detecting, quantifying, and monitoring freshwater water resources, identifying potential threats to water resources and agriculture, as well as other themes related to water resources and environmental management at various spatial scales. The Special Issue attracted over thirteen peer-reviewed scientific articles, with the majority of manuscripts originating from China. Most of the studies made use of satellite datasets, ranging from coarse spatial resolution data, such as the Gravity Recovery and Climate Experiment (GRACE) and GRACE Follow-On (GRACE-FO), to medium spatial resolution data, such as the Landsat series, ERA5, Modern-Era Retrospective Analysis for Research and Application Land version 2 reanalysis product (MERRA2), CLSM and NOAH ET, and MODIS (Moderate Resolution Imaging Spectroradiometer). Google Earth Engine (GEE) data, together with big data processing techniques, such as the remote sensing-based energy balance model (ALEXI/DisALEXI approach) and the STARFM data fusion technique, were used for analyzing geospatial datasets. Overall, this Special Issue demonstrated significant knowledge gaps in various big data image processing techniques and improved computing processes in assessing and monitoring water resources and the environment at various spatial and temporal scales.

Keywords: algorithms; climate change; droughts; environmental degradation; Earth observation; monitoring and assessment; sustainable management; water scarcity and security

1. Introduction

Although water resources are essential for life on Earth, the growing human population, combined with global environmental change, particularly climate change and variability, has put a strain on this finite resource and the environment [1]. Water scarcity caused by rapid population growth, droughts [2–4], erratic rainfall [5], overexploitation, and increased water pollution [6] from anthropogenic activities, such as mining and urbanization, pose a threat to the supply of usable quality and quantity of water to meet the demand [7]. This is especially concerning, given that approximately one-fifth of the world's population lives in water-stressed areas, and this trend is likely to worsen in the near future if the current trends, threats, and pressures are not fully understood and managed appropriately [2]. Fortunately, recent advances in geospatial modelling techniques, as well as an increase in readily available remotely sensed data with improved sensing characteristics (i.e., high spatial, spectral, and radiometric resolutions), have provided researchers with invaluable tools for not only monitoring and assessing the state of water resources, but the environment at various spatial scales [2]. In line with the United Nations Sustainable Development Goal (SDG) 6 of *"ensuring the availability and sustainable management of water and sanitation for all,"* this Special Issue invited scientific manuscripts that focus on innovative and cutting-edge practical applications of remotely sensed data and modelling techniques in water resources and environmental management at various spatial scales. Articles that used remote sensing to assess and monitor freshwater quality, quantity, availability, and management to ensure water security were published.

The goal of this Special Issue was to solicit scientific research articles that used cutting-edge remote sensing technologies to assess and monitor freshwater quality, quantity, availability and management in order to ensure water security. Articles that benefited from recent advances in satellite technologies, as well as the easy availability and accessibility to large volumes of archival spatial datasets and data analytic techniques for water resources and environmental management, were prioritized. Special consideration was given to scientific manuscripts that covered, but were not limited to, the development of geospatial techniques and remote sensing applications for detecting, quantifying, and monitoring freshwater water resources, identifying potential threats to water resources and agriculture, as well as other themes related to water resources and environmental management at various spatial scales. The Special Issue generated widespread global interest, with the majority of articles originating from China.

This Special Issue included thirteen scientific articles on land surface and plant water-use analysis, climate change impact assessment, agriculture, wetlands, water quality and quantity, soil moisture estimation, and land-use and land-cover analysis, as well as their implications for water resources and the environment. Several satellite datasets were used, ranging from coarse spatial resolution data, such as the Gravity Recovery and Climate Experiment (GRACE) and GRACE Follow-On (GRACE-FO), to medium spatial resolution data that include the Landsat series, ERA5, Modern-Era Retrospective Analysis for Research and Application Land version 2 reanalysis product (MERRA2), CLSM and NOAH ET, and MODIS (Moderate Resolution Imaging Spectroradiometer). Moderate- to high-resolution data were also used. The studies applied Google Earth Engine (GEE), as well as other advanced (big) data processing techniques, such as the remote sensing-based energy balance model (ALEXI/DisALEXI approach), and the STARFM data fusion technique. In addition, the studies applied machine learning algorithms such as the random forest (RF), support vector machine (SVM), CA–Markov model, and triple collocation (TC) method, as well as statistical methods, such as the Finite Volume Coastal Ocean Model (FVCOM). This editorial review divided the published scientific articles into six major themes, the contributions of which are summarized in detail in the following sections.

2. Analysis of Agriculture and Landcover Implications for Water Resources

Several studies focused on agriculture impacts on water resources and land-cover dynamics at the catchment scale, taking advantage of the various satellite imagery and advanced image processing techniques, such as the cellular automata and Markov chain (CA–Markov) model, random forest algorithms, and cloud computing Google Earth Engine platform [7,8]. The CA–Markov model was found to be the most appropriate change/time-series model, combining the standard Markov model and cellular automata (CA) for the prediction and modelling of future LULC changes at 30 m spatial resolution and linking these changes to the state of water resources at the catchment scale. This two-pronged approach has been shown to be viable because it allows for the integration of multidate satellite data in tracing the human footprint on the environment and, in particular, water resources. Cellular automata (CA), for example, has demonstrated the ability to change and control complex spatially distributed processes, as well as a strong ability to simulate the spatio-temporal characteristics of complex systems [7,9].

Climate variability, in addition to human drivers of LULC change, has important implications on moisture and temperature variations, which determine vegetation health and water availability. The decline in potential water sources may also exacerbate droughts and increase water supply challenges in an already water-stressed catchment. For instance, Magidi et al. [8] used the GEE platform and found that there has been an increase in irrigated areas over non-irrigated areas in semi-arid areas of Mpumalanga, South Africa, putting a strain on water resource demand in the agriculture sector. The study created the first detailed provincial irrigation map by leveraging moderately high-resolution Sentinel-2 and GEE computational power, as well as advanced machine learning algorithms, to demonstrate the scale of a shift from rainfed agriculture to irrigation in the area, as well as undocumented pressure on water resources. In addition, these irrigation activities have serious impacts on wetland systems due to nutrient accumulation. For example, Dlamini et al. [10] predicted seasonal variations in the concentrations of nine foliar biochemical elements in plant leaves of key floodplain wetland vegetation types and crops in the uMfolozi floodplain system, using Sentinel-2 multispectral data.

3. Plant Water Use and Its Impact on Water Resource Availability

Plant water use is critical to the survival, productivity, and spatial distribution of various vegetation functional types. Remote sensing of plant water-use estimation is, therefore, critical in water-controlled ecosystems or dryland areas with limited water resources, as well as in soil moisture dynamics [11,12]. This Special Issue features studies that used multi-source satellite datasets and advanced data analytics techniques to estimate plant water use at various scales. Carpintero et al. [11] used the ALEXI/DisALEXI model and the STARFM data fusion technique to estimate evapotranspiration (ET) dynamics in a Mediterranean oak savanna at fine spatial and temporal resolution (30 m, daily) from 2013 to 2015. DisALEXI energy fluxes (1 km, daily) and Landsat images (60–100 m, 16 days) were compared to in situ measurements from an eddy covariance flux tower system. Overall, the modelled fluxes performed reasonably well in comparison to field observations. The use of fine-scale ET maps (30 m, daily) provides critical information on plant water-use trends that are difficult to detect at a coarser spatial resolution over heterogeneous landscapes, and may aid management decisions at the field and farm scale.

The resulting high-resolution ET maps (daily, 30 m) revealed critical information for obtaining a better understanding of the hydrological functioning of various vegetation distributions that are typically difficult to detect at a 1 km pixel resolution. This can help with water-use decisions on the farm or in the community. Ochege et al. [12] performed the first detailed regional assessment of GLEAM, ERA5, MERRA2, CLSM, and NOAH ET products using the triple collocation (TC) method. Overall, these studies revealed that in water-stressed areas, gaps in regular evapotranspiration monitoring and evaluation pose a significant challenge to agricultural water resource allocation. As a result, the use of spatial explicit techniques in estimating plant water use is critical for improved water resource management and ensuring informed appropriate plant water allocation. Such efforts will greatly enhance global change and climate change research projects, particularly when these uncharted environments located in unique climate regions of the world are specifically captured in future studies.

4. Anthropogenic and Climate Change Impacts on Water Resource

Due to the rapid development of the coastal economy and continuous urbanization, more than 60% of the world's population now lives within 200 kilometers of a coastline. As a result, the impact of anthropogenic activity on the regional environment has increased, natural resources are being exploited on a larger scale, and natural coastal areas and marine ecosystems are being pushed back by intensive economic development and coastal zone projects. Land-derived pollutants (e.g., nitrogen and phosphorus) are continuously entering waters as a result of increasing urbanization, continuous mari-culture scale-up, and rapid development of the ocean economy, resulting in water quality deterioration and eutrophication in local areas. Based on Landsat images and long-term observation data, Zhu et al. [6] developed a nutrient retrieval model for Yueqing Bay to produce a long-term series of nutrient concentration products in Yueqing Bay from 2013 to 2020, combined with support vector machine learning and water temperature and satellite spectra as input parameters. Furthermore, the study analyzed the spatiotemporal variations and driving factors. The study found that natural factors, physical effects (e.g., seasonal variations in flow field) and biological effects (e.g., seasonal differences in the intensity of plankton photosynthesis) were the main causes of seasonal differences in nutrient concentration in Yueqing Bay. The study serves as an effective reference for remote sensing dynamic monitoring of water quality, as well as technical assistance for environmental protection and remediation of similar bays. Overall, these studies have demonstrated the unique utility of multi-source remotely sensed techniques for monitoring water quality variation in coastal areas with significant anthropogenic impact. However, in the future, an updated satellite-based algorithm for water quality assessment is required to better understand nutrient variation in coastal areas.

Furthermore, regional climate change has an impact on the state of inland and coastal water bodies, as well as their water balance, which is determined by a variety of hydrometeorological and hydrogeological factors [13]. The behavior of lake and reservoir levels is an integral characteristic of changes in the water balance, which not only determines the physical and ecological state of water bodies, but also has a significant impact on coastal infrastructure and socioeconomic development in the region. Using satellite altimetry data from 1993 to 2020, studies have been conducted to investigate the interannual variability in the level of the Ladoga and Onega lakes, Europe's largest lakes located in the northwest of Russia [13]. Three specialized altimetry databases were used for this purpose, including DAHITI, G-REALM, and HYDROWEB. These altimetry databases' water level data were compared to in situ records from water level gauge stations. The study demonstrated Google Earth as a useful tool for searching for critical areas in terms of the potential impact of Ladoga and Onega Lake water level rise on Russian railway infrastructure.

5. Conclusions

In conclusion, this Special Issue has demonstrated the utility of cutting-edge remote sensing data and advanced image processing techniques in dealing with large data sets at various scales and developing automated programs for operational water resource assessment and monitoring. This is an important step towards resolving water-related issues that are critical to achieving the UN Sustainable Development Goals. The applications shown range from freshwater quality, quantity, availability, assessment, and management to ensure water security, among others. Overall, this Special Issue has demonstrated that with the availability of advanced data processing techniques and improved computing processes, day-to-day, monthly, seasonal, and long-term water resource and environmental monitoring at high spatial resolution and covering large areas is now possible.

Author Contributions: Conceptualization, T.D.; methodology, T.D., M.D.S. and C.M.; writing—original draft preparation, T.D., M.D.S. and C.M., writing—review and editing, T.D., M.D.S. and C.M.; funding acquisition, T.D. All authors have read and agreed to the published version of the manuscript.

Funding: This work was supported by the WeMAST-Wetlands Monitoring and Assessment project, which is funded by the Africa Union and the European Union through the Global Monitoring for Environment and Security (GMES) and Africa Support Programme.

Data Availability Statement: Not applicable.

Acknowledgments: This work was supported by the WeMAST-Wetlands Monitoring and Assessment project, which is funded by the Africa Union and the European Union through the Global Monitoring for Environment and Security (GMES) and Africa Support Programme.

Conflicts of Interest: The author declares no known competing interest.

References

1. Yang, X.; Wang, N.; Chen, A.; Li, Z.; Liang, Q.; Zhang, Y. Impacts of Climate Change, Glacier Mass Loss and Human Activities on Spatiotemporal Variations in Terrestrial Water Storage of the Qaidam Basin, China. *Remote Sens.* **2022**, *14*, 2186. [CrossRef]
2. Bhaga, T.D.; Dube, T.; Shoko, C. Satellite monitoring of surface water variability in the drought prone Western Cape, South Africa. *Phys. Chem. Earth* **2021**, *124*, 102914. [CrossRef]
3. Maphanga, T.; Dube, T.; Shoko, C.; Sibanda, M. Advancements in the satellite sensing of the impacts of climate and variability on bush encroachment in savannah rangelands. *Remote Sens. Appl. Soc. Environ.* **2022**, *25*, 100689. [CrossRef]
4. Sharara, A.; Shekede, M.D.; Gwitira, I.; Masocha, M.; Dube, T. Fine-scale multi-temporal and spatial analysis of agricultural drought in agro-ecological regions of Zimbabwe. *Geomat. Nat. Hazards Risk* **2022**, *13*, 1342–1365. [CrossRef]
5. Varghese, D.; Radulović, M.; Stojković, S.; Crnojević, V. Reviewing the potential of sentinel-2 in assessing the drought. *Remote Sens.* **2021**, *13*, 3355. [CrossRef]
6. Zhu, B.; Bai, Y.; Zhang, Z.; He, X.; Wang, Z.; Zhang, S.; Dai, Q. Satellite Remote Sensing of Water Quality Variation in a Semi-Enclosed Bay (Yueqing Bay) under Strong Anthropogenic Impact. *Remote Sens.* **2022**, *14*, 550. [CrossRef]
7. Matlhodi, B.; Kenabatho, P.K.; Parida, B.P.; Maphanyane, J.G. Analysis of the future land use land cover changes in the gaborone dam catchment using ca-markov model: Implications on water resources. *Remote Sens.* **2021**, *13*, 2427. [CrossRef]
8. Magidi, J.; Nhamo, L.; Mpandeli, S.; Mabhaudhi, T. Application of the random forest classifier to map irrigated areas using google earth engine. *Remote Sens.* **2021**, *13*, 876. [CrossRef]
9. Guan, D.J.; Li, H.F.; Inohae, T.; Su, W.; Nagaie, T.; Hokao, K. Modeling urban land use change by the integration of cellular automaton and Markov model. *Ecol. Modell.* **2011**, *222*, 3761–3772. [CrossRef]
10. Dlamini, M.; Chirima, G.; Sibanda, M.; Adam, E.; Dube, T. Characterizing Leaf Nutrients of Wetland Plants and Agricultural Crops with Nonparametric Approach Using Sentinel-2 Imagery Data. *Remote Sens.* **2021**, *13*, 4249. [CrossRef]
11. Carpintero, E.; Anderson, M.C.; Andreu, A.; Hain, C.; Gao, F.; Kustas, W.P.; González-Dugo, M.P. Estimating evapotranspiration of mediterranean oak savanna at multiple temporal and spatial resolutions. Implications for water resources management. *Remote Sens.* **2021**, *13*, 3701. [CrossRef]
12. Ochege, F.U.; Shi, H.; Li, C.; Ma, X.; Igboeli, E.E.; Luo, G. Assessing satellite, land surface model and reanalysis evapotranspiration products in the absence of in-situ in central asia. *Remote Sens.* **2021**, *13*, 5148. [CrossRef]
13. Kostianoy, A.G.; Lebedev, S.A.; Kostianaia, E.A.; Prokofiev, Y.A. Interannual Variability of Water Level in Two Largest Lakes of Europe. *Remote Sens.* **2022**, *14*, 659. [CrossRef]

remote sensing

MDPI

Article

Integrating In Situ and Current Generation Satellite Data for Temporal and Spatial Analysis of Harmful Algal Blooms in the Hartbeespoort Dam, Crocodile River Basin, South Africa

Khalid Ali [1,2,*], Tamiru Abiye [3] and Elhadi Adam [2]

1 Department of Geology and Environmental Geosciences, College of Charleston, Charleston, SC 29424, USA
2 School of Geography, Archaeology and Environmental Studies, University of the Witwatersrand, Johannesburg 2050, South Africa
3 School of Geosciences, University of the Witwatersrand, Johannesburg 2050, South Africa
* Correspondence: alika@cofc.edu

Abstract: The Hartbeespoort Dam is a discharge point of a catchment that is characterized by urbanization, mining, industrial, and agricultural activities. These activities coupled with fluxes of heavily polluted wastewater from informal settlements, wastewater treatment works, as well as runoff from golf courses have led to the development of recurring harmful algal blooms (HABs). The predominant cyanobacteria scum that is largely covering the Dam water is toxic to fish and poses serious public health risks. Phosphorus is the limiting nutrient in terrestrial aquatic systems and excess concentration in the waters usually results in eutrophication. The productivity level in Hartbeespoort Dam is also a function of total phosphorous (TP) level, showing a positive correlation with chlorophyll-a, an index for phytoplankton which are predominantly HABs in this Dam. Analysis of long-term in situ water quality data (1980–2020) show that TP is not the only driver, changes in surface water temperatures also affect the productivity level, especially, when TP levels are below a threshold of approximately 0.4 mg/L. Chlorophyll-a was retrieved from current generation high resolution satellite (Landsat and Sentinel) at 5-year interval. Standard band ratio-based ocean color model applied to satellite data produced an accuracy of R^2 = 0.86 and RMSE of 5.56 μg/L. Time series analysis of in situ and satellite data show similar trends including capturing the effect of biocontrol on productivity levels between the late 1980s and the early 1990s, after which productivity increased with an increased flux of TP. Since 2015, the average annual surface temperature in the Dam has decreased leading to the decline in productivity level despite increasing levels of TP. The spatial dynamics of the HABs is a function of the discharges levels of the various rivers draining into the Dam as well as its geometry. Relatively higher concentrations are observed near river discharges and in areas of restricted water circulation.

Keywords: cyanobacteria; chlorophyll-a; water quality; HABs; remote sensing; Hartbeespoort Dam

1. Introduction

Climate change, urbanization, and agricultural activities have negative environmental impacts on freshwater bodies adding stress on the already scarce resource, especially, in developing countries such as South Africa. Currently, 68% of South Africa's population lives in urban areas, and this is expected to increase to 71% by 2030 (PMG, 2021; https://pmg.org.za/page/Urbanisation, accessed on 6 June 2022). Based on the current trend, it is also likely that the majority of future urban growth will be in informal areas as governments struggle to keep up with the increasing demand for basic services including water supply, sanitation, and formal housing. Such kinds of informal settlements are directly linked to the degradation of environmental quality, including water quality [1]. Runoff from agricultural land or other land uses with high application of fertilizers can also further exacerbate the water quality issue. As part of positive feedback process, deterioration in water quality can have significant economic impacts on the communities dependent on the streams, rivers, lakes, or other open freshwater bodies both for direct use, but also indirectly in terms of the water-dependent economic activities such as fishing [2].

Most of the reservoirs and impoundments in South Africa are located downstream of densely populated urban areas, thereby becoming nutrient enriched as a result of receiving untreated wastewater, raw sewage, and agricultural runoff [3]. Enrichment of nutrients, particularly total phosphorus (TP), which is the limiting nutrient in freshwater bodies, leads to eutrophication [2,4–6]. Studies of eutrophication in aquatic ecosystems has been extensively studied [7–9]; however, the significance of its impacts in South Africa has only recently been realized [10–13].

The Hartbeespoort Dam is a 195 million cubic meters reservoir with an average depth 10 m and is primarily used for irrigation and recreational activities. The reservoir is one of the highly productive eutrophic systems in the Gauteng Province of South Africa (Figure 1). It receives input from streams and rivers that drain the Johannesburg region (formal and informal settlements), industrial developments, agricultural activities, and golf courses. It also receives a significant amount of treated and partially treated water from Waste Water Treatment works (WWTW) at a rate of 5000 L/s [14,15]. Moreover, the poor performance of WWTW, the wide spread agricultural activities, and the lack of compliance monitoring in South Africa's water resources have triggered numerous discharges of inadequately treated wastewater effluents, industrial, and agricultural effluents into the country's already threatened water resources [16].

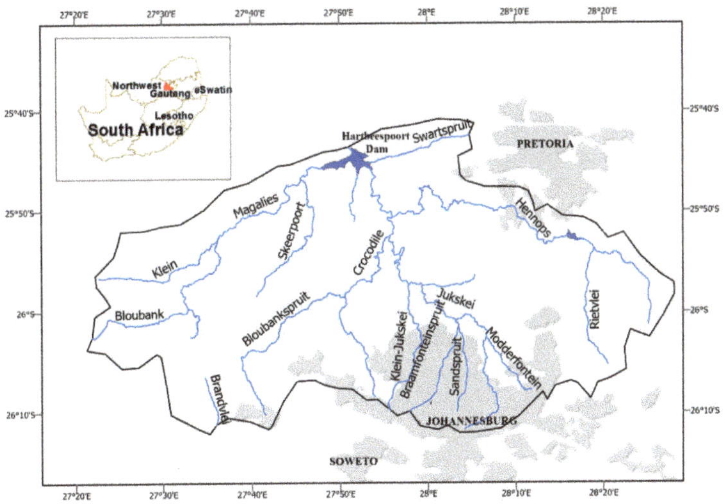

Figure 1. Map of the Hartbeespoort Dam and its watershed with drainage system indicated by lines.

These activities have led to the development of recurring algal blooms that are dominated by *Microcystis aeruginosa* characterized as harmful algal blooms (HABs) and the growth of invasive floating aquatic macrophyte, hyacinth [17–20]. Hartbeespoort Dam has been known for the occurrence of massive blooms, mostly exceeding the 20 μg/L chlorophyll-a cut-off trophic state level, of the potentially toxin producing cyanobacterium [21–25]. The cyanobacterial scum that is largely covering the Dam water is toxic to fish and poses serious public health risks [19,20,23].

Distributions of HABs in space and time have major implications for water quality and ecosystem function [2,23,26]. Water resource managers require robust water quality monitoring that is critical to address the question of how various natural and anthropogenic factors affect the health of these environments and to characterize the onset and the temporal and spatial scale of HABs. Currently, water quality assessment is based on in situ sampling and observation in the Hartbeespoort Dam at select locations. However, in-situ sampling, monitoring, and assessment of water quality require significant logistical and financial resources, which many developing countries such as South Africa lack. These kinds of conventional approaches of observation are challenging for large Dam water systems, where productivity levels change spatially and temporally. A remote sensing approach overcomes some of these challenges by providing a synoptic view of the aquatic bodies at high spatial and temporal resolution, with the potential to greatly improve our understanding of water quality variations. This method also allows for effective water quality monitoring, particularly in surveillance programs of HABs and provides researchers high resolution data for ecological and climate-related studies [7,27].

Current generation satellites have high spectral and radiometric resolution sensors that can detect variations in reflectance from open water surfaces [7,8,28,29]. Measured reflectance at multiple spectral windows carry spectral signatures that characterize various water quality parameters (WQP) such as chlorophyll-a and turbidity [30–33]. Chlorophyll-a, a universal pigment that is used as an index for phytoplankton or HABs, is an optically active constituent that alters irradiance light through absorption and scattering and changes the spectral signature of reflected radiance from water bodies [31,34]. These return signatures in the portion of visible and near-infrared bands of the electromagnetic spectrum are used to model chlorophyll-a concentration in the water body.

Several ocean color (OC) algorithms have been developed to retrieve concentrations of biochemical constituents in the upper euphotic zone using the reflectance from the water column. These algorithms relate measurements of satellite remote sensing reflectance Rrs (λ) to chlorophyll-a to provide e.g., phytoplankton biomass approximation [34–36].

In waters dominated by phytoplankton (Case I waters), blue and green wavelength-based OC algorithms have been successfully used to retrieve chlorophyll-a concentrations [7,37]. In coastal and inland waters (Case II waters), multiple biochemical constituents affect remote sensing reflectance, Rrs (λ), including suspended sediments, tripton, and the presence of HABs [8,26,38–40]. The various in-water constituents competingly absorb light in the blue spectral region and may add to uncertainties when using this region to detect specifically chlorophyll-a concentration [34,41–43]. The Rrs products have been evaluated in Case-2 waters using data from the Aerosol Robotic Network (AERONET-OC) and showed that the red-infrared band combinations can be effective for the Rrs based retrievals in optically complex (Case II) turbid waters [44]. In the recent study carried out by [11], in a nearby reservoir, the Vaal Dam, it was demonstrated that both the blue-green and red-infrared based OC chlorophyll-a algorithms worked effectively when applied to high resolution sensors. This may be attributed to the overwhelmingly strong signal from the excessively high chlorophyll-a concentrations in the water column, which is also the case in the Hartbeespoort Dam.

Current generation satellite sensors such as the Landsat 8 OLI and Sentinel 2 MSI are equipped with spectral windows that are sensitive to chlorophyll-a's absorption features. These sensors have relatively high spatio-temporal and radiometric resolution and therefore, allow for better spatial characterization of the water quality dynamics in large open waters bodies such as the Hartbeespoort Dam [11,28,37,39].

Application of satellite remote sensing to monitor water quality of large open waters in South Africa is primitive. The aim of the study is to evaluate long-term water quality changes in the Hartbeespoort Dam using an integrated approach of in situ data and multispectral satellite sensors. Specific goals of this study are to (a) evaluate long-term (1980–2021) nutrient loading into the Dam and the response in productivity, and (b) apply Landsat 8 OLI and Sentinel 2 MSI for spatio-temporal assessment of HABs, using chlorophyll-a as an index. For the satellite-based chlorophyll-a retrievals, we focused on satellite data that were collected in late summer, when productivity is at peak level, and the signal-to-noise ratio is high.

2. Study Area

The Hartbeespoort Dam is the outflow of the Upper Crocodile River Basin (UCRB) (Figure 1), which constitutes part of the Crocodile West and Marico Water Management Area as per the Department of Water and Sanitation (DWS) classification. The basin covers an area of approximately 6300 km^2 [12] within the Gauteng and North West Provinces of South Africa. Due to the presence of high topography in the south of the basin, the UCRB is generally drained in a northerly direction. The confluence of the Jukskei and Hennops rivers in the east of the basin produces the Crocodile River, which contributes to 90% of Hartbeespoort Dam's inflow [15]. The Magalies River flowing from the west contributes minor inflow to Hartbeespoort Dam. The Crocodile River flows northward from Hartbeespoort Dam into the Limpopo river.

The UCRB is characterized by a subtropical highland climate with a mean annual rainfall of 700 mm [45]. The basin generally receives most of the rainfall between the summer months of October and March, which occurs as convective rainfall in the form of afternoon thundershowers and occasional hailstorms [46,47]. The average temperatures in summer are a minimum of 15 °C and a maximum of 30 °C while winter temperatures range from 5 °C to 25 °C [48].

3. Materials and Methods

3.1. Water Quality Data

Water quality data were obtained from the water management system database of the Department of Water and Sanitation (DWS) in South Africa. The DWS maintains water quality data throughout the UCRB watershed including on sites within the Hartbeespoort Dam. Data are available since 1980 although a gap exists between 1983 and 1987. The data include major ion chemistry, total phosphorus concentrations (TP), and chlorophyll-a concentrations. Physical data include in situ temperature and dissolved oxygen (DO) measurements taken at multiple depths.

Time-series graphs for temperature (°C), DO concentrations (mg/L), TP concentrations (mg/L), and chlorophyll-a concentrations (μg/L) are plotted. Annual-based trend analysis is run on the time series data and a line plot is added to each time series to assess long-term changes of the parameters. Trend analyses were also run to show decadal changes using a 5 m depth integrated temperature, DO concentrations (mg/L), TP concentrations (mg/L), and chlorophyll-a concentrations (μg/L). The correlation between these parameters has also been evaluated.

3.2. Satellite Data

Level 2 Landsat OLI and Sentinel 2 MSI satellite data of the study area are acquired from USGS earth explorer https://earthexplorer.usgs.gov/, accessed on 5 May 2022) and Copernicus Open Access Hub (https://scihub.copernicus.eu/dhus/#/home, accessed on 5 May 2022, respectively. Data were acquired for the month of April at five-year intervals, as available between 1980 and 2020. This allows for understanding long-term changes without much influence from short-term effects including annual and inter-annual seasonal anomalies. We specifically focused on satellite data from late summer (April) as a period of reference, where chlorophyll-a and bloom events are at peak level. Landsat 5/7 data were obtained for the period 1995 to 2010. Both series have near identical sensor characteristics in-terms of resolution and band placements. For recent dates (2013 onwards) the Landsat 8 and Sentinel 2 data were used. Both have higher radiometric resolution of 12 bit which allows for the enhanced capability to discriminate very slight differences in the water column optics which is a function of WQPs. The new generation sensors are also equipped with more spectral bands that are narrower creating the opportunity to identify specific materials by their band absorption characteristics. The Sentinel 2 sensor has a higher spatial resolution of 10 m than the Landsat series (30 m).

NASA's standard ocean color processing software the SeaWiFS Data Analysis System (SEADAS; version 8) level −2 generator (l2 gen) is used to process the images, perform the atmospheric correction process, and retrieve remote sensing reflectance (R_{rs}) per unit steradian (sr) and the normalized water-leaving radiance (nLw) [49,50]. For this analysis, the multi-scattering with 2-band model selection and iterative near infra-red (NIR) correction method is applied [51]. This method is effective in turbid waters and attempts to separate the radiometric signal of suspended particulates from aerosol noise [52–56]. Cloud free images were downloaded but additional cloud masks are applied to remove haze related effects using the standard SeaDAS masking default thresholds from the NIR spectral region. A land mask was also applied to remove the land component and retrieve images of the open water body only.

Spectral band ratios are applied to both Sentinel and Landsat satellite images for modeling productivity levels of the Dam water. This study uses two standard OC chlorophyll-a retrieval models, specifically the two-band blue-green-based algorithm as presented in [36] or the NIR-red model given by [33] depending on performances. These ratio-based models are expressed as:

$$Chl\ a \approx \frac{R^1_{(\lambda_1)}}{R^1_{(\lambda_2)}} \tag{1}$$

The models are based on ratios of spectral regions that are sensitive to absorption and scattering features of chlorophyll-a. The spectral regions are matched with the corresponding satellites bands. The chlorophyll-a concentrations retrieved using the two-band models are used to present the spatial and temporal dynamics of the HABs.

4. Result and Discussion

Distribution analysis of biophysical parameters between 1980 and 2020 shows that concentration levels of chlorophyll-a in the Hartbeespoort Dam varied from approximately 3.5 µg/L to 390 µg/L, and TP levels ranged from 0.1 mg/L to 1.5 mg/L (Figure 2). During the same period, average temperature and DO were approximately 19 °C and 8.4 mg/L, respectively. The chlorophyll-a data is positively skewed with mean centered near the low range values. Figure 2 shows that for chlorophyll-a, 75% of the data fall below the 55 µg/L level. This trend is also observed in the TP values where the values range from 0–1.5 mg/L with the mean centered around 0.4 mg/L. Temperature and DO data have values that follow a relatively normal distribution with few outliers on both sides. The extremely low frequency outliers in temperature and DO may be attributed to systematic or random errors.

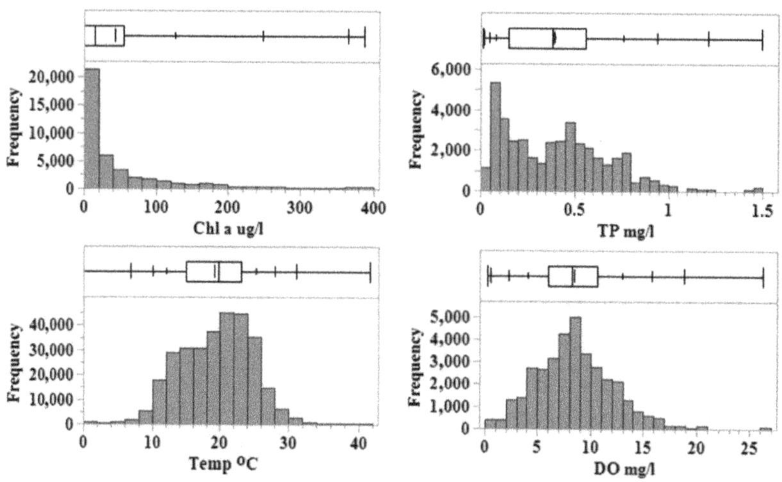

Figure 2. Distribution plot of various water quality parameters in the Hartbeespoort Dam; DWS.

Time series plot of chlorophyll-a, TP, DO, and temperature show seasonal, annual, and decadal variation of the water quality parameters (Figure 3). Concentrations of the TP nutrient were high between 1982 and 1986 with values consistently greater than 0.5 mg/L and reaching up to 1.5 mg/L. The chlorophyll-a was also high with an average concentration of 52 μg/L. These values put the waters in the hypereutrophic status [57]. TP values decreased significantly by the late 1980s and into the early 1990s, primarily in response to the bioremediation project [18]. Chlorophyll-a concentration in the Hartbeespoort Dam also decreased during the 1990s, and the average value dropped to 20 μg/L. However, when biocontrol project ended in late 1990s, due to lack of funding, chlorophyll-a concentrations spiked in the 2000s reaching values near 400 μg/L with an average of 64 μg/L between 2000 and 2010. In the past decade (2010–2018), there was a general decrease in productivity in the Dam but the average chlorophyll-a remained high at 45 μg/L, which is well above the 7 μg/L threshold for eutrophic systems (Figure 3) [25]. Although the amount of TP concentration continues to increase, a steep decline in the chlorophyll-a concentrations has been observed since 2015. This also correlates with drop in temperature during the same period (Figure 3). This indicates that temperature has strong influence on productivity and DO levels. Higher chlorophyll-a and DO values correspond to warmer seasons and vise-versa. The increase in TP over the past two decades can be linked to higher discharges of untreated wastewater along with other anthropogenic nutrient sources such as agricultural and golf course runoffs.

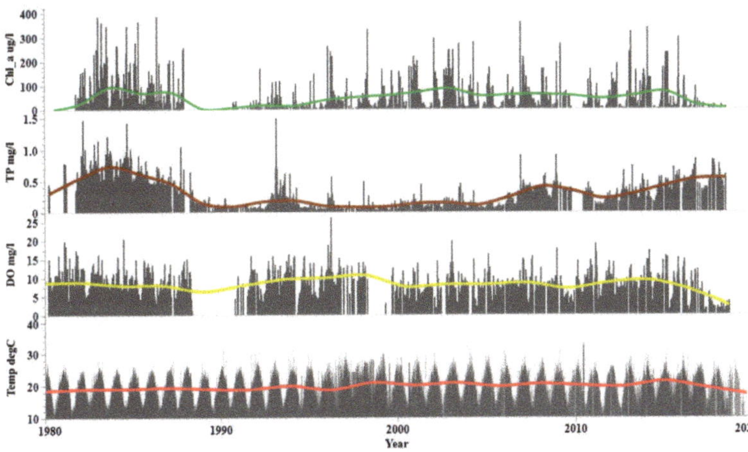

Figure 3. Time series of water quality parameters in the Hartbeespoort Dam; DWS.

Over the long-term (past 4 decades), temperature and DO show near constant values with a sharp decline in 2015 (Figure 3). Chlorophyll-a follows the 2015 declining trend while TP levels continue to rise. To understand these relationships at higher resolution, the data were categorized by decades. During the period from 1980 to 1990; chlorophyll-a, TP, and DO showed a decline while temperature increased by 0.8 °C (Figure 4a). The average annual concentration of chlorophyll-a decreased from ~62 μg/L to ~40 μg/L; the average DO levels changed from ~9.1 mg/L to 7.5 mg/L and the TP levels dropped from an average of 0.7 mg/L to 0.2 mg/L. The decline of TP and chlorophyll-a overlaps with the implementation period (late 1980s to the early 1990s) of the bioremediation project which ultimately reduced the chlorophyll-a levels by up to 80% [18]. The increase in measured temperature records during this decade could be attributed to the dominance of the El Nino effect over La Nina where the years 1982 and 1983 were characterized by drought throughout the country [58]. The effect of continued bioremediation into the 1990's is reflected in the further decline of TP levels between 1990 and 2000 (Figure 4b). However, during the same period, chlorophyll-a increased significantly to levels of up to 60 μg/L causing super-saturation in DO and reaching 11 mg/L. The increase in productivity may be attributed to the sudden increase in loading of TP into the Dam as recorded between 1992 and 1995 (Figure 4b). During this decade the Dam was also going through temperature stabilization. This shift in temperature trend could be linked to the weakening of the El Nino effect in the region leading to cooler temperatures over the decade (Figure 4b).

In the past two decades (2000–2020), as the result of increased urbanization and increased flux of untreated or partially treated wastewater, TP levels in the Hartbeespoort Dam have been significantly increasing (Figure 4c,d). However, despite this increase in TP concentrations of the limiting nutrient, the average chlorophyll-a and DO levels have been decreasing at a rate of up to 7.5 μg/L/year and 0.8 mg/L/year, respectively (Figure 4c,d). The biocontrol project kept TP levels relatively low between 1990 and 2010 compared to the 1980s where levels were consistently higher at TP > 0.7 mg/L. However, recent records of TP levels are approaching to that of the 1980s. A slightly higher average surface water temperature was recorded during the past two decades but inter-annual variability over the 10 year periods shows a decreasing trend by as much as 1.5 °C or at a rate of 0.15 °C/year (Figure 4c,d). This temperature gradient could be due to the El Nino effect that affected South Africa in 2015 and 2016 [58].

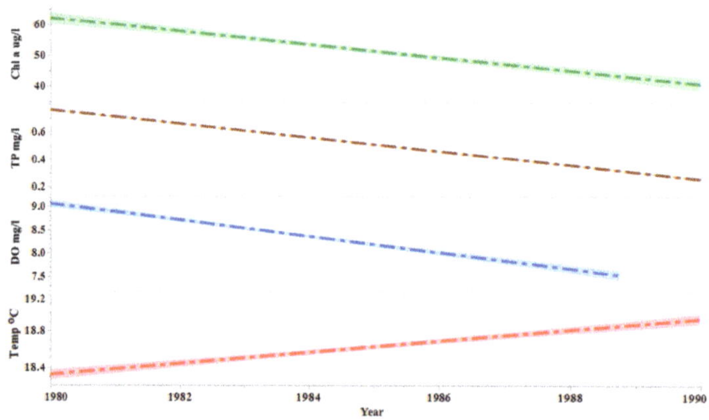

(a)

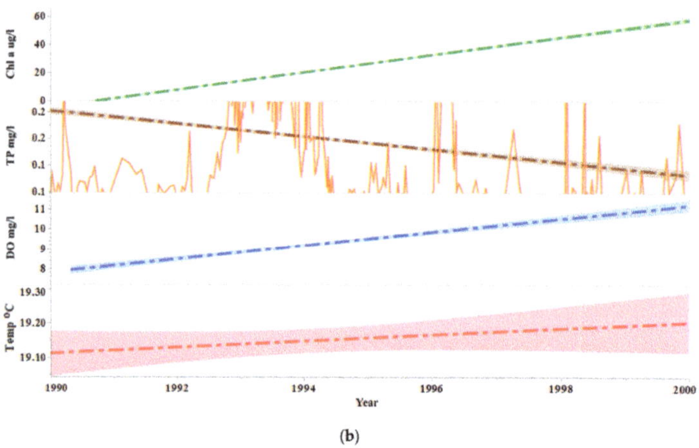

(b)

Figure 4. *Cont.*

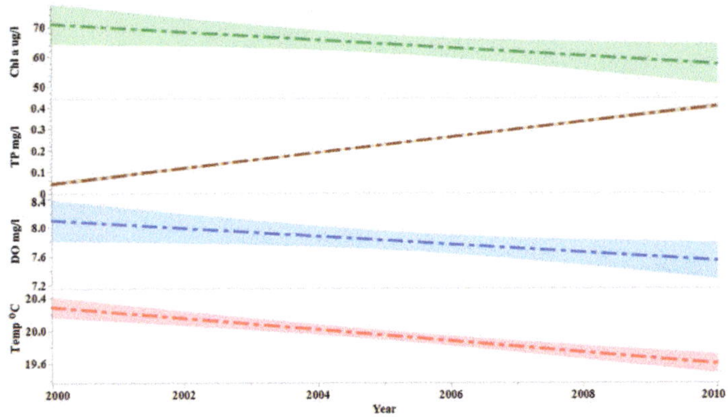

(c)

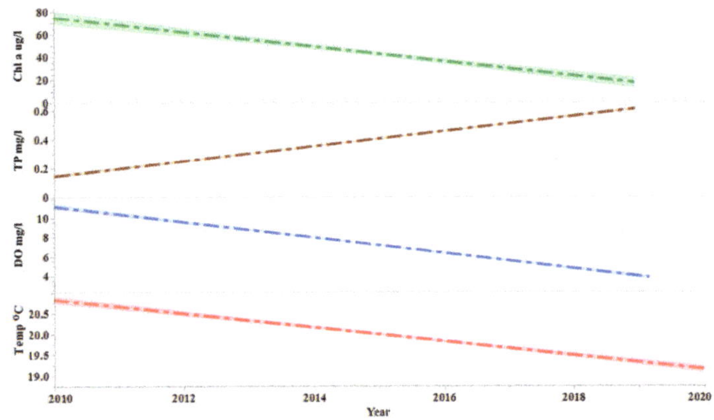

(d)

Figure 4. Decadal trends of chlorophyll-a, TP, DO and temperature in the Dam during (**a**) 1980–1990, (**b**) 1990–2000, (**c**) 2000–2010, (**d**) 2000–2010; DWS.

Productivity in the open water system is a function of multiple biophysical factors but limiting nutrients such as TP are considered the major drivers in many aquatic settings [12,13,18,31,59]. However, results in this study have shown that TP is not necessarily the only factor driving phytoplankton growth in the Hartbeespoort Dam. This is especially the case when TP levels are relatively low, in this case below 0.4 mg/L. Chlorophyll-a concentration in Hartbeespoort Dam showed a decreasing trend for the past two decades while the TP levels were increasing. The steep decline in the temperature slope, chlorophyll-a and DO as shown in Figures 3 and 4c,d clearly indicate that temperature plays a significant role in controlling productivity in the Hartbeespoort Dam. The effect of temperature on productivity is well documented in previous studies [60,61].

The regression analysis between the water quality parameters shows a positive correlation between chlorophyll-a and temperature; chlorophyll-a and DO; chlorophyll-a and TP; as well as between temperature and DO (Figure 5). The data between 1980 and 1990 showed productivity in the Dam was primarily influenced by TP when concentrations were high. But in recent decades, despite the increase in TP, chlorophyll-a has continued to decrease. As temperature changes, the productivity in the Dam also changes which in turn affects the net DO levels. The TP's positive correlation with chlorophyll-a indicates the effect of nutrient loading in the Dam, however, the near zero correlation with the physical parameters such as the DO suggests that TP may not necessarily be the main driver of the productivity in the Dam. These relationships confirm temperature as the main driver of the biophysical dynamics in the Dam's aquatic system. Our study shows that a TP level of above 0.4 mg/L is required to drive the primary productivity. However, these effects require further study.

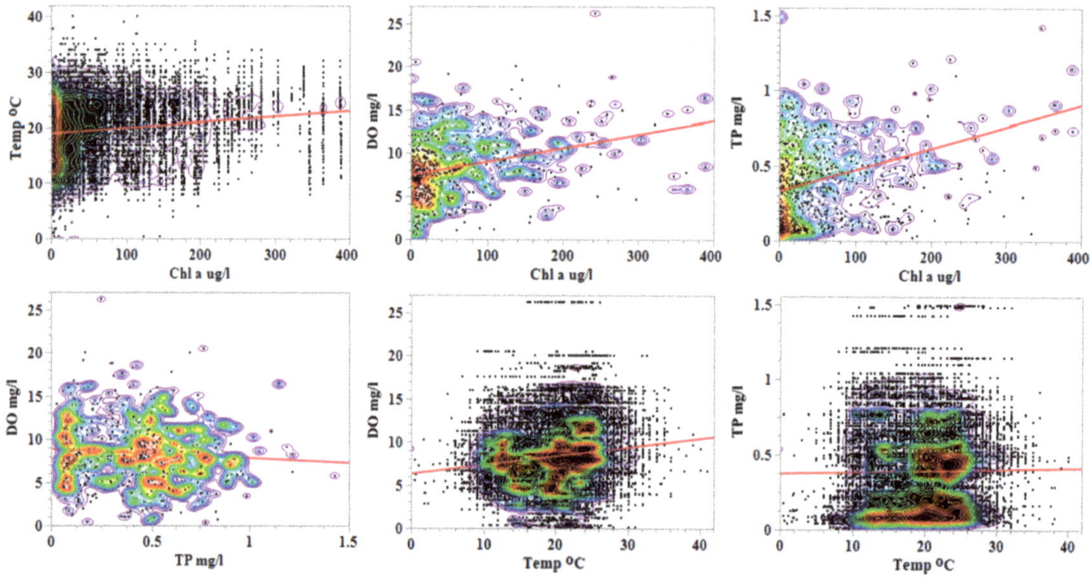

Figure 5. Scatter plots between the various water quality parameters; DWS.

Regression analysis between satellite-based and in-situ chlorophyll-a data show a strong correlation with an R^2 = 0.86 and RMSE of 5.56 µg/L (Figure 6). This demonstrates the applicability of global OC model to effectively retrieve chlorophyll-a in large productive open waters such as the Hartbeespoort Dam, where productivity levels are high. The OC model was used to predict chlorophyll-a level for 2020, where in situ data are not available.

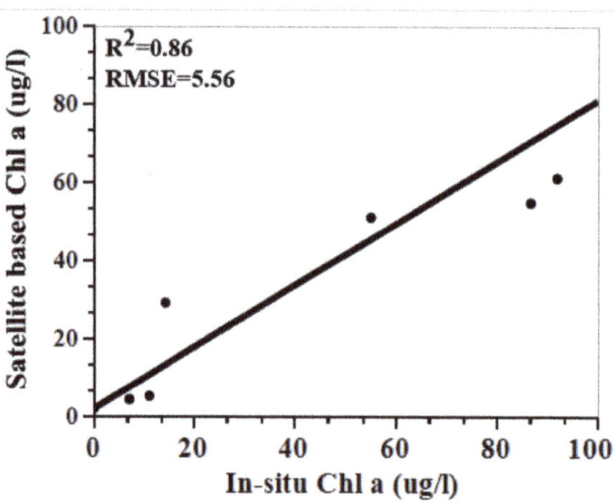

Figure 6. Satellite chlorophyll-a estimates comparisons against in-situ concentrations.

Time series plots of the 5-year interval satellite-based and the in situ chlorophyll-a data show the interdecadal variability of productivity levels in the Hartbeespoort Dam (Figure 7). The satellite data captures the major shifts in chlorophyll-a concentrations in the Dam. The steep decline in the satellite-based chlorophyll-a level during mid-to-late 1980s, and the continuous low chlorophyll-a level in the early 1990s reflects the effects of the bioremediation project. Satellite data also captured the increase in productivity level between the mid-1990s and early-2000s, which is attributed to increased urbanization and higher discharges of untreated wastewater along with other anthropogenic sources. The general decline in satellite-based chlorophyll-a levels between mid-2000 and 2015 mimics the observed in situ chlorophyll-a trend. As of 2015, there is a steep decline in the chlorophyll-a concentrations which is also captured by the satellite data.

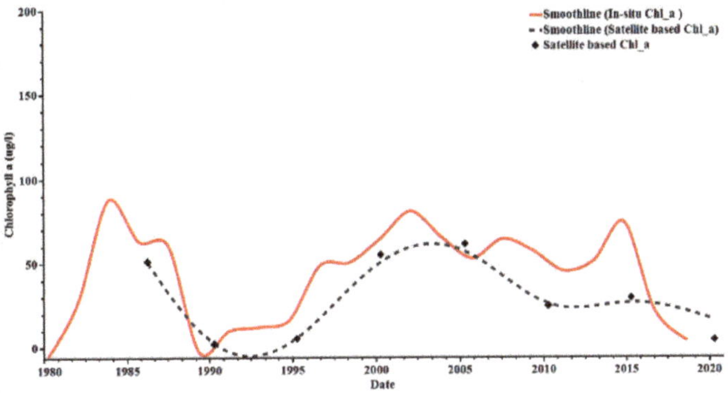

Figure 7. Time series of chlorophyll-a data retrieved from satellite and in-situ measurements.

The spatial and temporal changes of water quality and the dynamics of the HABs across the Hartbeespoort Dam is shown in Figure 8. The Landsat series data are primarily used because of its large temporal coverage and was supplemented with Sentinel 2 data for the recent years. In this study, a five-year interval satellite data were used to understand the long-term spatial and temporal dynamics of HABs in the reservoir by consistently comparing images from April (late summer), as a period of reference, where productivity levels are considered to be at maximum levels. This interval of data collection also minimized influence of annual and seasonal changes. A total 20 satellite images were processed and normalized to be able to compare the variation over time. The image that was taken in 1985 (Figure 8a) shows relatively high concentration levels of chlorophyll-a which is linked to high TP levels in the system (Figure 3). Relatively higher productivity is observed near the discharge areas of the crocodile river as well as near the western part of the dam where the Magalis river is discharged (Figure 8b). The relatively high level in the western parts could be attributed to direct runoff from golf courses and restricted water circulation in the reservoir.

The positive impact of bioremediation in the late 1980s can be visualized in the image taken in April of 1990, where productivity level was significantly low (Figure 8c,d). Since the bioremediation project took place in the early 1990s, productivity levels also remained relatively low in mid-1990s as shown in Figure 8e,f. This is consistent with the measured ground data as presented in Figure 3. With the end of the biocontrol measures coupled with an increase in TP levels, productivity in the reservoir increased between the late 1990s and the early 2010s (Figure 8g,h). The five-year interval images between 2000 and 2010 show a progressive increase in phytoplankton biomass (Figure 8i–l). This is also the period where TP levels started to increase due to increased human activities such as discharges of untreated or partially treated wastewater, runoff from urban areas, agricultural activities, and golf courses. Landsat data from 2015 onwards, and Sentinel 2 data for the past 2 years show a declining trend in the productivity in the Hartbeespoort Dam (Figure 8m–t). These changes are strongly correlated with the measured in-situ chlorophyll-a and DO levels. The spatial dynamics of the HABs is linked to fluxes of the various rivers draining into the Dam and its geometry.

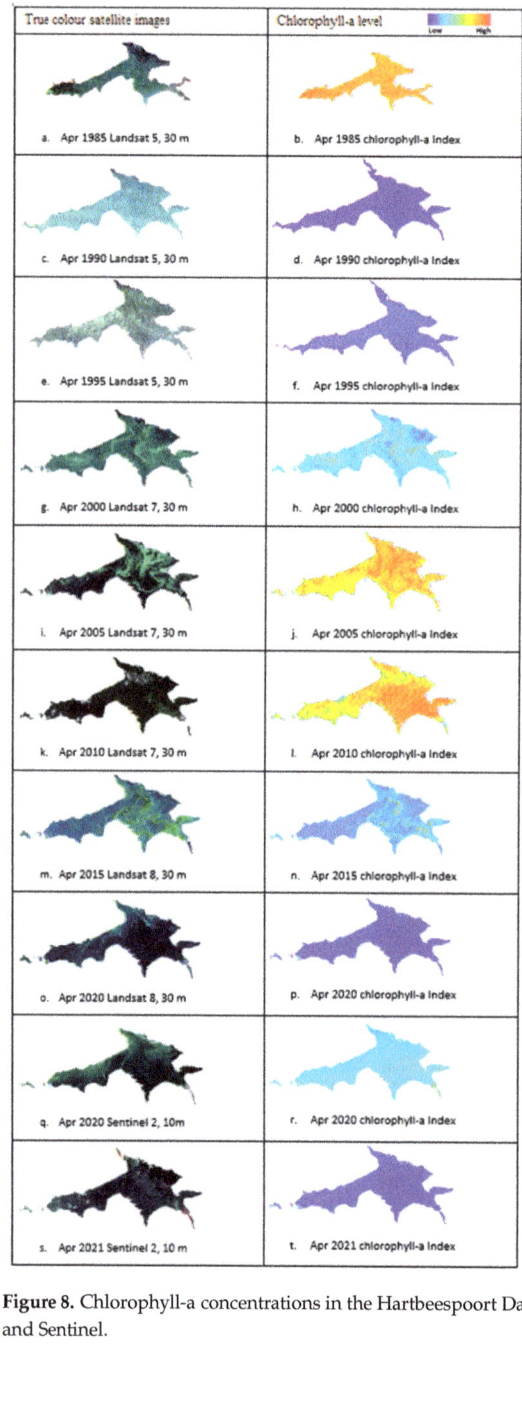

Figure 8. Chlorophyll-a concentrations in the Hartbeespoort Dam between 1980–2021 from Landsat and Sentinel.

5. Conclusions

The results obtained from the measured temporal water quality data revealed a high level of anthropogenic impact on the Hartbeespoort Dam which was designed to provide water for agricultural and recreational uses. The deterioration of the water quality over the years is exacerbated by the discharge of poor quality and nutrient-rich effluent of untreated wastewater, raw sewage, and agricultural runoff.

The study showed that the productivity levels in the Hartbeespoort Dam is primarily a function of TP levels showing relatively strong positive correlation with chlorophyll-a, an index for phytoplankton and in this case predominantly HABs. This finding is consistent with other studies that showed phosphorous is the limiting factor in freshwater bodies [4–6]. However, in this case, the analysis of long-term in situ water quality data showed that TP is not the only driver. Changes in surface water temperatures also affect the productivity levels, especially, when TP levels are below a threshold of approximately 0.4 mg/L. Since 2015, the average annual surface temperature in the Dam has decreased leading to the decline in productivity levels despite increasing levels of TP.

Chlorophyll-a retrieved from satellite images show the dynamics of HABs over the past four decades. The OC model applied to retrieve the chlorophyll-a in the Hartbeespoort Dam performed at an accuracy of $R^2 = 0.86$, RMSE = 5.56 µg/L. Relatively high chlorophyll-a concentrations were observed at river discharge points near the reservoir mouth, and in peripheral regions where nutrient rich waters runoff and water circulations are restricted due to low levels of mixing. Time series analysis of the 5-year interval satellite data captured the major shifts and trends of chlorophyll-a levels in the Dam. The data has showed the positive effect of biocontrol on productivity level between the late 1980s and the early 1990s. However, when biocontrol was terminated in mid-1990s, productivity increased with an increased flux of TP from uncontrolled disposal of wastewater. The bioremediation activity of the late 1980s and early 1990s was effective in reducing the TP and hence the eutrophication level of the reservoir. The effect of declining surface water temperature since 2015 on productivity levels is also evident from the satellite data showing decrease in productivity both spatially and temporally. This is consistent with the dynamics observed in measured data. Further studies will be conducted to understand short-term effects including annual, inter-annual, and seasonal changes in HAB dynamics.

Author Contributions: Conceptualization, K.A., E.A. and T.A.; Formal analysis, K.A.; Funding acquisition, E.A.; Investigation, K.A.; Methodology, K.A.; Writing—original draft, K.A.; Writing—review & editing, K.A., E.A. and T.A. All authors have read and agreed to the published version of the manuscript.

Funding: This research received no external funding.

Conflicts of Interest: The authors declare no conflict of interest.

References

1. Cullis, J.D.S.; Horn, A.; Rossouw, N.; Fisher-Jeffes, L.; Kunneke, M.M.; Hoffman, W. Urbanisation, climate change and its impact on water quality and economic risks in a water scarce and rapidly urbanising catchment: Case study of the Berg River Catchment. *H2Open J.* **2019**, *2*, 146–167. [CrossRef]
2. Alprol, A.E.; Heneash, A.M.M.; Soliman, A.M.; Ashour, M.; Alsanie, W.F.; Gaber, A.; Mansour, A.T. Assessment of water quality, eutrophication, and zooplankton community in lake Burullus, Egypt. *Diversity* **2021**, *13*, 268. [CrossRef]
3. Edokpayi, J.N.; Odiyo, J.O.; Durowoju, O.S. Impact of Wastewater on Surface Water Quality in Developing Countries: A Case Study of South Africa. In *Water Quality*; InTech: London, UK, 2017.
4. Blomqvist, S.; Gunnars, A.; Elmgren, R. Why the limiting nutrient differs between temperate coastal seas and freshwater lakes: A matter of salt. *Limnol. Oceanogr.* **2004**, *49*, 2236–2241. [CrossRef]
5. Cha, Y.K.; Alameddine, I.; Qian, S.S.; Stow, C.A. A cross-scale view of N and P limitation using a Bayesian hierarchical model. *Limnol. Oceanogr.* **2016**, *61*, 2276–2285. [CrossRef]
6. Elser, J.J.; Bracken, M.E.S.; Cleland, E.E.; Gruner, D.S.; Harpole, W.S.; Hillebrand, H.; Ngai, J.T.; Seabloom, E.W.; Shurin, J.B.; Smith, J.E. Global analysis of nitrogen and phosphorus limitation of primary producers in freshwater, marine and terrestrial ecosystems. *Ecol. Lett.* **2007**, *10*, 1135–1142. [CrossRef]

7. Ali, K.A.; Ortiz, J.; Bonini, N.; Shuman, M.; Sydow, C. Application of Aqua MODIS sensor data for estimating chlorophyll a in the turbid Case 2 waters of Lake Erie using bio-optical models. *GISci. Remote Sens.* **2016**, *53*, 483–505. [CrossRef]
8. Ali, K.; Witter, D.; Ortiz, J. Application of empirical and semi-analytical algorithms to MERIS data for estimating chlorophyll a in Case 2 waters of Lake Erie. *Environ. Earth Sci.* **2014**, *71*, 4209–4220. [CrossRef]
9. Drobac, D.; Tokodi, N.; Simeunović, J.; Baltić, V.; Stanić, D.; Svirčev, Z. Human exposure to cyanotoxins and their effects on health. *Arh. Hig. Rada Toksikol.* **2013**, *64*, 305–316. [CrossRef]
10. Coetzee, J.A.; Hill, M.P. The role of eutrophication in the biological control of water hyacinth, Eichhornia crassipes, in South Africa. *BioControl* **2012**, *57*, 247–261. [CrossRef]
11. Obaid, A.A.; Ali, K.A.; Abiye, T.A.; Adam, E.M. Assessing the utility of using current generation high-resolution satellites (Sentinel 2 and Landsat 8) to monitor large water supply dam in South Africa. *Remote Sens. Appl. Soc. Environ.* **2021**, *22*, 100521. [CrossRef]
12. Van Ginkel, C.E.; Silberbauer, M.J. Temporal trends in total phosphorus, temperature, oxygen, chlorophyll a and phytoplankton populations in Hartbeespoort Dam and Roodeplaat Dam, South Africa, between 1980 and 2000. *Afr. J. Aquat. Sci.* **2007**, *32*, 63–70. [CrossRef]
13. van Ginkel, C.E. Eutrophication: Present reality and future challenges for South Africa. *Water SA* **2011**, *37*, 693–702. [CrossRef]
14. Leketa, K.C. Holistic Approach to Groundwater Recharge Assessment in the Upper Crocodile River Basin, Johannesburg, South Africa. Ph.D. Thesis, University of the Witwatersrand, Johannesburg, South Africa, 2019.
15. Leketa, K.; Abiye, T.; Butler, M. Characterisation of groundwater recharge conditions and flow mechanisms in bedrock aquifers of the Johannesburg area, South Africa. *Environ. Earth Sci.* **2018**, *77*, 727. [CrossRef]
16. Naidoo, S.; Olaniran, A.O. Treated wastewater effluent as a source of microbial pollution of surface water resources. *Int. J. Environ. Res. Public Health* **2013**, *11*, 249–270. [CrossRef]
17. Ashton, P.J.; Robarts, R.D. Apparent predation of microcystis aeruginosa katz. Emend elenkin by a saprospira-like bacterium in a hypertrophic lake (Hartbeespoort Dam, South Africa). *J. Limnol. Soc. S. Afr.* **1987**, *13*, 44–47. [CrossRef]
18. Auchterlonie, J.; Eden, C.L.; Sheridan, C. The phytoremediation potential of water hyacinth: A case study from Hartbeespoort Dam, South Africa. *S. Afr. J. Chem. Eng.* **2021**, *37*, 31–36. [CrossRef]
19. Gumbo, R.J.; Ross, G.; Cloete, E.T. Biological control of Microcystis dominated harmful algal blooms. *Afr. J. Biotechnol.* **2008**, *7*, 4765–4773.
20. Ballot, A.; Sandvik, M.; Rundberget, T.; Botha, C.J.; Miles, C.O. Diversity of cyanobacteria and cyanotoxins in Hartbeespoort Dam, South Africa. *Mar. Freshw. Res.* **2014**, *65*, 175–189. [CrossRef]
21. Zohary, T. Hyperscums of the cyanobacterium Microcystis aeruginosa in a hypertrophic lake (Hartbeespoort Dam, South Africa). *J. Plankton Res.* **1985**, *7*, 399–409. [CrossRef]
22. Zohary, T.; Madeira, A.M.P. Structural, physical and chemical characteristics of Microcystis aeruginosa hyperscums from a hypertrophic lake. *Freshw. Biol.* **1990**, *23*, 339–352. [CrossRef]
23. Oberholster, P.J.; Botha, A.M. Use of remote sensing and molecular markers to detect toxic cyanobacterial hyperscum crust: A case study on Lake Hartbeespoort, South Africa. *Afr. J. Biotechnol.* **2010**, *9*, 8783–8790.
24. Conradie, K.R.; Barnard, S. The dynamics of toxic Microcystis strains and microcystin production in two hypertrofic South African reservoirs. *Harmful Algae* **2012**, *20*, 1–10. [CrossRef]
25. Carlson, R.E. A trophic state index for lakes 1. *Limnol. Oceanogr.* **1977**, *22*, 361–369. [CrossRef]
26. Pahlevan, N.; Smith, B.; Schalles, J.; Binding, C.; Cao, Z.; Ma, R.; Alikas, K.; Kangro, K.; Gurlin, D.; Hà, N. Seamless retrievals of chlorophyll-a from Sentinel-2 (MSI) and Sentinel-3 (OLCI) in inland and coastal waters: A machine-learning approach. *Remote Sens. Environ.* **2020**, *240*, 111604. [CrossRef]
27. Gurlin, D.; Gitelson, A.A.; Moses, W.J. Remote estimation of chl-a concentration in turbid productive waters–Return to a simple two-band NIR-red model? *Remote Sens. Environ.* **2011**, *115*, 3479–3490. [CrossRef]
28. Pahlevan, N.; Chittimalli, S.K.; Balasubramanian, S.V.; Vellucci, V. Sentinel-2/Landsat-8 product consistency and implications for monitoring aquatic systems. *Remote Sens. Environ.* **2019**, *220*, 19–29. [CrossRef]
29. Wu, M.; Zhang, W.; Wang, X.; Luo, D. Application of MODIS satellite data in monitoring water quality parameters of Chaohu Lake in China. *Environ. Monit. Assess.* **2009**, *148*, 255–264. [CrossRef]
30. Gitelson, A.A.; Dall'Olmo, G.; Moses, W.; Rundquist, D.C.; Barrow, T.; Fisher, T.R.; Gurlin, D.; Holz, J. A simple semi-analytical model for remote estimation of chlorophyll-a in turbid waters: Validation. *Remote Sens. Environ.* **2008**, *112*, 3582–3593. [CrossRef]
31. Gholizadeh, M.H.; Melesse, A.M.; Reddi, L. A comprehensive review on water quality parameters estimation using remote sensing techniques. *Sensors* **2016**, *16*, 1298. [CrossRef]
32. Schalles, J.F. Chapter 3 optical remote sensing techniques to estimate phytoplankton chlorophyll a concentrations in coastal waters with varying suspended matter and CDOM concentrations. In *Remote Sensing of Aquatic Coastal Ecosystem Processes*; Springer: Dordrecht, The Netherlands, 2006.
33. Moses, W.J.; Gitelson, A.A.; Berdnikov, S.; Povazhnyy, V. Estimation of chlorophyll-a concentration in case II waters using MODIS and MERIS data–Successes and challenges. *Environ. Res. Lett.* **2009**, *4*, 045005. [CrossRef]
34. Franz, B.A.; Bailey, S.W.; Kuring, N. Ocean Color Measurements from Landsat-8 OLI using SeaDAS. 2014. Available online: https://www.researchgate.net/publication/267679321_Ocean_Color_Measurements_from_Landsat-8_OLI_using_SeaDAS (accessed on 14 July 2022).

35. Binding, C.E.; Greenberg, T.A.; McCullough, G.; Watson, S.B.; Page, E. An analysis of satellite-derived chlorophyll and algal bloom indices on Lake Winnipeg. *J. Great Lakes Res.* **2018**, *44*, 436–446. [CrossRef]
36. O'Reilly, J.E.; Werdell, P.J. Chlorophyll algorithms for ocean color sensors–OC4, OC5 & OC6. *Remote Sens. Environ.* **2019**, *229*, 32–47. [CrossRef]
37. Hieronymi, M.; Hereon, H.-Z.; Müller, D.; Krasemann, H.; Brockmann, C.; Ruescas, A.; Stelzer, K.; Bouchra, N.; Ruddick, K.; Simis, S.; et al. *Ocean Colour Remote Sensing of Extreme Case-2 Waters Case 2 eXtreme View Project Space based Cyanobacteria Information & Service (CyanoAlert) View Project Ocean Colour Remote Sensing of Extreme Case-2 Waters*; European Space Agency, Special Publication: Amsterdam, The Netherlands, 2016.
38. Doerffer, R.; Fischer, J. Concentrations of chlorophyll, suspended matter, and gelbstoff in case II waters derived from satellite coastal zone color scanner data with inverse modeling methods. *J. Geophys. Res. Ocean.* **1994**, *99*, 7457–7466. [CrossRef]
39. Malahlela, O.E.; Oliphant, T.; Tsoeleng, L.T.; Mhangara, P. Mapping chlorophyll-a concentrations in a cyanobacteria-and algae-impacted Vaal Dam using Landsat 8 OLI data. *S. Afr. J. Sci.* **2018**, *114*, 1–9. [CrossRef]
40. Balasubramanian, S.V.; Pahlevan, N.; Smith, B.; Binding, C.; Schalles, J.; Loisel, H.; Gurlin, D.; Greb, S.; Alikas, K.; Randla, M. Robust algorithm for estimating total suspended solids (TSS) in inland and nearshore coastal waters. *Remote Sens. Environ.* **2020**, *246*, 111768. [CrossRef]
41. Gitelson, A.A.; Schalles, J.F.; Hladik, C.M. Remote chlorophyll-a retrieval in turbid, productive estuaries: Chesapeake Bay case study. *Remote Sens. Environ.* **2007**, *109*, 464–472. [CrossRef]
42. Gons, H.J.; Rijkeboer, M.; Ruddick, K.G. A chlorophyll-retrieval algorithm for satellite imagery (Medium Resolution Imaging Spectrometer) of inland and coastal waters. *J. Plankton Res.* **2002**, *24*, 947–951. [CrossRef]
43. Gons, H.J. Optical teledetection of chlorophyll a in turbid inland waters. *Environ. Sci. Technol.* **1999**, *33*, 1127–1132. [CrossRef]
44. Pahlevan, N.; Roger, J.-C.; Ahmad, Z. Revisiting short-wave-infrared (SWIR) bands for atmospheric correction in coastal waters. *Opt. Express* **2017**, *25*, 6015–6035. [CrossRef]
45. Abiye, T.A. Provenance of groundwater in the crystalline aquifer of Johannesburg area, South Africa. *Int. J. Phys. Sci.* **2011**, *6*, 98–111.
46. De Villiers, C.J.; Barnard, P. Environmental reporting in South Africa from 1994 to 1999: A research note. *Meditari Account. Res.* **2000**, *8*, 15–23. [CrossRef]
47. The Department of Water Affairs and Forestry (DWAF). *Crocodile (West) River Catchment: Internal Strategic Perspective*; Department of Water Affairs and Forestry: Pretoria, South Africa, 2004.
48. The Department of Water Affairs and Forestry (DWAF). *Hartbeespoort Dam Integrated Biological Remediation Program*; Department of Water Affairs and Forestry: Pretoria, South Africa, 2007; pp. 1–18.
49. Kerrigan, K.; Ali, K.A. Application of Landsat 8 OLI for monitoring the coastal waters of the US Virgin Islands. *Int. J. Remote Sens.* **2020**, *41*, 5743–5769. [CrossRef]
50. Pahlevan, N.; Sarkar, S.; Franz, B.A.; Balasubramanian, S.V.; He, J. Sentinel-2 MultiSpectral Instrument (MSI) data processing for aquatic science applications: Demonstrations and validations. *Remote Sens. Environ.* **2017**, *201*, 47–56. [CrossRef]
51. Franz, B.A.; Werdell, P.J. A Generalized Framework for Modeling of Inherent Optical Properties in Remote Sensing Applications. In Proceedings of the Ocean Optics, Anchorage, AK, USA, 25 September–1 October 2010.
52. Shi, W.; Wang, M. Detection of turbid waters and absorbing aerosols for the MODIS ocean color data processing. *Remote Sens. Environ.* **2007**, *110*, 149–161. [CrossRef]
53. Shi, W.; Wang, M.; Zhang, Y. Inherent optical properties in lake taihu derived from VIIRS satellite observations. *Remote Sens.* **2019**, *11*, 1426. [CrossRef]
54. Carswell, T.; Costa, M.; Young, E.; Komick, N.; Gower, J.; Sweeting, R. Evaluation of MODIS-Aqua atmospheric correction and chlorophyll products of Western North American coastal waters based on 13 years of data. *Remote Sens.* **2017**, *9*, 1063. [CrossRef]
55. Wang, M.; Shi, W.; Jiang, L. Atmospheric correction using near-infrared bands for satellite ocean color data processing in the turbid western Pacific region. *Opt. Express* **2012**, *20*, 741–753. [CrossRef]
56. Wang, D.; Ronghua, M.; Xue, K.; Li, J. Improved atmospheric correction algorithm for Landsat 8–OLI data in turbid waters: A case study for the Lake Taihu, China. *Opt. Express* **2019**, *27*, A1400. [CrossRef]
57. DWAF. *South African Water Quality Guidelines—Domestic Water Use*, 2nd ed.; Department of Water Affairs and Forestry: Pretoria, South Africa, 1996; Volume 1, pp. 86–87.
58. Abiye, T.A.; Leketa, K.C. Historic Climatic Variability and Change: The Importance of Managing Holocene and Late Pleistocene Groundwater in the Limpopo River Basin, Southern Africa. In *Climate Change and Water Resources in Africa*; Springer: Berlin/Heidelberg, Germany, 2021; pp. 131–143.
59. Welch, E.B.; Cooke, G.D. Internal phosphorus loading in shallow lakes: Importance and control. *Lake Reserv. Manag.* **2005**, *21*, 209–217. [CrossRef]
60. Mulholland, P.G.; Ronnie Best, G.; Coutant, C.C.; Hornberger, G.M.; Meyer, J.L.; Robinson, P.J.; Stenberg, J.R.; Eugene Turner, R.; Vera-Herrera, F.; Wetzel, R.G. Effects of climate change on freshwater ecosystems of the south-eastern United States and the Gulf Coast of Mexico. *Hydrol. Process.* **1997**, *11*, 949–970. [CrossRef]
61. Lürling, M.; Eshetu, F.; Faassen, E.J.; Kosten, S.; Huszar, V.L.M. Comparison of cyanobacterial and green algal growth rates at different temperatures. *Freshw. Biol.* **2013**, *58*, 552–559. [CrossRef]

remote sensing

MDPI

Article

Comparison of UAS and Sentinel-2 Multispectral Imagery for Water Quality Monitoring: A Case Study for Acid Mine Drainage Affected Areas (SW Spain)

Melisa A. Isgró [1,2], M. Dolores Basallote [1,*], Isabel Caballero [3] and Luis Barbero [2]

1 Research Center on Natural Resources, Health, and the Environment, University of Huelva, 21071 Huelva, Spain
2 Department of Earth Sciences, University of Cádiz, Av. República Saharaui s/n, Puerto Real, 11510 Cádiz, Spain
3 Instituto de Ciencias Marinas de Andalucía (ICMAN-CSIC), Campus Universitario Río San Pedro, s/n, Puerto Real, 11510 Cádiz, Spain
* Correspondence: maria.basallote@dct.uhu.es; Tel.: +34-959-219-835

Abstract: Uncrewed Aerial Systems (UAS) and satellites are used for monitoring and assessing the quality of surface waters. Combining both sensors in a joint tool may scale local water quality retrieval models to regional and global scales by translating UAS-based models to satellite imagery. The main objective of this study is to examine whether Sentinel-2 (S2) data can complement UAS data, specifically from the MicaSense RedEdge MX-Dual sensor, for inland water quality monitoring in mining environments affected by acid mine drainage (AMD). For this purpose, a comparison between UAS reflectance maps and atmospherically corrected S2 imagery was performed. S2 data were processed with Case 2 Regional Coast Colour (C2RCC) and Case 2 Regional Coast Colour for Complex waters (C2X) atmospheric correction (AC) processors. The correlation between the UAS data and the atmospherically corrected S2 data was evaluated on a band-by-band and a pixel-by-pixel basis, and the compatibility of the spectral data was analyzed through statistical methods. The results showed C2RCC and C2X performed better for acidic greenish-blue and non-acidic greenish-brown water bodies concerning the UAS data than for acidic dark reddish-brown waters. However, significant differences in reflectance between the UAS sensor and both S2 AC processors have been detected. The poor agreement between sensors should be considered when combining data from both instruments since these could have further consequences in developing multi-scale models.

Keywords: abandoned mine; acidic water; surface water monitoring; drone; Copernicus programme

1. Introduction

Remote sensing techniques have gained popularity in recent years and the improvement in terms of spectral, spatial, and temporal resolutions of the sensors have increased their use for analysis and monitoring in different fields of study [1–3]. Lately, Uncrewed Aerial Systems (UAS) have grown as an innovative platform for acquiring ultra-high-resolution images at low altitudes, providing both high spatial centimeter-scale and flexible temporal resolution, at an increasingly affordable price. At the same time, recently launched satellites withing the Copernicus programme, such as Sentinel-2A/B (S2), can also provide a medium spatial resolution (10–20–60 m) with a revisit frequency of five days (at the equator) but delivering global coverage. Each of these data platforms is generally used separately even though it is known that they can be complementary and have strong synergies [4]. While the main applications that exploit the UAS/Satellite complementarities are precision agriculture and ecology [5–8], water resources in inland waters remain an underexplored field [9].

UAS imagery can be used to complement satellite imagery with cloud cover [10] and increase spatial resolution. These characteristics are especially relevant to mapping small water bodies, where the satellite resolution does not allow the recognition of events occurring at a local scale [11–13] or in the case of highly dynamic scenarios such as coastal areas [14], and to avoid mixed pixels particularly near the water boundaries [15]. On the other hand, the use of satellite data can scale local models to regional and global scales by translating UAS-based models to satellite imagery [4,7]. This would be impossible using only UAS since they have only a few km^2 swaths, due to the short flight time and the national legislation regarding air traffic and people's privacy. Nevertheless, the differences in acquisition time, resolution, environmental conditions, spectral bands, sensor position, and atmospheric effects (among others factors), may result in discrepancies in the performances of the sensors [4]. Thus, it is imperative to study the consistency between UAS and satellite observations, which may affect the multi-scale model proficiency.

This study focuses on the comparison of reflectance data acquired by the commercial sensor Micasense RedEdge-MX Dual onboard an UAS (DJI Matrice 210 V2) and by S2 in small water bodies affected by acid mine drainage (AMD) at different extents. AMD is the formation and movement of highly acidic water rich in metals and sulfate due to the oxidation of sulfide-bearing minerals exposed to atmospheric weathering processes, which frequently occurs in metal mine (and coal mine) areas. The result is a highly toxic solution that causes harmful effects affecting animals, plants, and even humans [16–18]. Data acquisition was carried out in the Tharsis mine complex, at the Iberian Pyrite Belt (IPB) (SW Spain; Figure 1), where the historical mining activity has left a legacy of several abandoned sites, which became the sources of AMD in the area. The acid leachates have reached the aquifers and river streams by percolation and runoff, being responsible for the degradation of the water system and the formation of water bodies containing highly acidic and metal-polluted water [16,18,19]. This research will help with understanding the coupling between chemical and optical properties of water and how remote sensing is applied to resolving these AMD-affected waters cases. AMD-waters have a complex composition due to their physicochemical characteristics, which are also observed in their different colors. Ocean Colour remote sensing usually distinguish two water classes: Case 1 waters which are ocean waters whose optical properties are determined by phytoplankton and other covarying compounds and Case 2 waters which are coastal and inland waters whose optical properties are determined by other constituents, such as mineral particles and Chromophoric Dissolved Organic Matter (CDOM), and their concentrations do not covary with phytoplankton [20]. Few studies have focused on quantifying water quality parameters in mining areas and most of them have been based on satellite datasets [21–26], whereas the application of UAS is rather recent [27]. The novelty of this study consists in exploring the possibility of combining different scale sensors in order to achieve multi-scale models. Therefore, the main objective is to examine whether S2 data can complement UAS data, specifically from the MicaSense RedEdge MX-Dual sensor, for inland water quality monitoring in mining environments. For this purpose, a comparison between UAS reflectance maps and atmospherically corrected S2 imagery was performed.

In addition, S2 requires atmospheric correction (AC) to subtract the atmospheric and sunglint contribution from the top of the atmosphere (TOA) signal, and there are several AC processors available for inland and coastal waters [28]. In other water masses, these models have been satisfyingly applied to process from TOA to BOA (bottom of atmosphere) reflectance, a step required for monitoring terrestrial, emerged intertidal, benthic or coastal ecosystems [2,28–30]. Hence, for this study, two AC processors were selected to be applied on S2 data: Case 2 Regional Coast Colour (C2RCC) and Case 2 Regional Coast Colour for Complex waters (C2X).

Figure 1. Location map of sampling sites. Five water bodies (Filón Norte (FN), Filón Centro (FC), Sierra Bullones (SB), Embalse Grande (EG), and Th18) sampled in the Tharsis Mine (**A**), and five water bodies (Mina Lagunazo (ML), Embalse Lagunazo (EL), Laguna Lagunazo A, B, and C (LLA, LLB, and LLC, respectively)] in Lagunazo Mine site (**B**). (**C**) corresponds to aerial images of different inland water masses evaluated in this study.

2. Materials and Methods

2.1. Study Area and Water Physicochemical Parameters

The study area was the IPB (SW Spain), which contains one of the greatest concentrations of polymetallic massive sulfide mineralization on Earth, with more than 1700 Mt of sulfide ore estimated (mined and reserves) [17,31]. As a result of the historical mineral exploitation, there are numerous abandoned mining sites, such as the Tharsis and Lagunazo mines [32], which were the areas sampled for this study.

Figure 1 shows the water bodies selected for this study: Embalse Grande (EG), Sierra Bullones (SB), Filón Centro (FC), Filón Norte (FN), and TH18 from Tharsis mine (Figure 1A); and Mina Lagunazo (ML), Embalse Lagunazo (EL), Laguna Lagunazo A, B and C (LLA, LLB, LLC) from Lagunazo mine site (Figure 1B). EG and EL are relatively clean water reservoirs used for agricultural purposes. SB, FC, FN, and ML are pit lakes flooded with acidic waters, and TH18, LLA, LLB, and LLC are surficial AMD sites.

The study area covers a wide gradient of water compositions, from neutral to extremely acidic pH (0.01–7.76) and from low to high metal-enriched solutions (e.g., 0.05–60, 895 mg L^{-1} Fe). The water composition is also reflected in the watercolor: EG and ML are nonacidic water bodies and have a greenish-brown color, ML is acidic and presents greenish-blue color and the rest of them are dark reddish-brown acid waters with high iron concentrations. In situ physicochemical data were collected simultaneously with the UAS data acquisition. Detailed information about sampling can be found in [33]. Briefly, pH, electrical conductivity (EC), oxidation–reduction potential (ORP), and temperature (T) were measured at each sampling point with a CrisonMM40 þ multimeter, previously calibrated with certified solutions. Turbidity was measured using a Hanna HI-93703 portable turbidimeter at the water surface. Water samples were collected in HDPE bottles previously acid washed, filtered (0.45 mm pore size cellulose nitrate membrane), and HNO_3-acidified to pH < 2. Major elements (Al, Ba, Ca, Cu, Fe, K, Mg, Mn, Na, P, S, Si, Sr, and Zn) were determined by inductively coupled plasma-atomic emission spectroscopy (ICP-AES; PerkinElmer® Optima 3200 RL, Waltham, MA, USA) at the Institute of Environment Assessment and Water Research (IDAEA-CSIC, Barcelona, Spain).

2.2. Remotely Sensed Data

Two different multispectral sensors were used for this study: an airborne sensor (MicaSense RedEdge-MX Dual onboard a DJI Matrice 210 V2) and a spaceborne sensor (S2). The spatial resolution and bandwidths of the sensors used in this study are described in Table 1. Only the bands used for this study are shown in the table.

Table 1. Central wavelengths (nm) and bandwidths of the evaluated sensors bands.

Sensors	Coastal Aerosol	Blue	Green	Red	Red-Edge1	Red-Edge2	NIR
Sentinel-2							
Central Wavelength	443	490	560	665	704	740	865
Bandwith	20	60	36	30	15	15	21
MicaSense							
Central Wavelength	444	475	560	668	705	740	840
Bandwith	28	20	20	10	10	18	40

2.2.1. UAS Data Acquisition and Processing

The UAS DJI Matrice 210 V2, a rotary-wing quadcopter platform with vertical take-off and landing (VTOL), was used to collect aerial images at a constant flight altitude over water bodies. The sensor mounted on the UAS to acquire the imagery was a lightweight multispectral camera: MicaSense RedEdge-MX Dual, which has ten different channels. The sensor resolution is 1.2-megapixel for each of the multispectral bands and 3.6-megapixel for the RGB captures. The lens achieves a ground sampling distance (GSD) of 8 cm/pixel at 120 m above ground level (AGL). Autonomous aerial flights were performed using the DJI GS Pro planning software from 120 m AGL altitude with 80% frontal and 75% side image overlap, the grid was simple, and the speed was set at 10 m/s. It is assumed that all the multispectral imagery is in the nadir position due to the location of the camera. No Ground Control Points (GCPs) were added to the flights since the Micasesnse sensor has an integrated GPS that geo-tag each of the images acquired by the UAS. Furthermore, it was also equipped with a Downwelling Light Sensor (DLS 2), and a MicaSense's Calibrated

Reflectance Panel (CRP) to perform the radiometric calibration on the ambient light changes during the flight.

The DLS 2 and CRP are necessary for the radiometric calibration on the ambient light changes during the flight. For this purpose, a picture of the CRP was taken before and after the flight to capture the lighting conditions of the flight ensuring no shadows were covering the panel. The use of the CRP allows comparing the data acquisition over different dates or at different times of day, which is essential for a multiple flight analysis. On the other hand, the DLS 2 is located at the top of the UAS looking upwards towards the sky and measures the variations in light intensity throughout the flight by using 10 light sensors located in different planes to define the sun position and its effect on the images. This light information is stored in each TIFF image metadata and is used to correct the illumination variability during the mosaicking process [34]. By using both the DLS2 and CRP during the data collection, it is possible to choose the best calibration option to get a high-quality reflectance map during the image processing.

Pix4D mapper (Pix4D S.A., Lucerne, Switzerland) was the Structure from Motion (SfM) software used to process the multispectral images. Before starting the pre-processing, images were inspected and discarded if blurred. From all images captured by the UAS platform, the software creates orthomosaics of the surface reflectance values by image alignment, creation of a 3D point cloud and triangle mesh, and digital surface model (DSM), orthomosaic, and index generation (Figure S1). During these steps, the radiometric processing and calibration are performed, and different calibration methods can be applied depending on the input data and the light conditions. In this study, two methods were implemented: Camera and Sun Irradiance and Camera only [35]. For Camera and Sun Irradiance calibration, Pix4D uses the parameters written in the EXIF metadata and relate to the camera, as well as the CRP to calculate the absolute irradiance and the DLS data for normalizing each image for changes in the incoming radiation during the flight. For Camera only calibration, Pix4D also uses the parameters written in the EXIF metadata and relate to the camera and the CRP, but the DLS data is not considered. The last method is more appropriate when the sky is clear and there is no variation in the light conditions [36]. As a result of this process, ten single orthomosaics of the surface reflectance values with a GSD of 0.08 m were generated for all the flights. These reflectance maps contain reflectance values ranging between 0.0 and 1.0 for each pixel. The atmospheric correction was not considered for the UAS imagery, since it is a close-range remote sensing approach and the atmosphere layer between the UAS and the ground is so thin that it can be omitted [9,37]. Considering that the UAS data are a low altitude remote sensing tool with ultra-high spatial resolution that is not affected by the atmosphere and that field spectro-radiometric measurements were not performed, the UAS data were considered as the ground truth reference source for the remote sensing.

The UAS data acquisition was carried out on 7 October 2020 and the overpassing of S2 was the next day on 8 October 2020.

2.2.2. Sentinel-2 Data Acquisition and Processing

S2 Level-1C product was acquired from the ESA Copernicus Open Access Hub (https://scihub.copernicus.eu/, accessed on 1 March 2022). The study region is covered by one S2 tile (29SQB), which is a 100 km by 100 km squared orthoimage in UTM/WGS84 projection. The tile was acquired as close as possible to the date on which the UAS-based survey was carried out. Images with cloud cover and sun glint over the water bodies were discarded. S2 images were available on 8 October 2020, at 11:08:49 UTC.

Satellite imagery needs to be atmospherically corrected since the radiance reaching the sensor interacts with the atmosphere and is affected by several parameters, such as wind speed, sun glint, angle of the sun, aerosols, water vapor, and land adjacency effects, among others. AC processors can correct the "noise" from the radiance to get the water-leaving reflectance true value. Many AC processors have been tested and validated for coastal waters, but only a few studies have been performed over inland waters [20,28,38–40].

In this study, the images were processed to Level-2A (L2A) with the Case 2 Regional CoastColour processor (C2RCC) and Case 2 Regional Coast Colour for Complex waters (C2X) [40]. C2RCC is an AC processor based on artificial neural networks (ANN) that relate TOA water reflectance to radiative transfer simulations of water-leaving radiances. It uses a radiative transfer model for the atmospheric characteristics and a bio-optical model for the water characteristics. These water characteristics are parameterized considering the water reflectance in the blue spectrum (443 nm), based on bio-optical databases. The ANN are then trained based on TOA reflectance, these modelled water characteristics and environmental variables [41]. C2RCC was complemented with the CoastColour dataset to extend the range for coastal waters including extreme cases resulting in the C2X. C2X uses a special set of networks trained for extreme ranges of absorption and scattering, that is, for very turbid waters [42]. Previous studies with S2 imagery in inland waters have obtained good performance for both AC processors [19,43–46], even though for coastal waters ACOLITE has been proved to be more consistent [23]. The AC was performed through the ESA's Sentinel toolbox in the Sentinel Application Platform (SNAP).

All the bands were previously resampled to 10 m in SNAP using the bilinear upsampling method. The satellite imagery was then reprojected to WGS 84/UTM zone 29N with nearest neighbor interpolation to match the UAS data.

2.3. Spectral Band Comparison

The UAS data (8 cm) was resampled and aligned in the open-source geographic information system software QGIS to match the satellite pixel size of 10 m employing the bilinear interpolation method. To compare the results of both sensors datasets, for each water body the average of six 3 × 3 boxes were extracted from the UAS and S2. The comparison was based on pure pixels, by choosing central pixels of the water body when possible, meaning that pixels with only water cover were selected. The UAS bands were transformed into remote sensing reflectance (Rrs) with unit 1/sr by the following equation [47,48]:

$$Rrs = Rhow/\pi$$

where π has units of steradian and Rhow is the non-dimensional water-leaving reflectance. Then the UAS-imagery [47,48] were paired with the most contemporaneous available S2 image (8 October 2020) (see detailed information on the flowchart in Figure 2).

The correlation between the UAS data and the atmospherically corrected S2 data (C2RCC and C2X) was evaluated on a band-by-band basis to determine the strength of the relationship between both sensors. For this purpose, a linear model was fitted to the UAS-S2 image pair and the coefficient of determination (R^2) was calculated and compared. Moreover, a pixel-by-pixel and band-by-band comparison was performed as a graphical output of the analyzed reflectance values. To validate the consistency of both sensors, the band-by-band root mean square error (RMSE), mean absolute error (MAE), and bias between all the UAS resampled pixels and the corresponding S2 data were calculated. MAE and bias provide more robust output with non-Gaussian distributions and outliers [44].

Afterwards, the compatibility of the spectral data from the UAS bands and the S2 bands was analyzed through statistical methods. The non-parametric paired-sample Wilcoxon signed-rank test was applied to compare the mean difference of reflectance values from UAS data with the S2 data band-by-band. This test is an alternative to paired t-test since the data have no normal distribution according to the results obtained by the Shapiro-Wilk test. The test will indicate whether there is a statistical difference between the two means of UAS and S2 data in order to discern if the data combination of both sensors should be successful and can be complementary. In contrast, if significant differences exist, this should be taken into account when combining data from both sensors since these differences could affect the performance of models, leading to biases in the estimation of variables, such as water quality parameters.

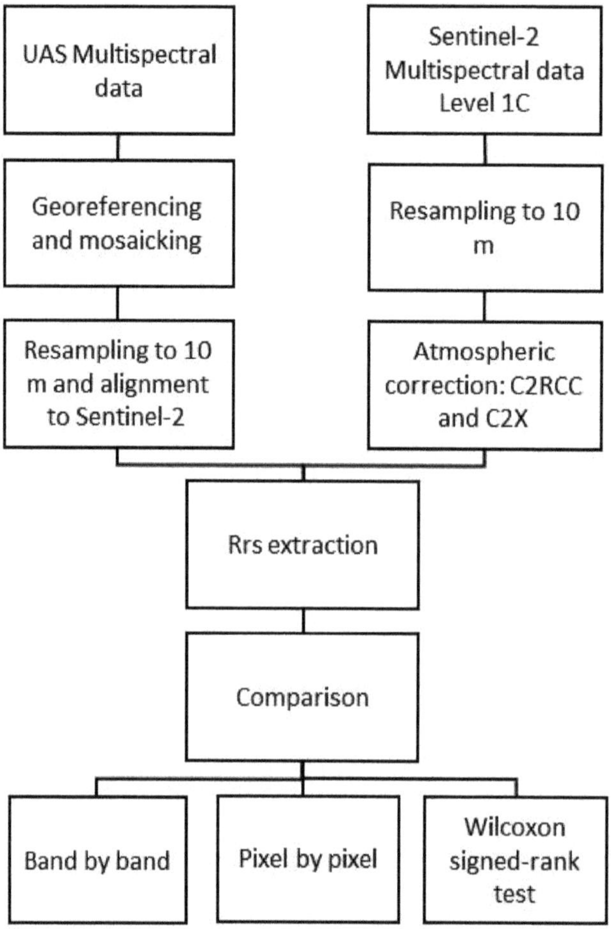

Figure 2. Flowchart of the methodological procedures for sensors comparison.

3. Results

3.1. Reflectance Spectra

The UAS and S2 sensors' reflectance spectra for each water body are shown in Figure 3. The S2 reflectances are atmospherically corrected with C2RCC and C2X processors. Based on the general shape of the spectra, greenish-blue and greenish-brown water bodies (ML, EG, and EL) showed a better agreement between sensors (Figure 3A,B,F). For dark reddish-brown waters (LLA, LLB, LLC, Th18, FC, FN, SB), the shape of the spectra seems to vary depending on the sensor and atmospheric correction procedure. The differences between UAS and S2 measurements are pronounced at the coastal blue and NIR bands, while reflectances are more consistent in the blue, green, red, red-edge 1, and red-edge 2 bands. Moreover, dark reddish-brown water reflectance is much lower than greenish-blue and greenish-brown waters due to the higher absorption and smaller backscattering of its water surfaces.

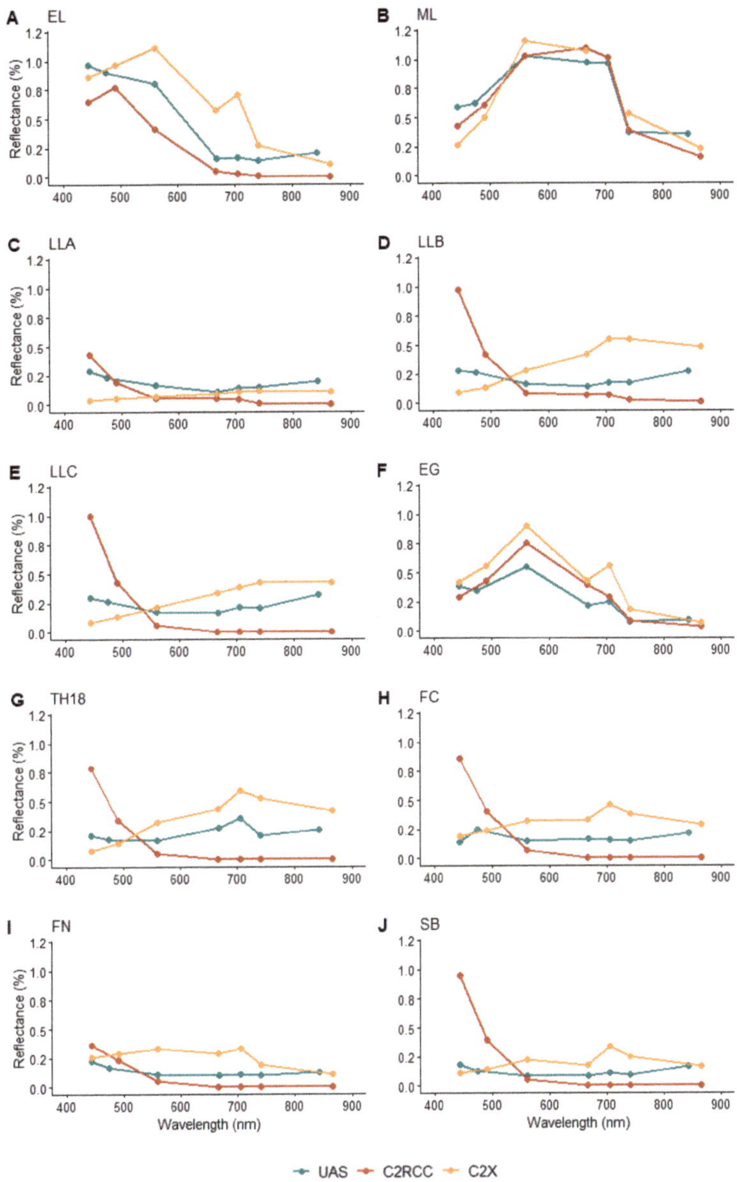

Figure 3. Reflectance spectra collected from Embalse Lagunazo (**A**), Mina Lagunazo (**B**), Laguna Lagunazo A (**C**), Laguna Lagunazo B (**D**), Laguna Lagunazo C (**E**), Embalse Grande (**F**), TH18 (**G**), Filón Centro (**H**), Filón Norte (**I**), and Sierra Bullones (**J**) in October with UAS and Sentinel-2 bands obtained with C2RCC and C2X atmospheric correction processing.

3.2. Spectral Band Comparison

The band-by-band and pixel-by-pixel comparison showed that the UAS and C2RCC data follow the same trend for the green, red, red-edge1, red-edge2, and NIR bands (Figure 4C–G)). Nevertheless, C2X showed a high variability of the pixel values in all the analyzed bands. The UAS surface reflectance data has low variability in the reflectance values of all the analyzed bands. The data showing higher reflectance in the coastal blue, blue, and green band correspond to EL, ML, and EG (Figure 4A–C) which are the greenish-blue and greenish-brown water bodies.

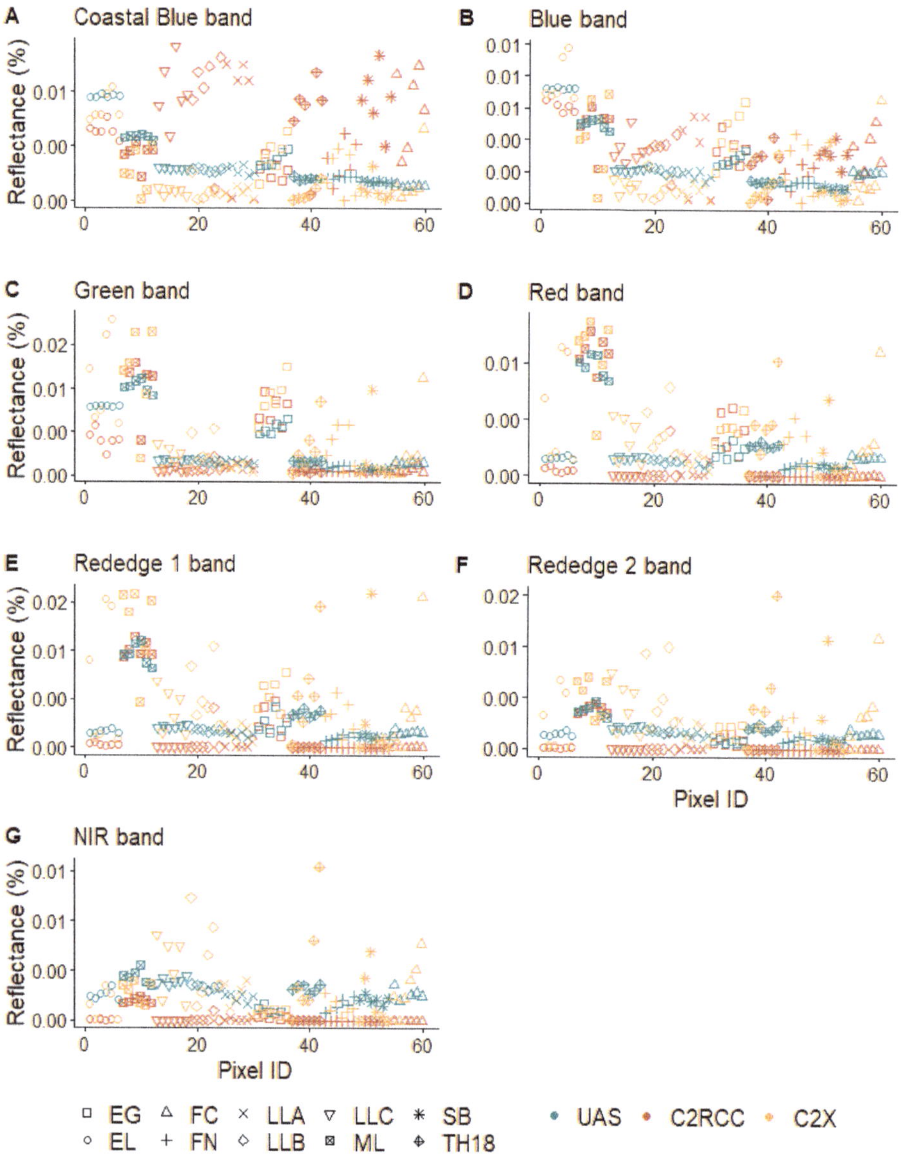

Figure 4. Pixel-by-pixel and band-by-band comparison of UAS remote sensing reflectance data with S2 atmospherically corrected products (CR2CC and C2X). The analyzed bands are costal blue (**A**), blue (**B**), green (**C**), red (**D**), rededge1 (**E**), rededge2 (**F**), and NIR (**G**). UAS data collected on 7 October and S2 data on 8 October 2020.

To assess the consistency of both sensors, Figures 5A and S2 shows the statistical analysis applied to all spectrum bands: R^2, RMSE, MAE, and bias, per band and per

AC. It can be observed that UAS vs. C2RCC performed better than UAS vs. C2X. The coefficients of determination obtained from comparing UAS orthomosaics and C2RCC were higher than those obtained with C2X for the green, red, red-edge1, red-edge2, and NIR bands (Figures 5A and S2). There is a weak relationship between the two sensors for both atmospheric corrections for the NIR band and there is an extremely weak relationship for the coastal blue band whose R^2 is the lowest of all bands. This may be because the coastal blue band usually has more residuals from the atmospheric correction than other visible bands. For the regression between UAS and C2RCC data, the results suggest that while more than 70% of the variation in the green, red and red-edge1 bands can be explained by the linear model, the remaining variation might be due to the differences such as acquisition dates, the spectral bands, bandwidth of the sensors, radiometric correction, and atmospheric residuals, among others [46,49].

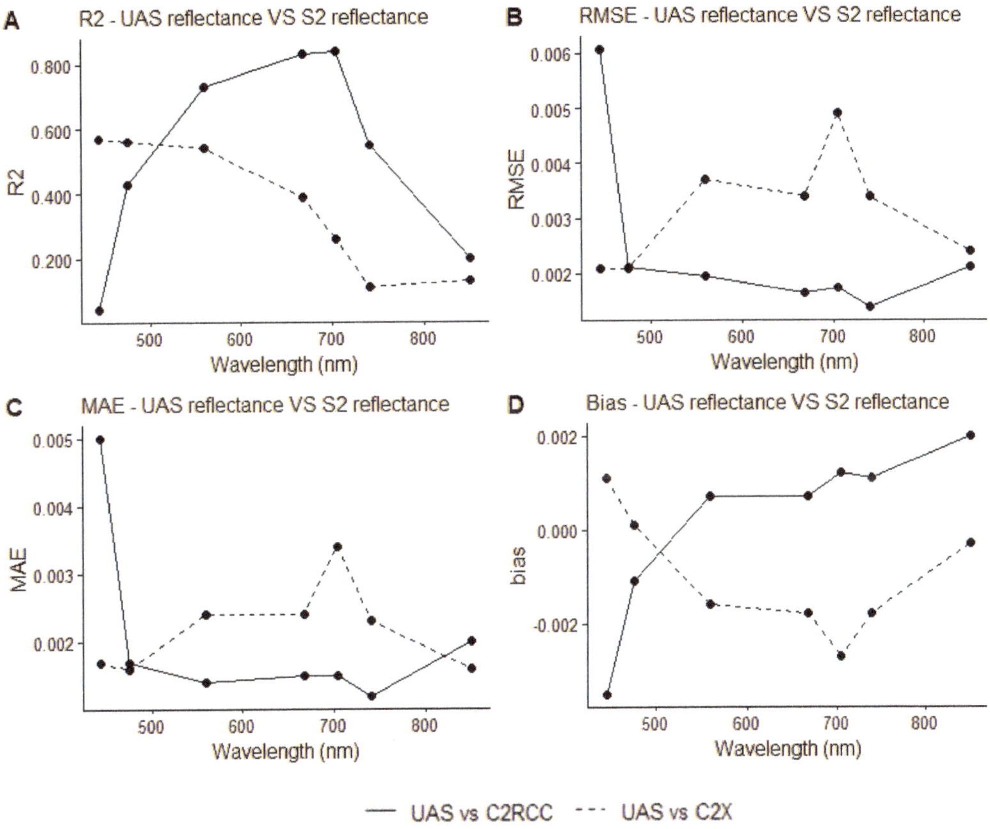

Figure 5. Band by band summary of statistics. (**A**) R2 (coefficient of determination); (**B**) RMSE (Root Mean Square Error) (1/sr), (**C**) MAE (Mean Absolute Error) (1/sr); (**D**) bias. UAS data of 7 October and S2 data of 8 October 2020.

Regarding the coastal blue band, the pixel-by-pixel comparison showed high variability between the UAS data and both satellite atmospheric corrections (Figure 4). Only in this case, C2X values were closer to the UAS data than C2RCC ($RMSE_{CB}$ = 0.0060 1/sr, MAE_{CB} = 0.050 1/sr for C2RCC and $RMSE_{CB}$ = 0.0021 1/sr, MAE_{CB} = 0.017 1/sr for C2X) (Figure 5B). The blue band has also high variability between the UAS data and both satellite atmospheric corrections and the RMSE values are the same ($RMSE_B$ = 0.0021 1/sr) for this band while the MAE values are similar (MAE_B = 0.017 1/sr for C2RCC and MAE_B = 0.016 1/sr for C2X). The bias values of C2RCC show negative values for these bands, indicating an underestimation of the reflectance obtained by this AC processor, while C2X shows positive values indicating an overestimation of the reflectance values comparing to the UAS data.

Regarding the green, red, red-edge1, and red-edge2 bands, C2RCC values were generally closer to the UAS data ($RMSE_G$ = 0.0019 1/sr, $RMSE_R$ = 0.0016 1/sr, $RMSE_{RE1}$ = 0.0017 1/sr, $RMSE_{RE2}$ = 0.0014 1/sr) than the C2X ($RMSE_G$ = 0.0036 1/sr, $RMSE_R$ = 0.0034 1/sr, $RMSE_{RE1}$ = 0.0049 1/sr, $RMSE_{RE2}$ = 0.0034 1/sr) (Figure 5B). The MicaSense remote sensing reflectance data were always greater than the satellite's for the green, red, and red-edge1, except in the pixels corresponding to ML and EG (Figure 4C–E). C2RCC show a low positive bias, while C2X has negative values (Figure 5).

Regarding the NIR band, the pixel-by-pixel plot indicates that C2RCC and UAS have a more similar trend than C2X (Figure 4G), but both atmospheric corrections have similar RMSE values ($RMSE_{NIR}$ = 0.0021 1/sr for C2RCC, $RMSE_{NIR}$ = 0.0024 1/sr for C2X) in this spectral region. Furthermore, C2RCC increases the bias in the NIR band and C2X decreases it.

The Wilcoxon signed-rank test from the paired two samples showed a statistically significant difference between mean reflectances of UAS and mean reflectances of S2 based on a band-by-band comparison (p-value < 0.001) (Table 2). The p-value indicates strong evidence against the null hypothesis, as there is less than a 5% probability the null is correct. Therefore, the null hypothesis is rejected, and the alternative hypothesis is accepted. The only non-significant difference from the whole analysis was found in the blue (p-value = 0.214) and NIR band (p-value < 0.680) for the C2X processor.

Table 2. Wilcoxon signed-rank test of UAS and S2 remote sensing reflectance data.

Bands	Coastal Blue	Blue	Green	Red	RedEdge1	RedEdge2	NIR
Observations	60	60	60	60	60	60	60
Spearman correlation	−0.243	0.546	0.685	0.338	0.277	0.000	−0.084
Wilcoxon signed-rank test for UAS and C2RCC data	<0.0001	<0.0001	0.001	0.001	<0.0001	<0.0001	<0.0001
Wilcoxon signed-rank test for UAS and C2X data	<0.0001	0.214	0.002	<0.0001	<0.0001	<0.0001	0.680

4. Discussion

The type of waters presented in this study corresponds to Case 2 waters, which have great variability in optical properties and can be classified as highly absorbing or highly scattering waters [38,50,51]. Dark reddish-brown acidic waters as can be seen in Figure 3C–E,G–J) are highly absorbing waters characterized by very low water-leaving reflectance and the maximum Rrs is typically <0.5%. While greenish-blue and greenish-brown waters, such as EL, ML, and EG, are highly scattering waters (Figure 3A,B,F).

The spatial resampling of the drone data from 8 cm to 10 m to match the satellite resolution provided reflectance values coherent with the values obtained from the S2 data, even though UAS data were more diverse in reflectance values than S2. The comparison of the UAS and S2 reflectance spectra for each water body revealed that C2RCC and C2X performed better for greenish-blue and greenish-brown water bodies concerning the UAS data, while for dark reddish-brown waters there is a lack of consistency in the form of the spectral signature obtained by both sensors (Figure 3). According to the band-by-band and pixel-by-pixel comparison and the statistics obtained, C2RCC data is more similar to the UAS data than C2X (Figure 4). However, despite the strong correlation found in the green, red, and red-edge1 bands for C2RCC, a significant difference in reflectance, between MicaSense RedEdge-MX Dual and S2 multispectral data, was found with the Wilcoxon signed-rank test (Table 2). This means that these differences in the reflectance values should be considered when combining data from both sensors since these could have further consequences in developing multi-scale models.

The poor agreement between sensors may arise from different causes such as the spectral bands of the sensors, the bandwidth of the sensors, the radiometric correction, sensor degradation over time and the atmospheric conditions [46,49,52]. In this case, it is important to mention that the bandwidths of S2 were much wider than the UAS in several important bands, maybe due to spectral contamination. Radiometric resolution (number of data bits) is often overlooked and can be a factor along with signal to noise characteristics of the sensor. In this case, the signal-to-noise ratio (SNR), might be a problem since the Micasense was developed for land-studies, not water quality (requiring a higher SNR due in part to factors of water leaving radiances). In addition, it is well known that SNR in Sentinel 2 is low in blue-green bands. Hence, that difference in radiometric resolution can be a factor along with signal to noise characteristics of the sensors. One important factor is the temporal matching, there is one day difference between the UAS and S2 data. It is known that the dynamics of small water masses can greatly impact the results especially at very high spatial resolutions [53]. The dynamics of the water bodies may appear calm, but it is recommended to take matchups within a two-hour window around crossover time. Finally, even though the sensor MicaSense RedEdge is provided with DLS 2 and CRP to accomplish factory radiometric calibration, the calibration parameters may change when the camera is utilized in real-world conditions. As a result, previous studies recommend regular calibration of the radiometric calibration coefficients, proving that there is an impact of the radiometric correction on the spectral consistency between UAS and satellite data [52,54].

Other studies have found a relatively good correlation of bands between MicaSense RedEge sensor and S2. For instance, it has found that the spectral response of MicaSense RedEge sensor matches the response of S2 in a forest environment, except for the NIR band [55]. A good agreement in the shape of the spectral signature of a drinking water reservoir affected by eutrophication for both sensors has been showed, making it possible to develop a multi-scale monitoring tool [9]. On the other hand, the authors of [46] performed the t-test and Wilcoxon signed-rank test to compare the reflectance values on a crop environment from MicaSense RedEge sensor and S2 on a band-by-band basis obtaining the same result as ours, a significant difference in reflectances. It is important to mention that the analyzed water bodies with similar spectral signatures to the one studied by [9] (EG, EL, ML) have shown coherence between the shapes of the spectral curves of both sensors, but are the red AMD-affected water bodies the ones that did not show coherence in the results and should be further study to determine the cause of the difference in the sensors' responses. It is suggested that the spectrum of the AMD-affected water was not within the training database of the AC processors or the spectrum is out-of-scope of the algorithm definition, thus giving a water leaving reflectance spectrum with large deviations from the UAS data [40]. In this sense, new AC should be designed for this complex type of water. A recommendation to improve the obtained results is to verify the sensors data with field spectroradiometric measurements or hyperspectral sensor mounted on a UAS, especially in the dark reddish-brown acidic waters where UAS and S2 data failed to obtain similar reflectance data. Field spectroradiometric measurement of surface reflectance is a useful source of information that is completely unaffected by the atmosphere and, thus, is suitable for assessing the accuracy of the data obtained with other sensors [55,56].

5. Conclusions

This study dealt with a comparison between images of small water bodies derived from two different platforms: an UAS and the satellite S2. The UAS is an ultra-high-resolution platform and S2 a free medium resolution platform. The comparison was mainly based to determine whether the S2 sensor can complement the UAS sensor, in this case, the MicaSense RedEdge MX-Dual sensor, for inland water quality monitoring in mining environments. For the S2 data, two AC processors were applied to the imagery:

C2RCC and C2X, which have a good performance in complex waters and extreme complex waters respectively. Thus, the overall comparison was made between the UAS, C2RCC, and C2X datasets resampled to the spatial resolution of 10 m. Considering that the UAS data are a low altitude remote sensing tool with ultra-high spatial resolution (8 cm) that is not affected by the atmosphere and that field spectroradiometric measurements were not performed, the UAS data were considered as the ground truth reference source for the remote sensing reflectance. The results suggest that even though C2RCC showed a better performance than C2X compared to the UAS data, there is a significant difference in reflectance between MicaSense RedEdge-MX Dual and both AC processors which should be considered when combining data from both sensors since these could have further consequences in developing multi-scale models. The results obtained may be considered preliminary due to the limited in situ database and the lack of handheld spectroradiometric measurements to validate the performance of the UAS sensor and the AC processors.

Supplementary Materials: The following supporting information can be downloaded at: https://www.mdpi.com/article/10.3390/rs14164053/s1, Figure S1: Simplified orthomosaics of the surface reflectance building process: (a) images acquisition planning, (b) image alignment and 3D point cloud generation in Pix4Dmapper, and (c) reflectance map, in Filón Centro water body (Tharsis mine). Figure S2: Regression models between S2 (atmospherically corrected with C2X and C2RCC) and UAV-derived datasets for each analyzed band.

Author Contributions: Conceptualization, M.A.I., M.D.B. and L.B.; methodology, M.A.I.; Sentinel-2 processing and correction, I.C.; UAS processing, M.A.I., M.D.B. and L.B.; validation, M.A.I.; writing—review and editing M.A.I., M.D.B., I.C. and L.B. All authors have read and agreed to the published version of the manuscript.

Funding: This study was supported in part by the Erasmus Mundus Joint Master Degree (EMJMD) in Water and Coastal Management (WACOMA) with the contribution of the Erasmus+ Programme of the European Union. This work was also supported by Plan Andaluz de Investigación RNM 166 Environmental radioactivity research group (LB) and partially by the FEDER UHU2020-21 Project. UAS equipment from University of Cádiz Drone Service was supported by MINECO infrastructure projects (EQC2018-00446-P and UNCA-2013-1969). M.D.B. thanks the Spanish Ministry of Science and Innovation for the Postdoctoral Fellowship granted under application reference IJC2018-035056-I. This research was also funded by the Spanish Ministry of Science and Innovation, the Spanish State Research Agency, the European Regional Development Fund MCIN/AEI/10.13039/501100011033 (Sen2Coast Project; RTI 2018-098784-J-I00), and the Consejería de Transformación Económica, Industria, Conocimiento y Universidades from Andalusian Government through the Andalusian FEDER operational program 2014–2020 (A1123060E0_PYC20 RE 032 UHU) and the call 2020 for collaborative interest projects in the field of the Innovation Ecosystems of the International Excellence Centers.

Data Availability Statement: Not applicable.

Acknowledgments: We thank to Tharsis Mining & Metallurgy for allowing us to collect the samples in the abandoned mining areas that belong to the company since 2018. The authors also gratefully acknowledge the support of the Guest Editor, the Assistant Editor and the three anonymous reviewers for their comments and positive criticisms, which notably improved the quality of the original paper.

Conflicts of Interest: The authors declare no conflict of interest.

References

1. Ritchie, J.C.; Zimba, P.V.; Everitt, J.H. Remote Sensing Techniques to Assess Water Quality. *Photogramm. Eng. Remote Sens.* **2003**, *69*, 695–704. [CrossRef]
2. Dörnhöfer, K.; Oppelt, N. Remote sensing for lake research and monitoring—Recent advances. *Ecol. Indic.* **2016**, *64*, 105–122. [CrossRef]
3. Chawla, I.; Karthikeyan, L.; Mishra, A.K. A review of remote sensing applications for water security: Quantity, quality, and extremes. *J. Hydrol.* **2020**, *585*, 124826. [CrossRef]
4. Emilien, A.-V.; Thomas, C.; Thomas, H. UAV & satellite synergies for optical remote sensing applications: A literature review. *Sci. Remote Sens.* **2021**, *3*, 100019.

5. Matese, A.; Toscano, P.; Filippo Di Gennaro, S.; Genesio, L.; Primo Vaccari, F.; Primicerio, J.; Belli, C.; Zaldei, A.; Bianconi, R.; Gioli, B. Intercomparison of UAV, aircraft and satellite remote sensing platforms for precision viticulture. *Remote Sens.* **2015**, *7*, 2971–2990. [CrossRef]
6. Elkind, K.; Sankey, T.; Munson, S.; Aslan, C. Invasive buffelgrass detection using high-resolution satellite and UAV imagery on Google Earth Engine. *Remote Sens. Ecol. Conserv.* **2019**, *5*, 318–331. [CrossRef]
7. Sagan, V.; Maimaitijiang, M.; Sidike, P.; Maimaitiyiming, M.; Erkbol, H.; Hartling, S.; Peterson, K.; Peterson, J., II; Burken, J.; Fritschi, F. UAV/satellite multiscale data fusion for crop monitoring and early stress detection. *ISPRS Int. Arch. Photogramm. Remote Sens. Spat. Inf. Sci.* **2019**, *XLII-2/W13*, 715–722. [CrossRef]
8. Messina, G.; Peña, J.M.; Vizzari, M.; Modica, G. A comparison of UAV and satellites multispectral imagery in monitoring onion crop. An application in the 'Cipolla Rossa di Tropea'(Italy). *Remote Sens.* **2020**, *12*, 3424. [CrossRef]
9. Cillero Castro, C.; Domínguez Gómez, J.A.; Delgado Martín, J.; Hinojo Sánchez, B.A.; Cereijo Arango, J.L.; Cheda Tuya, F.A.; Díaz-Varela, R. An UAV and Satellite Multispectral Data Approach to Monitor Water Quality in Small Reservoirs. *Remote Sens.* **2020**, *12*, 1514. [CrossRef]
10. Olivetti, D.; Roig, H.; Martinez, J.-M.; Borges, H.; Ferreira, A.; Casari, R.; Salles, L.; Malta, E. Low-cost unmanned aerial multispectral imagery for siltation monitoring in reservoirs. *Remote Sens.* **2020**, *12*, 1855. [CrossRef]
11. Michez, A.; Piégay, H.; Lisein, J.; Claessens, H.; Lejeune, P. Mapping of riparian invasive species with supervised classification of Unmanned Aerial System (UAS) imagery. *Int. J. Appl. Earth Obs. Geoinf. ITC J.* **2016**, *44*, 88–94. [CrossRef]
12. Gonçalves, G.; Andriolo, U.; Pinto, L.; Bess, F. Mapping marine litter using UAS on a beach-dune system: A multidisciplinary approach. *Sci. Total Environ.* **2020**, *706*, 135742. [CrossRef] [PubMed]
13. Taddia, Y.; Russo, P.; Lovo, S.; Pellegrinelli, A. Multispectral UAV monitoring of submerged seaweed in shallow water. *Appl. Geomat.* **2020**, *12*, 19–34. [CrossRef]
14. Andriolo, U.; Gonçalves, G.; Bessa, F.; Sobral, P. Mapping marine litter on coastal dunes with unmanned aerial systems: A showcase on the Atlantic Coast. *Sci. Total Environ.* **2020**, *736*, 139632. [CrossRef]
15. Choodarathnakara, A.L.; Kumar, T.A.; Koliwad, S.; Patil, C.G. Mixed Pixels: A Challenge in Remote Sensing Data Classification for Improving Performance. *Int. J. Adv. Res. Comput. Eng. Technol.* **2012**, *1*, 261.
16. Grande, J.; Valente, T.; Torre, M.; Santisteban, M.; Cerón, J.; Pérez-Ostalé, E. Characterization of acid mine drainage sources in the Iberian Pyrite Belt: Base methodology for quantifying affected areas and for environmental management. *Environ. Earth Sci.* **2014**, *71*, 2729–2738. [CrossRef]
17. Olías, M.; Nieto, J.M. Background Conditions and Mining Pollution throughout History in the Río Tinto (SW Spain). *Environments* **2015**, *2*, 295–316. [CrossRef]
18. Favas, P.J.C.; Sarkar, S.K.; Rakshit, D.; Venkatachalam, P.; Prasad, M.N.V. Chapter 17—Acid Mine Drainages From Abandoned Mines: Hydrochemistry, Environmental Impact, Resource Recovery, and Prevention of Pollution. In *Environmental Materials and Waste*; Prasad, M.N.V., Shih, K., Eds.; Academic Press: Cambridge, MA, USA, 2016; pp. 413–462.
19. Sánchez España, J.; Lopez Pamo, E.; Santofimia Pastor, E.; Diez Ercilla, M. The acidic mine pit lakes of the Iberian Pyrite Belt: An approach to their physical limnology and hydrogeochemistry. *Appl. Geochem.* **2008**, *23*, 1260–1287. [CrossRef]
20. Soomets, T.; Uudeberg, K.; Jakovels, D.; Brauns, A.; Zagars, M.; Kutser, T. Validation and comparison of water quality products in baltic lakes using sentinel-2 msi and sentinel-3 OLCI data. *Sensors* **2020**, *20*, 742. [CrossRef]
21. Buzzi, J.; Riaza, A.; García-Meléndez, E.; Weide, S.; Bachmann, M. Mapping changes in a recovering mine site with hyperspectral airborne HyMap imagery (Sotiel, SW Spain). *Minerals* **2014**, *4*, 313–329. [CrossRef]
22. Davies, G.E.; Calvin, W.M. Quantifying Iron Concentration in Local and Synthetic Acid Mine Drainage: A New Technique Using Handheld Field Spectrometers. *Mine Water Environ.* **2017**, *36*, 299–309. [CrossRef]
23. Modiegi, M.; Rampedi, I.T.; Tesfamichael, S.G. Comparison of multi-source satellite data for quantifying water quality parameters in a mining environment. *J. Hydrol.* **2020**, *591*, 125322. [CrossRef]
24. Pyankov, S.V.; Maximovich, N.G.; Khayrulina, E.A.; Berezina, O.A.; Shikhov, A.N.; Abdullin, R.K. Monitoring acid mine drainage's effects on surface water in the Kizel coal basin with Sentinel-2 satellite images. *Mine Water Environ.* **2021**, *40*, 606–621. [CrossRef]
25. Riaza, A.; Buzzi Marcos, J.; García-Meléndez, E.; Vázquez, I.; Bellido, E.; Carrère, V.; Muller, A. Pyrite mine waste and water mapping using Hymap and Hyperion hyperspectral data. *Environ. Earth Sci.* **2012**, *66*, 1957–1971. [CrossRef]
26. Tesfamichael, S.G.; Ndlovu, A. Utility of ASTER and Landsat for quantifying hydrochemical concentrations in abandoned gold mining. *Sci. Total Environ.* **2018**, *618*, 1560–1571. [CrossRef] [PubMed]
27. Flores, H.; Lorenz, S.; Jackisch, R.; Tusa, L.; Contreras, I.C.; Zimmermann, R.; Gloaguen, R. UAS-based hyperspectral environmental monitoring of acid mine drainage affected waters. *Minerals* **2021**, *11*, 182. [CrossRef]
28. Pereira-Sandoval, M.; Ruescas, A.; Urrego, P.; Ruiz-Verdú, A.; Delegido, J.; Tenjo, C.; Soria-Perpinyà, X.; Vicente, E.; Soria, J.; Moreno, J. Evaluation of atmospheric correction algorithms over Spanish inland waters for sentinel-2 multi spectral imagery data. *Remote Sens.* **2019**, *11*, 1469. [CrossRef]
29. Caballero, I.; Stumpf, R.P. Atmospheric correction for satellite-derived bathymetry in the Caribbean waters: From a single image to multi-temporal approaches using Sentinel-2A/B. *Opt. Express* **2020**, *28*, 11742–11766. [CrossRef]
30. Caballero, I.; Navarro, G. Monitoring cyanoHABs and water quality in Laguna Lake (Philippines) with Sentinel-2 satellites during the 2020 Pacific typhoon season. *Sci. Total Environ.* **2021**, *788*, 147700. [CrossRef]

31. Leistel, J.M.; Marcoux, E.; Thiéblemont, D.; Quesada, C.; Sánchez, A.; Almodóvar, G.R.; Pascual, E.; Sáez, R. The volcanic-hosted massive sulphide deposits of the Iberian Pyrite BeltReview and preface to the Thematic Issue. *Miner. Depos.* **1997**, *33*, 2–30. [CrossRef]
32. Pinedo Vara, I. *Piritas de Huelva: Su Historia, Minería y Aprovechamiento;* Suc. de Rivadeneyra: Madrid, Spain, 1963.
33. Isgró, M.A.; Basallote, M.D.; Barbero, L. Unmanned Aerial System-Based Multispectral Water Quality Monitoring in the Iberian Pyrite Belt (SW Spain). *Mine Water Environ.* **2022**, *41*, 30–41. [CrossRef]
34. Micasense. Capturing Multispectral Data in Ideal and Adverse Lighting Conditions. Available online: https://micasense.com/capturing-multispectral-data-in-ideal-and-adverse-lighting-conditions/ (accessed on 12 December 2021).
35. Micasense. How to Process MicaSense Sensor Data in Pix4D. 2022. Available online: https://support.micasense.com/hc/en-us/articles/115000831714-How-to-Process-MicaSense-Sensor-Data-in-Pix4D (accessed on 1 March 2022).
36. Micasense. Using Panels and/or DLS in Post-Processing. 2022. Available online: https://support.micasense.com/hc/en-us/articles/360025336894-Using-Panels-and-or-DLS-in-Post-Processing (accessed on 1 March 2022).
37. Burkhart, J.F.; Kylling, A.; Schaaf, C.; Wang, Z.; Bogren, W.; Storvold, R.; Solbø, S.; Pedersen, C.; Gerland, S. Unmanned aerial system nadir reflectance and MODIS nadir BRDF-adjusted surface reflectances intercompared over Greenland. *Cryosphere* **2017**, *11*, 1575–1589. [CrossRef]
38. Martins, V.S.; Barbosa, C.C.F.; De Carvalho, L.A.S.; Jorge, D.S.F.; Lobo, F.D.L.; Novo, E.M.L.d.M. Assessment of atmospheric correction methods for Sentinel-2 MSI images applied to Amazon floodplain lakes. *Remote Sens.* **2017**, *9*, 322. [CrossRef]
39. Warren, M.A.; Simis, S.; Martinez-Vicente, V.; Poser, K.; Bresciani, M.; Alikas, K.; Spyrakos, E.; Giardino, C.; Ansper, A. Assessment of atmospheric correction algorithms for the Sentinel-2A MultiSpectral Imager over coastal and inland waters. *Remote Sens. Environ.* **2019**, *225*, 267–289. [CrossRef]
40. Brockmann, C.; Doerffer, R.; Peters, M.; Kerstin, S.; Embacher, S.; Ruescas, A. Evolution of the C2RCC neural network for Sentinel 2 and 3 for the retrieval of ocean colour products in normal and extreme optically complex waters. In Proceedings of the Living Planet Symposium, Prague, Czech Republic, 9–13 May 2016; p. 54.
41. Tavares, M.H.; Martinez, J.; da Motta Marques, D.; Lins, R.; Fragoso, C., Jr.; Harmel, T. Atmospheric and sunglint correction for retrieving chlorophyll-a in a productive tropical estuarine-lagoon system using Sentinel-2 MSI imagery. *ISPRS J. Photogramm. Remote Sens.* **2021**, *174*, 215–236. [CrossRef]
42. Pan, Y.; Bélanger, S.; Huot, Y. Evaluation of Atmospheric Correction Algorithms over Lakes for High-Resolution Multispectral Imagery: Implications of Adjacency Effect. *Remote Sens.* **2022**, *14*, 2979. [CrossRef]
43. Soriano-González, J.; Urrego, P.; Sòria-Perpinyà, X.; Angelats, E.; Alcaraz, C.; Delegido, J.; Ruiz-Verdu, A.; Tenjo, C.; Vicente, E.; Moreno, J. Towards the Combination of C2RCC Processors for Improving Water Quality Retrieval in Inland and Coastal Areas. *Remote Sens.* **2022**, *14*, 1124. [CrossRef]
44. Seegers, B.N.; Stumpf, R.P.; Schaeffer, B.A.; Loftin, K.A.; Werdell, P.J. Performance metrics for the assessment of satellite data products: An ocean color case study. *Opt. Express* **2018**, *26*, 7404–7422. [CrossRef]
45. Mishra, D.R.; Ogashawara, I.; Gitelson, A.A. *Bio-Optical Modeling and Remote Sensing of Inland Waters*; Elsevier: Amsterdam, The Netherlands, 2017.
46. Zabala, S. Comparison of Multi-Temporal and Multispectral Sentinel-2 and Unmanned Aerial Vehicle Imagery for Crop Type Mapping, in Department of Physical Geography and Ecosystem Science. Master's Thesis, Lund University, Lund, Sweden, 2017.
47. Schaepman-Strub, G.; Schaepman, M.E.; Painter, T.H.; Dangel, S.; Martonchik, J.V. Reflectance quantities in optical remote sensing—Definitions and case studies. *Remote Sens. Environ.* **2006**, *103*, 27–42. [CrossRef]
48. Renosh, P.R.; Doxaran, D.; Keukelaere, L.D.; Gossn, J.I. Evaluation of atmospheric correction algorithms for sentinel-2-MSI and sentinel-3-OLCI in highly turbid estuarine waters. *Remote Sens.* **2020**, *12*, 1285. [CrossRef]
49. Sozzi, M.; Kayad, A.; Marinello, F.; Taylor, J.; Tisseyre, B. Comparing vineyard imagery acquired from Sentinel-2 and Unmanned Aerial Vehicle (UAV) platform. *OENO One* **2020**, *54*, 189–197. [CrossRef]
50. Hieronymi, M.; Krasemann, H.; Müller, D.; Brockmann, C.; Ruescas, A.B.; Stelzer, K.; Nechad, B.; Ruddick, K.; Simis, S.G.H.; Tilstone, G.; et al. Ocean colour remote sensing of extreme case-2 waters. *Spectrum* **2016**, *2*, 4.
51. Uudeberg, K.; Ansko, I.; Põru, G.; Ansper, A.; Noorma, A. Using optical water types to monitor changes in optically complex inland and coastal waters. *Remote Sens.* **2019**, *11*, 2297. [CrossRef]
52. Jiang, J.; Johansen, K.; Tu, Y.; McCabe, M.F. Multi-sensor and multi-platform consistency and interoperability between UAV, Planet CubeSat, Sentinel-2, and Landsat reflectance data. *GISci. Remote Sens.* **2022**, *59*, 936–958. [CrossRef]
53. Zeng, C.; Richardson, M.; King, D.J. The impacts of environmental variables on water reflectance measured using a lightweight unmanned aerial vehicle (UAV)-based spectrometer system. *ISPRS J. Photogramm. Remote Sens.* **2017**, *130*, 217–230. [CrossRef]
54. Mamaghani, B.; Salvaggio, C. Multispectral Sensor Calibration and Characterization for sUAS Remote Sensing. *Sensors* **2019**, *19*, 4453. [CrossRef]
55. Padró, J.-C.; Muñoz, F.J.; Ávila, L.A.; Pesquer, L.; Pons, X. Radiometric correction of Landsat-8 and Sentinel-2A scenes using drone imagery in synergy with field spectroradiometry. *Remote Sens.* **2018**, *10*, 1687. [CrossRef]
56. Warner, J.X.; Wei, L.Z.; Strow, L.; Dickerson, R.R.; Nowak, J.B. The global tropospheric ammonia distribution as seen in the 13-year AIRS measurement record. *Atmos. Chem. Phys.* **2016**, *16*, 5467–5479. [CrossRef]

remote sensing

MDPI

Article

Application of a PLS-Augmented ANN Model for Retrieving Chlorophyll-*a* from Hyperspectral Data in Case 2 Waters of the Western Basin of Lake Erie

Khalid A. Ali [1,*] and Wesley J. Moses [2]

1 Department of Geology and Environmental Geosciences, College of Charleston, University of Charleston, Charleston, SC 29424, USA
2 Naval Research Laboratory, Washington, DC 20375, USA; wesley.moses@nrl.navy.mil
* Correspondence: alika@cofc.edu

Abstract: We present results that demonstrate the utility of machine learning techniques that are based on partial least squares (PLS) and artificial neural networks (ANNs) for estimating low-moderate chlorophyll-*a* (chl-*a*) concentrations in the western basin of Lake Erie (WBLE). Previous ocean color studies have resulted in a large number of algorithms that are based on spectral indices to estimate water quality parameters (WQPs) such as chl-*a* concentration from remote sensing reflectance. However, these spectral index algorithms are based on reflectance features at specific wavelengths and do not take advantage of the wealth of spectral information that is contained in hyperspectral data, and are often not easily adaptable to waters with conditions that are different from those in the datasets that were used to originally calibrate the indices. Recently, there have been efforts to use machine learning techniques that are based on ANNs and PLS regression to exploit the spectral richness contained in hyperspectral data and retrieve WQPs. In this study, we have combined an ANN model with output from PLS regression to retrieve chl-*a* concentration from hyperspectral data in the WBLE. We compared the results from the PLS-ANN method to those that were obtained from a band-ratio algorithm that is based on reflectances in the blue and green spectral regions, a band ratio algorithm that is based on reflectances in the red and near-infrared (NIR) spectral regions, and a PLS-only approach. For a dataset that was collected in 2012, with chl-*a* concentrations ranging from 0.48 to 21.2 μg/L, the PLS-ANN method yielded a root mean square error (RMSE) of 1.22 μg/L, whereas the blue-green ratio algorithm yielded an RMSE of 1.75 μg/L, the NIR-red ratio algorithm yielded an RMSE of 1.95 μg/L, and the PLS-only approach yielded an RMSE of 1.95 μg/L. The PLS-ANN method takes advantage of the PLS regression to identify specific wavelengths that contain most information about the variation in chl-*a* concentration, minimize spectral collinearity and redundancy in the data, and simplify the neural network's input structure. The better performance of the PLS-ANN method can also be attributed to the neural network's ability to account for nonlinearity in the relationship between chl-*a* concentration and spectral reflectance. The results indicate that the PLS-ANN method can be reliably used to estimate and monitor low-moderate chl-*a* concentrations in optically complex waters.

Keywords: partial least squares; artificial neural networks; remote sensing; Lake Erie

1. Introduction

The spectral response of water in the visible-near-infrared (VNIR) range contains information on the composition of optically significant constituents in water and forms the basis for retrieving water quality parameters (WQPs) such as chlorophyll-*a* (chl-*a*) concentration from satellite data [1]. Historically, methods employed to quantitatively retrieve WQPs from remote sensing signals have included semi-empirical regression techniques that are based on ratios of or differences between reflectances at two or three spectral regions [2–5] and spectral inversion techniques that are based on the radiative transfer model e.g., [6,7]. In optically complex Case 2 waters [8] such as the Western Basin of Lake Erie (WBLE) [9], where the spectral signature is a function of multiple, non-covarying WQPs, retrieving estimates of various WQPs using reflectance at only a few select bands is challenging [10]. Approaches that are based on spectral feature selection and full spectral optimization to determine an optimal set of solutions for retrieving numerous WQPs from individual satellite-measured reflectance spectra have been attempted but found to be computationally too expensive and too time-consuming for operational implementation, especially when they are implemented on a per-pixel basis on local processors [11,12]. With the introduction of Application Programming Interfaces (APIs) that are run on cloud computing infrastructure such as the Google Earth Engine, the computational burden has become less critical of an issue, and there is renewed interest in using full spectral optimization as an operational tool to retrieve WQPs from hyperspectral data e.g., [13]. Several spaceborne hyperspectral sensors have been launched recently or are scheduled to be launched in the near future, resulting in an abundance of global hyperspectral data for the next several years, ready to be exploited for environmental monitoring and resource management purposes. The sensors include the recently launched Italian mission PRecursore IperSpettrale della Missione Applicativa (PRISMA), the recently launched German mission Environmental Mapping and Analysis Program (EnMAP), and NASA's future hyperspectral missions, the Plankton, Aerosol, Cloud, ocean Ecosystem (PACE) mission and the Surface Biology and Geology (SBG) mission, both of which are currently under development.

With the availability of large volumes of high dimensional remote sensing data and enhanced computational efficiency, linear methods such as the partial least squares (PLS) regression and machine learning techniques such as the artificial neural network (ANN), which is well suited for modeling non-linear relationships, are increasingly being applied in many remote sensing studies. In 1999, Schiller and Doerffer [14] developed a neural network (NN), which was later applied to data from the multispectral MEdium Resolution Imaging Spectrometer (MERIS) to retrieve concentrations of chl-*a* and suspended particulate matter (SPM) [15]. Several other NN-based approaches have also been developed to retrieve WQPs, especially chl-*a* concentration, from satellite data (e.g., [16–21]), with varying levels of success. In general, NN-based approaches tend to yield accurate estimates when they are applied to data from waters with biophysical conditions within the range of parameters that are used to train the NN model and fail when they are applied to conditions outside the training data range. This is a limitation, especially for multispectral NN-based algorithms that are based on modeled relationships between WQPs and reflectance at a few specific wavelengths. Hyperspectral data offer the ability to capture minor spectral variations in optical properties of water and its constituents and are well suited for optically complex waters. However, hyperspectral images contain large volumes of data that are inherently collinear, with often significant correlation amongst adjacent spectral channels. The large volume of the data and high correlation amongst data variables can introduce complexity to the neural network and can lead to overfitting of the model [21,22], thereby yielding inaccurate estimates. It is, therefore, necessary to reduce dimensionality and multicollinearity in hyperspectral data before applying NN-based algorithms. Data dimensionality and multicollinearity can be reduced by using methods such as the PLS, resulting in a low-dimensional dataset with a simplified structure, consisting of variables that are most significant for explaining variations in the dataset. PLS reduces the spectrally rich hyperspectral dataset to an optimal, low-dimensional set of variables that are most sensitive to parameters of interest to be retrieved. Once the redundancy and multicollinearity in the data are removed or minimized, ANNs are well suited to account for nonlinearity in the dataset and retrieve the parameter of interest. Optimized neural network algorithms that are based on feature dimension reduction have been successfully used for many applications [23].

Because of the complementary advantages of PLS and ANN, we combined both methods in a two-step process to estimate chl-*a* concentration in optically complex waters of the western basin of Lake Erie (WBLE) using hyperspectral data. A similar approach has been previously tested by Song et al. [24], who applied a PLS-ANN approach to datasets from several inland water bodies in Indiana, USA, and China for retrieving chl-*a* concentration and reported that the PLS-ANN approach yielded accurate results, marginally outperforming a three-band model that was based on reflectances in the red and near-infrared (NIR) spectral regions. The dataset that was used by Song et al. [24] encompassed a very wide range of chl-*a* concentrations, ranging over two orders of magnitude. While the overall accuracy reported by Song et al. [24] for the PLS-ANN algorithm is high for the entire range of chl-*a* concentrations, one may wonder whether similar accuracies may be obtained if the algorithm is applied to data from optically complex waters with a narrower range of low-moderate chl-*a* concentrations. Accurate retrieval of low-moderate chl-*a* concentrations (~5–15 ug/L) in optically complex, Case 2 waters [8] is often problematic [25–29] because it is at the high end of sensitivity for algorithms that rely on chl-*a*-specific features in the blue and green spectral regions and at the low end of sensitivity for algorithms that rely on chl-*a*-specific features in the red and NIR spectral regions. Moreover, at low-moderate chl-*a* concentrations, chl-*a*-specific spectral features at wavelengths of interest are affected, and in some cases overwhelmed, by contributions from other optically significant constituents such as SPM and colored dissolved organic matter (CDOM). The objective of this study is to test the suitability of the PLS-ANN approach for retrieving low-moderate chl-*a* concentrations, using as a case study the WBLE, which is an optically complex water body [9] with significant concentrations of SPM and CDOM, which do not co–vary with chl-*a* concentration. We focused on data that were collected in late spring and early summer, before chl-*a* concentrations attain peak values. First, we used the PLS framework to assess the relationship between hyperspectral reflectance measurements and chl-*a* concentrations and identify the spectral channels of importance that contribute significantly to explaining variations in chl-*a* concentration. We also used PLS to identify spectral channels that are highly correlated to each other and used the information to reduce dimensionality and multicollinearity in the hyperspectral dataset. Second, we used information from the PLS output in an ANN framework to model the nonlinear relationship between hyperspectral reflectance and chl-*a* concentrations, determine the relative importance and predictive value of the input wavelengths, and estimate chl-*a* concentration. We have also provided estimates of chl-*a* concentration retrieved through the PLS method alone and through well-established semi-empirical algorithms that are based on the blue-green and NIR-red reflectance ratios for the sake of comparison with PLS-ANN retrievals.

2. Data and Methods

2.1. Data

The data consisted of reflectance and chl-*a* concentrations that were collected from 18 stations in the WBLE (Figure 1) in June and July of 2012 using the research vessel *Gibraltar III*.

2.1.1. Reflectance Measurements

Hyperspectral data were collected at each station using a GER 1500 Visible and Near-Infrared (VIS-NIR) spectroradiometer at continuous wavelengths within the spectral range 350–1080 nm at 1.5 nm spectral intervals and 3.2 nm spectral resolution. Upwelling radiance, $L_u(\lambda)$, and downwelling irradiance, $E_d(\lambda)$, were collected simultaneously. To minimize adjacency effects from the vessel, the measurements were taken from the sun-lit side of the vessel using an optical fiber that was attached to an extendable pole, with the tip of the optical fiber placed a few centimeters below the water surface and pointed towards nadir. In order to minimize the effect of uncertainties due to noise, six spectra were taken and averaged for each station. Each average spectrum was then spectrally averaged at 10 nm intervals between 400 and 750 nm, resulting in 36 spectral channels of data for each spectrum. A white Spectralon® reflectance standard was used to calibrate the

GER 1500 spectra. The remote sensing reflectance (Rrs) was calculated by normalizing the upwelling radiance, $L_u(\lambda)$, by the downwelling irradiance, $E_d(\lambda)$, as follows [30]:

$$R_{rs} = \frac{L_u\,(\lambda)}{E_d\,(\lambda)}\frac{E_{cal}\,(\lambda)}{L_{cal}\,(\lambda)}\frac{\rho_{ref}\,(\lambda)}{\pi}\frac{t}{n^2}F(\lambda) \tag{1}$$

where $L_{cal}\,(\lambda)$ and $E_{cal}\,(\lambda)$ are the upwelling radiance from and the downwelling irradiance incident on the Spectralon® reflectance standard, respectively, $\rho_{ref}(\lambda)$ is the known irradiance reflectance of the Spectralon® reflectance standard, π is used to transform the irradiance reflectance into remote sensing reflectance, t is the water-to-air transmittance (t = 0.98), n is the refractive index of water relative to air (n = 1.34 at 20 °C), and $F(\lambda)$ is the spectral immersion factor, which accounts for the radiance transfer between water and air [31,32].

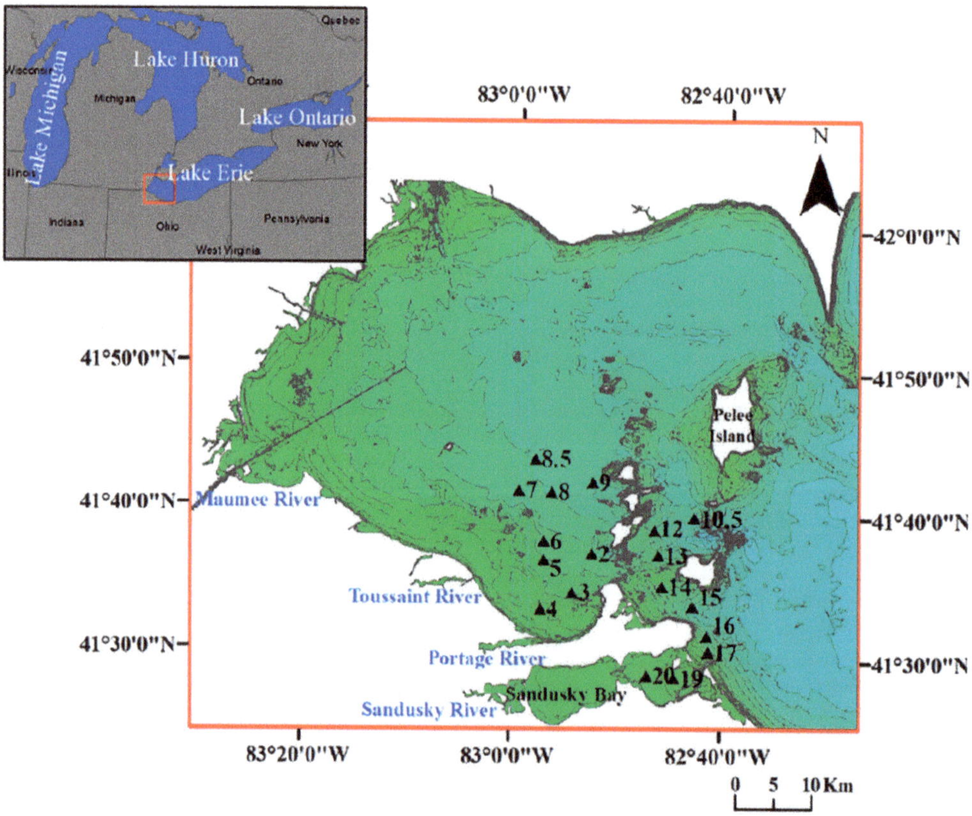

Figure 1. Sampling locations in the western basin of Lake Erie (after [3]).

Figure 2 shows hyperspectral data that were collected from the WBLE. The spectral signatures are characteristic of productive waters with absorption and scattering features across various spectral regions [30,33,34]. The features are indicative of the effects of optically significant in-water constituents, such as absorption and scattering of phytoplankton and other associated WQPs. Absorption minima near the green (560 nm) and NIR (710 nm) spectral regions and maxima near 440, 620, and 680 nm are clearly apparent in the data; these features indicate the presence of phytoplankton biomass [35–37]. The NIR reflectance peaks are much higher for chl-a concentrations > 1.5 ug/L. The absorption minima near 560 nm and the maxima near 680 nm are more distinct for samples from stations that are closer to stream and river inlets, such as in the Sandusky Bay, compared to those further offshore in the open waters of the WBLE (Figure 1).

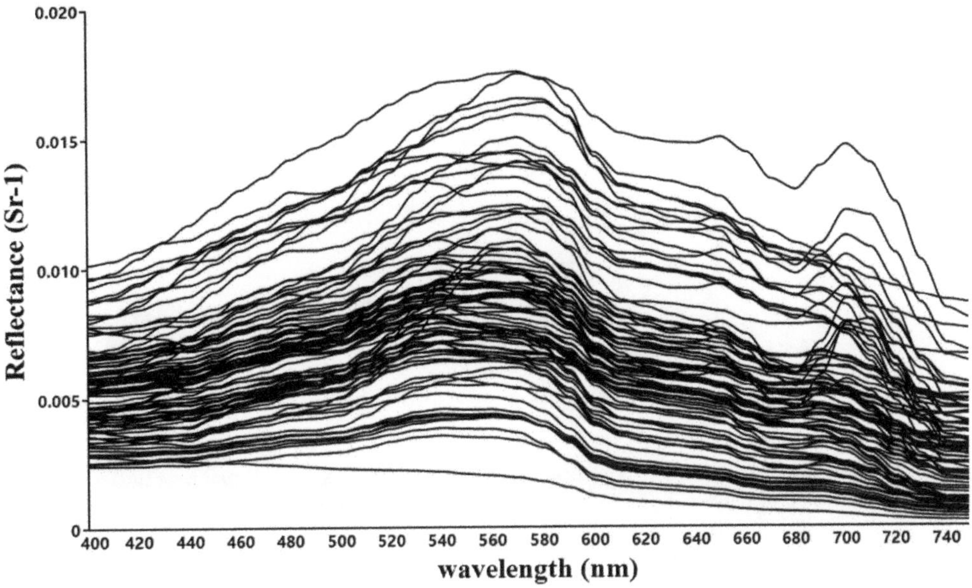

Figure 2. In situ reflectance spectra (400–750 nm) at 18 stations in the WBLE.

2.1.2. Chl-*a* Measurements

At each of the 18 stations, water samples were collected at 0.5 m depth, filtered using glass fiber filters (GF/F), and temporarily stored in a freezer at −20 °C for subsequent analysis in the laboratory. Chl-*a* was extracted from the samples and quantified following the U.S. Environmental Protection Agency's (EPA) 445 protocol [38]. The average *in situ* chl-*a* concentration for the whole dataset was 4.83 ug/L. The concentration tended to increase throughout the sampling period. The concentrations across the basin varied between 0.48 and 21.2 ug/L, with higher concentrations recorded at stations located in the Sandusky Bay (Figure 1). The standard deviation of the chl-*a* concentrations was 4 ug/L (Figure 3). The high standard deviation is attributed to the relatively higher seasonal algal density in the Sandusky Bay. During late spring and early summer, river discharges are high, and significant amounts of terrestrial matter, including nutrients, are transported into the bay. This increases the combined concentrations of allochthonous and autochthonous phytoplankton and, consequently, leads to higher chl-*a* concentrations in the bay (Stations 17, 19 and 20—see Figure 1), which results in significant differences in chl-*a* concentration between the waters in the Sandusky Bay and the central WBLE. By late summer, the river discharge decreases and the waters between the two sub-basins undergo mixing, resulting in a decrease in the overall variability of chl-*a* concentration in the WBLE [3].

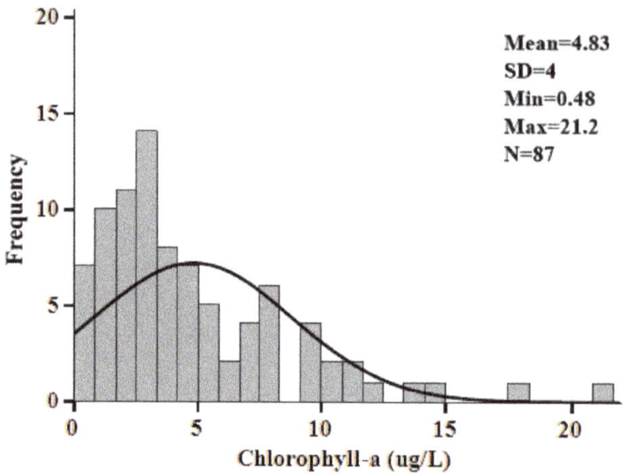

Figure 3. Distribution of chl-*a* concentrations in the WBLE during June–July 2012.

2.2. Methods

The data were analyzed to retrieve the chl-*a* concentrations using partial least squares, artificial neural network, and two well-established semi-empirical algorithms—a blue-green algorithm and an NIR-red algorithm.

2.2.1. Partial Least Squares (PLS)

PLS is a feature-reduction approach that has the ability to analyze data with a low signal-to-noise ratio and numerous, strongly collinear variables [39]. It projects relationships between two sets of matrices containing dependent and independent variables onto a latent space, extracting several principal independent factors while capturing most of the variance in the original data [40–42].

Suppose there are n observations of K feature variables, $X = x_1, x_2, \ldots, x_K$, and m dependent variables, $Y = y_1, y_2, \ldots, y_m$. In an aquatic remote sensing context, K represents the number of spectral channels in the reflectance dataset and m represents the number of WQPs that are being retrieved from the reflectance measurements. As with Principal Component Analysis (PCA), the data are mean-centered and scaled to ensure zero mean and unit variance. The preprocessed data matrix X is expressed as:

$$X = TP^T + E \quad (2)$$

Here, T is the score matrix, which contains the iterative projections of the observed data onto successive orthogonal axes or components of maximum variation (akin to the principal components in PCA) and has the dimension $n \times A$, where n is the number of observations in the dataset and A is the number of components; P is the load matrix, which consists of unit vectors, each describing the direction of the respective component, and has the dimension $K \times A$; E is the residual error matrix, which is an $n \times K$ matrix consisting of the perpendicular distance between each observation data point and the hyperplane, which encompasses all components. For a given component, i, the matrix multiplication TP^T can be expressed as the sum product of the score vector t_i (the ith column of the score matrix T) and the load vector p_i (the ith column of the loading matrix P). For a system with A principal components, Equation (2) can be rewritten as:

$$X = \sum_{i=1}^{A} t_i p_i^T + E, i = 1, 2, \ldots, A \quad (3)$$

Similarly, matrix Y is decomposed into:

$$Y = UQ^T + F \tag{4}$$

Here, U is the score matrix, Q is the loading matrix, and F is the residual error matrix. For a given component, i, the matrix multiplication UQ^T can be expressed as the sum product of the score vector u_i (the ith column of the score matrix U) and the loading vector q_i (the ith column of the loading matrix Q); Equation (4) can be rewritten as:

$$Y = \sum_{i=1}^{A} u_i q_i^T + F, i = 1,2,\ldots,A \tag{5}$$

PLS differs from PCA in that the score matrices are determined for both dependent and independent variables. Moreover, the score matrix T is determined based on not only X but also Y. For a given component i, each of the K columns in X_i is regressed against an initial estimate of u_i to determine a weight vector, w_i, which is normalized to unit length and regressed against every row in X_i to determine the score vector t_i. Similarly, every column in Y_i (each representing the m WQPs to be retrieved) is regressed against t_i to determine a weight vector for the Y data space, q_i, which is regressed against every row in Y_i to determine the score vector u_i. The process continues iteratively until u_i changes negligibly. At convergence, w_i, t_i, q_i, and u_i together define the ith component and are stored in the respective matrices W, T, Q, and U. One of the salient features of the PLS is that the score matrices T and U not only maximally explain the variance in the X and Y data space but they also maximize the covariance, and hence the correlation, between t_i and u_i. Thus, they represent not only the best explanations of variance in the X and Y data space but also the best explanation of the relationship between X and Y. The model predictions of the X and Y matrices at the ith component are determined as follows:

$$\hat{X}_i = t_i {p'}_i \tag{6}$$

where p_i is the loading vector that is determined by regressing the columns of X_i against t_i, and:

$$\hat{Y}_i = u_i {q'}_i \tag{7}$$

The data matrices are deflated and new matrices for the next component are determined as $X_{i+1} = X_i - \hat{X}_i$ and $Y_{i+1} = Y_i - \hat{Y}_i$. The process continues until successive components do not explain new variability. When used for predicting the dependent variables, the relationship between the score matrix T and the data matrix X is exploited to predict Y. Given that the weight vectors for successive components are calculated based on deflated X matrices and not the original data matrix, W cannot be directly used to define the relationship between T and X. The loading matrix, P, and the weight matrix with deflation, W, can be combined as $R = W\left(P^T W\right)^{-1}$, which relates T and X as:

$$T = XR \tag{8}$$

The weight matrix R is informative as it indicates the importance of the dependent variables in X space for explaining the variation in Y. In the aquatic remote sensing context, the matrix R helps to identify spectral channels of importance for estimating the WQPs of interest in Y. Spectral channels containing similar information will have similar weights. Thus, R also provides a means of minimizing data redundancy.

Following Equation (7), Y can be calculated as:

$$\hat{Y} = TQ^T \tag{9}$$

Using Equation (8) in Equation (9), we get:

$$\hat{Y} = XRQ^T \tag{10}$$

$B = RQ^T$ is the regression coefficient of the PLS model.

PLS accounts for spectral correlation amongst the various spectral channels, assigns weights to the spectral channels depending on their predictive capability, and enhances the potential discrimination of retrieved constituents. PLS does not need to use all of the input hyperspectral channels to establish maximum correlation with WQPs. It only needs information from select spectral channels. As such, it provides a convenient way to minimize multicollinearity and reduce dimensionality in the dataset.

2.2.2. Artificial Neural Network (ANN)

ANNs provide a powerful means to get insight into and uncover underlying nonlinear relationships and structures that are existing in datasets [43]. ANN imitates the biological neural network learning process and possesses powerful functions for dealing with nonlinear processes with complex underlying structures. The model is formed by artificial neurons on multiple layers including input/output that emulate biological neurons and the synaptic connections.

The feed-forward back-propagation multi-layer perceptron (MLP) is the most frequently used ANN in remote sensing applications [44]. The basic structure is comprised of three distinctive layers: (i) the input layer, where the remote sensing reflectance data are introduced into the model, represented by the array $x = [x_1, x_2, \ldots, x_n]$; (ii) the hidden layer, where the activation function $\alpha = \sigma(x_i w_i)$ is applied to x after multiplying it by the corresponding synaptic weight vector, w, for data transformation; and (iii) the output layer, where the results, $y = f(\alpha)$, of the ANN are produced [45,46]. The ANN is trained through a backpropagation algorithm to minimize the error (E) between the predicted and expected outputs. The error is redistributed by recomputing w for the previous neuron layers from back to front [47]. The backward iteration processes and the step size that is taken to update w is controlled by the ANN learning rate, η, expressed as:

$$w \leftarrow w - \eta \frac{\partial E}{\partial w} \tag{11}$$

2.2.3. Semi-Empirical Algorithms

For the sake of comparison with the PLS and PLS-ANN methods, we retrieved chl-*a* concentrations using two well-established semi-empirical algorithms, one that is based on reflectance in the blue and green spectral regions and the other based on reflectance in the red and NIR regions.

Blue-Green Chl-*a* Algorithm

Recently, O'Reilly and Werdell [4] used an extensive collection of global datasets to re-parameterize the blue-green algorithm that was originally developed decades ago by O'Reilly et al. [48]. The blue-green chl-*a* algorithm is a fourth-degree polynomial equation of the form:

$$\log_{10}(Chl - a) = a_0 + a_1 X + a_2 X^2 + a_3 X^3 + a_4 X^4 \tag{12}$$

where, $X = \log_{10}\left(\frac{max[Rrs_{440}, Rrs_{490}, Rrs_{510}]}{Rrs_{560}}\right)$, and a_0, a_1, a_2, a_3, and a_4 are polynomial coefficients obtained by regressing the maximum blue-green ratio values against measured chl-*a* concentrations. O'Reilly et al. [4] developed regression coefficients for 25 spaceborne sensors, taking into consideration minor variations across the sensors in the wavelength locations of respective spectral channels in the blue and green regions.

Considering the wavelengths of the GER 1500 spectrometer, we used a 4th order polynomial function [4] for a wavelength combination that was closest to spectral channels centered at 440 nm, 490 nm, and 510 nm in the numerator and 560 nm in the denominator. The specific optimized algorithm was:

$$\log_{10}(Chl-a) = 0.4254 - 3.21679X + 2.86907X^2 - 0.62628X^3 - 1.09333X^4 \quad (13)$$

where: $X = \log_{10}\left(\frac{max[Rrs_{440}, Rrs_{490}, Rrs_{510}]}{Rrs_{560}}\right)$.

NIR-Red Chl-*a* Algorithm

The two-band NIR-red model that is based on the chl-*a* absorption feature in the red spectral region and the reflectance peak in the NIR region due to the combined effect of decreasing absorption of light by chl-*a* and increasing absorption by water [49,50] has yielded accurate retrievals of chl-*a* concentrations in various inland and coastal waters around the world when applied to in situ, airborne, and spaceborne remote sensing data.

Based on the wavelength locations of the red and NIR spectral channels of the GER 1500 spectrometer, an optimized two-band NIR-red algorithm, with the coefficients re-parameterized based on in situ measurements of chl-*a* concentration from the WBLE, was formalized as follows:

$$Chl - a = 0.5431 \times \left[R_{670}^{-1} \times R_{710}\right] + 2.315 \quad (14)$$

It must be noted that the wavelengths at the red and NIR spectral channels of the GER 1500 spectrometer used in this study, namely 670 and 710 nm, are slightly different from the original red and NIR wavelengths that were used in the development of the two-band NIR-red model [51], namely 665 and 708 nm. This could have an impact on the algorithm performance, although we expect the impact of the wavelength difference to be negligible.

3. Results and Discussions

Chl-*a* concentrations were estimated using the PLS method, PLS-ANN method, the blue-green algorithm, and the two-band NIR-red algorithm. As the dataset was rather small, it was not split into separate calibration and validation datasets. Instead, each of the algorithms was applied to the entire dataset and the retrievals were assessed by comparing them to in situ measurements. The goal of this study is not to develop universally applicable PLS-ANN algorithms but to rather test the suitability of the approach for a small dataset with low-moderate chl-*a* concentrations.

3.1. Error Metrics

The retrieved chl-*a* concentrations were evaluated based on the correlation coefficient (R^2), root mean square error (RMSE), and three additional error metrics that were recommended by Seegers et al. [52], namely, mean absolute error (MAE), bias, and 'percent wins'. It must be noted that R^2 is limited in its utility as a validation metric because a high R^2 does not necessarily mean high accuracy if there is significant bias in the estimates. Nevertheless, it can be used as a measure of consistency in the algorithm's behavior across the range of measurements. The MAE and bias are expressed as multiplicative metrics that were calculated by first log-transforming the retrievals to determine the residuals as log-transformed measures and back-transforming them. This approach is well suited for quantities such as chl-*a* concentration for which the error varies proportionally with the magnitude of the quantity [52]. The calculated MAE and bias are to be interpreted as multiplicative measures. For example, an MAE of 1.25 would indicate a relative measurement error of 25% and a bias of 1.3 would mean that the estimate is on average 30% more than the observed value, with bias values that are less than unity indicative of negative bias. The RMSE, MAE, and bias are expressed as follows:

$$\text{RMSE} = \sqrt{\frac{\sum_{i=1}^{n}(E_i - M_i)^2}{n}} \quad (15)$$

$$\text{Mean Absolute Error (MAE)} = 10^{\frac{\sum_{i=1}^{n} |log_{10}(E_i) - log_{10}(M_i)|}{n}} \quad (16)$$

$$\text{Bias} = 10^{\frac{\sum_{i=1}^{n} (log_{10}(E_i) - log_{10}(M_i))}{n}} \quad (17)$$

where E_i and M_i are the estimated and measured values, respectively, and n is the number of observations. It must be noted that while the MAE reported here is a unitless measure that is based on log-transformed values, the RMSE is in physical units of chl-*a* concentration determined based on untransformed values and, as such, the numerical magnitudes of MAE and RMSE cannot be compared against each other.

The 'percent wins' was calculated according to the steps that were described by Seegers et al. [52], in which each data point was examined individually, and for each data point, the algorithm that yielded the closest estimate to the observed measurement was declared the 'winner'. For each algorithm, the 'percent wins' is the percentage of data points for which its estimate was the closest in value to the measured value in comparison to estimates from the other algorithms. This is a valuable diagnostic metric that provides additional information on the algorithms' performance, such as identifying algorithms that yield very accurate estimates for some data points but fail significantly at other data points, which might provide insights on optimal conditions for those algorithms. It also helps to identify algorithms that may not yield the best or the worst estimates but may perform consistently well across all data points in the dataset.

3.2. Estimation of Chl-a Concentration Using the PLS Method

Distributional transformation was applied to the original input data to get the characteristic input variables in matrix X (hyperspectral remote sensing data) and output variables in matrix Y (measured chl-*a* concentrations). The data were mean-centered and standardized with a distribution of N(0,1). This ensures data normality, eliminates intrinsic numerical influence, overcomes model learning problems, and allows easy data comparison. PLS was then applied to reduce multicollinearity in the dataset and extract the leading components t and u from X and Y, respectively, while maximizing their correlation with each other and projecting the regression of X on the t axis and Y on the u axis. The new components represent relevant information that is present in the reflectance spectra that were used to estimate the chl-*a* concentration. Similar methods have been previously applied [53,54] for extracting information about spatial and temporal variations in sediment composition. The method works well for optically significant in-water constituents, allowing the extraction of absorption features that are correlated to the presence of chl-*a*. In many cases, spectral patterns identified by PLS can also be interpreted physically based on known spectral signatures [10,12,55].

The application of the PLS model to the input hyperspectral data, X, generated three principal components, $t_i (i = 1,2,3)$, which explained more than 96% (Figure 4) of the optical variance recorded in the data. The three principal components represent latent spectral structures that are obtained by maximizing the covariance between linear combinations of spectral bands and chl-*a* concentrations in the WBLE. The first component explained about 63% of the observed chl-*a* variability, and the second and third factors explained 9% and 24% of the chl-*a* variability, respectively. This suggests that the spectral bands that have relatively higher communality in the PLS components are the most sensitive to chl-*a* variability. These include bands that are centered at 690, 700, 710, and 720 nm in PLS factor 1 (components); spectral regions near 440, 480, 490, 560, 580, 620, 650, 670, 680, 700, and 710 nm were identified as the most sensitive to chl-*a* variations in PLS factors 2 and 3 (Figure 4), consistent with previous studies [5,10,55]. The correlation between the retrieved individual principal components, t_i, and chl-*a*, u_i, as shown in Figure 5, indicates the linear relationship of each component to the PLS-based chl-*a* model. Comparison between the measured chl-*a* concentrations and estimates from the PLS model yielded an R^2 of 0.76, an RMSE of 1.95 ug/L, an MAE of 1.41, a bias of 1.16, and a percent-wins measure of 16.28% (Table 1).

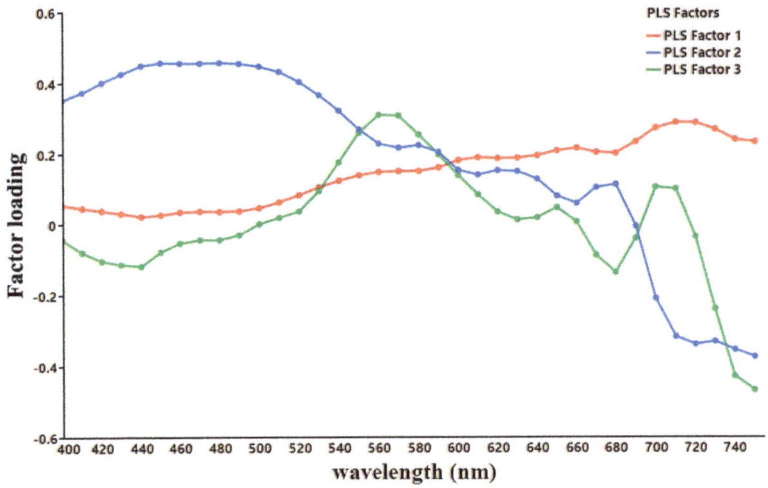

Figure 4. Three primary PLS component loadings.

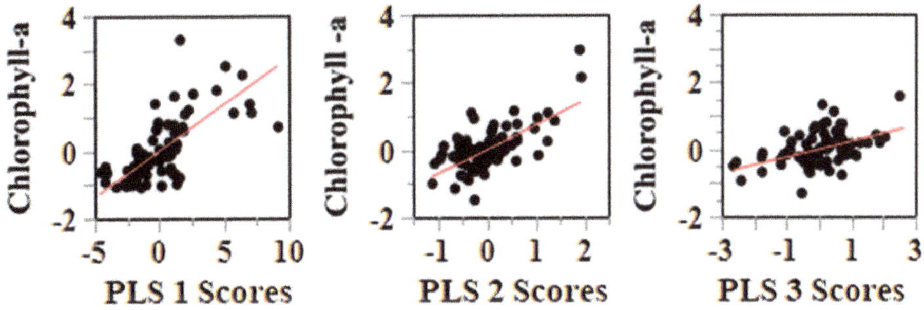

Figure 5. Linear relationship between the PLS components and the chlorophyll-a scores.

Table 1. Performance metrics for the four algorithms in retrieving chl-*a* concentration in the WBLE.

Algorithm	n	Bias	MAE	Percent Wins (%)	R^2	RMSE
PLS	87	1.16	1.41	16.28	0.76	1.95
PLS-ANN	87	1.1	1.31	58.14	0.92	1.22
Blue-green Model	87	1.21	1.73	12.79	0.61	1.75
NIR-red Model	87	1.19	1.74	12.79	0.56	1.95

3.3. Estimation of Chl-a Concentration Using the PLS-ANN Method

The dimension-reduced output from the PLS process, t and u, are supplied as input into the ANN as the predictor and response variables, respectively. The network employed in this study is a two-layer perceptron that is based on backpropagation (Figure 6). The first layer has two nodes, and the second layer has three nodes. Each of the two nodes in the first layer is a function of all three nodes in the second layer. All three nodes in the second layer are a function of the reduced-dimension input spectra, t. The estimated chl-*a* concentration (i.e., u) is a function of both nodes in the first layer. The PLS-ANN model was initially trained using a randomly selected training data subset representing 75% of the dataset. The remaining 25% was used to assess the model's capability to estimate chl-*a* concentration in the WBLE.

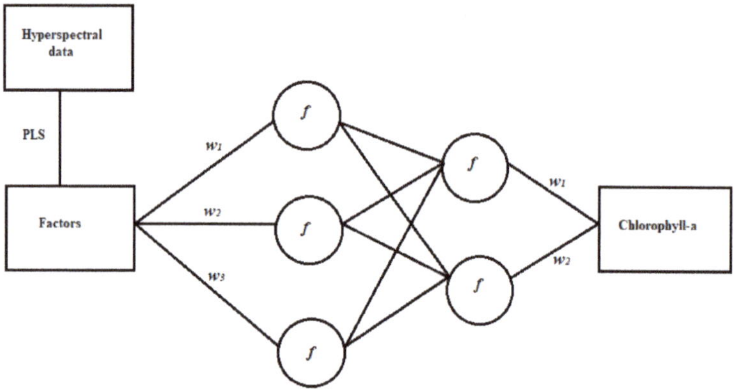

Figure 6. Optimized PLS-ANN algorithm.

Augmenting the PLS model with an ANN helps model nonlinearities that may exist between the spectral-based PLS factors and chl-*a* concentration, which the PLS may not necessarily capture. Figure 7 shows the linear and nonlinear components of the relationship between the three PLS factors (PLS X, scores 1–3) and chl-*a* concentrations on the y axis. The slopes of the relationships are indicative of the extent of variability in chl-*a* concentration that is captured by each factor, with steeper slopes indicative of higher proportions of the variability captured by the model. In this case, PLS Factor 1 captures significantly more linear and nonlinear components than do PLS Factors 2 and 3. A comparison between the measured chl-*a* concentrations and PLS-ANN estimates (Figure 8b) yielded an R^2 of 0.92, an RMSE of 1.22 ug/L, an MAE of 1.31, a bias of 1.1, and a percent-wins measure of 58.14% (Table 1), all of which are better than the corresponding figures from the PLS-only approach.

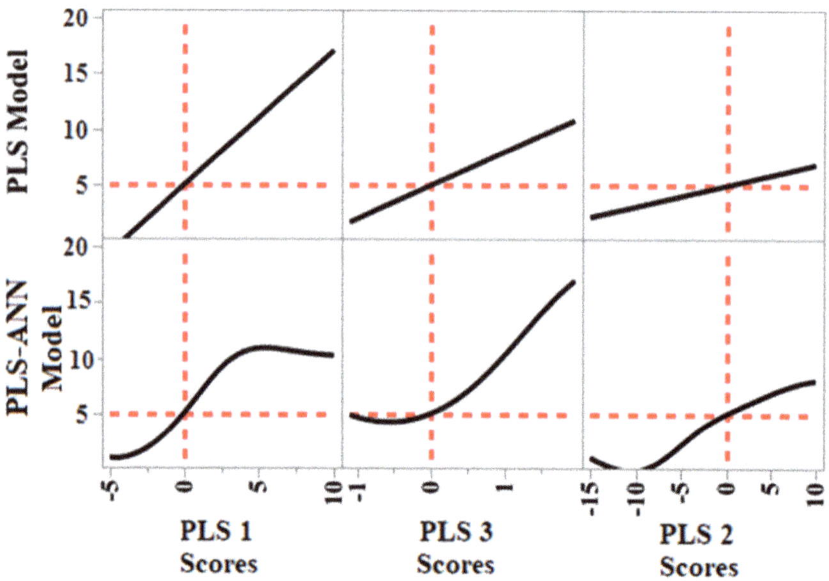

Figure 7. Linear and non-linear relationship of PLS components and chlorophyll-a.

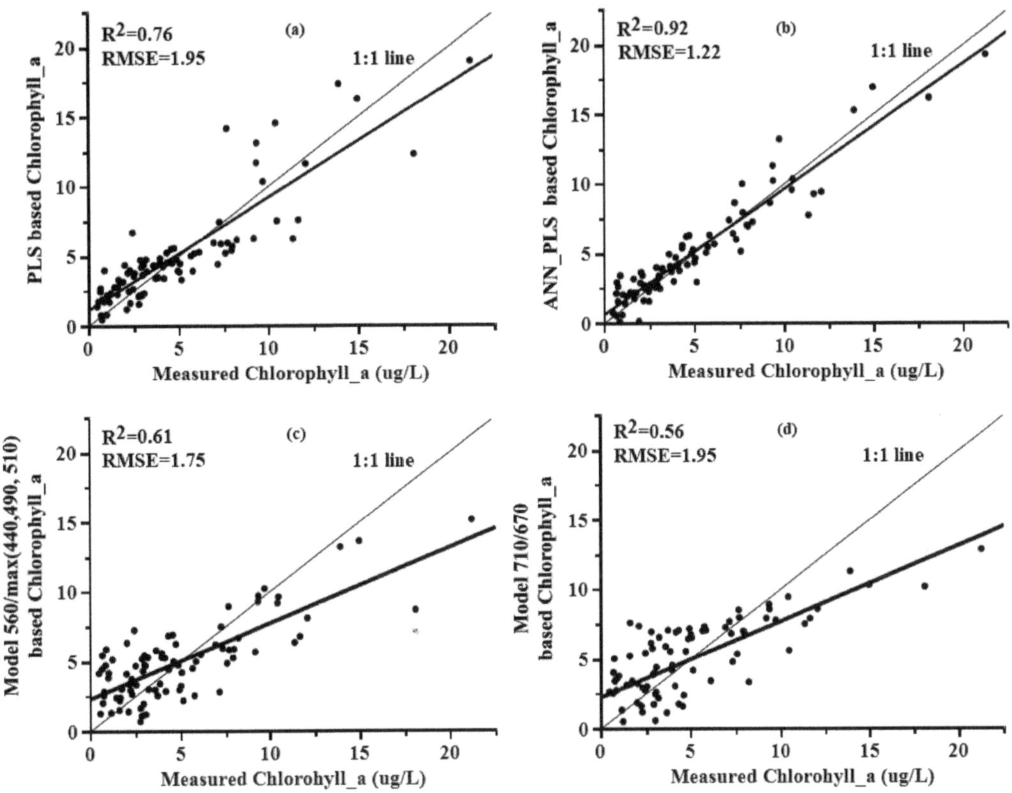

Figure 8. Scatter-and model line fit plots of in situ measured chl-*a* concentrations and estimates from the (**a**) PLS-only, (**b**) PLS-ANN, (**c**) blue-green band ratio, and (**d**) NIR-red band ration algorithms.

3.4. Estimation of Chl-a Concentration Using the Blue-Green Algorithm

The blue-green algorithm yielded reasonably accurate estimates of chl-*a* concentration (Figure 8c). The algorithm performed fairly well at low chl-*a* concentrations, although the plot of measured vs. estimated concentrations shows some scatter around the 1:1 line. The algorithm underestimated moderate chl-*a* concentrations, which is consistent with previous observations regarding the limitations of the blue-green algorithm at moderate-to-high chl-*a* concentrations [50].

The chl-*a* estimates from the blue-green algorithm had an R^2 of 0.61, an RMSE of 1.75 ug/L, an MAE of 1.73, a bias of 1.21, and a percent-wins measure of 12.79%.

3.5. Estimation of Chl-a Concentration Using the NIR-Red Algorithm

The two-band NIR-red algorithm also yielded reasonably accurate estimates of chl-*a* concentration (Figure 8d). Unsurprisingly, the algorithm performed poorer at low chl-*a* concentrations (<5 ug/L), where the chl-*a*-specific spectral features are weaker in the red and NIR spectral regions and the effect of uncertainty due to low signal in the NIR region becomes more pronounced. The algorithm performed fairly well at moderate chl-*a* concentrations, although the estimates were biased low. The chl-*a* estimates from the two-band NIR-red algorithm had an R^2 of 0.56, an RMSE of 1.95 ug/L, an MAE of 1.74, a bias of 1.19, and a percent-wins measure of 12.79% (Table 1).

The performance metrics for the algorithms are summarized in Table 1. Among the algorithms, the PLS-ANN algorithm performed the best in estimating the chl-*a* concentration. This is further confirmed using the regression plots that are presented in Figure 8, with show the hybrid model with the highest R^2 value and the least RMSE value.

4. Conclusions

This study shows that the PLS-ANN model, which combines linear (i.e., PLS) and nonlinear modeling (i.e., ANN), can be used to retrieve accurate and reliable estimates of chl-*a* concentration from hyperspectral data. The PLS-ANN model produced the best result in each of the five error metrics that were used to assess the chl-*a* estimates, followed by the PLS model and the band ratio algorithms. The results are consistent with those obtained by Song et al. [24], who compared chl-*a* estimates from the PLS-ANN method with those from the PLS-only and NIR-red band ratio models. As stated earlier, this study differed from that of Song et al. [24] in that we focused on low-moderate chl-*a* concentrations. One of the challenges with algorithms parameterized using datasets with a very wide range of chl-*a* concentrations is that the published overall performance metrics of the algorithm may be heavily influenced by values towards the extreme ends of the data range, and one may not be sure if the algorithm will yield the same accuracy when it is applied to a dataset representing a narrower range of conditions. This study demonstrates that the PLS-ANN method works reliably well at low-moderate chl-*a* concentrations, which is a critical range of transition from mesotrophic to eutrophic waters, which is outside the typical chl-*a* range for algorithms that are meant for oligotrophic waters and at the low end of the chl-*a* range for algorithms that are parameterized for eutrophic waters.

One of the unique advantages of the PLS-ANN model is the ability to model both linear and nonlinear components of the relationship between chl-*a* concentration and hyperspectral reflectance, which is critically important for optically complex, Case 2 waters. The PLS approach helps minimize collinearity in the dataset and reduce the input hyperspectral dataset into fewer components that explain most of the variation in the parameter of interest. In this study, the first three PLS components (factors) altogether explained 96% of the variation in chl-*a* concentration. By reducing data dimensionality and multicollinearity, the PLS approach helps to minimize potential overfitting problems, which can be a significant concern with hyperspectral data.

As demonstrated in this study, the PLS-ANN model is well-suited to take full advantage of spectral information that is contained in hyperspectral data to retrieve WQPs in optically complex water bodies such as the WBLE. The goal of this study was not to develop a universally applicable PLS-ANN algorithm for estimating chl-*a* concentration but to test the suitability of the approach for a specific case of low-moderate concentrations. A much larger dataset encompassing a wide range of biophysical conditions is needed for developing algorithms for broader application to multiple water bodies without the need for site-specific calibration. A key consideration in developing such a universal algorithm would be consistency in the spectral channels that are associated with the most significant PLS factors.

As hyperspectral data become increasingly common around the world, tools that are based on approaches such as the PLS-ANN model, which makes effective use of spectrally rich data, will be critical for monitoring environmental changes in optically complex inland, estuarine, and coastal waters.

Author Contributions: Conceptualization: K.A.A.; methodology: K.A.A. and W.J.M.; software: K.A.A.; validation: K.A.A.; formal analysis: K.A.A. and W.J.M.; investigation: K.A.A. and W.J.M.; resources: K.A.A. and W.J.M.; data curation: K.A.A.; writing—original draft preparation: K.A.A.; writing—review and editing: K.A.A. and W.J.M.; visualization: K.A.A.; supervision: K.A.A. and W.J.M.; project administration: K.A.A. and W.J.M.; funding acquisition: K.A.A. and W.J.M. All authors have read and agreed to the published version of the manuscript.

Funding: This research was funded by NASA's R3 funding, grant number SC-80NSSC20M0148 and basic research funds from the U.S. Naval Research Laboratory.

Acknowledgments: We acknowledge NASA's EPSCoR program and the U.S. Naval Research Laboratory for funding this research.

Conflicts of Interest: The authors declare no conflict of interest. The funders had no role in the design of the study; in the collection, analyses, or interpretation of data; in the writing of the manuscript; or in the decision to publish the results.

References

1. Albert, A.; Mobley, C. An analytical model for subsurface irradiance and remote sensing reflectance in deep and shallow case-2 waters. *Opt. Express* **2003**, *11*, 2873–2890. [CrossRef]
2. Ali, K.A.; Ortiz, J.; Bonini, N.; Shuman, M.; Sydow, C. Application of Aqua MODIS sensor data for estimating chlorophyll a in the turbid Case 2 waters of Lake Erie using bio-optical models. *GISci. Remote Sens.* **2016**, *53*, 483–505. [CrossRef]
3. Ali, K.; Witter, D.; Ortiz, J. Application of empirical and semi-analytical algorithms to MERIS data for estimating chlorophyll a in Case 2 waters of Lake Erie. *Environ. Earth Sci.* **2014**, *71*, 4209–4220. [CrossRef]
4. O'Reilly, J.E.; Werdell, P.J. Chlorophyll algorithms for ocean color sensors—OC4, OC5 & OC6. *Remote Sens. Environ.* **2019**, *229*, 32–47. [CrossRef]
5. Moses, W.J.; Gitelson, A.A.; Berdnikov, S.; Bowles, J.H.; Povazhnyi, V.; Saprygin, V.; Wagner, E.J.; Patterson, K.W. HICO-Based NIR–Red Models for Estimating Chlorophyll-α Concentration in Productive Coastal Waters. *IEEE Geosci. Remote Sens. Lett.* **2013**, *11*, 1111–1115. [CrossRef]
6. Lee, Z. *Visible-Infrared Remote Sensing Model and Applications for Ocean Waters*; University of South Florida: Tampa, FL, USA, 1994.
7. Roesler, C.S.; Perry, M.J. In situ phytoplankton absorption, fluorescence emission, and particulate backscattering spectra determined from reflectance. *J. Geophys. Res. Ocean.* **1995**, *100*, 13279–13294. [CrossRef]
8. Morel, A.; Prieur, L. Analysis of variations in ocean color1. *Limnol. Oceanogr.* **1977**, *22*, 709–722. [CrossRef]
9. O'Donnell, D.M.; Effler, S.W.; Strait, C.M.; Leshkevich, G.A. Optical characterizations and pursuit of optical closure for the western basin of Lake Erie through in situ measurements. *J. Great Lakes Res.* **2010**, *36*, 736–746. [CrossRef]
10. Ali, K.A.; Witter, D.L.; Ortiz, J.D. Multivariate approach to estimate colour producing agents in Case 2 waters using first-derivative spectrophotometer data. *Geocarto Int.* **2013**, *29*, 102–127. [CrossRef]
11. Doerffer, R.; Fischer, J. Concentrations of chlorophyll, suspended matter, and gelbstoff in case II waters derived from satellite coastal zone color scanner data with inverse modeling methods. *J. Geophys. Res. Earth Surf.* **1994**, *99*, 7457–7466. [CrossRef]
12. Ryan, K.; Ali, K. Application of a partial least-squares regression model to retrieve chlorophyll-a concentrations in coastal waters using hyper-spectral data. *Ocean Sci. J.* **2016**, *51*, 209–221. [CrossRef]
13. Van Nguyen, M.; Lin, C.-H.; Chu, H.-J.; Jaelani, L.M.; Syariz, M.A. Spectral Feature Selection Optimization for Water Quality Estimation. *Int. J. Environ. Res. Public Health* **2019**, *17*, 272. [CrossRef]
14. Schiller, H.; Doerffer, R. Neural network for emulation of an inverse model operational derivation of Case II water properties from MERIS data. *Int. J. Remote Sens.* **1999**, *20*, 1735–1746. [CrossRef]
15. Doerffer, R.; Schiller, H. The MERIS Case 2 water algorithm. *Int. J. Remote Sens.* **2007**, *28*, 517–535. [CrossRef]
16. Choi, J.-H.; Kim, J.; Won, J.; Min, O. Modelling Chlorophyll-a Concentration using Deep Neural Networks considering Extreme Data Imbalance and Skewness. In Proceedings of the 2019 21st International Conference on Advanced Communication Technology (ICACT), Pyeongchang-gun, Korea, 17–20 February 2019; pp. 631–634. [CrossRef]
17. Ioannou, I.; Foster, R.; Gilerson, A.; Gross, B.; Moshary, F.; Ahmed, S. Neural network approach for the derivation of chlorophyll concentration from ocean color. In Proceedings of the SPIE 8724, Ocean Sensing and Monitoring V, Baltimore, MD, USA, 29 April–3 May 2013. [CrossRef]
18. Zhan, H.; Shi, P.; Chen, C. Inversion of oceanic chlorophyll concentrations by neural networks. *Chin. Sci. Bull.* **2001**, *46*, 158–161. [CrossRef]
19. Vilas, L.G.; Spyrakos, E.; Palenzuela, J.M.T. Neural network estimation of chlorophyll a from MERIS full resolution data for the coastal waters of Galician rias (NW Spain). *Remote Sens. Environ.* **2011**, *115*, 524–535. [CrossRef]
20. Syariz, M.A.; Lin, C.-H.; Van Nguyen, M.; Jaelani, L.M.; Blanco, A.C. WaterNet: A Convolutional Neural Network for Chlorophyll-a Concentration Retrieval. *Remote Sens.* **2020**, *12*, 1966. [CrossRef]
21. Keiner, L.E.; Yan, X.-H. A Neural Network Model for Estimating Sea Surface Chlorophyll and Sediments from Thematic Mapper Imagery. *Remote Sens. Environ.* **1998**, *66*, 153–165. [CrossRef]
22. Krasnopolsky, V.M. Neural network emulations for complex multidimensional geophysical mappings: Applications of neural network techniques to atmospheric and oceanic satellite retrievals and numerical modeling. *Rev. Geophys.* **2007**, *45*, RG3009. [CrossRef]
23. Garlik, B.; Křivan, M. Identification of type daily diagrams of electric consumption based on cluster analysis of multi-dimensional data by neural network. *Neural Netw. World* **2013**, *23*, 271–283. [CrossRef]
24. Song, K.; Shao, T.; Li, L.; Li, S.; Tedesco, L.; Duan, H.; Li, Z.; Shi, K.; Du, J.; Zhao, Y. Using Partial Least Squares-Artificial Neural Network for Inversion of Inland Water Chlorophyll-a. *IEEE Trans. Geosci. Remote Sens.* **2014**, *52*, 1502–1517. [CrossRef]

25. Smith, M.; Lain, L.R.; Bernard, S. An optimized Chlorophyll a switching algorithm for MERIS and OLCI in phytoplankton-dominated waters. *Remote Sens. Environ.* **2018**, *215*, 217–227. [CrossRef]
26. Gilerson, A.A.; Gitelson, A.A.; Zhou, J.; Gurlin, D.; Moses, W.; Ioannou, I.; Ahmed, S.A. Algorithms for remote estimation of chlorophyll-a in coastal and inland waters using red and near infrared bands. *Opt. Express* **2010**, *18*, 24109–24125. [CrossRef]
27. Moses, W.J.; Gitelson, A.A.; Berdnikov, S.; Povazhnyy, V. Satellite Estimation of Chlorophyll-*a* Concentration Using the Red and NIR Bands of MERIS—The Azov Sea Case Study. *IEEE Geosci. Remote Sens. Lett.* **2009**, *6*, 845–849. [CrossRef]
28. Yacobi, Y.Z.; Moses, W.; Kaganovsky, S.; Sulimani, B.; Leavitt, B.C.; Gitelson, A.A. NIR-red reflectance-based algorithms for chlorophyll-a estimation in mesotrophic inland and coastal waters: Lake Kinneret case study. *Water Res.* **2011**, *45*, 2428–2436. [CrossRef]
29. Odermatt, D.; Gitelson, A.; Brando, V.E.; Schaepman, M. Review of constituent retrieval in optically deep and complex waters from satellite imagery. *Remote Sens. Environ.* **2012**, *118*, 116–126. [CrossRef]
30. Dall'Olmo, G.; Gitelson, A.A. Effect of bio-optical parameter variability on the remote estimation of chlorophyll-a concentration in turbid productive waters: Experimental results. *Appl. Opt.* **2005**, *44*, 412–422. [CrossRef]
31. Mueller, J.L.; Austin, R.W. Ocean optics protocols for SeaWiFS validation, revision 1. *Oceanogr. Lit. Rev.* **1995**, *42*, 805.
32. Ohde, T.; Siegel, H. Derivation of immersion factors for the hyperspectral TriOS radiance sensor. *J. Opt. A Pure Appl. Opt.* **2003**, *5*, L12–L14. [CrossRef]
33. Gitelson, A.A.; Yacobi, Y.Z.; Schalles, J.F.; Rundquist, D.C.; Han, L.; Stark, R.; Etzion, D. Remote estimation of phytoplankton density in productive waters. *Arch. Hydrobiol. Spec. Issues Advanc. Limnol.* **2000**, *55*, 121–136.
34. Schalles, J.F.; Rundquist, D.C.; Schiebe, F.R. The influence of suspended clays on phytoplankton reflectance signatures and the remote estimation of chlorophyll. *Int. Ver. Theor. Angew. Limnol. Verh.* **2001**, *27*, 3619–3625. [CrossRef]
35. Gitelson, A.A. Nature of the peak near 700 nm on the radiance spectra and its application for remote estimation of phytoplankton pigments in inland waters. In Proceedings of the SPIE 1971, 8th Meeting on Optical Engineering in Israel: Optical Engineering and Remote Sensing, Tel Aviv, Israel, 14–16 December 1992; p. 170.
36. Gower, J.; King, S.; Borstad, G.; Brown, L. Detection of intense plankton blooms using the 709 nm band of the MERIS imaging spectrometer. *Int. J. Remote Sens.* **2005**, *26*, 2005–2012. [CrossRef]
37. Gower, J.F.R.; Doerffer, R.; Borstad, G.A. Interpretation of the 685 nm peak in water-leaving radiance spectra in terms of fluorescence, absorption and scattering, and its observation by MERIS. *Int. J. Remote Sens.* **1999**, *20*, 1771–1786. [CrossRef]
38. Arar, E.J.; Collins, G.B. *Method 445.0: Chlorophyll a*; U.S. Environmental Protection Agency: Washington, DC, USA, 1997; pp. 1–22.
39. Esbensen, K.H.; Guyot, D.; Westad, F.; Houmoller, L.P. Multivariate Data Analysis-In Pactice. An Introduction to multivariate data analysis and experimental design (4th edn), Kim H. Esbensen, CAMO, OSLO, 2000, ISBN 82-9933302-4, xviii + 600pp, US$230.00. *J. Chemom.* **2002**, *16*, 117–118. [CrossRef]
40. Guo, G.; Mu, G. Simultaneous dimensionality reduction and human age estimation via kernel partial least squares regression. In Proceedings of the CVPR 2011, Colorado Springs, CO, USA, 20–25 June 2011; pp. 657–664.
41. Krishnan, A.; Williams, L.J.; McIntosh, A.R.; Abdi, H. Partial Least Squares (PLS) methods for neuroimaging: A tutorial and review. *NeuroImage* **2011**, *56*, 455–475. [CrossRef]
42. Chen, C.; Cao, X.; Tian, L. Partial Least Squares Regression Performs Well in MRI-Based Individualized Estimations. *Front. Neurosci.* **2019**, *13*, 1282. [CrossRef]
43. Park, Y.-S.; Lek, S. *Artificial Neural Networks: Multilayer Perceptron for Ecological Modeling*; Elsevier: Amsterdam, The Netherlands, 2016; Volume 28. [CrossRef]
44. Siripatrawan, U.; Linz, J.E.; Harte, B.R. Electronic sensor array coupled with artificial neural network for detection of Salmonella Typhimurium. *Sens. Actuators B Chem.* **2006**, *119*, 64–69. [CrossRef]
45. Bishop, C.M. *Building Neural Network for Pattern Recognition*; Oxford University Press: Oxford, UK, 2004; pp. 357–361. [CrossRef]
46. Huang, Y.; Kangas, L.J.; Rasco, B.A. Applications of Artificial Neural Networks (ANNs) in Food Science. *Crit. Rev. Food Sci. Nutr.* **2007**, *47*, 113–126. [CrossRef]
47. Ham, F.M.; Kostanic, I. *Principles of Neurocomputing for Science and Engineering*; McGraw-Hill Higher Education: New York, NY, USA, 2000.
48. O'Reilly, J.E.; Maritorena, S.; Mitchell, B.G.; Siegel, D.A.; Carder, K.L.; Garver, S.A.; Kahru, M.; McClain, C. Ocean color chlorophyll algorithms for SeaWiFS. *J. Geophys. Res. Earth Surf.* **1998**, *103*, 24937–24953. [CrossRef]
49. Moses, W.J.; Gitelson, A.A.; Berdnikov, S.; Saprygin, V.; Povazhnyi, V. Operational MERIS-based NIR-red algorithms for estimating chlorophyll-a concentrations in coastal waters—The Azov Sea case study. *Remote Sens. Environ.* **2012**, *121*, 118–124. [CrossRef]
50. Dall'Olmo, G.; Gitelson, A.A.; Rundquist, D.C.; Leavitt, B.; Barrow, T.; Holz, J.C. Assessing the potential of SeaWiFS and MODIS for estimating chlorophyll concentration in turbid productive waters using red and near-infrared bands. *Remote Sens. Environ.* **2005**, *96*, 176–187. [CrossRef]
51. Moses, W.J.; Gitelson, A.A.; Berdnikov, S.; Povazhnyy, V. Estimation of chlorophyll-a concentration in case II waters using MODIS and MERIS data—successes and challenges. *Environ. Res. Lett.* **2009**, *4*, 045005. [CrossRef]
52. Seegers, B.N.; Stumpf, R.P.; Schaeffer, B.; Loftin, K.A.; Werdell, P.J. Performance metrics for the assessment of satellite data products: An ocean color case study. *Opt. Express* **2018**, *26*, 7404–7422. [CrossRef] [PubMed]
53. Harris, S.E.; Mix, A.C. Pleistocene Precipitation Balance in the Amazon Basin Recorded in Deep Sea Sediments. *Quat. Res.* **1999**, *51*, 14–26. [CrossRef]

54. Deaton, B.C.; Balsam, W.L. Visible spectroscopy; a rapid method for determining hematite and goethite concentration in geological materials. *J. Sediment. Res.* **1991**, *61*, 628–632. [CrossRef]
55. Ortiz, J.D.; Witter, D.L.; Ali, K.A.; Fela, N.; Duff, M.; Mills, L. Evaluating multiple colour-producing agents in Case II waters from Lake Erie. *Int. J. Remote Sens.* **2013**, *34*, 8854–8880. [CrossRef]

remote sensing

MDPI

Article

Improving Soil Moisture Estimation via Assimilation of Remote Sensing Product into the DSSAT Crop Model and Its Effect on Agricultural Drought Monitoring

Hongkui Zhou [1,*], Guangpo Geng [2], Jianhua Yang [3,4], Hao Hu [1], Li Sheng [1] and Weidong Lou [1]

[1] Institute of Digital Agriculture, Zhejiang Academy of Agricultural Sciences, Hangzhou 310021, China; huh@zaas.ac.cn (H.H.); shengli@zaas.ac.cn (L.S.); louwd@zaas.ac.cn (W.L.)
[2] College of Geomatics, Xi'an University of Science and Technology, Xi'an 710054, China; gengguangpo@xust.edu.cn
[3] State Key Laboratory of Remote Sensing Science, Beijing Normal University, Beijing 100875, China; yangjh15@mail.bnu.edu.cn
[4] Center for Drought and Risk Research, Beijing Normal University, Beijing 100875, China
* Correspondence: zhouhongkui@zaas.ac.cn

Abstract: Accurate knowledge of soil moisture is crucial for agricultural drought monitoring. Data assimilation has proven to be a promising technique for improving soil moisture estimation, and various studies have been conducted on soil moisture data assimilation based on land surface models. However, crop growth models, which are ideal tools for agricultural simulation applications, are rarely used for soil moisture assimilation. Moreover, the role of data assimilation in agricultural drought monitoring is seldom investigated. In the present work, we assimilated the European Space Agency (ESA) Climate Change Initiative (CCI) soil moisture product into the Decision Support System for Agro-technology Transfer (DSSAT) model to estimate surface and root-zone soil moisture, and we evaluated the effect of data assimilation on agricultural drought monitoring. The results demonstrate that the soil moisture estimates were significantly improved after data assimilation. Root-zone soil moisture had a better agreement with in situ observation. Compared with the drought index based on soil moisture modeled without remotely-sensed observations, the drought index based on assimilated data could improve at least one drought level in agricultural drought monitoring and performed better when compared with winter wheat yield. In conclusion, crop growth model-based data assimilation effectively improves the soil moisture estimation and further strengthens soil moisture-based drought indices for agricultural drought monitoring.

Keywords: soil moisture; data assimilation; agricultural drought; remote sensing; DSSAT model

1. Introduction

Drought is a fairly complex and disastrous climate disaster, which can cause tremendous damages to the economy and environment, especially for the agriculture sector. Typically, meteorological, agricultural, hydrological, and socioeconomic droughts are recognized as the four common types of drought [1]. Compared with other types of drought, agricultural drought provides a more immediate and obvious influence [2] and threatens regional and even national food security [3,4]. Under the scenario of global climate change, future drought may be exacerbated [5], and along with other climate hazards, drought may still pose a heightened threat to humanity [6]. Thus, accurate and effective agricultural drought monitoring is of great significance for impact assessment and targeted adaption planning.

Agricultural drought involves insufficient soil moisture due to the imbalance between water supply and crop water demand, which detrimentally affects crop growth and yield [7]. Hence, soil moisture becomes an essential indicator for assessing agricultural drought [2]. Several soil moisture-based drought indices have been proposed, which are generally divided into two categories: (1) based on soil moisture time series (e.g., SMP [8], NSM [9], ESSMI [10]); and (2) based on soil properties and other parameters (e.g., SMI [11,12], SWDI [2], RSMI [13]). Although the rationales of these indices are not completely identical, the accuracy of soil moisture is a critical factor affecting their performances for agricultural drought monitoring. Currently, methods for estimating soil moisture include field measurement, model simulation, and remote sensing retrieval. Field measurement is accurate but time-consuming and limited in coverage, making it unsuitable for large-scale and frequent soil moisture collection. The main advantage of using a land surface model to simulate soil moisture is that it has a clear physical mechanism, has good spatiotemporal consistency, and can obtain soil moisture throughout the soil profile; the drawback is that it requires many input parameters, and its accuracy depends largely on the model structure, model parameters and forcings. Over the past few decades, great advances in remote sensing techniques have enabled users to obtain soil moisture over large areas. Soil moisture can be estimated using remote sensing techniques with different spectra from optical to microwave ranges, i.e., the visible/near infrared (VIS/NIR, 350~2500 nm), thermal infrared (TIR, 3500~14,000 nm) and microwave (P—band, L—band, C—band, and X—band) [14–16]. Li et al. [15] reviewed the soil moisture retrieval from satellite remote sensing measurements, with emphasis on estimations from the visible/near infrared (VIS/NIR) and thermal infrared (TIR) observations. Babaeian et al. [16] elaborated on recent advances in soil moisture estimation using ground, proximal, and airborne and spaceborne remote sensing techniques, and they summarized the available soil moisture data sources. According to the above related literature, the satellite-based microwave remote sensing method is still the mainstream of soil moisture estimation. However, remote sensing measurements are limited to a few centimeters at the top of the soil and do not provide information on the moisture of the deep soil [17,18]. In fact, root-zone soil moisture is more appropriate than that from a fixed soil depth for assessing agricultural drought [19].

Given that each method has its own advantages and limitations in estimating soil moisture, data assimilation (DA) is a promising approach for improving soil moisture estimation by integrating a process model and observation. Benefiting from the rapid progress of remote sensing techniques, the accuracy of satellite soil moisture products has gradually increased in recent years, and they have become the core observations for soil moisture data assimilation. Many previous studies have assimilated remotely sensed products into hydrological or land surface models [17,20–23] and agricultural drought estimation [24–29]. For example, Baldwin et al. [17] assimilated the AMSR-E soil moisture into the Soil Moisture Analytical Relationship (SMAR) infiltration model for predicting root-zone soil moisture. De Lannoy and Reichle [28] conducted the assimilation of SMOS products into the Goddard Earth Observing System Model to improve estimates of surface and root-zone soil moisture. Seo et al. [29] assimilated the SMAP and ASCAT soil moisture into the JULES land surface model for improving soil moisture and drought monitoring. Kumar et al. [24] presented the assimilation of three remotely sensed soil moisture products into the Noah land surface model for soil moisture and drought estimation. The results of these studies showed that data assimilation contributes to the improvement of soil moisture estimation, but significant room for improvement exists for each component (model, observation, and algorithm) of the data assimilation system [25]. For the time being, land surface models are still the dominant model operators for moisture assimilation. As far as crops are concerned, the occurrence of agricultural drought has a close bearing on soil moisture, crop water demand, growth stage, and irrigation condition. Most land surface models, however, rely on observations or empirical functions to simulate the effects of vegetation and lack a more mechanistic representation [30], while generally ignoring the effect of human activities (e.g., irrigation) on soil moisture change. In contrast, crop models are more capable of explicitly characterizing the entire growth process and simulating the interactions between the root-zone soil moisture and crop water uptake during the growing season. Hence, crop growth models are ideal tools to characterize the relationship between crop growth and environmental factors, and drought indices combined with the agronomic parameters from the model output can better monitor agricultural drought. In current studies, crop growth model-based data assimilation is generally used for leaf area index, as well as yield or biomass estimation and prediction [31–35]. The linkage between crop models and remote sensing observations for improving surface and root-zone soil moisture is relatively rare, and evaluations of the impact of DA-based soil moisture estimates on agricultural drought monitoring are still lacking. Consequently, the goals of this study were: (1) to assimilate remote sensing data into a crop model to estimate soil moisture; (2) to investigate the impacts of different strategies (simultaneous state-parameter estimation, ensemble size, and assimilation interval) on soil moisture data assimilation; and (3) to evaluate the extent to which soil moisture assimilation enhances the level of agricultural drought monitoring.

2. Study Area and Data

2.1. Study Area

The focus of this article is on the Huang-Huai-Hai (HHH) Plain, the second largest plain in China, which is located between 32° and 41°N and 112° and 123°E, covering 390,000 km^2 (Figure 1). The HHH Plain is an important food-producing base in China, accounting for approximately 16% of the total cultivated land area [36–38]. The prevailing cropping system is the rotation of winter wheat and summer corn, and wheat and corn production account for 70% and 30% of the national total production, respectively [39]. The HHH Plain is affected by a warm-temperate monsoon climate, and the average temperatures in winter and summer are approximately 0 °C and 26 °C, respectively. In terms of precipitation, significant differences exist in time and space dimensions. The average annual precipitation ranges from 400 to 1000 mm, increasing from north to south, but 60–80% of the annual precipitation is concentrated in the summer corn growing season. Since winter wheat is more susceptible to drought than corn, the experimental period was selected during the winter wheat growing season. Considering the differences in topography, climate, hydrology, and agricultural production conditions, the HHH Plain consists of four divisions, as shown in Figure 1. Eight agro-meteorological sites were selected in this study, which were distributed in the four agricultural subregions.

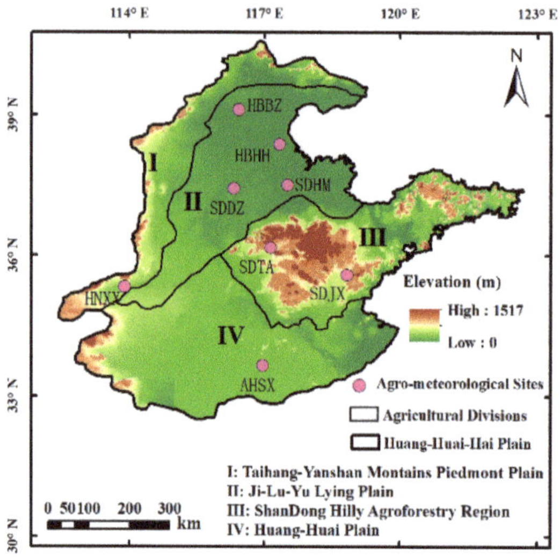

Figure 1. Study area and distribution of agro-meteorological sites.

2.2. Data

2.2.1. Meteorological Data

Meteorological variables of daily precipitation, air temperature, wind speed, and solar radiation were required to drive the crop growth simulation. Solar radiation was calculated from sunshine duration using the Ångström equation [40]. The other four meteorological variables and sunshine duration were observed by meteorological stations in China (http://data.cma.cn/ (accessed on 1 June 2022)).

2.2.2. Agro-Meteorological Data

Agro-meteorological data, including wheat phenology, yield, soil moisture, and soil properties, were also collected in this study; these data were derived from the observations of agro-meteorological sites affiliated with the China Meteorological Administration. Wheat phenology and yield were primarily used for crop growth model calibration. The phenology was measured every 10 days during the growing season, and yield was measured after harvest. Soil moisture was used for model initialization and data assimilation evaluation. We collected soil moisture at 10, 20, 30, 40 and 50 cm depths at the representative HBBZ site, as well as at 10, 20 and 50 cm depths at seven other sites. Soil moisture was measured every 10 days during the growing season. Gravimetric measurement was adopted with four replicates, and the averaged value was used. The soil moisture obtained in this way is gravimetric water content. In order to match the satellite soil moisture data, it was then multiplied by soil bulk density and converted to volumetric water content. The soil moisture used in this study refers to volumetric water content. Soil properties included the soil texture, bulk density, field water-holding capacity, and wilting point at the corresponding layers, and these properties were used for model input and soil moisture index calculation. Soil properties were measured every 5–10 years depending on the actual condition of each site.

2.2.3. Satellite Soil Moisture

Soil moisture from the ESA (European Space Agency) CCI (Climate Change Initiative) was used as the observation for soil moisture data assimilation. ESA CCI soil moisture contains the passive, active and combined products based on various microwave sensors, which have been extensively applied in a great deal of earth system applications, e.g., climate change, land-atmosphere interactions, drought monitoring and assessment, and land surface modeling [18]. The latest version is v07.1, which can be obtained from http://www.esa-soilmoisture-cci.org (accessed on 1 June 2022). This product covers the time range from November 1978 to December 2021, with spatial and temporal resolutions of 0.25° and 1 day, respectively. The combined product during the period from 2002 to 2011 was used in this study. For more details regarding this product, please refer to Liu et al. [41], Dorigo et al. [18] and Gruber et al. [42].

3. Methods

3.1. Crop Model Description and Verification

The Decision Support System for Agro-technology Transfer (DSSAT)—Cropping System Model (CSM)—Wheat model (herein referred to as the DSSAT model [43]) was used as the model operator for data assimilation. The DSSAT model was developed by the US Department of Agriculture, and it can simulate the vegetative and reproductive growth, physiological and ecological processes, and soil water balance on a daily basis. Input data include four categories: weather; crop; soil; and management data. The weather data should at least include daily maximum and minimum temperature, solar radiation, and precipitation. Crop data indicates the genetic and ecological parameters, which describe crop photosynthesis, respiration, phenological stage and other physiological processes. Soil data represent the soil physical, chemical, and morphological parameters throughout the soil profile. Management data include a number of parameters regarding planting, irrigation, fertilization, etc.

The soil moisture simulation in the DSSAT model was implemented using the soil water balance module [44]. Soil water balance involves the following processes: infiltration; evapotranspiration (ET); runoff; and drainage. Soil water infiltration is calculated as the sum of rainfall and irrigation minus surface runoff. Surface runoff is computed based on the SCS-CN model [45] developed by the US Department of Agriculture. The soil-plant-atmosphere module (SPAM) was developed to compute the daily ET, and it provides two methods—the Priestley–Taylor [46] and FAO-56 Penman–Monteith [40] methods, to calculate the potential ET. Actual ET is calculated based on potential ET and soil water stress factors [47].

Before using the DSSAT model, the model parameters should be calibrated according to the local climate, soil conditions, management methods, and cropping system, in order to enable the model to accurately measure the characteristics of local crops. Crop phenology and yield were used to calibrate seven critical cultivar parameters (P1V, P1D, P5, G1, G2, G3, and PHINT; for a detailed description, see Jones et al. [43]). The suitable parameters in the study area were well calibrated and validated in our previous study; for more details about the model verification process, please refer to Zhou et al. [13].

3.2. Assimilation Algorithms

The Augmented Ensemble Kalman Filter (AEnKF) [30,48] was used as the assimilation algorithm in the present paper. On the basis of EnKF (Ensemble Kalman Filter; [49]), AEnKF also takes model parameters as state variables to achieve the aim of simultaneously estimating these two categorical parameters (state variables and model parameters). Hence, EnKF is the foundation of AEnKF. The implementation of EnKF involves the following steps: initialization; forecast; and analysis.

Suppose $X = [x_1, x_2, \cdots, x_n]^T$ as the state variable, where x_i represents the soil moisture in the different soil layers and crop model parameters in this study. First, the state ensemble is initialized as follows (Equation (1)):

$$X^a_{i,0} = X^a_0 + w_i,\ \ w_i \sim N(0, P) \tag{1}$$

where $X^a_{i,0}$ is the i-th initial state member (i = 1, 2, ..., n; n is the ensemble size), X^a_0 is the expectation of the background field, w_i is the random error, N indicates the Gaussian distribution, and P is background covariance matrix. Note that the initial state ensemble can also be generated by perturbing model parameters.

Second, each ensemble member is forecasted by the model operator as follows (Equation (2)):

$$X^f_{i,t+1} = M(X^a_{i,t}, \alpha_{t+1}, \beta_{t+1}) + \mu_i,\ \ \mu_i \sim N(0, Q) \tag{2}$$

where $X^f_{i,t+1}$ is the forecasted state variable of the i-th ensemble member at the time t + 1; $M(\bullet)$ is the model operator, α represents the model forcing data, β is model parameter; μ_i is the model error, and Q is the model covariance matrix.

Third, when the observations are available, each ensemble member is analyzed based on the following linear combination of forecasted state and observation (Equation (3)).

$$X^a_{i,t+1} = X^f_{i,t+1} + K_{t+1}\left[(Y_{t+1} + v_i) - H\left(X^f_{i,t+1}\right)\right], v_i \sim N(0, R) \tag{3}$$

where $X^a_{i,t+1}$ is the analysis state variable, Y_{t+1} is the observation, v_i is the observation error, R is the observation covariance matrix, $H(\bullet)$ is the observation operator, and K_{t+1} is the Kalman gain matrix defined as follows:

$$K_{t+1} = P^f_{t+1}H^T\left(HP^f_{t+1}H^T + R\right)^{-1} \tag{4}$$

$$P_{t+1}^{f} = \frac{1}{n-1}\sum_{i=1}^{n}(X_{i,t+1}^{f} - \overline{X_{t+1}^{f}})\left(X_{i,t+1}^{f} - \overline{X_{t+1}^{f}}\right)^{T} \quad (5)$$

$$P_{t+1}^{f}H^{T} = \frac{1}{n-1}\sum_{i=1}^{n}(X_{i,t+1}^{f} - \overline{X_{t+1}^{f}})\left[H\left(X_{i,t+1}^{f}\right) - H\left(\overline{X_{t+1}^{f}}\right)\right]^{T} \quad (6)$$

$$HP_{t+1}^{f}H^{T} = \frac{1}{n-1}\sum_{i=1}^{n}[H(X_{i,t+1}^{f}) - H(\overline{X_{t+1}^{f}})][H(X_{i,t+1}^{f}) - H(\overline{X_{t+1}^{f}})]^{T} \quad (7)$$

$$\overline{X_{t+1}^{f}} = \frac{1}{n}\sum_{i=1}^{n}X_{i,t+1}^{f} \quad (8)$$

where P_{t+1}^{f} is the forecasted covariance matrix, and $\overline{X_{t+1}^{f}}$ is the average value of forecasted ensemble members.

Finally, the analysis state variable is calculated by averaging each ensemble member (Equation (9)). When the next observation occurs, the state variable continues to repeat the above process until the end of data assimilation.

$$X_{t+1}^{a} = \frac{1}{n}\sum_{i=1}^{N}X_{i,t+1}^{a} \quad (9)$$

3.3. Data Assimilation Implementation

Data assimilation framework includes three key components: model operator; observation; and algorithm. The soil water balance module of the DSSAT model was used as the model operator for soil moisture data assimilation. Soil moisture simulations of nine winter wheat growing seasons from 2002 to 2011 were conducted at the eight agro-meteorological sites in the HHH Plain. Management parameters, such as planting date, density, depth, soil properties, fertilization, and irrigation, were set according to the records of the agro-meteorological sites.

The ESA CCI soil moisture product was used as observation data of the assimilation framework. Given the differences in the climatology and spatial resolution of in situ and satellite observations, bias correction of satellite soil moisture was required prior to data assimilation. In this study, the cumulative distribution function matching approach was applied for bias correction, which has been widely applied in data assimilation studies (e.g., [24,29,50]). Referring to existing studies [18,24,34,51], the observation error was set to 0.06 $m^3 \cdot m^{-3}$.

The AEnKF was selected as the algorithm for soil moisture data assimilation, which was described in detail in Section 3.2. In this study, the state variable vector consists of soil moisture at different depths and model parameters that were sensitive to soil moisture. Soil parameters such as lower limit (SLLL), drained upper limit (SDUL), saturated limit (SSAT), drainage rate (SLDR) and runoff curve number (SLRO) in the DSSAT model have significant impacts on soil moisture estimation [47]. Thus, these soil parameters were also included as state variables, and the state variable vector was as follows:

$$X_t = [SM_1\ SM_2 \cdots\ SM_n\ \text{SLLL SDUL SSAT SLDR SLRO}]^T \quad (10)$$

where X_t is the state variable vector at the time t, SM_i is the soil moisture at soil layer i, and n is the number of soil layers. Since the ESA CCI data only reflects soil moisture information for the top layer, the observation operator H was set to $[1\,0\,0\,0\,0\,0\,0\,0\,0\,0]^T$. Following the study of Ines et al. [34], an uncertainty level of 10% of the model soil parameters was introduced to perturb the initial state variables.

The ensemble size was set to 20 by considering the estimation accuracy and computational time. Additionally, with the increase in assimilation time, the standard EnKF algorithm tends to exhibit a "filter divergence" phenomenon; that is, the analysis value gradually approaches the model forecast and ignores the observation. According to existing studies [33,34], when the observation error covariance is greater than four times the model forecast error covariance ($R/P^f_{t+1} > 4$), an inflation factor of 1.05 was used to enlarge the Kalman gain to reduce this phenomenon. In addition, the impacts of ensemble size, model parameter, and assimilation interval on data assimilation are discussed in Section 4.2.

3.4. Agricultural Drought Index

The root-weighted soil moisture index (*RSMI* [13]) was used as an agricultural drought index to evaluate the role of data assimilation. There are two main reasons for choosing the *RSMI* as an indicator: (a) *RSMI* considers the influence of soil properties and root distribution on crop water uptake, and (b) *RSMI* does not require long-term time series data. *RSMI* is as follows:

$$RSMI = \sum_{i=1}^{n} RW_i \times SMI_i \tag{11}$$

$$RW_i = \frac{z_i \times RLD_i}{\sum_{i=1}^{n} z_i \times RLD_j} \tag{12}$$

$$SMI = 10 \times \frac{\theta - \theta_{wp}}{\theta_{fc} - \theta_{wp}} - 5 \tag{13}$$

where *SMI* is the soil moisture index, *RW* is the weight value, n is the number of soil layers, in which roots reach the deepest part of the soil, which varies with the crop growth stage, z is the soil thickness, *RLD* is the root length density, and θ, θ_{fc} and θ_{wp} represent soil moisture, field capacity and wilting point, respectively. The drought classifications of the *RSMI* are shown in Table 1. For more details regarding the *RSMI*, please refer to Zhou et al. [13].

Table 1. Drought categories of the *RSMI*.

RSMI Value	Drought Level
−1 to 0	Less intense
−2 to −1	Moderate
−3 to −2	High intense
−4 to −3	Severe
−4 or less	Extreme

3.5. Evaluation Criteria

The root-mean-square error (*RMSE*, $m^3 \cdot m^{-3}$), correlation coefficient (*R*), and effectiveness (*Eff*, %) [52] were used to quantitatively measure the capability of data assimilation. These indicators were calculated as follows:

$$RMSE = \sqrt{\frac{1}{n}\sum_{i=1}^{n}(S_i - O_i)^2} \tag{14}$$

$$R = \frac{\sum_{i=1}^{n}(S_i - \overline{S})(O_i - \overline{O})}{\sqrt{\sum_{i=1}^{n}(S_i - \overline{S})^2 \sum_{i=1}^{n}(O_i - \overline{O})^2}} \tag{15}$$

$$Eff = 100 \times \left(1 - \frac{\sum_{i=1}^{n}(S_i^a - O_i)^2}{\sum_{i=1}^{n}\left(S_i^f - O_i\right)^2}\right) \tag{16}$$

where S_i is the estimated (simulated or assimilated) soil moisture at the time i, O_i is the measured soil moisture; n is the number of measurements, $\overline{S}$ and $\overline{O}$ represent the average value of the estimated and measured soil moisture, respectively, S_i^f and S_i^a are the soil moisture before and after data assimilation, respectively, and *Eff* indicates the efficiency of the data assimilation. If *Eff* is greater than 0, this situation indicates that the soil moisture accuracy is improved after data assimilation; otherwise, a negative value denotes a degraded performance.

4. Results

4.1. Comparisons of Open-Loop and Assimilated Soil Moisture

4.1.1. Soil Moisture at Different Depths

A data assimilation experiment was first conducted and validated in the representative Bazhou (HBBZ) agro-meteorological site. The experimental period incorporated the growing seasons of winter wheat from 2002 to 2011. Figure 2 shows the comparison of the open-loop and assimilated soil moisture at the HBBZ site. Open-loop soil moisture is the direct simulation from the DSSAT model without data assimilation. As shown, relatively large errors were present in the open-loop simulations compared with the measured soil moisture, and open-loop soil moisture was significantly lower than observed data in the late growth stage of winter wheat. In contrast, assimilated results were in good agreement with the observations. Quantitatively, the goodness-of-fit statistics (*RMSE*, *R*, and *Eff*) at the HBBZ site are shown in Table 2. The *RMSEs* of open-loop soil moisture in the 10–50 cm depth during the 2002–2011 growing season were 0.084, 0.087, 0.088, 0.092, and 0.066 $m^3 \cdot m^{-3}$, respectively, and the *RMSEs* of assimilated soil moisture were 0.048, 0.031, 0.074, 0.060, and 0.043 $m^3 \cdot m^{-3}$, respectively. Except for the depths of 10 and 50 cm, the correlation coefficients of soil moisture at other soil layers increased slightly compared with those in the open-loop simulations. The effectiveness *Eff* values of soil moisture at a depth of 10–50 cm were 66.9%, 87.4%, 27.9%, 57.9%, and 56.0%, respectively, indicating that the soil moisture at each layer improved after data assimilation.

Table 2. Statistics of open-loop (OL) and assimilated (DA) soil moisture at the HBBZ site.

	10 cm	20 cm	30 cm	40 cm	50 cm	Root Zone
-RMSE **(root mean square error, $m^3 \cdot m^{-3}$)**						
OL	0.084	0.087	0.088	0.092	0.066	0.067
DA	0.048	0.031	0.074	0.060	0.043	0.038
-R **(correlation coefficient)**						
OL	0.685	0.531	0.546	0.503	0.514	0.664
DA	0.598	0.603	0.644	0.650	0.499	0.671
-Eff **(effectiveness, %)**						
	66.875	87.389	27.947	57.882	56.037	67.336

We also compared the open-loop and assimilated soil moisture with in situ observations at all eight agro-meteorological sites. Except for the HBBZ site, measured soil moisture at the other sites was collected at three depths: 10, 20, and 50 cm. Figure 3 shows the scatter plots and histograms of soil moisture estimates and observations. The results indicated that assimilated soil moisture matched better with in situ observations than open-loop simulations. Additionally, compared with open-loop simulations, the histograms of assimilated soil moisture were more similar to these of observations. For the eight sites, the average *RMSEs* of open-loop simulations at depths of 10, 20, and 50 cm were 0.088, 0.091, and 0.056 $m^3 \cdot m^{-3}$, respectively, and the average *RMSEs* of assimilated soil moisture were 0.047, 0.046, and 0.045 $m^3 \cdot m^{-3}$, respectively (Table 3). The correlations between the assimilated soil moisture with observations at each layer were improved. The *Eff* values of the soil moisture at each layer were greater than zero, indicating that moisture estimates throughout the soil profile were improved via data assimilation.

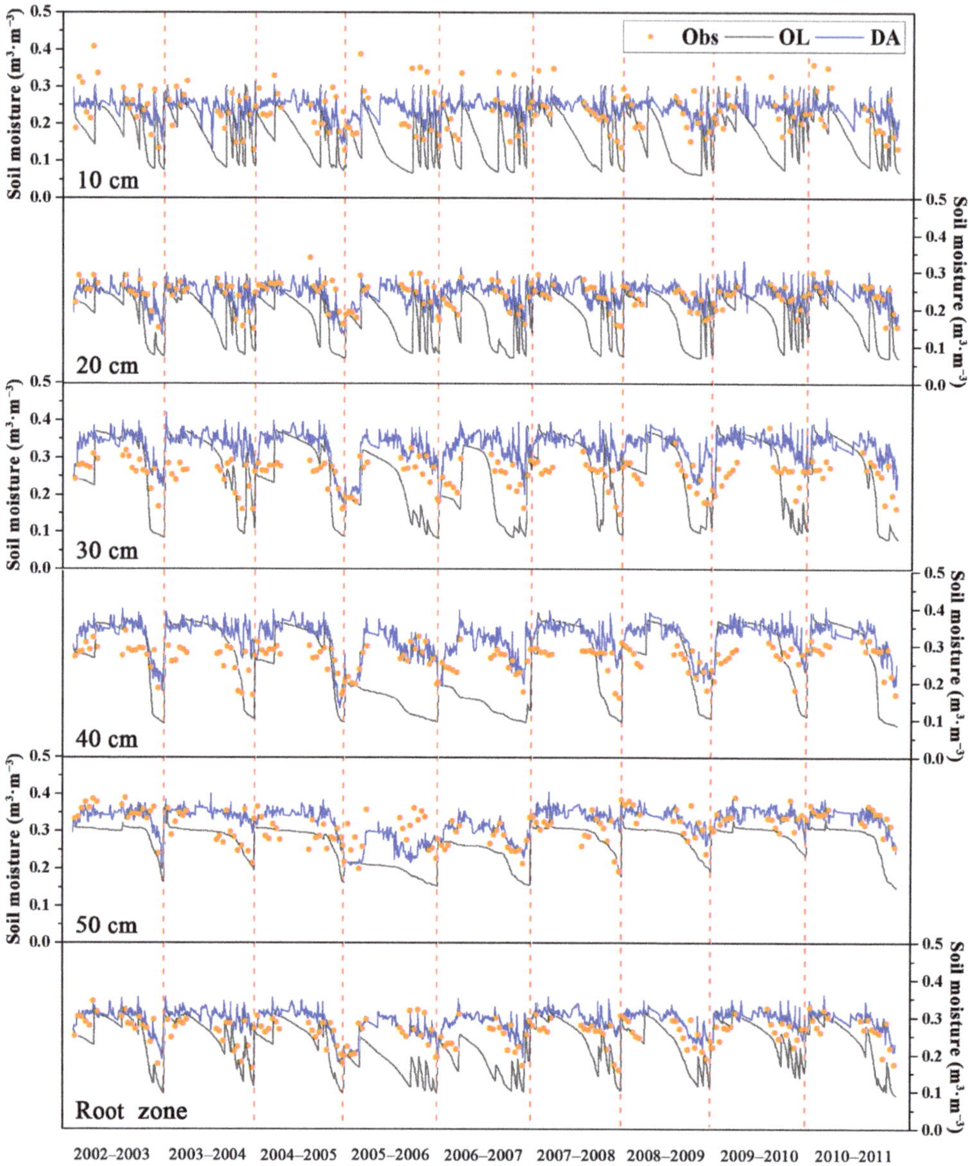

Figure 2. Comparison of open loop (OL) and assimilated (DA) soil moisture with observations (Obs) during the winter wheat growing seasons (2002–2011) at the HBBZ site.

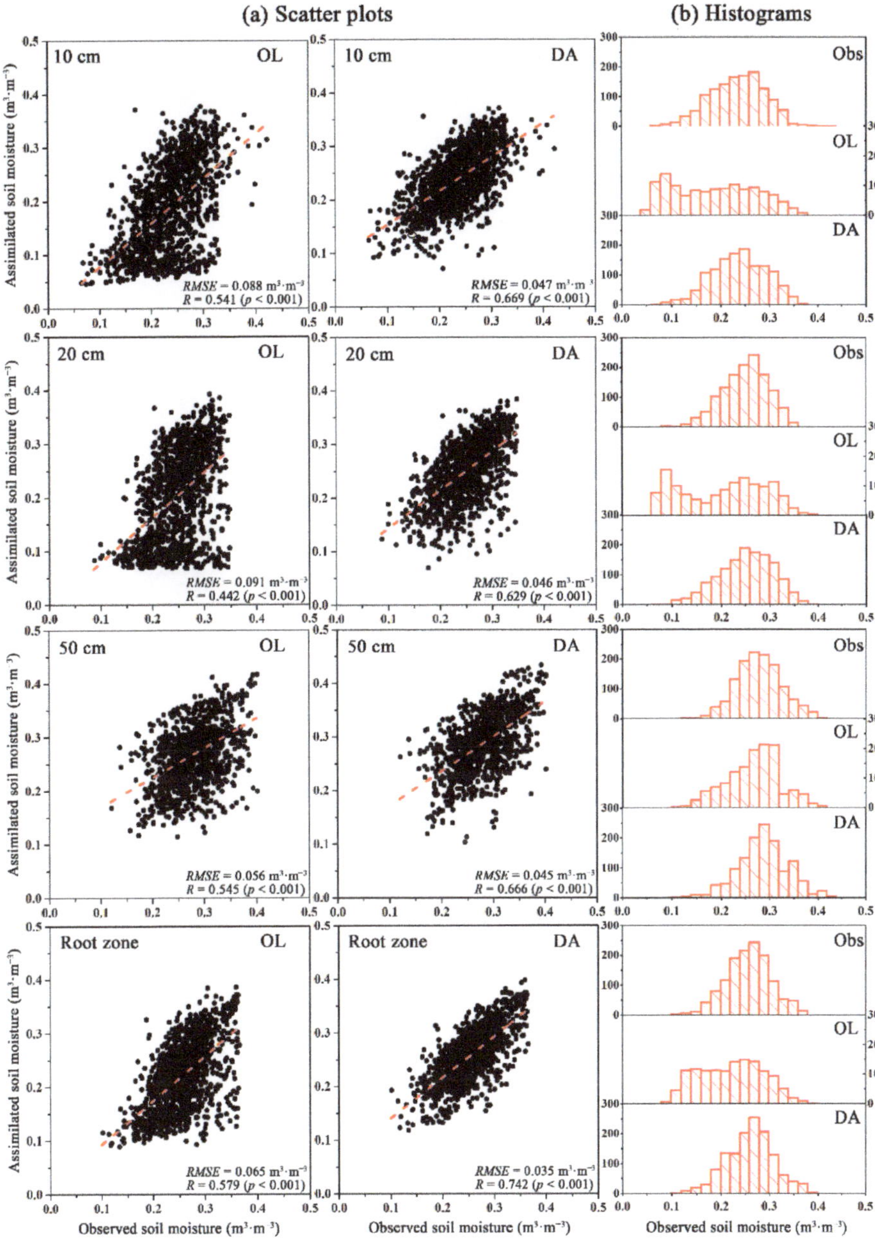

Figure 3. Comparisons of open-loop (OL) and assimilated (DA) soil moisture with observations (Obs) in the different soil depths: (**a**) scatter plots between OL (DA) and Obs; (**b**) histograms of OL, DA, and Obs. The red lines represent the linear regressions between open-loop/assimilated soil moisture and observed soil moisture.

Table 3. Statistics of open-loop (OL) and assimilated (DA) soil moisture at all eight sites.

Sites	OL				DA			
	10 cm	20 cm	50 cm	Root zone	10 cm	20 cm	50 cm	Root Zone
-*RMSE* ($m^3 \cdot m^{-3}$)								
AHSX	0.056	0.066	0.059	0.048	0.050	0.038	0.040	0.032
HBBZ	0.084	0.087	0.066	0.067	0.048	0.031	0.043	0.038
HBHH	0.101	0.090	0.024	0.062	0.053	0.046	0.027	0.031
HNXX	0.073	0.066	0.044	0.045	0.046	0.047	0.039	0.033
SDDZ	0.083	0.093	0.052	0.062	0.044	0.056	0.050	0.039
SDHM	0.129	0.140	0.074	0.105	0.046	0.054	0.050	0.040
SDJX	0.094	0.089	0.045	0.059	0.043	0.045	0.042	0.031
SDTA	0.058	0.077	0.064	0.048	0.040	0.041	0.063	0.035
All sites	0.088	0.091	0.056	0.065	0.047	0.046	0.045	0.035
-*R*								
AHSX	0.688	0.681	0.537	0.738	0.715	0.688	0.527	0.758
HBBZ	0.685	0.531	0.514	0.664	0.598	0.603	0.499	0.671
HBHH	0.502	0.583	0.908	0.717	0.643	0.698	0.880	0.810
HNXX	0.526	0.357	0.183	0.513	0.564	0.374	0.260	0.531
SDDZ	0.695	0.524	0.378	0.628	0.708	0.530	0.354	0.617
SDHM	0.355	0.174	0.399	0.312	0.366	0.326	0.473	0.487
SDJX	0.582	0.538	0.212	0.639	0.670	0.566	0.222	0.656
SDTA	0.692	0.577	0.168	0.658	0.763	0.647	0.076	0.652
All sites	0.541	0.442	0.545	0.579	0.669	0.629	0.666	0.742

Sites	10 cm	20 cm	50 cm	Root Zone
-*Eff* (%)				
AHSX	19.270	67.425	54.866	56.319
HBBZ	66.875	87.389	56.037	67.336
HBHH	73.014	74.341	−25.365	74.744
HNXX	60.265	50.033	22.175	46.646
SDDZ	71.841	63.894	9.073	60.075
SDHM	87.311	85.031	54.002	85.467
SDJX	78.943	74.336	13.335	73.261
SDTA	53.083	71.068	4.587	46.338
All sites	71.346	74.348	33.450	71.197

4.1.2. Root-Zone Soil Moisture

Root-zone soil moisture plays a significant role in agricultural drought monitoring. For cereal crops, roots are mainly distributed at a depth of 0–50 cm; thus, soil moisture within these depths has often been used to monitor and predict agricultural drought [2,53]. In this study, the average of soil moisture in each layer within 0–50 cm was regarded as the root-zone value. The assimilation results and error statistics of the root-zone soil moisture at HBBZ site are also shown in Figure 2 and Table 2. The *RMSEs* of open-loop and assimilated soil moistures were 0.067 and 0.038 $m^3 \cdot m^{-3}$, respectively. Similarly, Figure 3 and Table 3 display the results of the root-zone soil moisture at the eight sites. In comparison with the open-loop simulations, the *RMSE* of the root-zone soil moisture was reduced from 0.065 to 0.035 $m^3 \cdot m^{-3}$, and *R* was increased from 0.579 to 0.742. Furthermore, compared with the soil moisture at each layer, root-zone soil moisture also had a better agreement with in situ observations.

4.2. Impacts of Different Strategies on Data Assimilation

4.2.1. Impact of Simultaneous State-Parameter Estimation

During the progress of data assimilation, initial conditions, model parameters, external observations and algorithm are the main factors affecting the accuracy of state variable estimation. Hence, the accuracy of the model parameters is important for improving state variable estimation when other conditions are the same. In the AEnKF algorithm, state variables and model parameters are updated at the same time according to the Kalman gain, which is theoretically expected to enhance the soil moisture measurement. Here we compared the utilities of the results based on the AEnKF algorithm (simultaneous state–parameter estimation) and EnKF algorithm (state-only estimation) at the HHBZ site during the growing seasons of 2004–2005 and 2005–2006.

Figures 4 and 5 display the time series and *RMSEs* of soil moisture estimates based on the AEnKF and EnKF algorithms, respectively. The results show that, influenced by the model structure, the error of the soil moisture estimates based on EnKF were still large in the late growth season. The *RMSEs* of the assimilated soil moisture at depths of 10–50 cm and root zone based on the EnKF (AEnKF) algorithm were 0.048 (0.048), 0.030 (0.029), 0.085 (0.061), 0.084 (0.047), 0.071 (0.057), and 0.049 (0.029), respectively. The errors of assimilated soil moisture at each soil layer based on AEnKF were less than those based on the EnKF algorithm. According to the time series data and error statistics, the performance of the AEnKF-based soil moisture data assimilation was better, which also indicated that the simultaneous state-parameter estimation was effective for improving state variable estimation in data assimilation.

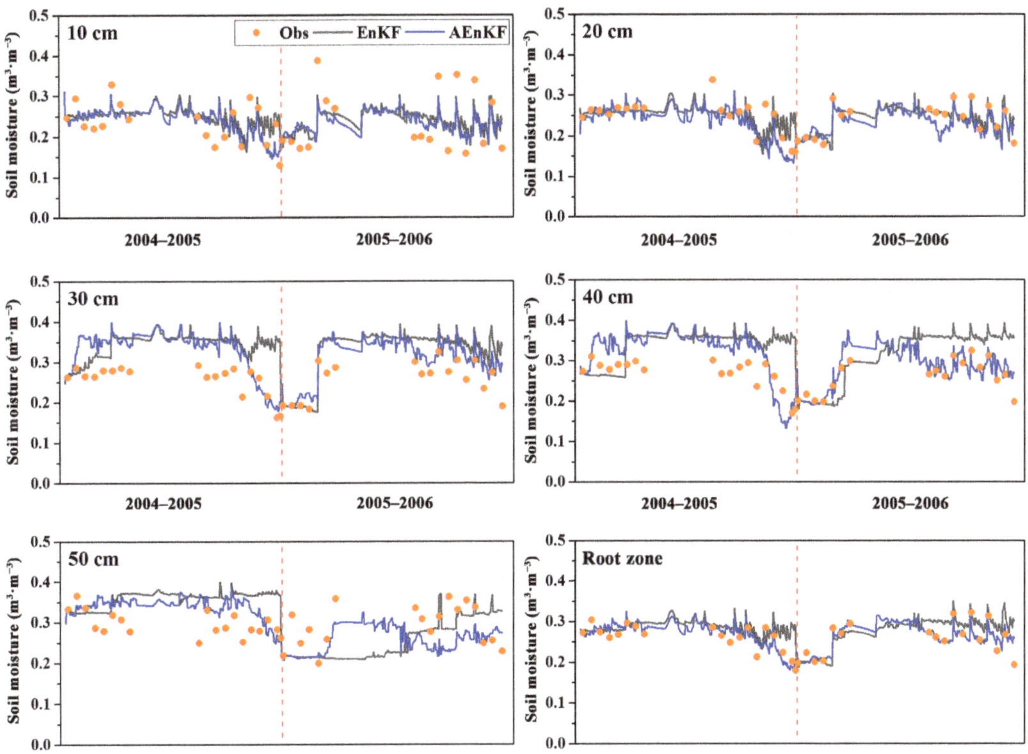

Figure 4. Soil moisture assimilation estimates using AEnKF and EnKF algorithms at different soil depths during the growing seasons of 2004–2005 and 2005–2006.

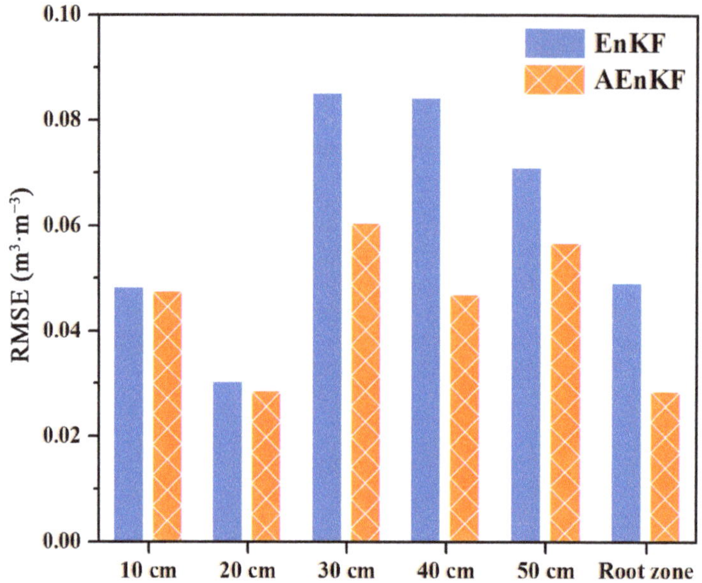

Figure 5. Root-mean-square errors (*RMSEs*) of assimilated soil moisture at different soil depths using AEnKF and EnKF algorithms.

4.2.2. Impact of Ensemble Size

The EnKF algorithm uses a Monte Carlo approach to generate the model error distribution. Theoretically, as ensemble size increases, the model error distribution can be better characterized. However, a significant expansion of ensemble size leads to a dramatic increase in model calculation costs. Therefore, ensemble size is often selected with the compromise of accuracy and time cost in practical applications, especially in the case of complex models with many parameters, which can not only ensure accuracy but also reduce calculation time. In this section, we compared the accuracy of the assimilated soil moisture with five, ten, twenty, and fifty ensemble members during the growing seasons of 2004–2005 and 2005–2006.

Figure 6 shows the time series of soil moisture estimates with different ensemble members. Generally, the fluctuations of the assimilated soil moisture time series under different ensemble members were similar. When the ensemble size was small, the fluctuations in soil moisture were relatively large; conversely, when the ensemble size was large, the change was relatively small. As ensemble size increased, the error of assimilated soil moisture decreased to some extent (Figure 7). However, overall, no obvious difference in the error of assimilation results occurred under the different ensemble members. The error of the root-zone soil moisture was the smallest, with *RMSEs* of 0.033, 0.029, 0.029, and 0.027 $m^3{\cdot}m^{-3}$ under 5, 10, 20, and 50 ensemble members, respectively. In previous studies, ensemble size was mostly set between 10 and 50 [24,33,34,54]. According to the error statistics, the *RMSEs* of the assimilated soil moisture under 20 and 50 members were similar. Compared with the observation data and model structure error, the influence of ensemble size on the soil moisture data assimilation was marginal. In view of the high dimensionality of the state variables in this study, the calculation cost of larger ensemble size was relatively expensive. Thus, considering the calculation accuracy and time cost, 20 ensemble members were deemed more appropriate for soil moisture assimilation at the other sites.

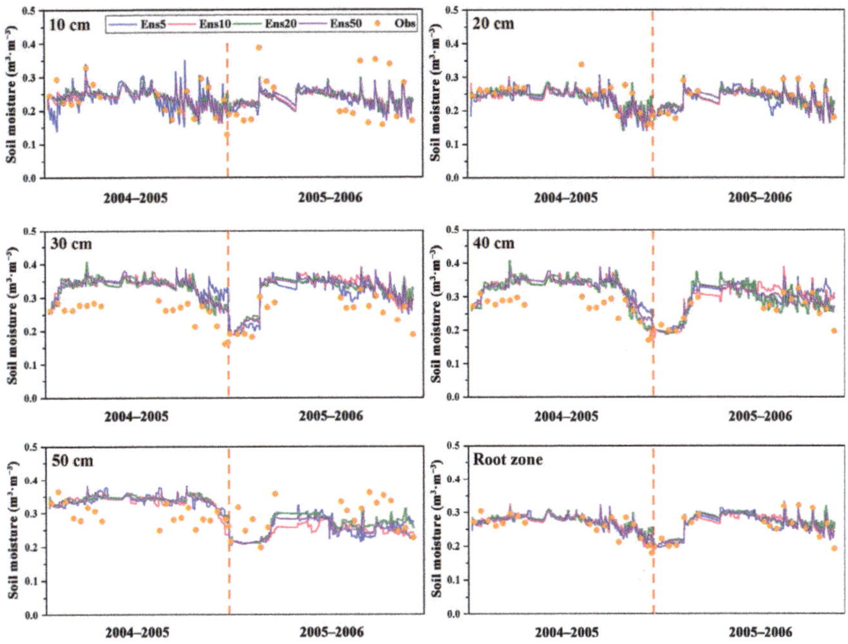

Figure 6. Soil moisture assimilation estimates with different ensemble sizes at different soil depths during the growing seasons of 2004–2005 and 2005–2006.

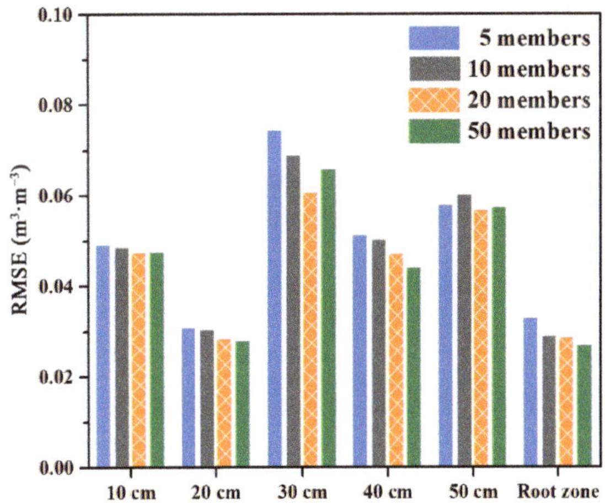

Figure 7. Root-mean-square errors (*RMSEs*) of assimilated soil moisture with five, ten, twenty and fifty members at different soil depths.

4.2.3. Impact of Temporal Interval

Although the ESA CCI soil moisture product has a one-day temporal resolution, the acquisition of continuous daily data during the growing season is difficult. This problem can be caused by many reasons, including satellite orbit changes, radio frequency interference (*RFI*), underlying surface (e.g., snow and ice), and physical limitations of satellite sensors [18,55,56]. Under the circumstances of incomplete observations, an exploration of the influence of observation frequency on soil moisture data assimilation is necessary.

An assimilation experiment was conducted with observation data at intervals of three, five and ten days for two winter wheat growing seasons from 2004 to 2006. The results showed that the fluctuation of soil moisture estimated with observation data at a ten-day interval was more obvious than that at three- and five-day intervals (Figure 8). When the temporal interval of observations was large, the assimilated soil moisture tended to approach the model simulation results. The model simulation values were lower in the late growth season, which led to an increase in soil moisture fluctuation. The detailed *RMSEs* of soil moisture with different assimilation intervals are shown in Figure 9. Generally, as the frequency of observation increased, the error of assimilated soil moisture decreased, but the difference in soil moisture error at intervals of three and five days was small. Therefore, the interval of observation data within five days was identified as the ideal time frequency for obtaining soil moisture with high accuracy, in agreement with the conclusion of Walker and Houser [57].

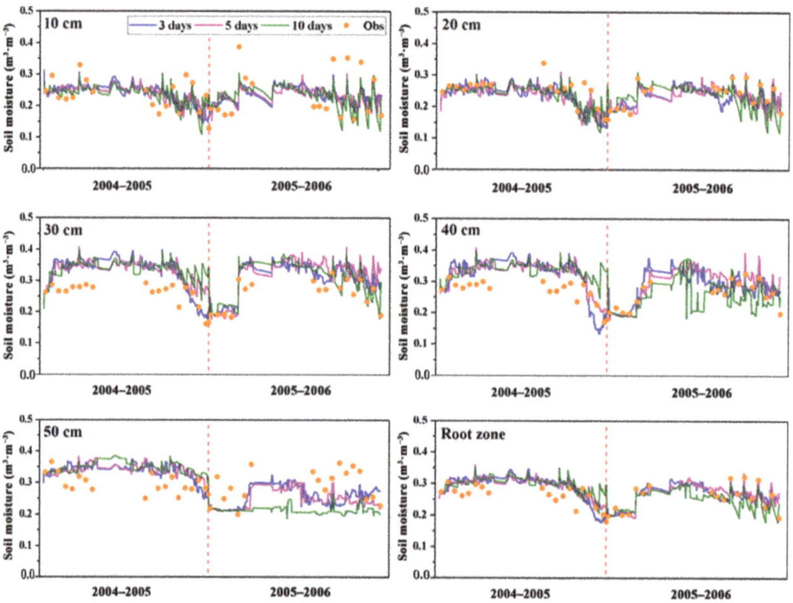

Figure 8. Soil moisture assimilation estimates with the observation intervals of three, five, and ten days at different soil depths during the growing seasons of 2004–2005 and 2005–2006.

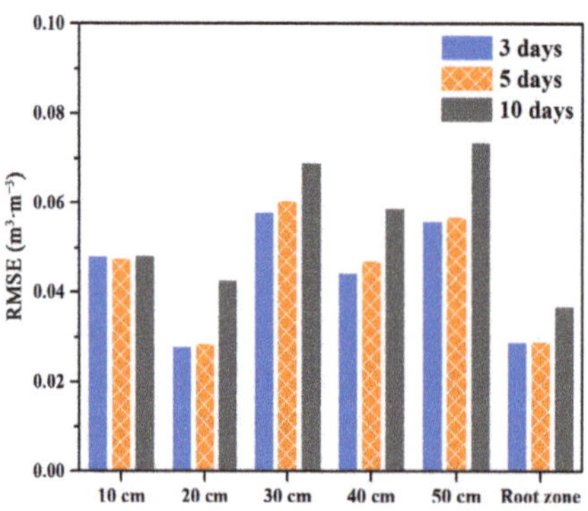

Figure 9. Comparison of the *RMSEs* of assimilated soil moisture with observation intervals of three, five, and ten days at different soil depths.

4.3. Effect of Data Assimilation on Agricultural Drought Monitoring

4.3.1. Comparison with Observed Soil Moisture-Based Drought Index

In this study, the *RSMI* was used as an indicator for agricultural drought monitoring. We compared the *RSMI* based on assimilated soil moisture ($RSMI_{DA}$) and open-loop simulations ($RSMI_{OL}$) with the RSMI based on in situ observations ($RSMI_{Obs}$), with the aim of investigating the extent to which soil moisture data assimilation improved the level of agricultural drought monitoring.

Firstly, from the perspective of qualitative assessment, time series graphs and scatter plots were used to show the fluctuations and consistency of the RSMI. Figure 10 displays the RSMI time series during the growing seasons from 2002 to 2011 at eight sites. The results illustrate that the $RSMI_{DA}$ and $RSMI_{obs}$ had similar fluctuations during the growing seasons. In the HHH Plain, due to the huge water demand for water in the late growing season of winter wheat, a decrease in soil moisture often occurred during this period. Both the $RSMI_{DA}$ and $RSMI_{obs}$ showed this characteristic. $RSMI_{OL}$ showed a good match with $RSMI_{obs}$ in the early growth season of winter wheat, however, in the late growth season, the values for $RSMI_{OL}$ were significantly lower than those for $RSMI_{obs}$. According to the drought levels of *RSMI*, the low value of $RSMI_{OL}$ in the late growth season indicated a severe or extreme drought, which might be inconsistent with the reality.

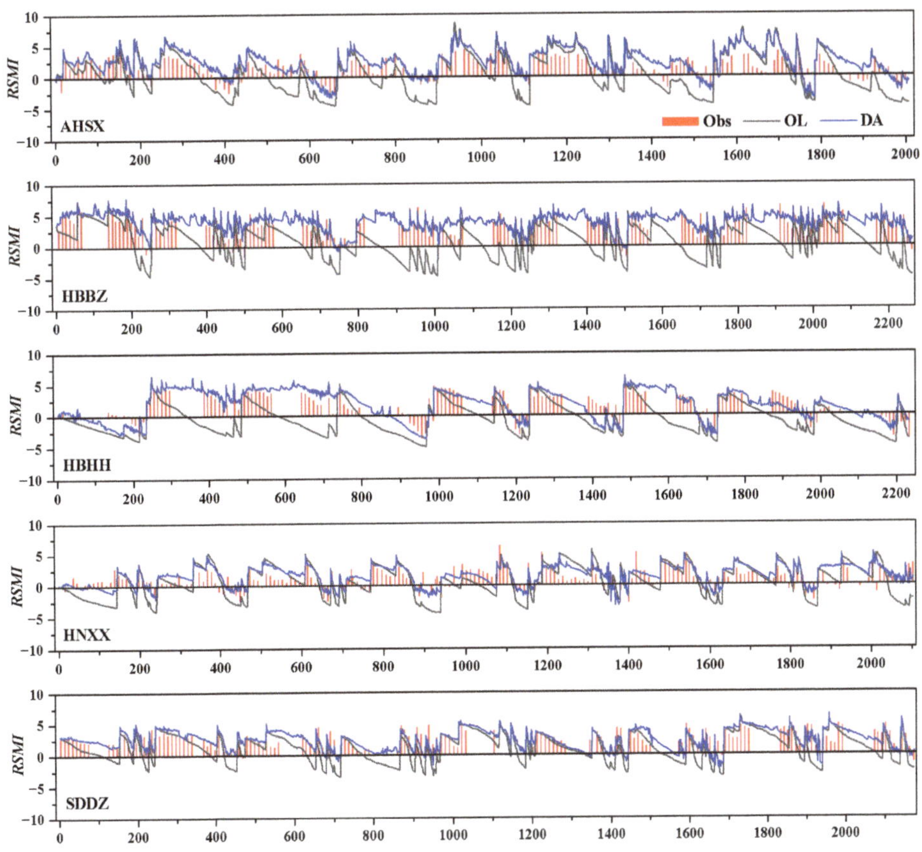

Figure 10. *Cont.*

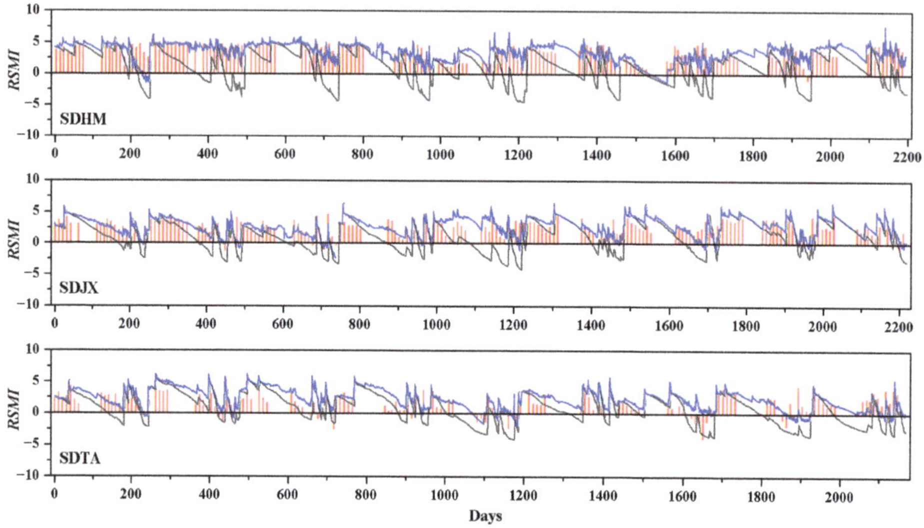

Figure 10. *RSMI* time series calculated from the observed, simulated and assimilated soil moisture for nine winter wheat growing seasons at the eight sites.

Secondly, from the perspective of quantitative assessment, Table 4 displayed the error statistics of *RSMI* based on open-loop and assimilated soil moisture at eight agro-meteorological sites. The average *RMSEs* of $RSMI_{DA}$ and $RSMI_{OL}$ at the eight sites were 2.60 and 1.38, respectively, and the corresponding average *R* values were 0.62 and 0.68, respectively. Compared with the $RSMI_{OL}$, the $RSMI_{DA}$ of all sites had a higher correlation at the 0.01 significance level, and the *RMSE* of $RSMI_{DA}$ was smaller than that of $RSMI_{OL}$ for all sites. According to the drought category of *RSMI* (Table 1), a value of less than zero indicated that drought occurred; when the *RSMI* value was decreased by one, the drought intensity was increased by one level. Therefore, the difference between the average *RMSEs* of $RSMI_{OL}$ and $RSMI_{DA}$ was 1.22, which implied that, compared with the drought index based on open-loop simulations, the index based on assimilated data could improve at least one drought level in agricultural drought monitoring.

Table 4. Root-mean-square errors (*RMSEs*) and correlation coefficients (*R*) for the *RSMI* based on open-loop (OL) and assimilated soil moisture (DA) at the eight sites. *R* values are shown in brackets.

	AHSX	HBBZ	HBHH	HNXX	SDDZ	SDHM	SDJX	SDTA	Mean
OL	2.31 (0.74)	3.49 (0.65)	2.81 (0.70)	2.12 (0.50)	2.19 (0.63)	3.61 (0.76)	2.35 (0.31)	1.95 (0.66)	2.60 (0.62)
DA	1.53 (0.75)	1.31 (0.74)	1.54 (0.80)	1.52 (0.55)	1.3 (0.63)	1.29 (0.78)	1.15 (0.50)	1.43 (0.66)	1.38 (0.68)

4.3.2. Comparison with Winter Wheat Yield

Crop yield reduction is a major manifestation of crop growth threatened by drought. Here, we discuss the relationship between agricultural drought and crop yield reduction to further evaluate the effect of data assimilation on agricultural drought monitoring. Yield loss ratio (*YLR*) was utilized to quantify the degree of crop yield reduction, calculated as follows: $YLR = \left(Y - \overline{Y}\right)/\overline{Y}$, where Y represents the annual crop yield, and $\overline{Y}$ is the multiyear average yield. According to the disaster record dataset of agro-meteorological sites from 2002 to 2011, six sites (AHSX, HBHH, SDDZ, SDHM, SDJX, and SDTA) experienced agricultural droughts. Figure 11 shows the relationship between the *RSMI* ($RSMI_{OL}$ and $RSMI_{DA}$) and crop yield loss ratio (*YLR*). Overall, the *RSMI* matched well with the *YLR* during the nine growing seasons; that is, when the *RSMI* value was small (drought), the crop yield was reduced, and when the *RSMI* value was large (wetness), the crop yield increased.

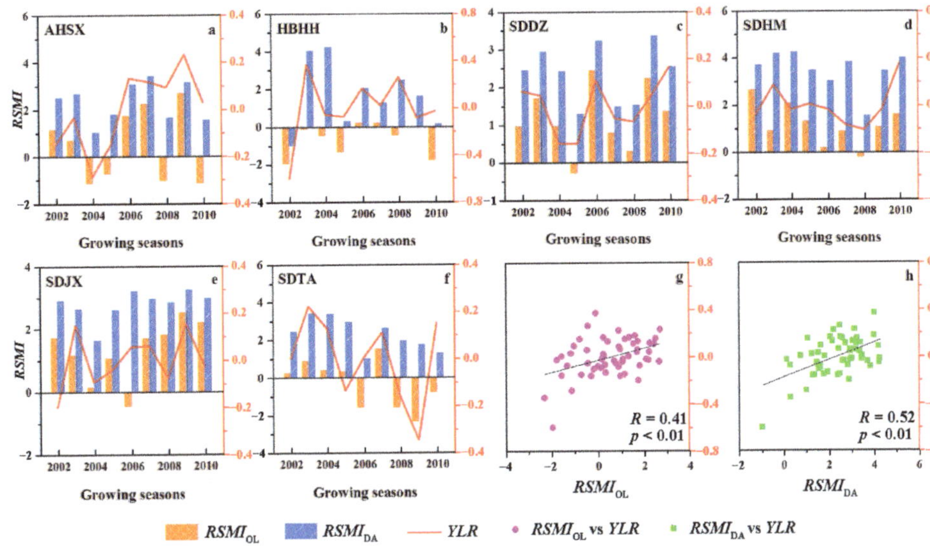

Figure 11. The relationship between the *RSMI* and yield loss ratio (*YLR*) during the growing seasons from 2002 to 2011: (**a–f**) the dynamics of *RSMI* and *YLR* at each site; (**g**,**h**) the scatter plots between *RSMI* and *YLR* at the six sites. $RSMI_{OL}$, *RSMI* based on open-loop soil moisture; $RSMI_{DA}$, *RSMI* based on assimilated soil moisture.

As a result of the small sample size per site, the *RSMI*s from the six sites were used together to statistically investigate the correlation with YLR (Figure 11g,h). The correlation coefficients between the *RSMI* ($RSMI_{OL}$ and $RSMI_{DA}$) and YLR were 0.41 and 0.52 ($p < 0.01$), respectively. The results demonstrated that the *RSMI* was significantly correlated with YLR, and the *RSMI* based on assimilated soil moisture performed better than that based on open-loop simulations. Therefore, data assimilation could further strengthen the skill of the drought index to reflect the yield losses by improving soil moisture estimation.

5. Discussion

In this study, we established a data assimilation scheme to estimate surface and root zone soil moisture, and we evaluated the effect of data assimilation on agricultural drought monitoring. The main advantage of this scheme is that, combined with a crop models and remote sensing observations, the process of crop growth and development can be simulated, and soil moisture information can also be obtained [19,58]. Such results are very helpful for constructing a drought index with agronomic traits [13,19]. The assimilated results demonstrate that soil moisture estimates were significantly improved after data assimilation (Figures 2 and 3; Tables 2 and 3). Although the observation data in the data assimilation framework was surface satellite soil moisture, deeper layers also showed the improved results after assimilation. From the perspective of different soil depths, the improvement of surface soil moisture accuracy after data assimilation was greater than that of the deep layers. As the soil depth increased, the ability of assimilating surface soil moisture to improve the accuracy of deep soil moisture decreased. These findings also agree with previous studies [20,30]. We can also find that the results of soil moisture assimilation at depths of 30 and 40 cm were less accurate. Soil water balance is a very complex physical process, and the possible reasons for the poor performance include the following: (1) the inability of the model structure to perfectly describe the soil water balance process; (2) the uncertainties of the input data (soil attribute data and management data) [58]. Due to the influence of many factors (models, observations, and forcings), the estimation of soil moisture at a fixed depth may have some uncertainties. Root-zone soil moisture (averaged at each layer within the root zone) was observed to be a better estimate (Table 3). Root-zone soil moisture plays a significant role in agricultural drought monitoring [2,53]. Our study found that compared with the drought index based on open-loop soil moisture, the index based on assimilated data could improve at least one drought level in agricultural drought monitoring. Therefore, it is an important and feasible approach to improve the accuracy of root-zone soil moisture through data assimilation, thereby enhancing the estimation of agricultural drought.

Model and observation are key components for data assimilation. In this study, we found that the performance of the AEnKF-based soil moisture data assimilation was better, which also indicated that the simultaneous state-parameter estimation was more effective for improving soil moisture estimation. This conclusion highlights the importance of simultaneous estimation of model parameters and state variables, which is consistent with related research [30,48,59,60]. In addition to model parameters, observation has a crucial impact on data assimilation. Errors caused by observations include the bias between model and observation and the error of the observation itself. Discrepancies between model and observation are caused by a variety of factors, including uncertainty of model parameters, a difference in spatial scales. Bias correction is effective to reduce the discrepancies between model and observation [59,60]. In this study, the CDF matching approach was used to reduce the impact of scale difference between the gridded satellite observation and point-scale model prior to data assimilation. The error of observation itself is generally set based on empirical values [24,34,59,60]. In our study, we set it based on the relevant studies on soil moisture validation [18,51], and we obtained good estimation results. Recently, several studies have proposed new methods for bias correction and model parameter estimation to further improve the performance of data assimilation. Instead of bias correction prior to data assimilation, Qin et al. [59] proposed on-line bias correction during data assimilation and obtained a good estimate. Tian et al. [60] developed a novel assimilation method to estimate both model and observation errors to improve soil moisture retrievals, in which model state, model parameters, model error, and observation error were estimated simultaneously. Overall, model parameters and observation errors are still worthy of continued research to further improve the estimation accuracy of state variables.

Through this study, we found that the crop growth model-based data assimilation scheme effectively improved soil moisture estimation and was significantly beneficial for agricultural drought monitoring. It is worth noting that this study also had some limitations that need to be further discussed. On one hand, the study was conducted on a point scale, as accurate regional crop model parameterization and high-precision gridded data (meteorological forcings, soil data, and field management data) used for crop modeling are still challenging. Some studies have investigated the gridded data assimilation scheme [32,33,61,62], however, generally gridded meteorological data are used, whereas soil data, and management data (e.g., irrigation), and crop parameters over large areas are still replaced by empirical values or coarse-resolution data. Therefore, observations at higher spatial and temporal resolution (e.g., SMAP product [63]) and more accurate model parameterization schemes can further improve the gridded data assimilation framework, help analyze the variations in drought characteristics in different regions, and better understand the mechanism of drought formation. On the other hand, a single remote sensing data source was used in this study. Multi-source remote sensing can provide land surface monitoring from different perspectives. Some scholars carried out data assimilation based on various remote sensing data sources [24,61,62,64,65]; however, so far collaborative applications based on multi-source remote sensing data are still rare in agricultural drought monitoring. With the launch of subsequent remote sensing satellites, data quality and spatial-temporal resolution will be further improved. The combination of multi-source remote sensing data and crop models will help to further strengthen the application of drought indicators in agricultural drought monitoring.

6. Conclusions

In this study, the DSSAT crop model and satellite soil moisture were integrated for data assimilation to improve soil moisture estimation. The assimilated results were validated using in situ observations of agro-meteorological sites. The impacts of different strategies (simultaneous state-parameter estimation, ensemble size and temporal interval) on data assimilation were also discussed. Furthermore, we evaluated the effect of data assimilation on agricultural drought monitoring through comparisons with the observed soil moisture-based drought index and winter wheat yield. The main conclusions are summarized as follows:

(1) Compared with open-loop simulations, crop growth model-based data assimilation effectively improved soil moisture estimation throughout the soil profile. Soil moisture in the root zone was more consistent with in situ observation ($RMSE$ = 0.035 $m^3 \cdot m^{-3}$, R = 0.742) than that in other soil layers;
(2) Simultaneous state-parameter estimation (based on AEnKF algorithm) performed better than state-only estimation (based on EnKF algorithm). As the observation frequency and ensemble size increased, the accuracy of soil moisture estimates also increased. In the case of a comprehensive consideration of estimation accuracy and calculation cost, observation frequency within five days with twenty ensemble members could meet the requirement of soil moisture accuracy;
(3) Compared with the drought index based on open-loop soil moisture, the drought index based on assimilated results improves at least one drought level in agricultural drought monitoring. Additionally, the drought index based on assimilated soil moisture was better correlated with winter wheat yield than that based on open-loop simulation.

Author Contributions: Conceptualization, H.Z.; methodology, H.Z. and J.Y.; software, H.Z. and J.Y.; validation, H.Z., G.G. and J.Y.; writing—original draft preparation, H.Z., G.G. and J.Y.; writing—review and editing, H.Z., G.G., J.Y., H.H., L.S. and W.L.; funding acquisition, H.Z. and L.S. All authors have read and agreed to the published version of the manuscript.

Funding: This research was funded by the National Natural Science Foundation of China (Grant No. 41907394, 41901360); Zhejiang Provincial Natural Science Foundation of China (Grant No. LY21D010004).

Data Availability Statement: The data presented in this study are available on request from the corresponding author.

Acknowledgments: We are grateful to those who participated in the data processing and manuscript revisions.

Conflicts of Interest: The authors declare no conflict of interest.

References

1. Mishra, A.K.; Singh, V.P. A review of drought concepts. *J. Hydrol.* **2010**, *391*, 202–216. [CrossRef]
2. Martínez-Fernández, J.; González-Zamora, A.; Sánchez, N.; Gumuzzio, A. A soil water based index as a suitable agricultural drought indicator. *J. Hydrol.* **2015**, *522*, 265–273. [CrossRef]
3. AghaKouchak, A.; Farahmand, A.M.; Melton, F.S.; Teixeira, J.P.; Anderson, M.; Wardlow, B.; Hain, C.R. Remote sensing of drought: Progress, challenges and opportunities. *Rev. Geophys.* **2015**, *53*, 452–480. [CrossRef]
4. Liu, X.; Pan, Y.; Zhu, X.; Yang, T.; Bai, J.; Sun, Z. Drought evolution and its impact on the crop yield in the North China Plain. *J. Hydrol.* **2018**, *564*, 984–996. [CrossRef]
5. Samaniego, L.; Thober, S.; Kumar, R.; Wanders, N.; Rakovec, O.; Pan, M.; Zink, M.; Sheffield, J.; Wood, E.F.; Marx, A. Anthropogenic warming exacerbates European soil moisture droughts. *Nat. Clim. Chang.* **2018**, *8*, 421–426. [CrossRef]
6. Mora, C.; Spirandelli, D.; Franklin, E.C.; Lynham, J.; Kantar, M.B.; Miles, W.; Smith, C.Z.; Freel, K.; Moy, J.; Louis, L.V.; et al. Broad threat to humanity from cumulative climate hazards intensified by greenhouse gas emissions. *Nat. Clim. Chang.* **2018**, *8*, 1062–1071. [CrossRef]

7. Crow, W.T.; Kumar, S.V.; Bolten, J.D. On the utility of land surface models for agricultural drought monitoring. *Hydrol. Earth Syst. Sci.* **2012**, *16*, 3451–3460. [CrossRef]
8. Sheffield, J.; Goteti, G.; Wen, F.; Wood, E. A simulated soil moisture based drought analysis for the United States. *J. Geophys. Res. D Atmos.* **2004**, *109*, D24108. [CrossRef]
9. Dutra, E.; Viterbo, P.; Miranda, P.M.A. ERA-40 reanalysis hydrological applications in the characterization of regional drought. *Geophys. Res. Lett.* **2008**, *35*, L19402. [CrossRef]
10. Carrão, H.; Russo, S.; Sepulcre-Canto, G.; Barbosa, P. An empirical standardized soil moisture index for agricultural drought assessment from remotely sensed data. *Int. J. Appl. Earth Obs. Geoinf.* **2016**, *48*, 74–84. [CrossRef]
11. Sridhar, V.; Hubbard, K.G.; You, J.; Hunt, E.D. Development of the Soil Moisture Index to Quantify Agricultural Drought and Its "User Friendliness" in Severity-Area-Duration Assessment. *J. Hydrometeorol.* **2008**, *9*, 660–676. [CrossRef]
12. Hunt, E.D.; Hubbard, K.G.; Wilhite, D.A.; Arkebauer, T.J.; Dutcher, A.L. The development and evaluation of a soil moisture index. *Int. J. Climatol.* **2009**, *29*, 747–759. [CrossRef]
13. Zhou, H.; Wu, J.; Geng, G.; Li, X.; Wang, Q.; Lei, T.; Mo, X.; Liu, L. Enhancing the Ability of a Soil Moisture-based Index for Agricultural Drought Monitoring by Incorporating Root Distribution. *JAWRA J. Am. Water Resour. Assoc.* **2017**, *53*, 1409–1423. [CrossRef]
14. Peng, J.; Albergel, C.; Balenzano, A.; Brocca, L.; Cartus, O.; Cosh, M.H.; Crow, W.T.; Dabrowska-Zielinska, K.; Dadson, S.; Davidson, M.W.; et al. A roadmap for high-resolution satellite soil moisture applications—Confronting product characteristics with user requirements. *Remote Sens. Environ.* **2020**, *252*, 112162. [CrossRef]
15. Li, Z.-L.; Leng, P.; Zhou, C.; Chen, K.-S.; Zhou, F.-C.; Shang, G.-F. Soil moisture retrieval from remote sensing measurements: Current knowledge and directions for the future. *Earth-Sci. Rev.* **2021**, *218*, 103673. [CrossRef]
16. Babaeian, E.; Sadeghi, M.; Jones, S.B.; Montzka, C.; Vereecken, H.; Tuller, M. Ground, Proximal, and Satellite Remote Sensing of Soil Moisture. *Rev. Geophys.* **2019**, *57*, 530–616. [CrossRef]
17. Baldwin, D.; Manfreda, S.; Keller, K.; Smithwick, E. Predicting root zone soil moisture with soil properties and satellite near-surface moisture data across the conterminous United States. *J. Hydrol.* **2017**, *546*, 393–404. [CrossRef]
18. Dorigo, W.; Wagner, W.; Albergel, C.; Albrecht, F.; Balsamo, G.; Brocca, L.; Chung, D.; Ertl, M.; Forkel, M.; Gruber, A.; et al. ESA CCI Soil Moisture for improved Earth system understanding: State-of-the art and future directions. *Remote Sens. Environ.* **2017**, *203*, 185–215. [CrossRef]
19. Mishra, A.K.; Ines, A.V.M.; Das, N.N.; Prakash Khedun, C.; Singh, V.P.; Sivakumar, B.; Hansen, J.W.; Mishra, A.K.; Sivakumar, B.; Singh, V.P. Anatomy of a local scale drought; application of assimilated remote sensing products, crop model, and statistical methods to an agricultural drought study. *J. Hydrol.* **2015**, *526*, 15–29. [CrossRef]
20. Sahoo, A.K.; De Lannoy, G.J.; Reichle, R.H.; Houser, P.R. Assimilation and downscaling of satellite observed soil moisture over the Little River Experimental Watershed in Georgia, USA. *Adv. Water Resour.* **2013**, *52*, 19–33. [CrossRef]
21. Parrens, M.; Mahfouf, J.-F.; Barbu, A.L.; Calvet, J.-C. Assimilation of surface soil moisture into a multilayer soil model: Design and evaluation at local scale. *Hydrol. Earth Syst. Sci.* **2014**, *18*, 673–689. [CrossRef]
22. Sabater, J.M.; Jarlan, L.; Calvet, J.-C.; Bouyssel, F.; De Rosnay, P. From Near-Surface to Root-Zone Soil Moisture Using Different Assimilation Techniques. *J. Hydrometeorol.* **2007**, *8*, 194–206. [CrossRef]
23. Huang, C.; Li, X.; Lu, L. Retrieving soil temperature profile by assimilating MODIS LST products with ensemble Kalman filter. *Remote Sens. Environ.* **2008**, *112*, 1320–1336. [CrossRef]
24. Kumar, S.; Peters-Lidard, C.; Mocko, D.; Reichle, R.; Liu, Y.; Arsenault, K.; Xia, Y.; Ek, M.; Riggs, G.; Livneh, B.; et al. Assimilation of Remotely Sensed Soil Moisture and Snow Depth Retrievals for Drought Estimation. *J. Hydrometeorol.* **2014**, *15*, 2446–2469. [CrossRef]
25. Han, E.; Crow, W.T.; Holmes, T.; Bolten, J. Benchmarking a Soil Moisture Data Assimilation System for Agricultural Drought Monitoring. *J. Hydrometeorol.* **2014**, *15*, 1117–1134. [CrossRef]
26. Bolten, J.D.; Crow, W.T.; Zhan, X.; Jackson, T.J.; Reynolds, C.A. Evaluating the Utility of Remotely Sensed Soil Moisture Retrievals for Operational Agricultural Drought Monitoring. *IEEE J. Sel. Top. Appl. Earth Obs. Remote Sens.* **2009**, *3*, 57–66. [CrossRef]
27. Sun, X.; Lai, P.; Wang, S.; Song, L.; Ma, M.; Han, X. Monitoring of Extreme Agricultural Drought of the Past 20 Years in Southwest China Using GLDAS Soil Moisture. *Remote Sens.* **2022**, *14*, 1323. [CrossRef]
28. De Lannoy, G.J.M.; Reichle, R.H. Assimilation of SMOS brightness temperatures or soil moisture retrievals into a land surface model. *Hydrol. Earth Syst. Sci.* **2016**, *20*, 4895–4911. [CrossRef]
29. Seo, E.; Lee, M.-I.; Reichle, R.H. Assimilation of SMAP and ASCAT soil moisture retrievals into the JULES land surface model using the Local Ensemble Transform Kalman Filter. *Remote Sens. Environ.* **2020**, *253*, 112222. [CrossRef]
30. Monsivais-Huertero, A.; Graham, W.; Judge, J.; Agrawal, D. Effect of simultaneous state–parameter estimation and forcing uncertainties on root-zone soil moisture for dynamic vegetation using EnKF. *Adv. Water Resour.* **2010**, *33*, 468–484. [CrossRef]
31. Dong, T.; Liu, J.; Qian, B.; Zhao, T.; Jing, Q.; Geng, X.; Wang, J.; Huffman, T.; Shang, J. Estimating winter wheat biomass by assimilating leaf area index derived from fusion of Landsat-8 and MODIS data. *Int. J. Appl. Earth Obs. Geoinf.* **2016**, *49*, 63–74. [CrossRef]
32. Huang, J.; Sedano, F.; Huang, Y.; Ma, H.; Li, X.; Liang, S.; Tian, L.; Zhang, X.; Fan, J.; Wu, W. Assimilating a synthetic Kalman filter leaf area index series into the WOFOST model to improve regional winter wheat yield estimation. *Agric. For. Meteorol.* **2016**, *216*, 188–202. [CrossRef]

33. Huang, J.; Tian, L.; Liang, S.; Ma, H.; Becker-Reshef, I.; Huang, Y.; Su, W.; Zhang, X.; Zhu, D.; Wu, W. Improving winter wheat yield estimation by assimilation of the leaf area index from Landsat TM and MODIS data into the WOFOST model. *Agric. For. Meteorol.* **2015**, *204*, 106–121. [CrossRef]
34. Ines, A.V.; Das, N.N.; Hansen, J.W.; Njoku, E.G. Assimilation of remotely sensed soil moisture and vegetation with a crop simulation model for maize yield prediction. *Remote Sens. Environ.* **2013**, *138*, 149–164. [CrossRef]
35. Jiang, Z.; Chen, Z.; Chen, J.; Ren, J.; Li, Z.; Sun, L. The Estimation of Regional Crop Yield Using Ensemble-Based Four-Dimensional Variational Data Assimilation. *Remote Sens.* **2014**, *6*, 2664–2681. [CrossRef]
36. Ren, S.; Guo, B.; Wu, X.; Zhang, L.; Ji, M.; Wang, J. Winter wheat planted area monitoring and yield modeling using MODIS data in the Huang-Huai-Hai Plain, China. *Comput. Electron. Agric.* **2021**, *182*, 106049. [CrossRef]
37. National Bureau of Statistics of China. *China Statistical Yearbook*; National Bureau of Statistics of China: Beijing, China, 2021.
38. Hou, M. Study of Agricultural Structure Adjustment and Agricultural Development Stratagem on HuangHuaiHai Plain. Ph.D. Thesis, China Agricultural University, Beijing, China, 2004.
39. Xu, J.; Ju, H.; Mei, X.; Liu, Q.; Yang, J. Simulation on potential effects of drought on winter wheat in Huang-Huai-Hai Plain from 1981 to 2010. *Trans. Chin. Soc. Agric. Eng.* **2015**, *31*, 150–158. (In Chinese) [CrossRef]
40. Allen, R.G.; Pereira, L.S.; Raes, D.; Smith, M. *Crop Evapotranspiration—Guidelines for Computing Reference Crop Evapo-Transpiration*; FAO Irrigation and Drainage Paper No. 56; FAO: Rome, Italy, 1998; pp. 1–15.
41. Liu, Y.Y.; Dorigo, W.A.; Parinussa, R.M.; de Jeu, R.A.M.; Wagner, W.; McCabe, M.F.; Evans, J.P.; van Dijk, A.I.J.M. Trend-preserving blending of passive and active microwave soil moisture retrievals. *Remote Sens. Environ.* **2012**, *123*, 280–297. [CrossRef]
42. Gruber, A.; Dorigo, W.A.; Crow, W.; Wagner, W. Triple Collocation-Based Merging of Satellite Soil Moisture Retrievals. *IEEE Trans. Geosci. Remote Sens.* **2017**, *55*, 6780–6792. [CrossRef]
43. Jones, J.W.; Hoogenboom, G.; Porter, C.H.; Boote, K.J.; Batchelor, W.D.; Hunt, L.A.; Wilkens, P.W.; Singh, U.; Gijsman, A.J.; Ritchie, J.T. The DSSAT cropping system model. *Eur. J. Agron.* **2003**, *18*, 235–265. [CrossRef]
44. Ritchie, J.T. Model for predicting evaporation from a row crop with incomplete cover. *Water Resour. Res.* **1972**, *8*, 1204–1213. [CrossRef]
45. SCS. *National Engineering Handbook, Section 4: Hydrology*; Soil Conservation Service: Washington, DC, USA, 1972.
46. Priestley, C.H.B.; Taylor, R.J. On the assessment of surface heat flux and evaporation using large-scale parameters. *Mon. Weather Rev.* **1972**, *100*, 81–92. [CrossRef]
47. Soldevilla-Martinez, M.; Quemada, M.; López-Urrea, R.; Muñoz-Carpena, R.; Lizaso, J. Soil water balance: Comparing two simulation models of different levels of complexity with lysimeter observations. *Agric. Water Manag.* **2014**, *139*, 53–63. [CrossRef]
48. Chen, W.; Huang, C.; Shen, H.; Li, X. Comparison of ensemble-based state and parameter estimation methods for soil moisture data assimilation. *Adv. Water Resour.* **2015**, *86*, 425–438. [CrossRef]
49. Evensen, G. The Ensemble Kalman Filter: Theoretical formulation and practical implementation. *Ocean. Dyn.* **2003**, *53*, 343–367. [CrossRef]
50. Nair, A.S.; Indu, J. Improvement of land surface model simulations over India via data assimilation of satellite-based soil moisture products. *J. Hydrol.* **2019**, *573*, 406–421. [CrossRef]
51. An, R.; Zhang, L.; Wang, Z.; Quaye-Ballard, J.A.; You, J.; Shen, X.; Gao, W.; Huang, L.; Zhao, Y.; Ke, Z. Validation of the ESA CCI soil moisture product in China. *Int. J. Appl. Earth Obs. Geoinf.* **2016**, *48*, 28–36. [CrossRef]
52. Aubert, D.; Loumagne, C.; Oudin, L. Sequential assimilation of soil moisture and streamflow data in a conceptual rainfall–runoff model. *J. Hydrol.* **2003**, *280*, 145–161. [CrossRef]
53. Narasimhan, B.; Srinivasan, R. Development and evaluation of Soil Moisture Deficit Index (SMDI) and Evapotranspiration Deficit Index (ETDI) for agricultural drought monitoring. *Agric. For. Meteorol.* **2005**, *133*, 69–88. [CrossRef]
54. Chakrabarti, S.; Bongiovanni, T.; Judge, J.; Zotarelli, L.; Bayer, C. Assimilation of SMOS Soil Moisture for Quantifying Drought Impacts on Crop Yield in Agricultural Regions. *IEEE J. Sel. Top. Appl. Earth Obs. Remote Sens.* **2014**, *7*, 3867–3879. [CrossRef]
55. Almendra-Martín, L.; Martínez-Fernández, J.; Piles, M.; González-Zamora, Á. Comparison of gap-filling techniques applied to the CCI soil moisture database in Southern Europe. *Remote Sens. Environ.* **2021**, *258*, 112377. [CrossRef]
56. Zhang, L.; Liu, Y.; Ren, L.; Teuling, A.J.; Zhang, X.; Jiang, S.; Yang, X.; Wei, L.; Zhong, F.; Zheng, L. Reconstruction of ESA CCI satellite-derived soil moisture using an artificial neural network technology. *Sci. Total Environ.* **2021**, *782*, 146602. [CrossRef]
57. Walker, J.P.; Houser, P.R. Requirements of a global near-surface soil moisture satellite mission: Accuracy, repeat time, and spatial resolution. *Adv. Water Resour.* **2004**, *27*, 785–801. [CrossRef]
58. Jin, X.; Kumar, L.; Li, Z.; Feng, H.; Xu, X.; Yang, G.; Wang, J. A review of data assimilation of remote sensing and crop models. *Eur. J. Agron.* **2018**, *92*, 141–152. [CrossRef]
59. Qin, J.; Tian, J.; Yang, K.; Lu, H.; Li, X.; Yao, L.; Shi, J. Bias correction of satellite soil moisture through data assimilation. *J. Hydrol.* **2022**, *610*, 127947. [CrossRef]
60. Tian, J.; Qin, J.; Yang, K.; Zhao, L.; Chen, Y.; Lu, H.; Li, X.; Shi, J. Improving surface soil moisture retrievals through a novel assimilation algorithm to estimate both model and observation errors. *Remote Sens. Environ.* **2022**, *269*, 112802. [CrossRef]
61. Huang, J.; Ma, H.; Sedano, F.; Lewis, P.; Liang, S.; Wu, Q.; Su, W.; Zhang, X.; Zhu, D. Evaluation of regional estimates of winter wheat yield by assimilating three remotely sensed reflectance datasets into the coupled WOFOST–PROSAIL model. *Eur. J. Agron.* **2019**, *102*, 1–13. [CrossRef]

62. Zhang, Z.; Li, Z.; Chen, Y.; Zhang, L.; Tao, F. Improving regional wheat yields estimations by multi-step-assimilating of a crop model with multi-source data. *Agric. For. Meteorol.* **2020**, *290*, 107993. [CrossRef]
63. Colliander, A.; Jackson, T.J.; Bindlish, R.; Chan, S.; Das, N.; Kim, S.B.; Cosh, M.H.; Dunbar, R.S.; Dang, L.; Pashaian, L.; et al. Validation of SMAP surface soil moisture products with core validation sites. *Remote Sens. Environ.* **2017**, *191*, 215–231. [CrossRef]
64. Jin, N.; Tao, B.; Ren, W.; He, L.; Zhang, D.; Wang, D.; Yu, Q. Assimilating remote sensing data into a crop model improves winter wheat yield estimation based on regional irrigation data. *Agric. Water Manag.* **2022**, *266*, 107583. [CrossRef]
65. Rahman, A.; Maggioni, V.; Zhang, X.; Houser, P.; Sauer, T.; Mocko, D.M. The Joint Assimilation of Remotely Sensed Leaf Area Index and Surface Soil Moisture into a Land Surface Model. *Remote Sens.* **2022**, *14*, 437. [CrossRef]

remote sensing

Article

Impacts of Climate Change, Glacier Mass Loss and Human Activities on Spatiotemporal Variations in Terrestrial Water Storage of the Qaidam Basin, China

Xuewen Yang [1,2], Ninglian Wang [1,2,3,*], An'an Chen [1,2], Zhijie Li [1,2], Qian Liang [1,2] and Yujie Zhang [1,2]

1 Shaanxi Key Laboratory of Earth Surface System and Environmental Carrying Capacity, College of Urban and Environmental Sciences, Northwest University, Xi'an 710127, China; xuewenyang@stumail.nwu.edu.cn (X.Y.); aachen@nwu.edu.cn (A.C.); lizhijie@stumail.nwu.edu.cn (Z.L.); liangqian1126@stumail.nwu.edu.cn (Q.L.); zhanyujie@stumail.nwu.edu.cn (Y.Z.)
2 Institute of Earth Surface System and Hazards, College of Urban and Environmental Sciences, Northwest University, Xi'an 710127, China
3 Institute of Tibetan Plateau Research, Chinese Academy of Sciences, Beijing 100101, China
* Correspondence: nlwang@nwu.edu.cn

Abstract: Monitoring the variations in terrestrial water storage (TWS) is crucial for understanding the regional hydrological processes, which helps to allocate and manage basin-scale water resources efficiently. In this study, the impacts of climate change, glacier mass loss, and human activities on the variations in TWS of the Qaidam Basin over the period of 2002−2020 were investigated by using Gravity Recovery and Climate Experiment (GRACE) and GRACE Follow-On (GRACE-FO) data, and other hydrological and meteorological data. The results indicate that TWS anomalies (TWSA) derived from five GRACE solutions experienced significant increasing trends over the study period, with the change rates ranging from 4.85 to 6.90 mm/year (1.37 to 1.95 km^3/year). The GRACE TWSA averaged from different GRACE solutions exhibited an increase at a rate of 5.83 $\pm$ 0.12 mm/year (1.65 $\pm$ 0.03 km^3/year). Trends in individual components of TWS indicate that the increase in soil moisture (7.65 mm/year) contributed the most to the variations in TWS. Through comprehensive analysis, it was found that the temporal variations in TWS of the Qaidam Basin were dominated by the variations in precipitation, and the spatial variations in TWS of the Qaidam Basin were mostly driven by the increase in glacier meltwater due to climate warming, particularly in the Narin Gol Basin. In addition, the water consumption associated with human activities had relatively fewer impacts.

Keywords: terrestrial water storage; GRACE; climate change; glacier meltwater; water balance

1. Introduction

Terrestrial water storage (TWS) refers to an integrated estimate of water stored on and beneath the land surface, including soil moisture, groundwater, surface water (lakes, rivers, and reservoirs), snow, glaciers, and plant canopy water, and is of great significance to the terrestrial and global hydrological cycles [1,2]. The variations in TWS can effectively indicate the balance or imbalance of water fluxes strongly affected by regional climate conditions [2]. The recent global warming has exacerbated the variability of TWS in many arid and semi-arid regions, as well as alpine regions [3–5]. Hence, accurate estimation of TWS is crucial for understanding the regional hydrological process, which helps to allocate and manage basin-scale water resources efficiently.

The Qaidam Basin, an arid-to-semi-arid endorheic basin in the Tibetan Plateau, is facing the problems of water scarcity and a fragile ecological environment [6]. The Qaidam Basin is considered to be the most susceptible region with the most significant warming in the Tibetan Plateau [7], which has a profound impact on water resources in this region, such as glacier retreat and drastic changes in the lake area. It is thus of vital significance to understand the impacts of climate change, glacier mass loss, as well as human activities on spatiotemporal variations in TWS of the Qaidam Basin. However, observation stations are sparse in high mountain regions, and the in situ observations only provide information on individual hydrological variables at point scales [8], which limits our understanding of TWS variations in this region.

The Gravity Recovery and Climate Experiment (GRACE) mission, a twin satellite-to-satellite tracking mission launched in March 2002, enables us to measure the mean and time-varying components of the Earth's gravity field [9]. Since the change of the Earth's gravity field is dominated by hydrological signals, the variations in TWS can be accurately quantified using GRACE data [10–12]. Despite the coarser spatial resolution (~300 km), GRACE can reveal the water storage changes with a precision of 1–1.5 cm across the globe [13]. After the retirement of the GRACE mission in October 2017, its follow-on mission, launched in May 2018, continues to provide consecutive data for the hydrological applications [14].

GRACE observations have been used to assess the variations in TWS of the Qaidam Basin in several studies. Jiao et al. [15] and Wei et al. [16] investigated the variations in TWS of the Qaidam Basin and the impacts of climate indices and meteorological changes on TWS variations. Bibi et al. [17] and Hu and Jiao [18] estimated the variations in groundwater (GW) of the Qaidam Basin by using TWS, soil moisture (SM), surface water storage (SWS), and snow water equivalent (SWE) based on the water storage equation (GW = TWS − SM − SWS − SWE). The above studies only selected the mass concentration (Mascon) solutions of GRACE data; the uncertainties between spherical harmonics (SH) solutions and mascon solutions were not taken into consideration. In addition, they mostly focused on some individual components of TWS, and the variations in surface water and glacier mass balance were generally ignored. Moreover, the analysis of the impacts of glacier mass loss and human water consumption on the variations in TWS is still lacking.

In this study, we intended to present a comprehensive analysis of the impacts of climate change, glacier mass loss, and human activities on spatiotemporal variations in TWS of the Qaidam Basin. The objectives of this study were as follows: (1) to evaluate the spatial and temporal variations in TWS of the Qaidam Basin over the past two decades; (2) to quantify the inter-annual variations in all components of TWS and their contributions to the variations in TWS; (3) to explore the potential driving factors affecting the spatiotemporal variations in TWS of the Qaidam Basin from the perspectives of climate change, glacier mass loss and human activities.

2. Study Area

The Qaidam Basin is a closed inland basin located in the northwestern Qinghai Province and the northeastern Tibetan Plateau, China, with a spatial range of approximately 34°30′–39°30′ N, 87°30′–99°30′ E (Figure 1). It is surrounded by the Kunlun Mountains in the south, the Qilian Mountains in the northeast, and the Altyn Tagh Mountains in the northwest [17]. The elevation of the basin ranges from 2636–6835 m, with an average elevation of over 4000 m in the surrounding mountains and approximately 2800 m in the basin interior. The water resources in the Qaidam Basin mainly originate from glaciers and snowfields in the surrounding mountainous areas. According to the second glacial catalog dataset of China [19], there were 1939 glaciers distributed in the basin in 2006, covering an area of 1654.37 ± 74.05 km^2.

The climate condition in the Qaidam Basin is typically alpine-arid continental, and there are obvious discrepancies between surrounding mountains and central plains. The annual mean temperature is about 4.8 °C in the central plains, while it is only about 1.5 °C in the surrounding mountains [18]. The spatial distribution of precipitation is uneven, which decreases from the periphery to the interior of the basin. The annual total precipitation ranges from approximately 16 mm in the central part of the basin to about 190 mm in the surrounding mountains.

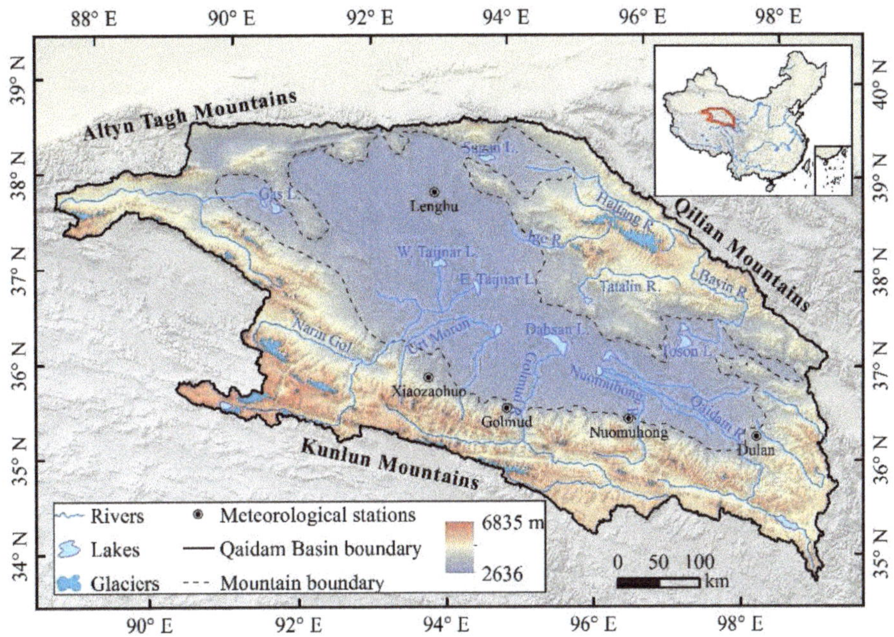

Figure 1. Location map of the Qaidam Basin. The mountain boundary of the basin is also indicated.

3. Data and Methods

3.1. Data

3.1.1. GRACE/GRACE-FO Data

The GRACE/GRACE-FO data are available in two forms, including spherical harmonics (SH) solutions and mass concentration (Mascon) solutions [20]. The SH solutions have been the standard for the first decade of GRACE observations, which utilize global SH basis functions to parameterize the earth gravity field model [16,21]. The monthly GRACE/GRACE-FO SH solutions are released by the Scientific Data System composed of three processing centers: Center for Space Research, University of Texas (CSR), Jet Propulsion Laboratory (JPL), and Geo-Forschungs-Zentrum (GFZ). The mascon solutions utilize mass concentration blocks to parameterize the earth gravity field model, which makes the implementation of geophysical constraints much easier and can improve the GRACE gravity information for the study of Earth surface processes [22–24].

In this study, the latest Release-06 (RL06) Level-2 SH solutions from CSR (CSR-SH), JPL (JPL-SH), and GFZ (GFZ-SH) (available at: ftp://isdcftp.gfz-potsdam.de/grace/, accessed on 1 June 2021) and the gridded RL06 version 02 mascon data processed by CSR (CSR-M) (http://www2.csr.utexas.edu/grace/RL06_mascons.html, accessed on 1 July 2021) and JPL (JPL-M) (https://grace.jpl.nasa.gov/data/get-data/jpl_global_mascons, accessed on 1 July 2021) (Table 1) were selected to estimate the variations in TWSA of the Qaidam Basin. The data covered the period of April 2002 to December 2020, for a total of 192 months, including 163 months from GRACE (April 2002 to June 2017) and 29 months from GRACE-FO (June 2018 to December 2020). The missing data of GRACE were interpolated by average values from adjacent months, and the 11-month data gap between GRACE and GRACE-FO was not remedied since it was caused by the discontinuation of satellites [25,26].

Table 1. The GRACE/GRACE-FO data used in this study.

Data Sources	Abbreviation	Spatial Resolution	Spatial Resolution (Resampled)	Temporal Resolution	Period
CSR spherical harmonics	CSR-SH	$1° \times 1°$	$0.25° \times 0.25°$	Monthly	2002–2020
JPL spherical harmonics	JPL-SH	$1° \times 1°$	$0.25° \times 0.25°$	Monthly	2002–2020
GFZ spherical harmonics	GFZ-SH	$1° \times 1°$	$0.25° \times 0.25°$	Monthly	2002–2020
CSR mascon	CSR-M	$0.25° \times 0.25°$	$0.25° \times 0.25°$	Monthly	2002–2020
JPL mascon	JPL-M	$0.5° \times 0.5°$	$0.25° \times 0.25°$	Monthly	2002–2020

3.1.2. GLDAS Data

The Global Land Data Assimilation System (GLDAS) is a land surface modeling system that aims to generate optimal fields of land surface states and fluxes based on satellite- and ground-based observations, using advanced land surface modeling and data assimilation techniques [27]. GLDAS includes five different land surface models, namely Noah, Mosaic, the Community Land Model (CLM), the Variable Infiltration Capacity (VIC), and Catchment (CLSM). Several studies have confirmed that the Noah model is the most robust global land surface model in GLDAS [28]. Therefore, the soil moisture, snow water equivalent, and plant canopy water data over the period of 2002–2020 were obtained from the latest GLDAS-2.1 Noah model outputs (https://disc.gsfc.nasa.gov, accessed on 1 June 2021), with a spatial resolution of 0.25°. The soil moisture is the sum of soil moisture in all soil profiles, including soil moisture content in 0–10 cm, 10–40 cm, 40–100 cm, and 100–200 cm underground, and root zone soil moisture.

3.1.3. Glacier Data

The glacier data of the Qaidam Basin in 2006 were taken from the second glacial catalog dataset of China (v1.0) published by the National Cryosphere Desert Data Center (http://www.ncdc.ac.cn, accessed on 1 December 2019) [19]. This dataset was compiled based on 218 Landsat TM/ETM+ images collected mainly from 2006 to 2010, and the glaciers in the study area were mainly concentrated in 2006. The widely used band ratio segmentation method was applied as the first step in extracting the glacier outlines, and intensive manual improvements were performed then [29]. This dataset provides several attributes for individual glaciers, and the area and volume of glaciers were used in this study.

The glacier data in 2016 were derived in this study. Since many debris-covered glaciers distributed in the Qaidam Basin, which are difficult to identify using automatic classification method, the glacier outlines were manually delineated here. A total of 26 Landsat OLI images, 2016, obtained from the United States Geological Survey (USGS) (http://earthexplorer.usgs.gov, accessed on 1 December 2019) were mainly used to delineate glacier outlines. Moreover, Google Earth images and SAR images [30–32] were used as reference images to help identify the debris-covered glaciers as well as illegible glaciers in Landsat images. After glacier delineation, the area of individual glaciers was calculated on the WGS84 spheroid in an Albers equal-area map projection. The glacier volume was estimated using the volume-area (V-A) scaling method based on empirical parameters proposed by Radić and Hock [33], which is the same as that used in the second glacial catalog dataset of China.

3.1.4. Meteorological Data

The monthly precipitation data from five meteorological stations in the Qaidam Basin (Figure 1, Table 2) for the period 2002–2020 were obtained from the National Meteorological Information Center of the China Meteorological Administration (CMA) (available at: http://data.cma.cn/, accessed on 1 July 2021). The meteorological stations are sparse in the Qaidam Basin, especially in alpine areas. Therefore, monthly precipitation data from GLDAS-2.1 Noah model outputs from 2002 to 2020 were also used in this study. To ensure the accuracy of the precipitation data, we compared GLDAS data with five CMA meteorological stations located in the Qaidam Basin. In addition, it is quite difficult to obtain the actual evapotranspiration from the observation stations, especially in the Qaidam Basin. Therefore, the evapotranspiration data during the period 2002 to 2020 were also obtained from GLDAS-2.1 Noah model outputs [8,34].

Table 2. The location of meteorological stations in the Qaidam Basin.

Station Code	Name	Latitude (°N)	Longitude (°E)	Altitude (m)
52602	Lenghu	38.45	93.20	2770.0
52707	Xiaozaohuo	36.48	93.41	2767.0
52818	Golmud	36.25	94.55	2807.6
52825	Nuomuhong	36.26	96.26	2790.4
52836	Dulan	36.18	98.06	3189.0

3.1.5. Other Relevant Data

The water resources data of the Qaidam Basin during the period 2002–2020 were collected from the Qinghai Water Resources Bulletin, including the amount of surface water resources, groundwater resources, as well as water consumption. The water consumption includes the total water consumption and the water consumption for agricultural irrigation, forestry, husbandry, fishery, industry, resident, urban public, and eco-environment.

3.2. Methods

The main methods of this study include the processing of GRACE/GRACE-FO data, quantification of the contribution of TWS components to the variations in TWS based on the water storage equation, and assessment of the influencing factors of the variations in TWS based on water balance elements, glacier mass balance, and data for human water consumption.

3.2.1. GRACE/GRACE-FO Data Processing

The formula of converting SH solutions observed by GRACE and GRACE-FO to the variations in TWS expressed by equivalent water height (EWH) is as follows [35]:

$$\mathrm{EWH}(\theta,\varphi) = \frac{\alpha\rho_{ave}}{3\rho_{water}}\sum_{l=0}^{\infty}\sum_{m=0}^{l}\frac{2l+1}{1+k_l}\overline{P}_{lm}(cos\theta)\times(\Delta C_{lm}\cos(m\varphi)+\Delta S_{lm}\sin(m\varphi)) \quad (1)$$

where ρ_{ave} is the mean density of the Earth (the constant is 5517 kg/m^3), ρ_{water} is the density of water (the constant is 1000 kg/m^3), α is the equatorial radius (the constant is 6378136.33 m), θ and φ are the geocentric latitude and longitude, ΔC_{lm} and ΔS_{lm} are the SH coefficients anomalies, k_l is the Loew coefficient, and $\overline{P}_{lm}$ is the normalized Legendre function, with maximum degree l and order m expanded to 60.

Five basic processing standards were applied to SH coefficients, including: (a) C_{20} terms replacement by solutions from Satellite Laser Ranging (SLR) [36], (b) the first-degree terms replacement by estimates from the atmosphere and ocean models [37], (c) Glacial Isostatic Adjustment (GIA) correction based on the model proposed by Geruo et al. [38], (d) mean baseline value subtraction from January 2004 to December 2009, and (e) grid value of equivalent water height with a spatial resolution of 0.25° transformation from SH coefficients. SH coefficients in the original data show obvious strip errors in space, generally oriented north to south, and there also exists large error in the high-degree terms of SH coefficients. Therefore, to reduce the north-south strip error and high-frequency noise, the SH solutions were smoothed with a Swenson destriping combined with a 300 km Gaussian smoothing [39,40]. Previous studies have shown that a Gaussian filter radius of 300 km is more appropriate, and it is commonly used [41,42].

The GRACE/GRACE-FO mascon solutions are provided in the form of equivalent water height. The post-processing was applied to the mascon solutions by following the above-mentioned steps (a)–(d), except that the GIA correction was applied based on the ICE-6G_D model from Peltier et al. [43] rather than the model from Geruo et al. [38] in step (c). The CSR-M and JPL-M solutions are provided at the spatial resolution of 0.25° and 0.5°, respectively. The JPL-M solution was then resampled at 0.25° by using nearest neighbor approach to keep consistent with TWSA from CSR-M and other SH solutions.

In this study, TWSA obtained from five GRACE solutions, including CSR-SH, JPL JPL-SH, GFZ-SH, CSR-M, and JPL-M, were compared during the period 2002 to 2020. TWSA within the basin was spatially averaged to obtain its monthly time series of it. To improve the signal-to-noise, GRACE TWSA was calculated by averaging the TWSA from five GRACE solutions [44,45]. When assuming the solution uncertainty as the major contributor, the standard deviation of TWSA among different solutions was calculated as the solution error [46]. In addition, the calibrated error of TWSA provided by JPL-M was also considered [47]. The total uncertainty of GRACE TWSA was calculated as:

$$\sigma_{total} = \sqrt{\sigma_{solution}{}^2 + \sigma_{calibrated}{}^2} \quad (2)$$

where σ_{total} represents the total uncertainty, $\sigma_{solution}$ and $\sigma_{calibrated}$ are the solution error and calibrated error, respectively.

3.2.2. Water Storage Equation

TWS includes six major components, as shown in the following formula:

$$\text{TWS} = \text{SWS} + \text{GWS} + \text{SMS} + \text{SWE} + \text{CWS} + \text{Glacier} \quad (3)$$

where TWS refers to the terrestrial water storage, SWS is surface water storage, GWS is groundwater storage, SMS is soil moisture storage, SWE is snow water equivalent, CWS is plant canopy water storage, and Glacier refers to glacier mass balance. The SWS, GWS, SMS, SWE, and CWS were calculated by using the same way as GRACE TWSA, calculating the basin-averaged monthly time series, and then subtracting the average values for the period of January 2004 to December 2009 to obtain the monthly time series anomalies. The components of TWS were represented as the equivalent water height (mm) or as equivalent water volume (km^3).

3.2.3. Water Balance Equation

Precipitation (P), evapotranspiration (ET), and net water flux (P–ET) are key components of both atmospheric and terrestrial water balances, which have a vital impact on the variations of TWS [41,48]. For an endorheic basin, changes in TWS (ΔTWS) can be estimated using the following formula:

$$\Delta\text{TWS} = \text{P} - \text{ET} \quad (4)$$

In this study, the variations in P, ET, and P–ET were compared with the variation in TWS to assess the impacts of climate change on the variations in TWS of the Qaidam Basin.

4. Results

4.1. Variations in TWSA from GRACE

4.1.1. Spatial Pattern of TWSA Trends from GRACE

The spatial pattern of linear trends of TWSA derived from five GRACE solutions over the period of 2002 to 2020 is presented in Figure 2. TWSA from different GRACE solutions in the Qaidam Basin indicated significant increasing trends ($p < 0.01$) during the entire study period. The variation trends in TWSA from CSR-SH, JPL-SH, and GFZ-SH were relatively small, with the values ranging from 1.81 to 6.70 mm/year, 1.92 to 6.93 mm/year, and 4.01 to 7.90 mm/year, respectively (Figure 2a–c). However, obviously larger trends in TWSA were observed by CSR-M and JPL-M, especially in the southwest of the basin (Figure 2d,e). Although the spatial resolution of mascon solutions released by CSR and JPL are 0.25° and 0.5°, respectively, their native resolutions are 1° (∼120 km) and 3° (330 km), which is the reason for the differences in TWSA from CSR-M and JPL-M [22].

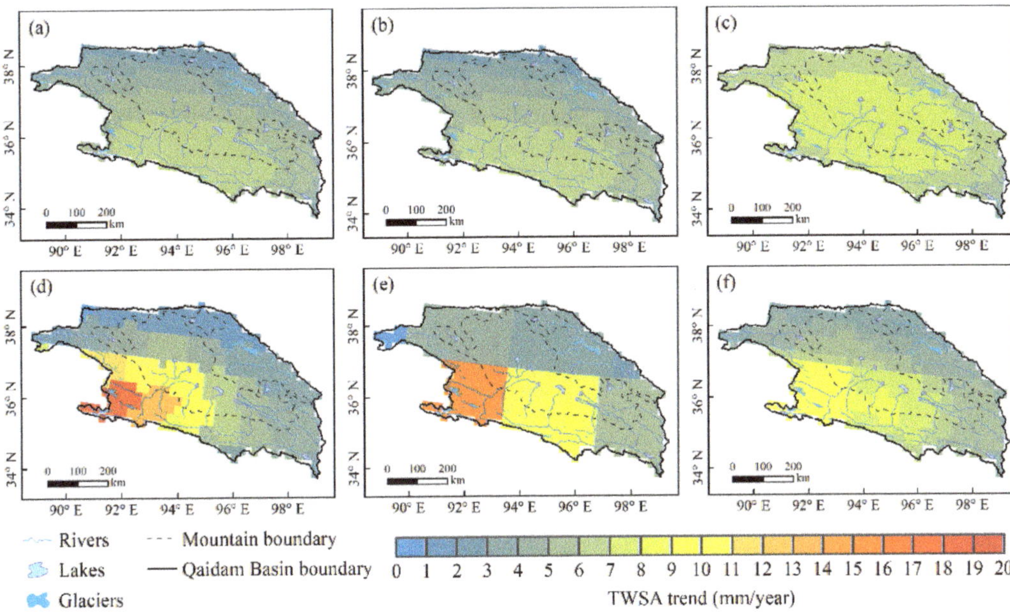

Figure 2. The spatial variation trends in TWSA of the Qaidam Basin derived from CSR-SH (**a**), JPL-SH (**b**), GFZ-SH (**c**), CSR-M (**d**), JPL-M (**e**), and the variation trend in the average TWSA from five GRACE solutions (**f**), during the period of 2002–2020.

The GRACE TWSA was computed as the average of TWSA from five GRACE solutions, and the spatial variation of it in the Qaidam Basin is shown in Figure 2f. GRACE TWSA of the entire basin exhibited a significant increase during the period from 2002 to 2020, with the spatial variation trend ranging from 2.64 to 10.70 mm/year. As the rivers flow into the central basin from the mountainous area, the variation trend in GRACE TWSA generally increased from upstream mountains to the interior of the basin in space. The largest variation trend was mostly concentrated in the southwest and the smallest in the north of the basin.

4.1.2. Temporal Variations in TWSA from GRACE

The monthly time series of TWSA derived from different GRACE solutions during the period of 2002 to 2020 are displayed in Figure 3. TWSA from the five GRACE solutions agreed well in terms of amplitude and dynamics and were significantly correlated with the correlation coefficients larger than 0.80. TWSA from different GRACE solutions in the Qaidam Basin indicated significant increasing trends ($p < 0.01$) during the entire study period, with the change rates ranging from 4.85 to 6.90 mm/year, equivalent to the increase in water storage of 1.37 to 1.95 km^3/year (Table 3). The increasing trends revealed by JPL-M (6.90 mm/year) and GFZ-SH (6.68 mm/year) were much larger than those of other solutions.

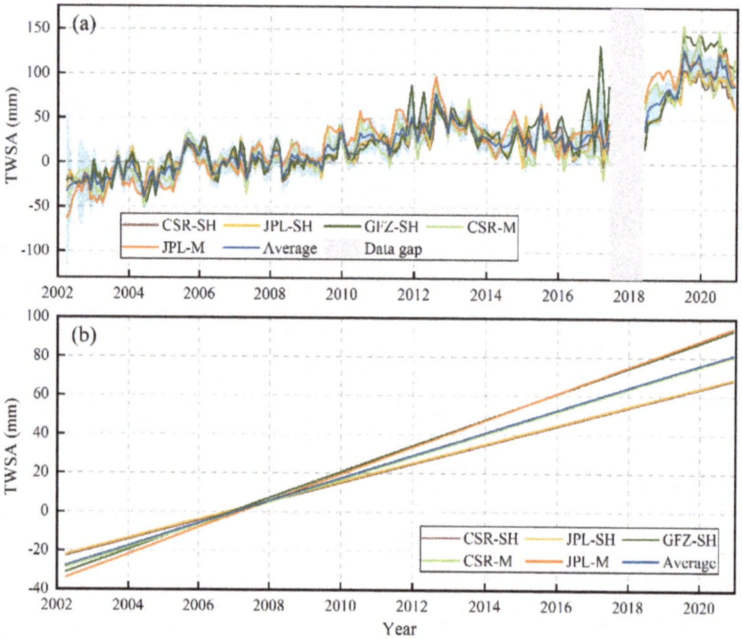

Figure 3. The monthly time series of TWSA (**a**) and linear trends (**b**) derived from five GRACE solutions in the Qaidam Basin over the period of 2002–2020. In (**a**), the average TWSA series and the uncertainty bound are also shown with a blue solid line and light blue shaded area, respectively. The gray shaded region represents the gap of 11 months between GRACE and GRACE-FO data.

Table 3. Trends in TWSA of the Qaidam Basin derived from five different GRACE solutions.

Solutions	**Trends**		***p*-Value**
	mm/Year	**km^3/Year**	
CSR-SH	4.86	1.37	$p < 0.01$
JPL-SH	4.85	1.37	$p < 0.01$
GFZ-SH	6.68	1.89	$p < 0.01$
CSR-M	5.86	1.66	$p < 0.01$
JPL-M	6.90	1.95	$p < 0.01$
Average	5.83 ± 0.12	1.65 ± 0.03	$p < 0.01$

By averaging TWSA from five GRACE solutions, the GRACE TWSA exhibited an increase at a rate of 5.83 ± 0.12 mm/year, equivalent to the increase in water storage of 1.65 ± 0.03 km^3/year. Specifically, the study period can be divided into two sub-periods (2002–2009 and 2009–2020). GRACE TWSA was mainly negative in the former sub-period and positive in the latter sub-period, indicating that the water storage of the Qaidam Basin was in deficit during the period of 2002–2009 and in surplus during the period of 2009–2020. Moreover, GRACE TWSA experienced a rapid increase after 2017, with a change rate of 22.82 ± 3.22 mm/year (6.46 ± 0.91 km^3/year). The maximum positive TWSA of 128.52 ± 22.20 mm was observed in July 2019.

The annual cycles of TWSA derived from five GRACE solutions over the study period were also analyzed, as shown in Figure 4. The average TWSA of each month from different solutions was positive, indicating that water storage of the Qaidam Basin was mainly in surplus each month. However, TWSA of the Qaidam Basin displayed significant seasonal variabilities, with the minimum and maximum values of TWSA occurring in spring (March–May) and autumn (September–November), respectively. The variations of mean GRACE TWSA in different seasons from 2002 to 2020 are presented in Figure 5. It can be found that the increasing rate of TWSA was the largest in summer and the smallest in spring, with the values of 6.14 ± 0.41 mm/year and 5.48 ± 0.89 mm/year, equivalent to the increase in water storage of 1.74 ± 0.12 km^3/year and 1.55 ± 0.25 km^3/year, respectively.

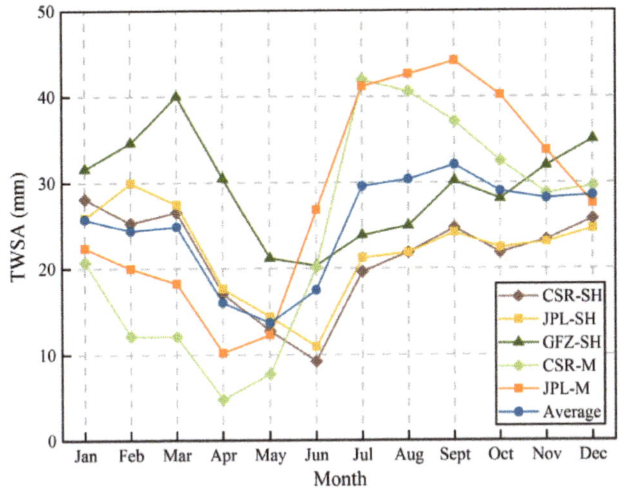

Figure 4. The annual cycles of TWSA derived from five GRACE solutions in the Qaidam Basin over the period of 2002–2020.

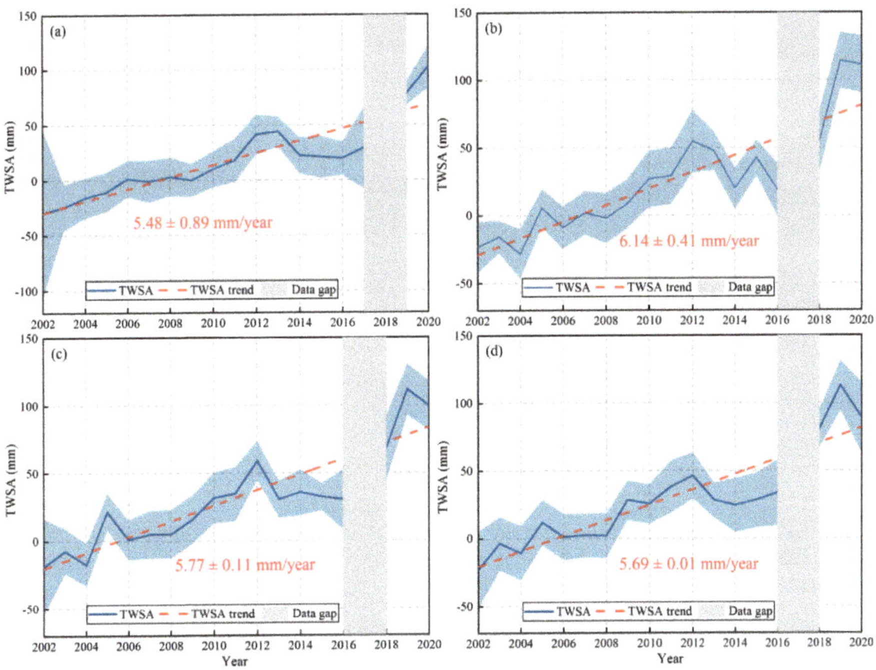

Figure 5. The variations of mean TWSA in spring (March–May) (**a**), summer (June–August) (**b**), autumn (September–November) (**c**), and winter (December–February) (**d**) in the Qaidam Basin over the period of 2002–2020. TWSA here was the average TWSA from different GRACE solutions.

4.2. Variations in All Components of TWS

In this section, the variations in all components of TWS, including soil moisture, snow water equivalent, glacier mass balance, surface water, groundwater, and plant canopy water, were investigated. The variations in soil moisture anomalies, snow water equivalent anomalies, and plant canopy water anomalies of the Qaidam Basin derived from GLDAS are shown in Figure 6a–c. The soil moisture anomalies experienced an increase at a rate of 7.65 mm/year (2.16 km^3/year) during the period from 2002 to 2020, which contributes ~131.17% to the variation in GRACE TWSA. This indicates a significant role of soil moisture in the variations in TWS of the Qaidam Basin. The snow water equivalent anomalies exhibited a slight increase at the rate of 0.06 mm/year (0.02 km^3/year), and the plant canopy water remained basically unchanged (0.00 mm/year), suggesting a negligible role of the two components in TWS variations.

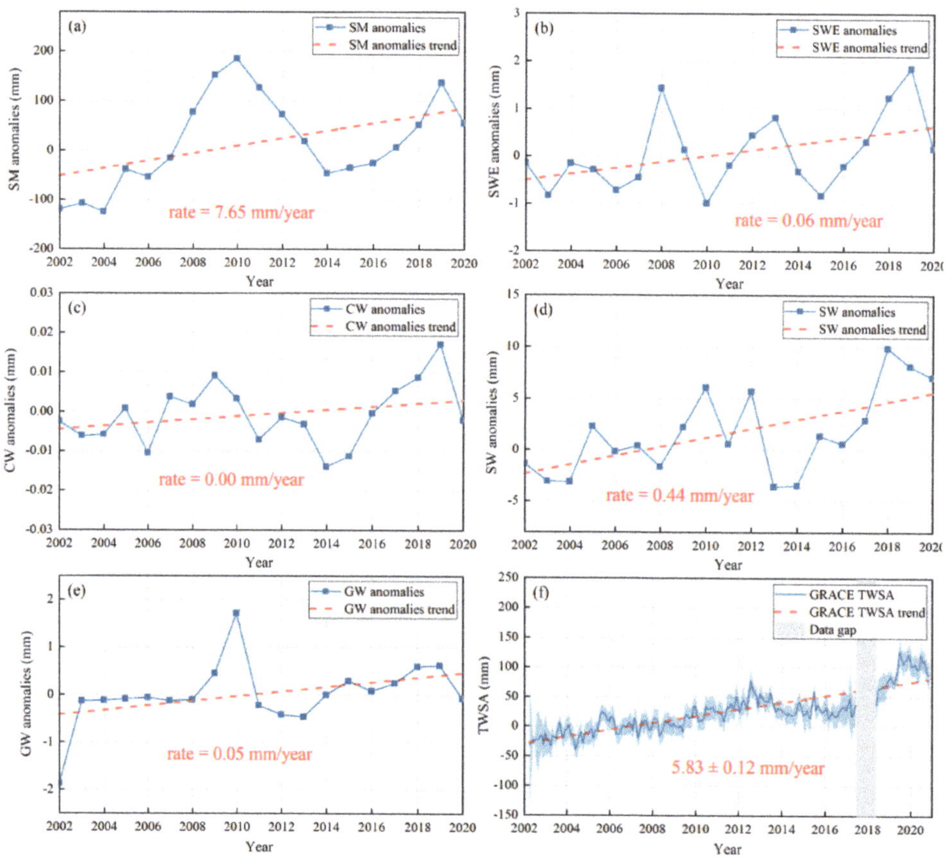

Figure 6. The variations in soil moisture (SM) anomalies (**a**), snow water equivalent (SWE) anomalies (**b**), plant canopy water (CW) anomalies (**c**), surface water (SW) anomalies (**d**), groundwater (GW) anomalies (**e**), and GRACE TWSA (**f**) of the Qaidam Basin during the period of 2002 to 2020.

The variations in surface water anomalies and groundwater anomalies of the Qaidam Basin obtained from the Qinghai Water Resources Bulletin are presented in Figure 6d,e. The surface water of the Qaidam Basin increased at a rate of 0.44 mm/year over the period from 2002 to 2020, equivalent to the increase in water storage of 0.12 km^3/year. The groundwater exhibited a slight increase at a rate of 0.05 mm/year (0.01 km^3/year). The contributions of surface water and groundwater to the variation in TWSA of the Qaidam Basin were approximately 7.54% and 0.86%, respectively, indicating little effect of the two components on the variations in TWS.

Based on the second glacial catalog dataset of China and the glacier data derived in this study, we found that the glacier area in the Qaidam Basin reduced from 1654.37 ± 74.05 km^2 in 2006 to 1587.66 ± 48.75 km^2 in 2016, which reduced by 66.71 ± 88.65 km^2 (4.03 ± 5.36%), at an average rate of 0.40 ± 0.54%. The glacier volume decreased by approximately 7.25 ± 5.22 km^3 (7.10 ± 5.12%) over the 2006–2016 period, equivalent to the total loss of glacier mass of 6.16 ± 4.44 km^3 w.e., by assuming a volume to mass conversion factor of 850 ± 60 kg/m^3. Based on the glacier mass loss during 2006–2016, the annual glacier mass balance of the Qaidam Basin over the period of 2002–2020 was estimated to be −0.62 ± 0.44 km^3 w.e./year (−2.18 ± 1.57 mm/year). For endorheic basins, glacier meltwater flows to rivers and eventually into the interior of the basin, which is an essential water resource supply in the basin. The glacier mass loss in the Qaidam Basin had a negative contribution of −37.38 ± 26.92% in the variation of TWS, indicating that the increase in glacier meltwater may lead to an increase in other components of TWS in this region.

Table 4 lists trends in all TWS components of the Qaidam Basin as well as their contributions to GRACE TWSA over the period of 2002 to 2020. The increase in TWS in the Qaidam Basin was mainly reflected by the increase in soil moisture and the loss of glacier mass. The increase in soil moisture contributed the most (~131.17%) to the variations in TWS. Despite that, the glacier mass loss had a negative contribution (−37.38 ± 26.92%) to the variations in TWS; the increase in glacier meltwater also contributed to the increase in other TWS components, particularly the soil moisture. In addition, other components of TWS had relatively little effect on the variations in TWS of the Qaidam Basin.

Table 4. Trends in soil moisture, snow water equivalent, glacier mass balance, surface water, groundwater and plant canopy water and their contributions to GRACE TWSA during the period of 2002–2020.

Variables	Trends		Contributions (%) to TWSA
	mm/Year	km^3/Year	
Soil moisture anomalies	7.65	2.16	131.17
Snow water equivalent anomalies	0.06	0.02	1.03
Plant canopy water anomalies	0.00	0.00	0.00
Glacier mass balance	−2.18 ± 1.57	−0.62 ± 0.44	−37.38 ± 26.92
Surface water	0.44	0.12	7.54
Groundwater	0.05	0.01	0.86
GRACE TWSA	5.83 ± 0.12	1.65 ± 0.03	–

5. Discussion

5.1. Impacts of Climate Change on the Variations in TWS

Climate change can affect the variations in TWS of the Qaidam Basin directly through precipitation and evapotranspiration. Figure 7 presents the variations in annual total precipitation from five meteorological stations during the period from 2002 to 2020. Four stations in the Qaidam Basin exhibited increases with the change rate ranging from 0.02 to 0.40 mm/year. Only Dulan station showed a decreasing trend (−0.47 mm/year). The meteorological stations in the Qaidam Basin are mainly distributed in plain areas, while the water resources in the basin are mainly affected by precipitation in the upstream mountains. Therefore, the precipitation data from the GLDAS model were selected here. Figure 8 displays the comparison of precipitation data from GLDAS Noah model outputs and meteorological stations from 2002 to 2020. Although the GLDAS model overestimates or underestimates precipitation at the point scale, there is a high consistency and correlation (r = 0.88) between basin-averaged precipitation from GLDAS and the average precipitation from five meteorological stations at the basin scale.

Since the water storage of the Qaidam Basin was in deficit during the period of 2002–2009 and in surplus during the period of 2009–2020, the temporal variations in precipitation (P), evapotranspiration (ET), and net water flux (P–ET) over the two sub-periods were analyzed, as shown in Figure 9. P and ET both exhibited tremendous increases during the period of 2002–2009, and the much larger increasing rate of ET indicates that ET was the dominant factor leading to the loss of TWS over this period. P and ET showed relatively small increasing trends from 2009 to 2020, suggesting that the increase in TWS over this period was mostly attributed to an increase in P. Overall, TWS of the Qaidam Basin exhibited a significant increase over the entire study period, which demonstrates that the increase in precipitation played a dominant role on the temporal variations in TWS.

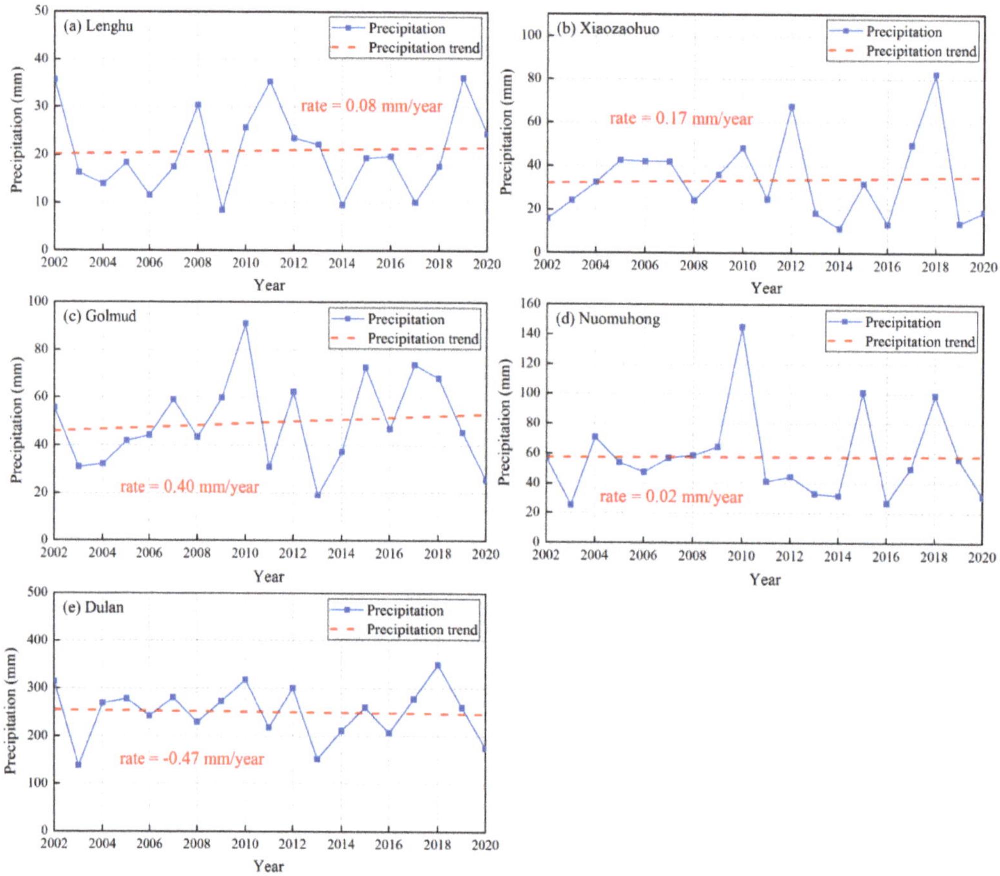

Figure 7. The variations in annual total precipitation of the Qaidam Basin from Lenghu station (**a**), Xiaozaohuo station (**b**), Golmud station (**c**), Nuomuhong station (**d**), and Dulan station (**e**) during the period of 2002–2020.

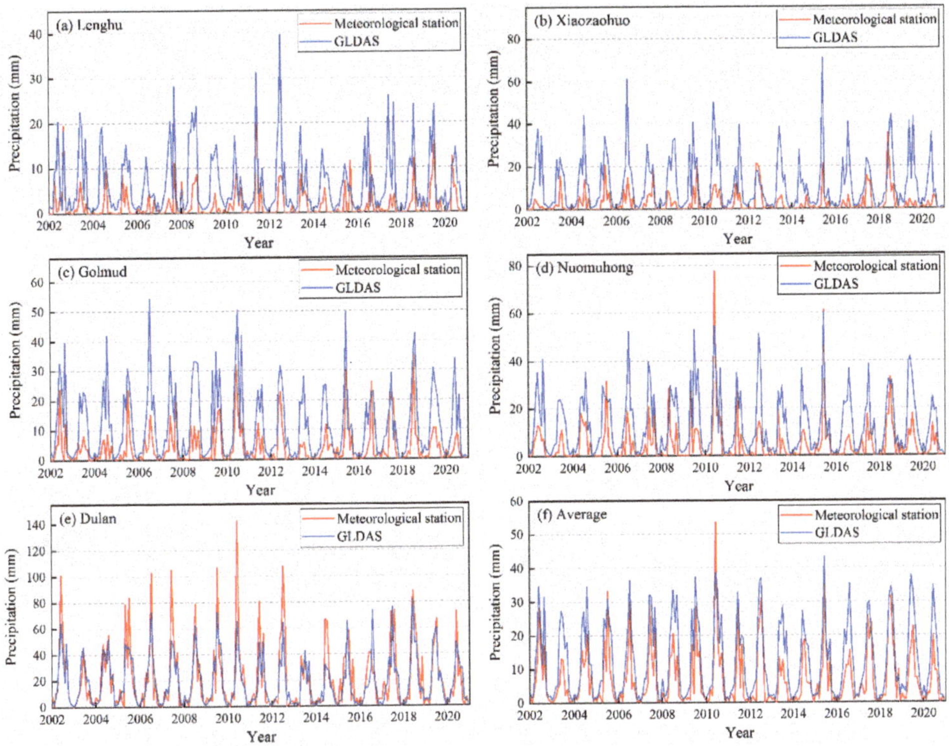

Figure 8. The comparison of precipitation data from GLDAS Noah model outputs and meteorological stations in Lenghu station (**a**), Xiaozaohuo station (**b**), Golmud station (**c**), Nuomuhong station (**d**), Dulan station (**e**) and the average precipitation from five meteorological stations (**f**) during the period of 2002–2020.

Figure 10 presents the spatial distribution of the annual total P and ET, as well as the spatial variation trends of P, ET, and P–ET over the period of 2002–2020. The mean annual total P and ET both decreased from east to west, with the maximum value mostly concentrated in the southeast and the minimum value in the north of the basin (Figure 10a,b). P, ET, and P–ET generally exhibited increases during the period of 2002–2020 (Figure 10c–e). The increase in P–ET in the mountainous area of the basin will lead to the increase in TWS in the downstream plain. For example, P–ET, in the upstream of the Qaidam River Basin and Golmud River Basin, showed significant increasing trends, and accordingly, TWS of the downstream plains also increased significantly. However, the opposite trends of P–ET in mountainous areas and TWS in the corresponding downstream plains were observed in several regions, e.g., a significant decrease in P–ET in the upstream of the Narin Gol Basin and a tremendous increase in TWS in the downstream of the basin, suggesting that precipitation had a relatively small contribution to river runoff and generally had little impact on the spatial variations in TWS of the Qaidam Basin.

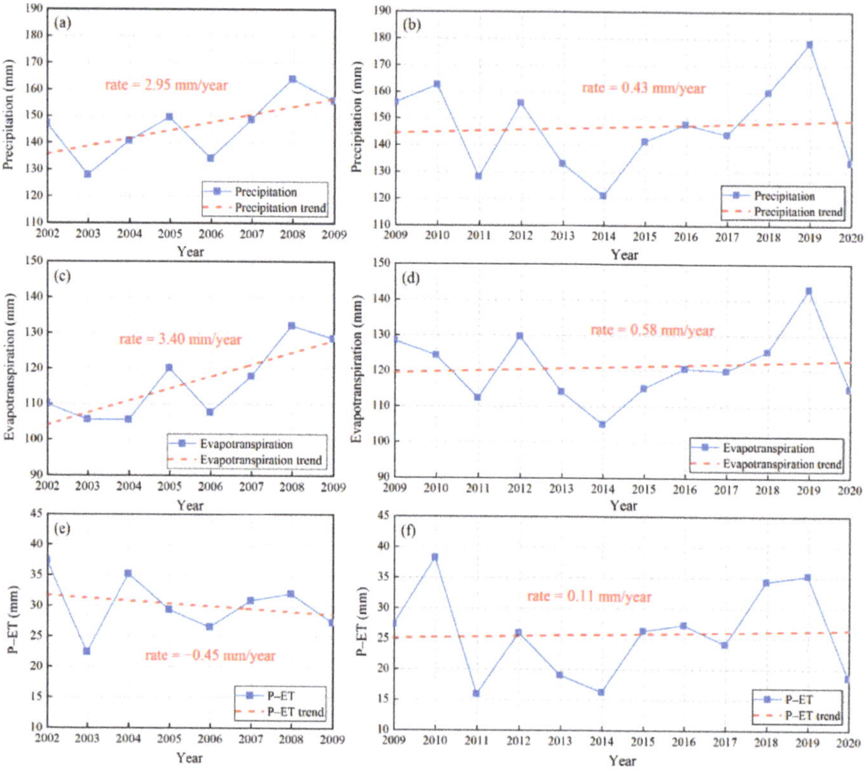

Figure 9. The temporal variations in precipitation (P) (**a**), evapotranspiration (ET) (**c**) and net water flux (P–ET) (**e**) of the Qaidam Basin from GLDAS during the period of 2002–2009, and the temporal variations in P (**b**), ET (**d**) and P–ET (**f**) of the Qaidam Basin during the period of 2009–2020.

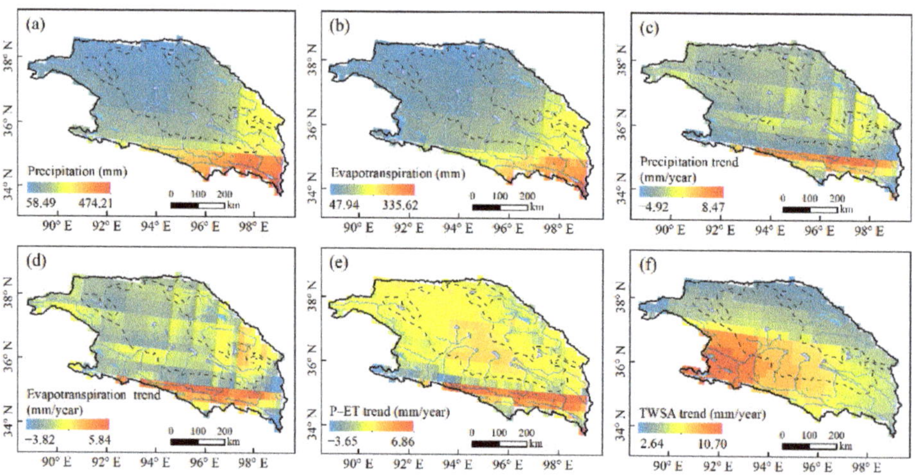

Figure 10. The spatial distribution of the annual total P (**a**) and ET (**b**), and the spatial variation trends of P (**c**), ET (**d**), P–ET (**e**) and GRACE TWSA (**f**) of the Qaidam Basin during the period of 2002–2020.

5.2. Impacts of Glacier Mass Loss on the Variations in TWS

Glaciers are essential contributors to river runoff in the Qaidam Basin. According to the glacier data derived in this study, there were 2242 glaciers distributed in the basin in 2016, with a total area of 1587.66 ± 48.75 km^2, an average area of 0.71 ± 0.02 km^2, the minimum area of 0.01 ± 0.0038 km^2, the maximum area of 82.22 ± 0.55 km^2, and a total volume of 94.89 ± 2.68 km^3. Since the Qaidam Basin is an endorheic basin, glacier change mainly affects the spatial variations in TWS. The area and volume of glaciers in the Qaidam Basin reduced by about 66.71 ± 88.65 km^2 (4.03 ± 5.36%) and 7.25 ± 5.22 km^3 (7.10 ± 5.12%) respectively from 2006 to 2016. Changes in glacier volume of sub-basins in the Qaidam Basin during the period of 2006–2016 were used to estimate the changes during the period of 2002–2020, which are presented in Figure 11.

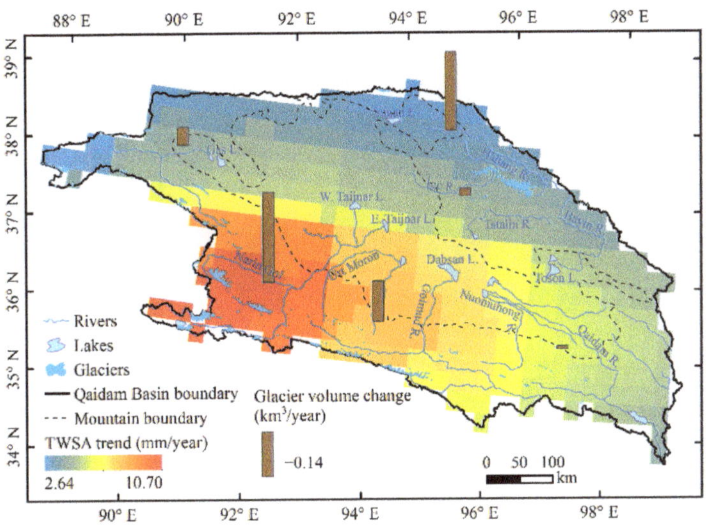

Figure 11. Changes in glacier volume of sub-basins in the Qaidam Basin over the period of 2006–2016.

The glacier volume in the Narin Gol Basin and Haltang River Basin decreased a lot at the rates of 0.29 ± 0.10 km^3/year and 0.25 ± 0.06 km^3/year, equivalent to the negative glacier mass balance of 0.24 ± 0.09 km^3 w.e./year (7.52 ± 2.71 mm/year) and 0.22 ± 0.05 km^3 w.e./year (9.64 ± 2.25 mm/year), respectively. Since P–ET exhibited decreasing trends in the upstream of the two sub-basins, the increase in glacier meltwater played a dominant role in the increase in TWS in the downstream plains, especially the downstream of the Narin Gol Basin. In addition, the runoff infiltration of the Narin Gol reaches more than 70% [49], which might be the reason for the larger increase in TWS in the upstream than downstream of the basin.

The glacier mass loss was also observed in the upstream of the Golmud River Basin and Qaidam River Basin, where TWS and P–ET both increased significantly. This indicates that the increase in TWS in the downstream of the two sub-basins is affected by the increase in precipitation and glacier meltwater in the upstream mountains, and the latter contributed less than the former. For other sub-basins in the Qaidam Basin, glacier meltwater exhibited slight increasing trends, while P–ET showed a slight decrease in the upstream mountains, suggesting the important role of glacier meltwater on the variations in TWS of the downstream plains. In a word, the increase in glacier meltwater was the dominant factor for the spatial variations in TWS of the Qaidam Basin.

5.3. Impacts of Human Activities on the Variations in TWS

The water consumption associated with human activities in oasis areas is the main component of the total water consumption of the Qaidam Basin, which has a significant impact on the variations in TWS. Figure 12 presents the variations in total water consumption and the water consumption for individual terms, including agricultural irrigation, forestry, husbandry, fishery, industry, resident, urban public, and eco-environment in the Qaidam Basin over the period of 2002 to 2020. Water withdrawal for irrigation accounts for a large proportion of total water consumption in this region and experienced a decreasing trend (−0.01 km^3/year) over the study period. The water consumption for forestry, husbandry, and fishery exhibited an increase at a rate of 0.01 km^3/year, while water consumption for other purposes remained basically unchanged.

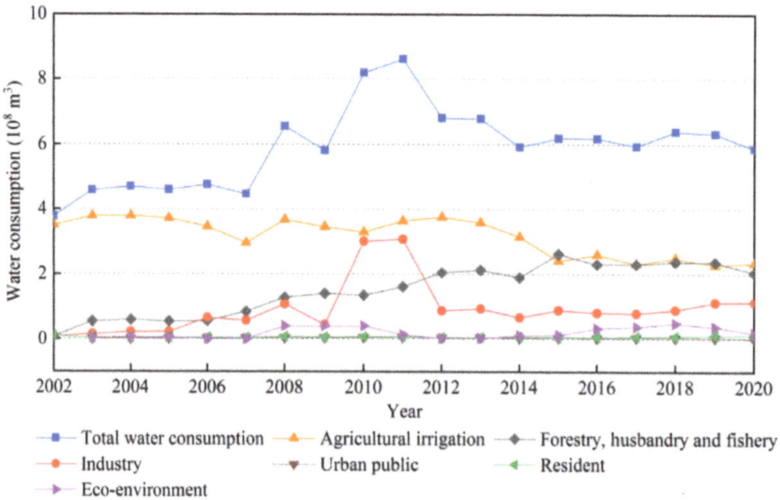

Figure 12. The variations in total water consumption and the water consumption for agricultural irrigation, forestry, husbandry, fishery, industry, resident, urban public and eco-environment in the Qaidam Basin, during the period of the 2002–2020.

The total water consumption in the Qaidam Basin increased at a rate of 0.05 km^3/year (0.17 mm/year) during the period of 2002 to 2011 and decreased at a rate of 0.02 km^3/year (0.06 mm/year) during the period of 2011 to 2020. The opposite variation trends can explain the reason why TWS in this region was in deficit from 2002 to 2009 and in surplus after 2009. However, the total water consumption showed a slight increase, and the TWSA increased significantly over the period of 2002–2020, suggesting that the impacts of human activities on the variations in TWS of the Qaidam Basin were relatively small.

In summary, the precipitation played a dominant role in the temporal variations in TWS of the Qaidam Basin. The increase in glacier meltwater due to climate warming was the driving factor affecting the spatial variations in TWS of the Qaidam Basin, particularly in the Narin Gol Basin. In addition, the water consumption associated with human activities had relatively fewer impacts on the variations in TWS.

6. Conclusions

In this study, the variations in TWS and individual TWS components of the Qaidam Basin over the period of 2002 to 2020 were investigated by using GRACE and GRACE-FO data, GLDAS Noah model outputs, as well as data from Qinghai Water Resources Bulletin. The impacts of climate change, glacier mass loss, and human activities on the variations in TWS were further analyzed based on water balance elements, glacier mass balance, and data for human water consumption. The major results of this study are summarized as follows:

(1) TWSA of the Qaidam Basin derived from different GRACE solutions agreed well in terms of amplitude and dynamics and were significantly correlated. The significant increasing trends were observed by TWSA from five GRACE solutions over the period of 2002–2020, with the change rates ranging from 4.85 to 6.90 mm/year (1.37 to 1.95 km^3/year). The increasing trends revealed by JPL-M (6.90 mm/year) and GFZ-SH (6.68 mm/year) were much larger than those of other solutions.

(2) The GRACE TWSA averaged from different GRACE solutions exhibited an increase at a rate of 5.83 $\pm$ 0.12 mm/year (1.65 $\pm$ 0.03 km^3/year) during the period of 2002–2020. The variation trend in GRACE TWSA generally increased from upstream mountains to the interior of the basin in space, with the largest increasing trend mostly concentrated in the southwest and the smallest in the north of the basin. TWSA of the Qaidam Basin displayed significant seasonal variabilities, with the maximum values occurring in autumn and the largest increasing trend observed in summer (6.14 $\pm$ 0.41 mm/year).

(3) The variation trends in components of TWS, including soil moisture, snow water equivalent, glacier mass balance, surface water, groundwater, and plant canopy water, were 7.65 mm/year, 0.06 mm/year, -2.18 ± 1.57 mm/year, 0.44 mm/year, 0.05 mm/year and 0.00 mm/year, respectively. Among these components of TWS, soil moisture contributed the most to the variations in TWS of the Qaidam Basin. Despite that, the glacier mass loss had a negative contribution to the variations in TWS; the increase in glacier meltwater also contributed to the increase in TWS.

(4) The temporal variations in TWS of the Qaidam Basin were primarily affected by the variations in precipitation. The spatial variations in TWS of the Qaidam Basin were mostly driven by the increase in glacier meltwater due to climate warming, particularly in the Narin Gol Basin. In addition, the water consumption associated with human activities had relatively fewer impacts on the variations in TWS.

Author Contributions: Conceptualization, X.Y. and N.W.; methodology, X.Y., A.C. and Q.L.; software, Y.Z.; validation, Z.L.; formal analysis, X.Y. and A.C.; investigation, Z.L. and Q.L.; resources, X.Y.; data curation, X.Y.; writing—original draft preparation, X.Y.; writing—review and editing, N.W., A.C., Z.L., Q.L. and Y.Z.; Visualization, A.C.; Supervision, Y.Z.; Project Administration, N.W.; Funding Acquisition, N.W. All authors have read and agreed to the published version of the manuscript.

Funding: This research was funded by the Strategic Priority Research Program of the Chinese Academy of Sciences (XDA20060201, XDA19070302), the National Natural Science Foundation of China (42130516), and the Second Tibetan Plateau Scientific Expedition and Research Program (2019QZKK020102).

Institutional Review Board Statement: Not applicable.

Informed Consent Statement: Not applicable.

Data Availability Statement: Not applicable.

Acknowledgments: The authors are very grateful to CSR, JPL and GFZ for providing GRACE RL06 Level 2 product and mascon solutions, and the Goddard Earth Sciences Data and Information Services Center, NASA, for providing GLDAS-2.1 outputs. We thank the National Cryosphere Desert Data Center for providing the glacier data, and the China Meteorological Administration for providing the precipitation data. We appreciate the editors and the three anonymous reviewers for their valuable suggestions and advice.

Conflicts of Interest: The authors declare no conflict of interest.

References

1. Rodell, M.; Famiglietti, J.S. Detectability of variations in continental water storage from satellite observations of the time dependent gravity field. *Water Resour. Res.* **1999**, *35*, 2705–2723. [CrossRef]
2. Zhu, Y.; Liu, S.Y.; Yi, Y.; Xie, F.; Grünwald, R.; Miao, W.F.; Wu, K.P.; Qi, M.M.; Gao, Y.P.; Singh, D. Overview of terrestrial water storage changes over the Indus River Basin based on GRACE/GRACE-FO solutions. *Sci. Total Environ.* **2021**, *799*, 149366. [CrossRef]
3. Chang, L.L.; Yuan, R.Q.; Gupta, H.V.; Winter, C.; Niu, G.Y. Why Is the Terrestrial Water Storage in Dryland Regions Declining? A Perspective Based on Gravity Recovery and Climate Experiment Satellite Observations and Noah Land Surface Model with Multiparameterization Schemes Model Simulations. *Water Resour. Res.* **2020**, *56*, e2020WR027102. [CrossRef]
4. Wang, J.D.; Song, C.Q.; Reager, J.T.; Yao, F.F.; Famiglietti, J.S.; Sheng, Y.W.; Macdonald, G.M.; Brun, F.; Schmied, H.M.; Marston, R.A. Recent global decline in endorheic basin water storages. *Nat. Geosci.* **2018**, *11*, 926–932. [CrossRef]
5. Shugar, D.H.; Jacquemart, M.; Shean, D.; Bhushan, S.; Upadhyay, K.; Sattar, A.; Schwanghart, W.; McBride, S.; de Vries, M.V.W.; Mergili, M.; et al. A massive rock and ice avalanche caused the 2021 disaster at Chamoli, Indian Himalaya. *Science* **2021**, *373*, 300–306. [CrossRef] [PubMed]
6. Xiao, Y.; Shao, J.; Frape, S.K.; Cui, Y.; Dang, X.; Wang, S.; Ji, Y. Groundwater origin, flow regime and geochemical evolution in arid endorheic watersheds: A case study from the Qaidam Basin, northwestern China. *Hydrol. Earth Syst. Sci.* **2018**, *22*, 4381–4400. [CrossRef]
7. Wang, X.; Yang, M.; Liang, X.; Pang, G.; Wan, G.; Chen, X. The dramatic climate warming in the Qaidam Basin, northeastern Tibetan Plateau, during 1961–2010. *Int. J. Climatol.* **2014**, *34*, 1524–1537. [CrossRef]
8. Jing, W.L.; Zhao, X.D.; Yao, L.; Jiang, H.; Xu, J.H.; Yang, J.; Li, Y. Variations in terrestrial water storage in the Lancang-Mekong river basin from GRACE solutions and land surface model. *J. Hydrol.* **2020**, *580*, 124258. [CrossRef]
9. Tapley, B.D.; Bettadpur, S.; Ries, J.C.; Thompson, P.F.; Watkins, M.M. GRACE Measurements of Mass Variability in the Earth System. *Science* **2004**, *305*, 503–505. [CrossRef]
10. Tapley, B.D.; Watkins, M.M.; Flechtner, F.; Reigber, C.; Bettadpur, S.; Rodell, M.; Sasgen, I.; Famiglietti, J.S.; Landerer, F.W.; Chambers, D.P.; et al. Contributions of GRACE to understanding climate change. *Nat. Clim. Chang.* **2019**, *9*, 358–369. [CrossRef]
11. Rodell, M.; Famiglietti, J.S.; Wiese, D.N.; Reager, J.T.; Beaudoing, H.K.; Landerer, F.W.; Lo, M.H. Emerging trends in global freshwater availability. *Nature* **2018**, *557*, 651–659. [CrossRef] [PubMed]
12. Ahmed, M.; Sultan, M.; Wahr, J.; Yan, E. The use of GRACE data to monitor natural and anthropogenic induced variations in water availability across Africa. *Earth-Sci. Rev.* **2014**, *136*, 289–300. [CrossRef]
13. Wahr, J.; Swenson, S.; Zlotnicki, V.; Velicogna, I. Time-variable gravity from GRACE: First results. *Geophys. Res. Lett.* **2004**, *31*, L11501. [CrossRef]
14. Landerer, F.W.; Flechtner, F.M.; Save, H.; Webb, F.H.; Bandikova, T.; Bertiger, W.I.; Bettadpur, S.V.; Byun, S.H.; Dahle, C.; Dobslaw, H.; et al. Extending the Global Mass Change Data Record: GRACE Follow-On Instrument and Science Data Performance. *Geophys. Res. Lett.* **2020**, *47*, e2020GL088306. [CrossRef]
15. Jiao, J.J.; Zhang, X.; Liu, Y.; Kuang, X. Increased Water Storage in the Qaidam Basin, the North Tibet Plateau from GRACE Gravity Data. *PLoS ONE* **2015**, *10*, e0141442. [CrossRef] [PubMed]
16. Wei, L.Y.; Jiang, S.H.; Ren, L.L.; Tan, H.B.; Ta, W.Q.; Liu, Y.; Yang, X.L.; Zhang, L.Q.; Duan, Z. Spatiotemporal changes of terrestrial water storage and possible causes in the closed Qaidam Basin, China using GRACE and GRACE Follow-On data. *J. Hydrol.* **2021**, *598*, 126274. [CrossRef]
17. Bibi, S.; Wang, L.; Li, X.P.; Zhang, X.T.; Chen, D.L. Response of Groundwater Storage and Recharge in the Qaidam Basin (Tibetan Plateau) to Climate Variations From 2002 to 2016. *J. Geophys. Res. Atmos.* **2019**, *124*, 9918–9934. [CrossRef]
18. Hu, L.T.; Jiao, J.J. Calibration of a large-scale groundwater flow model using GRACE data: A case study in the Qaidam Basin, China. *Hydrogeol. J.* **2015**, *23*, 1305–1317. [CrossRef]
19. Liu, S.Y.; Guo, W.Q.; Xu, J.L. *The Second Glacial Catalogue Data Set of China (V1.0)*; National Cryosphere Desert Data Center: Lanzhou, China, 2019.
20. Soltani, S.S.; Ashtiani, B.A.; Simmons, C.T. Review of assimilating GRACE terrestrial water storage data into hydrological models: Advances, challenges and opportunities. *Earth-Sci. Rev.* **2021**, *213*, 103487. [CrossRef]
21. Landerer, F.W.; Swenson, S.C. Accuracy of scaled GRACE terrestrial water storage estimates. *Water Resour. Res.* **2012**, *48*, 4531. [CrossRef]
22. Scanlon, B.R.; Zhang, Z.; Save, H.; Wiese, D.N.; Landerer, F.W.; Long, D.; Longuevergne, L.; Chen, J. Global Evaluation of New GRACE Mascon Products for Hydrologic Applications. *Water Resour. Res.* **2016**, *52*, 9412–9429. [CrossRef]
23. Wiese, D.N.; Landerer, F.W.; Watkins, M.M. Quantifying and reducing leakage errors in the JPL RL05M GRACE mascon solution. *Water Resour. Res.* **2016**, *52*, 7490–7502. [CrossRef]
24. Save, H.; Bettadpur, S.; Tapley, B.D. High-resolution CSR GRACE RL05 mascons. *J. Geophys. Res. Solid Earth* **2016**, *121*, 7547–7569. [CrossRef]
25. Huang, W.J.; Duan, W.L.; Chen, Y.N. Rapidly declining surface and terrestrial water resources in Central Asia driven by socio-economic and climatic changes. *Sci. Total Environ.* **2021**, *784*, 147193. [CrossRef]
26. Liu, X.; Hu, L.T.; Sun, K.N.; Yang, Z.Q.; Sun, J.C.; Yin, W.J. Improved Understanding of Groundwater Storage Changes under the Influence of River Basin Governance in Northwestern China Using GRACE Data. *Remote Sens.* **2021**, *13*, 2672. [CrossRef]

27. Rodell, M.; Houser, P.R.; Jambor, U.; Gottschalck, J.; Mitchell, K.; Meng, C.J.; Arsenault, K.; Cosgrove, B.; Radakovich, J.; Bosilovich, M.; et al. The global land data assimilation system. *Bull. Am. Meteorol. Soc.* **2004**, *85*, 381–394. [CrossRef]
28. Yang, T.; Wang, C.; Yu, Z.B.; Xu, F. Characterization of spatio-temporal patterns for various GRACE- and GLDAS-born estimates for changes of global terrestrial water storage. *Glob. Planet. Chang.* **2013**, *109*, 30–37. [CrossRef]
29. Guo, W.Q.; Liu, S.Y.; Xu, J.L.; Wu, L.Z.; Shangguan, D.H.; Yao, X.J.; Wei, J.F.; Bao, W.J.; Yu, P.C.; Liu, Q.; et al. The second Chinese glacier inventory: Data, methods and results. *J. Glaciol.* **2015**, *61*, 357–372. [CrossRef]
30. van Zyl, J.J.; Zebker, H.A.; Elachi, C. Imaging radar polarization signatures: Theory and observation. *Radio Science* **1987**, *22*, 529–543. [CrossRef]
31. Muhuri, A.; Manickam, S.; Bhattacharya, A. Scattering Mechanism Based Snow Cover Mapping Using RADARSAT-2 C-Band Polarimetric SAR Data. *IEEE J. Sel. Top. Appl. Earth Obs. Remote Sens.* **2017**, *10*, 3213–3224. [CrossRef]
32. Touzi, R. Target Scattering Decomposition in Terms of Roll-Invariant Target Parameters. *IEEE Trans. Geosci. Remote Sens.* **2007**, *45*, 73–84. [CrossRef]
33. Radić, V.; Hock, R. Regional and global volumes of glaciers derived from statistical upscaling of glacier inventory data. *J. Geophys. Res. Earth Surf.* **2010**, *115*, F01010. [CrossRef]
34. Wang, F.; Chen, Y.; Li, Z.; Fang, G.; Li, Y.; Wang, X.; Zhang, X.; Kayumba, P.M. Developing a Long Short-Term Memory (LSTM)-Based Model for Reconstructing Terrestrial Water Storage Variations from 1982 to 2016 in the Tarim River Basin. *Remote Sens.* **2021**, *13*, 889. [CrossRef]
35. Swenson, S.; Wahr, J. Methods for inferring regional surface-mass anomalies from Gravity Recovery and Climate Experiment (GRACE) measurements of time-variable gravity. *J. Geophys. Res. Solid Earth* **2002**, *107*, 2193. [CrossRef]
36. Cheng, M.K.; Tapley, B.D. Variations in the Earth's oblateness during the past 28 years. *J. Geophys. Res. Solid Earth* **2004**, *109*, B03406. [CrossRef]
37. Swenson, S.; Chambers, D.; Wahr, J. Estimating geocenter variations from a combination of GRACE and ocean model output. *J. Geophys. Res. Solid Earth* **2008**, *113*, B08410. [CrossRef]
38. Geruo, A.; Wahr, J.; Zhong, S.J. Computations of the viscoelastic response of a 3-D compressible Earth to surface loading: An application to Glacial Isostatic Adjustment in Antarctica and Canada. *Geophys. J. Int.* **2013**, *192*, 557–572.
39. Swenson, S.; Wahr, J. Post-processing removal of correlated errors in GRACE data. *Geophys. Res. Lett.* **2006**, *33*, L08402. [CrossRef]
40. Wahr, J.; Molenaar, M.; Bryan, F. Time variability of the Earth's gravity field: Hydrological and oceanic effects and their possible detection using GRACE. *J. Geophys. Res. Solid Earth* **1998**, *103*, 30205–30229. [CrossRef]
41. Yang, X.; Wang, N.; Liang, Q.; Chen, A.; Wu, Y. Impacts of Human Activities on the Variations in Terrestrial Water Storage of the Aral Sea Basin. *Remote Sens.* **2021**, *13*, 2923. [CrossRef]
42. Deng, H.; Chen, Y. Influences of recent climate change and human activities on water storage variations in Central Asia. *J. Hydrol.* **2017**, *544*, 46–57. [CrossRef]
43. Peltier, W.R.; Argus, D.F.; Drummond, R. Comment on "An Assessment of the ICE-6G_C (VM5a) Glacial Isostatic Adjustment Model" by Purcell et al. *J. Geophys. Res. Solid Earth* **2018**, *123*, 2019–2028. [CrossRef]
44. Sakumura, C.; Bettadpur, S.; Bruinsma, S.L. Ensemble prediction and intercomparison analysis of GRACE time-variable gravity field models. *Geophys. Res. Lett.* **2014**, *41*, 1389–1397. [CrossRef]
45. Wang, Q.; Zheng, W.; Yin, W.; Kang, G.; Zhang, G.; Zhang, D. Improving the Accuracy of Water Storage Anomaly Trends Based on a New Statistical Correction Hydrological Model Weighting Method. *Remote Sens.* **2021**, *13*, 3583. [CrossRef]
46. Abhishek; Kinouchi, T.; Sayama, T. A comprehensive assessment of water storage dynamics and hydroclimatic extremes in the Chao Phraya River Basin during 2002-2020. *J. Hydrol.* **2021**, *603*, 126868. [CrossRef]
47. Liu, J.; Jiang, L.; Zhang, X.; Druce, D.; Kittel, C.M.M.; Tøttrup, C.; Bauer-Gottwein, P. Impacts of water resources management on land water storage in the North China Plain: Insights from multi-mission earth observations. *J. Hydrol.* **2021**, *603*, 126933. [CrossRef]
48. Meng, F.C.; Su, F.G.; Li, Y.; Tong, K. Changes in terrestrial water storage during 2003-2014 and possible causes in Tibetan Plateau. *J. Geophys. Res. Atmos.* **2019**, *124*, 2909–2931. [CrossRef]
49. Tan, H.B.; Lu, S.C.; Ta, W.Q.; Rao, W.B.; Guo, H.Y. Water resource and mineral source significances of Nalenggele River Catchment in Qaidam Basin. *J. Hohai Univ. Nat. Sci.* **2020**, *48*, 392–397.

remote sensing

MDPI

Article

Assessment of Three Long-Term Satellite-Based Precipitation Estimates against Ground Observations for Drought Characterization in Northwestern China

Hao Guo [1], Min Li [1], Vincent Nzabarinda [2], Anming Bao [2,*], Xiangchen Meng [1], Li Zhu [1] and Philippe De Maeyer [3]

[1] School of Geography and Tourism, Qufu Normal University, Rizhao 276825, China; guohao@qfnu.edu.cn (H.G.); qfnulimin@163.com (M.L.); xiangchenmeng@qfnu.edu.cn (X.M.); zhuli@qfnu.edu.cn (L.Z.)
[2] State Key Laboratory of Desert and Oasis Ecology, Xinjiang Institute of Ecology and Geography, Chinese Academy of Sciences, Urumqi 830011, China; vincentnzabarinda@mails.ucas.ac.cn
[3] Department of Geography, Ghent University, 9000 Ghent, Belgium; Philippe.DeMaeyer@UGent.be
* Correspondence: baoam@ms.xjb.ac.cn; Tel.: +86-0991-7885378

Abstract: Long-term satellite-based precipitation estimates (LSPE) play a significant role in climatological studies like drought monitoring. In this study, three popular LSPEs (Precipitation Estimation from Remotely Sensed Information using Artificial Neural Networks–Climate Data Record (PERSIANN-CDR), Rainfall Estimates from Rain Gauge and Satellite Observations (CHIRPS) and Multi-Source Weighted-Ensemble Precipitation (MSWEP)) were evaluated on a monthly scale using ground-based stations for capturing drought event characteristics over northwestern China from 1983 to 2013. To reflect dry or wet evolution, the Standardized Precipitation Index (SPI) was adopted, and the Run theory was used to identify drought events and their characteristics. The conventional statistical indices (relative bias (RB), correlation coefficient (CC), and root mean square error (RMSE)), as well as categorical indices (probability of detection (POD), false alarm ratio (FAR), and missing ratio (MISS)) are used to evaluate the capability of LSPEs in estimating precipitation and drought characteristics. We found that: (1) three LSPEs showed generally satisfactory performance in estimating precipitation and characterizing drought events. Although LSPEs have acceptable performance in identifying drought events with POD greater than 60%, they still have a high false alarm ratio (>27%) and a high missing ratio (>33%); (2) three LSPEs tended to overestimate drought severity, mainly because of an overestimation of drought duration; (3) the ability of CHIRPS to replicate the temporal evolution of precipitation and SPI values is limited; (4) in severe drought events, PERSIANN-CDR tends to overestimate precipitation, and drought severity, as well as drought area; (5) among the three LSPEs, MSWEP outperformed the other two in identifying drought events (POD > 66%) and characterizing drought features. Finally, we recommend MSWEP for drought monitoring studies due to its high accuracy in estimating drought characteristics over northwestern China. In drought monitoring applications, the overestimation of PERSIANN-CDR for drought peak value and area, as well as CHIRPS's inferiority in capturing drought temporal evolution, must be considered.

Keywords: MSWEP; PERSIANN-CDR; CHIRPS; satellite-based precipitation estimates; drought monitoring; evaluation

1. Introduction

Even though precipitation is expected to increase in the future, evaporation is expected to increase significantly as well, and precipitation variation is expected to be extremely variable [1]. The semi-arid and arid regions are still facing high drought risk [2–4].

Droughts have significant impacts on both natural ecosystems and human society, and they are particularly severe in semi-arid and arid regions with fragile ecosystems, such as the arid zones of Central Asia. Drought indices are commonly used to quantify the drought condition and its impacts on global and regional scales [5,6]. Many drought indices have been developed in recent years to characterize drought conditions [7]. Among them, SPI, Standardized Precipitation Evapotranspiration Index (SPEI), and Palmer Drought Severity Index (PDSI) are the three most commonly used drought indices for evaluating meteorological droughts. The SPI is a standardized drought index calculated based on probabilities derived from long-term precipitation records [8]. It is widely used and recommended by the World Meteorological Organization (WMO) because of its simple calculation, flexible time scales, and limited variable requirements [9]. Even though SPEI and PDSI consider the effect of temperature; however, variables like radiation, relative humidity, and the temperature required to calculate potential evapotranspiration (PET) based on the Penman or Penman-Monteith algorithm are usually not available in these areas. While calculating potential evapotranspiration based on temperature, like the Thornthwaite algorithm, has been shown to overestimate drought conditions in semi-arid and arid regions [10–12].

Long-term data records with adequate spatial and temporal coverage are required for drought characterization and drought index calculation using ground gauges [13]. However, drought monitoring in Central Asia's arid zones, like many other arid and semi-arid countries around the world, is hampered by a lack of reliable precipitation observations. The climate stations are sparse and unevenly distributed in many drought-prone areas, resulting in large spatial gaps. Overall, the long latency of data acquisition, limited spatial representation, and data gaps in ground observations limit meteorological drought monitoring [13,14]. For example, many studies have lamented the scarcity of observations in Central Asia's arid zones [15,16]. Furthermore, to ensure the reliability of results, drought monitoring and event characterization require a long-term data period of at least 30 years [17].

Satellite-based precipitation estimates (SPEs) are a very valuable alternative resource for drought monitoring due to their consistent and homogeneous data with quasi-global coverage, near-real-time estimates, increasing data records, and high spatial resolution [17–19]. In recent years, great efforts have been devoted to the development of SPEs such as the Tropical Rainfall Measuring Mission (TRMM), Multisatellite Precipitation Analysis (TMPA), Precipitation Estimation from Remotely Sensed Information using Artificial Neural Networks (PERSIANN), Rainfall Estimates from Rain Gauge and Satellite Observations (CHIRPS), Global Satellite Mapping of Precipitation (GSMaP), Climate Prediction Center morphing technique (CMORPH), Multi-Source Weighted-Ensemble Precipitation (MSWEP), Climate Prediction Center (CPC), MORPHing technique (CMORPH), and Integrated Multi-satellite Retrievals for Global Precipitation Mission (IMERG). However, most of them are limited by the short record length in meteorological drought monitoring, which requires long-term data records ($\geq$30 years) [20]. According to AghaKouchak et al. [17], short-term data records will have an impact on drought monitoring accuracy. WMO reported that a reliable SPI calculation needs at least 30-years of data records [9].

Fortunately, three long-term LSPEs with records larger than 30 years are identified among the listed SPEs: PERSIANN-CDR, CHIRPS and MSWEP. These long-term SPEs (LSPEs) provide an opportunity to assess drought from a climatological perspective, especially for gauge-sparse regions like northwestern China. However, the inherent uncertainties of LSPEs in estimating precipitation may affect the accuracy of drought characterization. For example, PERSIANN-CDR has been reported to overestimate precipitation in mountainous regions [21]. Therefore, it is important to evaluate their suitability and reliability before employing them in drought monitoring.

Satellite-based precipitation estimates have already been put to the test in some regions for drought monitoring. Among these studies, TMPA products have been thoroughly evaluated in many regions, including Mexico [22], Africa [23], and different river basins [24–26]. The assessment of LSPEs for drought monitoring has also gotten a lot of attention in recent years. For example, the drought characterization performance of

CHIRPS has been evaluated in China [27], Mekong River Basin [28], and the Brazilian Midwest [29]. PERSIANN-CDR's capability for drought monitoring has also been assessed in China [30,31], India [32], and Iran [30]. The works on evaluating MSWEP for drought characterization have been limited because of the late release [16,30]. However, rather than original observations, the majority of existing evaluation works rely on interpolated grids or gridded reanalysis. The interpolation or re-grid process may introduce significant errors in gauge sparse regions.

This paper aims to inter-compare the performance of three LSPEs (i.e., PERSIANN-CDR, CHIRPS, and MSWEP) against ground-based gauges for meteorological drought monitoring in northwestern China from 1983 to 2013. The rest of this paper is organized as follows. Section 2 introduces the study area, datasets, and evaluation metrics. Section 3 focuses on the assessment of LSPEs in capturing precipitation and droughts. A summary and conclusions are given in Section 4.

2. Materials and Methods

2.1. Study Area

Northwestern China is located in the central part of Asia, in the mid-latitude regions of the northern hemisphere. The terrain of northwestern China varies greatly, with DEM ranging from <−190 m in the Turpan Basin to over 7900 m in the Kunlun Mountain (Figure 1). The majority of northwestern China is characterized by typical semi-arid and arid climates with relatively lower precipitation and limited water resources. The local climate is mainly influenced by the westerly winds from the mid-latitudes, which bring moisture from the North Atlantic Ocean [33]. Despite the high aridity, precipitation fluctuates a lot, which could lead to more drought events. Severe drought events occurred in 1983–1984, 1990–1991, 1997, 2000, 2010 [34–37]. Because of the region's high aridity and limited water resources, drought events have posed a serious threat to local agricultural production and natural ecosystems [38]. The great impacts of drought events on both society and the fragile ecology urge improvements in drought event monitoring and analysis.

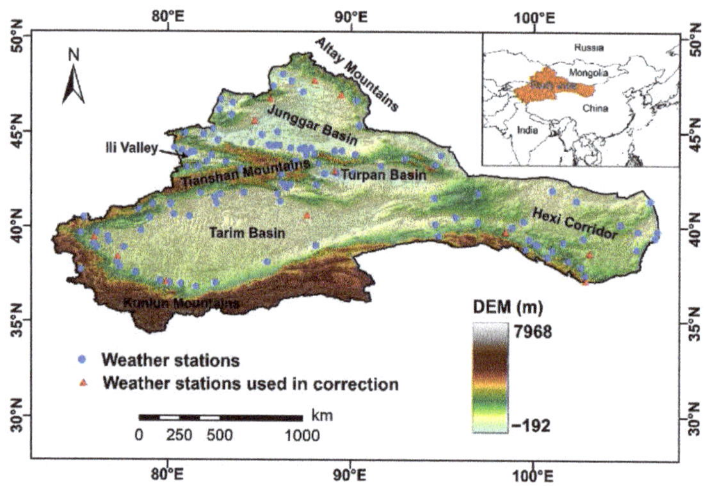

Figure 1. Topographic features and locations of weather stations in northwestern China.

2.2. Datasets

2.2.1. Ground-Based Observations

The spatial distribution of rain gauge locations is shown in Figure 1. The monthly precipitation data from 139 in situ rainfall gauges were collected from the National Meteorological Information Center (NMIC) and China Meteorological Administration (CMA, http://data.cma.cn/en, accessed on 1 October 2020) from 1983 to 2013. The evaluation period is selected based on the availability of both LSPEs and the gauge dataset. Seven of these gauges were eliminated because they had missing data for any of the 31 years. To ensure a relatively independent evaluation, 13 gauges involved in the Global Telecommunication System (GTS) were not considered in this study as they were used in the correction of the LSPEs' generation [39–42]. Despite the removal of GTS gauges, the current evaluation may not be completely independent, as some gauges may be involved in LSPEs' correction through other ways [39–41]. Finally, for the drought monitoring evaluation of LSPEs, precipitation observations from 119 in situ gauges marked with blue circles were used in this study. All of the in situ precipitation data used in this study went through a four-step quality control process: (1) Checking for extreme values, (2) checking for internal consistency, (3) checking for spatial consistency, and (4) checking for missing value [43]. The evaluation was conducted on a point scale to avoid potential errors introduced by the limited number and uneven distribution of rain gauges in this region. It is noted that some uncertainties may be introduced when the results of this study are extended to station-lacking regions, like the central part of the Tarim basin.

2.2.2. Long-Term Satellite Precipitation Datasets

From 1983 to 2013, three long-term satellite-based precipitation products are evaluated against the ground-based gauges for meteorological drought monitoring. The three long-term satellite-based precipitation products are summarized in Table 1.

Table 1. Summary of three long-term satellite-based precipitation products used in the study.

Dataset	Data Range	Highest Resolution	Coverage	Reference
PERSIANN-CDR V1R1	1983–current	0.25°-monthly	60° S–60° N	[40]
CHIRPS2.0	1983–current	0.05°-daily	60° S–60° N	[41]
MSWEP2.8	1979–current	0.1°-3 hourly	90° S–90° N	[39]

PERSIANN-CDR is developed by the U.S National Climatic Data Center (NCDC) Climate Data Record program. It covers 60° S–60° N and provides 0.25° daily satellite-based precipitation estimates from 1983 to the near present. The Artificial Neural Network algorithm was used to generate the uncorrected estimates using multiple infrared satellite data and the National Centers for Environmental Prediction (NCEP) stage IV hourly precipitation data. After that, the monthly product from the Global Precipitation Climatology Project (GPCP) was used to correct the initial estimates on a scale of 2.5 degrees. Finally, the corrected 2.5-degree estimates were downscaled to 0.25° to generate the final PERSIANN-CDR product [40]. The daily PERSIANN-CDR Version 1 Revision 1 (hereafter referred to as PERSIANN-CDR) data from 1983 to 2013 were obtained from the U.S. National Oceanic and Atmospheric Administration (NOAA) (https://www.ncei.noaa.gov/data/precipitation-persiann/access/, accessed on 15 August 2020). In this study, daily precipitation was accumulated to monthly to keep the consistency of gauge data and the other two LSPEs.

The CHIRPS is a quasi-global satellite and observation-based precipitation estimate. It has provided 0.05° precipitation estimates since 1981. It covers the land from 50°N to 50°S and is available from daily to annual time intervals. The generation of CHIRPS mainly includes two steps: the preliminary version of CHIRPS was generated based on multiple global cold cloud duration precipitation estimates with the correction of the Tropical Rainfall Measuring Mission Multi-Satellite Precipitation Analysis version 7 (TMPA 3B42 v7) and sparse World Meteorological Organization's Global Telecommunication System (GTS) gauge data; then, the final CHIRPS product was built by blending more available gauge datasets from around the world. Because of the different latency of obtaining various gauge data, the preliminary and final CHIRPS have a latency of ~2 days and ~3 weeks, respectively [41]. In this study, the monthly CHIRPS version 2 (hereafter referred to as CHIRPS) data from 1983 to 2013 were obtained from the University of California (https://www.chc.ucsb.edu/data/chirps, accessed on 10 August 2020).

MSWEP is a recently released global precipitation dataset from 1979 to the near-present with a spatial resolution of 0.1 degrees and temporal resolution of 3 hours [39,44]. The hourly, daily, monthly, and annual time scale data are available. MSWEP version 2 (hereafter referred to as MSWEP) is a current and improved version of MSWEP. It is a combination of several precipitation data resources, including satellite-based precipitation estimates, reanalysis estimates, and ground-based precipitation observations. The satellite-based estimates mainly include TRMM 3B42RT V7, GSMaP Moving Vector with Kalman products, Gridded Satellite (GridSat) B1 infrared data, and CMORPH estimates. Reanalysis includes European Centre for Medium-Range Weather Forecasts (ECMWF) interim reanalysis, Japanese 55-year Reanalysis (JRA-55), Global Precipitation Climatology Centre (GPCC) product, and WorldClim V2.0 monthly climatic dataset. It should be noted that final MSWEP estimates were generated using a daily gauge correction algorithm based on various gauge sources like Global Historical Climatology Network–Daily (GHCN-D) and Global Summary of Day (GSOD). In this study, the monthly MSWEP data from 1983 to 2013 were obtained from the official website (http://www.gloh2o.org/mswep/, accessed on 3 February 2021).

2.3. Statistics Evaluation Metrics

It is well known that there is a scale mismatch problem between the site-scale gauge and grid-scale satellite-based precipitation estimates. Conventionally, the rain gauges could be interpolated to the same scale with the SPEs based on different spatial interpolation techniques, such as the Inverse Distance Weighting (IDW) and Kriging method [45,46]. However, these interpolation methods may introduce large uncertainties and errors when the gauges are sparse and unevenly distributed with complex terrain [47,48]. Previous studies proved that spatial resolution differences may introduce small uncertainties on a monthly scale [47]. Therefore, ignoring the uncertainties from the different spatial resolutions of three LSPEs, the evaluation was conducted at a point scale without using any interpolation technique.

A series of conventional statistical metrics were used to evaluate the performance of three LSPEs in capturing monthly precipitation and drought characteristics. These statistical metrics used include relative bias (RB), Pearson linear correlation coefficient (CC) and root mean square error (RMSE). These statistical indices have been used in many evaluation studies [15,16,49]. The equations are listed in Appendix A.

2.4. Drought Index and Drought Characteristics

In this study, SPI [8] was used to assess the capability of both LSPEs to reflect the drought conditions and drought event characteristics against the ground observations. The drought index was used in this study for the following reasons: (1) The SPI is calculated solely based on precipitation variable, which is consistent with our aim of determining the ability of LRPEs to capture drought characterizations and the fact that other variables like radiation and relative humidity are difficult to collect in the arid zones of Central Asia. (2) SPI is a probability-based index in which the SPI value is transformed from a long-term mean condition's deviation degree [50]. Table A1 lists the SPI classes as well as their relative probabilities. (3) SPI is dimensionless and normalized, the SPI values in different regions could be directly compared [51,52]. (4) SPI can be used to investigate various types of droughts and quantify drought conditions with flexible timescales [53,54]. In this study, four typical timescales (i.e., 1-month, 3-month, 6-month and 12-month) are considered.

To better evaluate the potential of LSPEs for characterizing drought event features, the Run theory [55] was used to define drought events and quantify their characteristics like drought duration, intensity, and severity. Based on the study of McKee et al. [8], a drought event occurs when the SPI is negative, the peak value is less than -1, and the duration is greater than 2 months (Figure 1). Drought duration (DD) is the number of months between the start and end of a drought event. Drought severity (DS) is defined as the absolute cumulative SPI values during a drought event, which is the grey area in Figure 2. Drought intensity (DI) is the average value of absolute SPI during a drought event (the slash texture area).

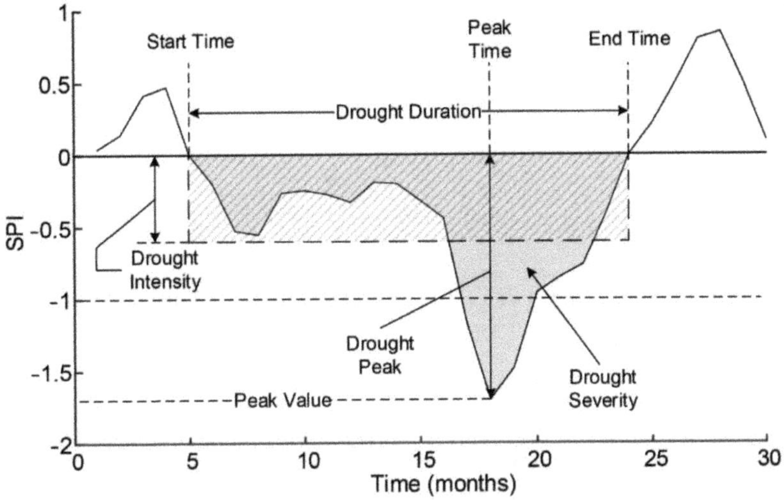

Figure 2. Definition of drought event and illustration of drought characteristics based on run theory.

2.5. Statistics for Drought Characteristic Evaluation

To quantify the ability of LSPEs to capture drought events, the probability of detection (*POD*), false alarm ratio (*FAR*), and missing event ratio (*MISS*) are used in the current study [56]. *POD* is used to quantify the hit ratio of drought events identified by both LSPEs and gauges. *FAR* is designed to measure the ratio of drought events falsely identified by LSPEs. The goal of *MISS* is to determine what percentage of drought events are detected by gauge but not by LSPEs [57]. These three categorical indices have been widely used in many studies [49,58,59]. The formulas are as follows:

$$POD(\%) = \frac{N(S = 1 \,\&\, G = 1)}{N(G = 1)} \times 100 \quad (1)$$

$$FAR(\%) = \frac{N(S = 1 \,\&\, G = 0)}{N(S = 1)} \times 100 \quad (2)$$

$$MISS(\%) = \frac{N(S = 0 \,\&\, G = 1)}{N(G = 1)} \times 100 \quad (3)$$

where S and G represent the LSPEs and Gauges, respectively. N means the number of drought events under different conditions. S = 1 (S = 0) indicates drought event identified by LSPE (or not); G = 1 (G = 0) illustrates a gauge-identified drought event (or not).

The mean value of drought duration (*MDD*), drought severity (*MDS*), and drought intensity (*MDI*) are also calculated throughout 1983–2013 to assess the ability of LSPEs to quantify drought characteristics. The following are the formulas for drought characteristics and their mean values:

$$MDD = \frac{\sum_{i=1}^{N} DD_i}{N} \quad (4)$$

$$MDS = \frac{\sum_{j=1}^{N} DS_j}{N},\ DS = \sum_{i=1}^{DD} |SPI_i| \quad (5)$$

$$MDI = \frac{\sum_{j=1}^{N} DI_j}{N},\ DI = \frac{\sum_{i=1}^{DD} |SPI_i|}{DD} \tag{6}$$

$$MDP = \frac{\sum_{j=1}^{N} DP_j}{N},\ DP = \max_{1 \le i \le DD} |SPI_i| \tag{7}$$

where i is a month during a drought event, SPI_i denotes the SPI value for the specified month i; N means the total number of drought events identified throughout the study period, and j is the index of the drought event.

3. Results

3.1. Evaluation of LSPE for Estimating Precipitation

The performance of LSPEs in estimating precipitation is firstly assessed by comparing their values with gauges. Both spatial pattern and temporal evolution are considered in this section.

Figure 3a–c present the spatial distribution of the multiyear mean precipitation derived from the three LSPEs and gauge stations during the period from 1983 to 2013. All three LSPEs could generally capture the spatial precipitation pattern. Because of its higher spatial resolution (0.05°), CHIRPS shows a smoother spatial precipitation pattern than MSWEP and PERSIANN-CDR. PERSIANN-CDR seems to have difficulty in accurately capturing local terrain precipitation, missing the high precipitation amounts in the Tianshan Mountain and the Kunlun Mountain area. This could be explained by the fact that cloud-top infrared estimates are used in the generation of PERSIANN-CDR [60,61]. CHIRPS performs best in terms of standard deviation, correlation, and RMSE, with the highest CC (0.89) and the lower RB (−0.02%) and RMSE (4.75 mm/month) (Figure 3b). PERSIANN-CDR tends to overestimate precipitation with RB of 6.78% (Figure 3a) while MSWEP tends to underestimate it by about 4.27% (Figure 3c).

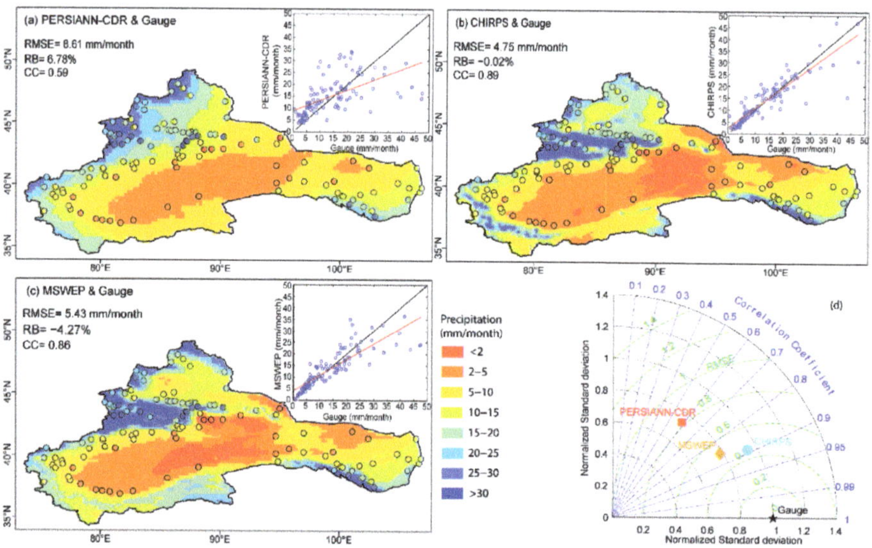

Figure 3. (**a**–**c**) Spatial pattern of multiyear monthly mean and the corresponding scatter plots for gauge and three satellite-based precipitation products; (**d**) Taylor diagram based on correlation, standard deviation, and RMSE of multiyear monthly mean precipitation between LSPEs and gauges. The blue circles represent pairs of monthly precipitation values between Gauge and LSPEs, while the black and red lines represent equally 1:1 and linear regression lines.

Temporal variation of precipitation is extremely significant for drought monitoring. To measure the performance of LSPEs in simulating temporal precipitation evolution throughout 1983–2013, the CC and RMSE are calculated at a point scale, as shown in Figure 4. In Xinjiang, all three LSPEs exhibit a high average CC ($\geq$0.63) and a low average RMSE (0.62), especially in the eastern Hexi corridor. PERSIANN-CDR and CHIRPS exhibit lower correlation and higher RMSE with gauges in both the Tianshan Mountain and the Kunlun Mountain area. Despite the better performance in replicating spatial patterns, CHIRPS shows some inferiority in capturing the temporal evolution of precipitation, with relatively lower CC (0.63) (Figure 4c). With average CC up to 0.73, MSWEP performed the best in simulating temporal precipitation variation.

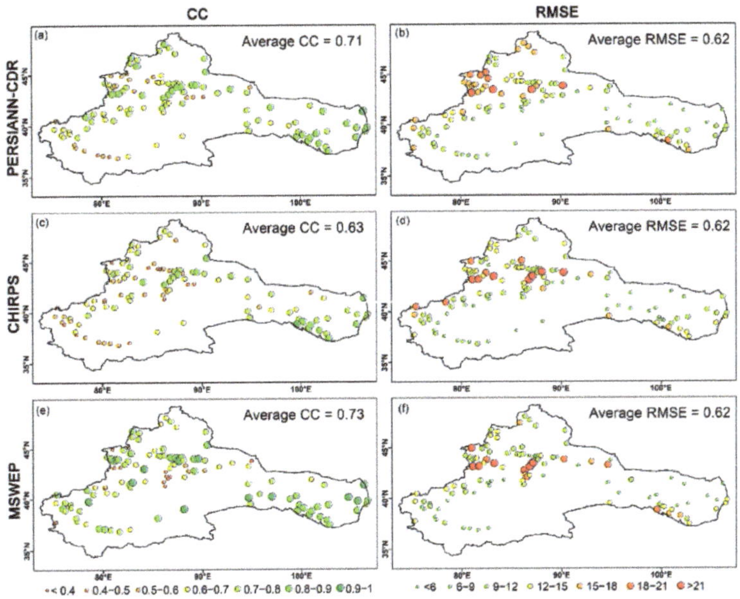

Figure 4. Spatial statistics of 31-year mean precipitation ranging from 1983 to 2013 derived from (**a**,**b**) PERSIANN-CDR, (**c**,**d**) CHIRPS, and (**e**,**f**) MSWEP over northwestern China.

3.2. Evaluation of LSPE for Estimating SPI

The accuracy of drought index estimation may directly affect the ability of drought characterization estimation. The point-scale SPI values based on LSPEs (SPI_{CDR}, SPI_{CHIRPS}, and SPI_{MSWEP}) and gauges (SPI_{Gauge}) are calculated at 1, 3, 6, and 12 months. The correlation (CC) and RMSE of SPI values based on LSPEs and gauges are shown in Figures 5 and 6.

As expected, SPI_{CHIRPS} has a relatively lower CC and a higher RMSE than those of SPI_{CDR} and SPI_{MSWEP}. This is mostly due to CHIRPS' inadequate ability to predict the temporal evolution of precipitation (Figure 4). Regardless of the time scales, SPI_{MSWEP} shows the highest correlation and lowest RMSE with SPI_{Gauge}, with average CC ranging from 0.66 for SPI12 to 0.72 for SPI1, among all three LSPEs. According to the CC and RMSE, there is no noticeable difference in performance for PERSIANN-CDR and CHIRPS across the four timescales. However, MSWEP's performance becomes poorer as the time scale increases. In the north Xinjiang region, PERSIANN-CDR seems to perform better in estimating SPI variation, while in the South Xinjiang and Hexi Corridor regions, it shows lower correlations with gauges. CHIRPS has relatively higher RMSE values in the Tianshan Mountain region.

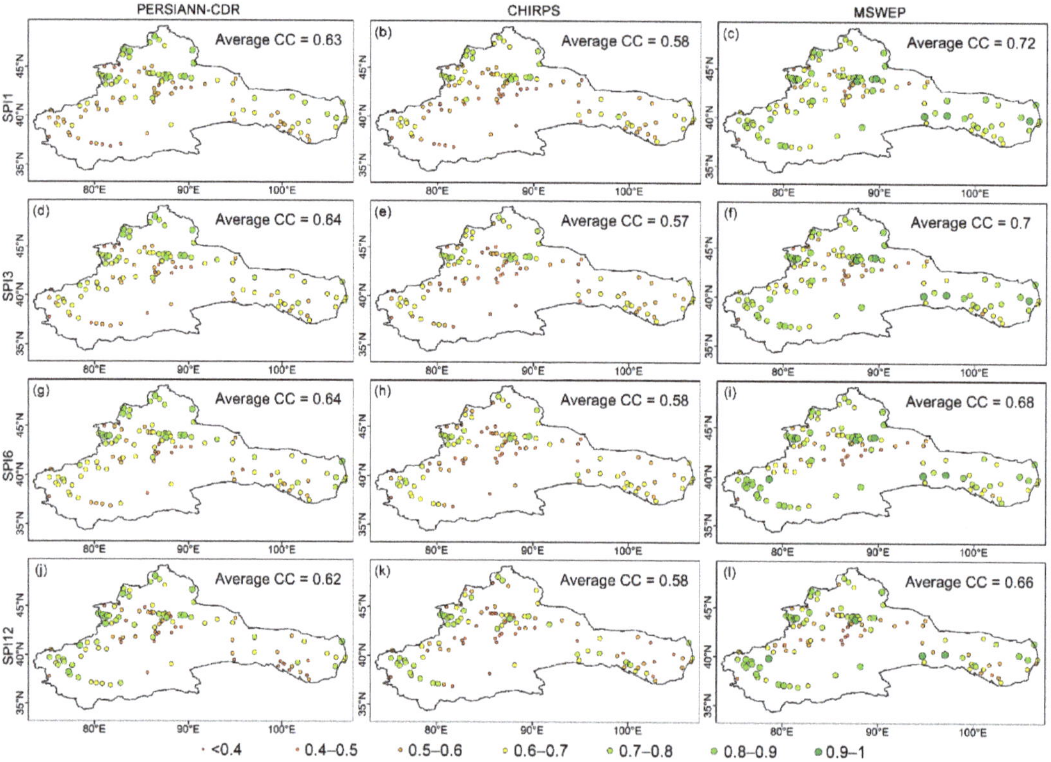

Figure 5. (**a**–**l**) Spatial distribution of CC between satellite products (i.e., PERSIANN_CDR, CHIRPS, and MSWEP) and gauge observations for SPI in four timescales.

Figure 7 shows the SPI temporal evolution of LSPEs and gauges. To quantify the uncertainties of LSPEs for capturing temporal variation in SPI, the RB, CC, and RMSE are also calculated for both wet (SPI > 0) and dry (SPI < 0) months.

It is revealed that the SPI values in short time scales vary frequently between positive and negative over time. This causes dry or wet events to be discontinuous, which is also found by other studies [37,62]. At multiple time scales, the performance of LSPEs in capturing SPI temporal evolution is generally satisfactory. Even though the frequently varied SPI values on short time scales, LSPEs could well capture the dry or wet fluctuations with some overestimation and underestimation (Figure 7). All three LSPEs show an overestimation during dry months, which decreases as the time scale increases.

MSWEP appears to perform better than the other two LSPEs in terms of reflecting dry or wet temporal evolution at multiple scales and has a higher correlation with gauges (CC > 0.8) during dry months. However, it also suffers from overestimation from 11.69% for SPI12 to 37.31% for SPI1. Short time scales have a higher SPI overestimation than long time scales, and vice versa. Surprisingly, MSWEP shows a great overestimation between 1992 and 1995, and the SPI overestimation is accumulated and becomes greater with the increase in time scale. Figure A1 shows an additional comparison of SPIs based on both LSPEs and gauges from 1992 to 1995. From 1992 to 1995, MSWEP greatly overestimates SPI against observations with RB from 51.56% for SPI1 to 1303.11% for SPI12 and CC lower than 0.64, while PERSIANN-CDR and CHIRPS appear to have performed normally with reasonable RB values (from −213% to 0%) and CC values (≥0.74). This is mainly because summer precipitation was consistently overestimated during these years. The anomalies during this period may have resulted in lower MSWEP and gauge correlations over longer time scales.

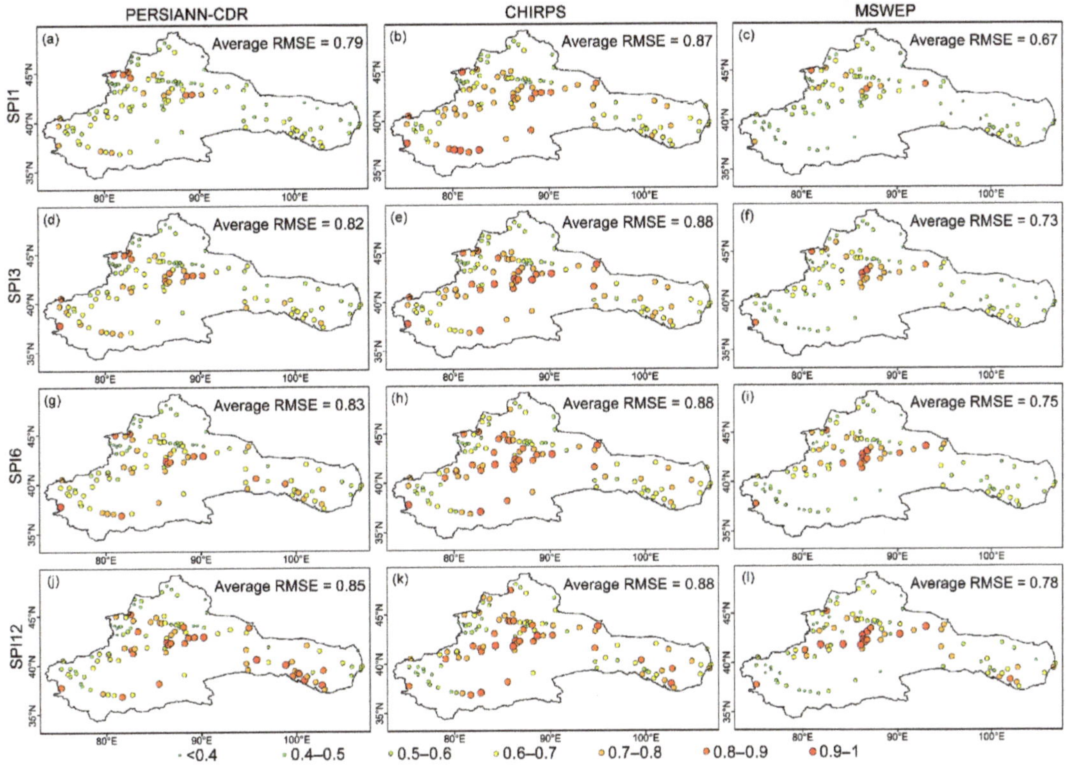

Figure 6. (**a–l**) Spatial distribution of RMSE between satellite products (i.e., PERSIANN_CDR, CHIRPS, and MSWEP) and gauge observations for SPI in four timescales.

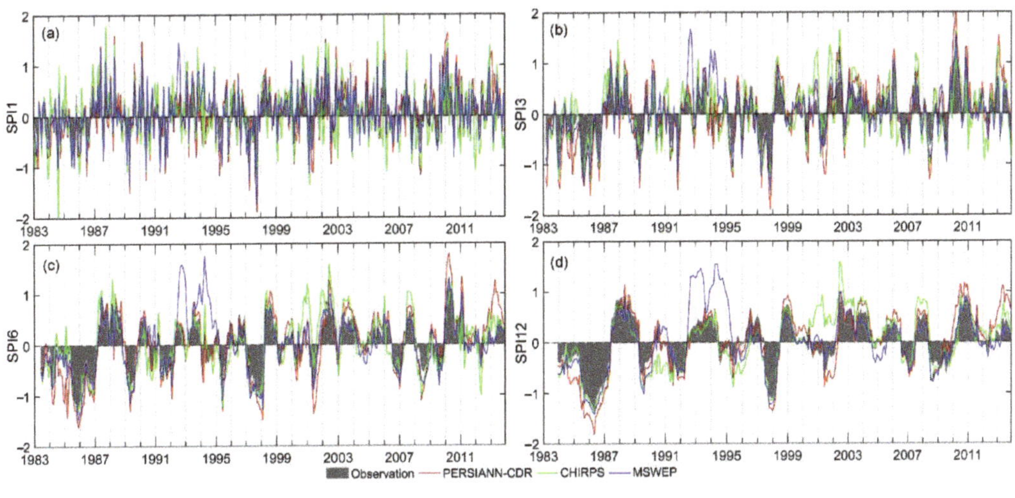

Figure 7. Temporal evolution of SPIs for LSPEs (i.e., PERSIANN-CDR, CHIRPS, MSWEP) and ground-based gauges. (**a**) SPI1; (**b**) SPI3; (**c**) SPI6 and (**d**) SPI12.

PERSIANN-CDR shows significant overestimation for both dry (SPI < 0) and wet (SPI > 0) months, with a relative bias of over 16% for wet conditions. The overestimation is more noticeable in dry months, with SPI1 at 66.34%, SPI3 at 42.21%, SPI6 at 33.99%, and SPI12 at 27.78%. CHIRPS shows a medium and stable performance among these three LSPEs. For dry months, the correlation between CHIRPS and gauges ranges from 0.63 to 0.86, with overestimation ranging from 19.78% (SPI12) to 62.96% (SPI1) (Table 2).

Table 2. Statistics between domain-averaged SPI between LSPEs and gauges from 1983 to 2013.

Scale	Name	Wet Months (SPI > 0)			Dry Months (SPI < 0)		
		RB (%)	CC	RMSE	RB (%)	CC	RMSE
SPI1	PERSIANN-CDR	16.72	0.86	0.20	66.34	0.81	0.31
	CHIRPS	12.05	0.70	0.29	62.96	0.63	0.34
	MSWEP	−3.14	0.81	0.20	37.31	0.89	0.20
SPI3	PERSIANN-CDR	23.87	0.85	0.22	42.21	0.84	0.29
	CHIRPS	17.98	0.49	0.34	21.07	0.74	0.24
	MSWEP	−0.10	0.66	0.26	17.93	0.89	0.16
SPI6	PERSIANN-CDR	24.27	0.82	0.22	33.99	0.84	0.27
	CHIRPS	16.15	0.54	0.31	12.66	0.67	0.25
	MSWEP	0.63	0.51	0.31	12.97	0.88	0.16
SPI12	PERSIANN-CDR	21.41	0.75	0.20	27.78	0.89	0.24
	CHIRPS	2.55	0.64	0.23	19.78	0.86	0.20
	MSWEP	−1.49	0.30	0.38	11.69	0.89	0.17

3.3. *Performance of LSPE in Estimating Drought Characteristics*

In this section, we would like to evaluate the performance of LSPEs in capturing drought events and estimating drought characteristics. The ability to capture drought events was assessed using categorical indexes (i.e., POD, FAR and MISS), whereas the ability to capture drought characteristics was assessed using the difference between three LSPEs and gauges for mean drought duration (MDD), mean drought severity (MDS), and mean drought intensity (MDI). To save space, only the 3-month SPI (SPI3), which is more suitable for identifying seasonal drought events [36,63], was selected as a representative case in the following sections.

Figure 8 depicts the spatial pattern of categorical indexes calculated between three LSPEs and gauges. The percentage of stations for each class is shown on the bar plots in the upper-right corner of each frame.

Three LSPEs have a good ability to capture drought events, with an average POD percentage of more than 60%. These LSPEs perform better in capturing drought events in the north of Xinjiang but have a lower POD score in the southern part. Among them, MSWEP has the highest average POD, at 66.34%, indicating significant superiority. MSWEP could capture more than 60% of drought events in more than 71% of stations. PERSIANN-CDR and CHIRPS perform poorly in capturing drought events with slightly lower POD values (61.46% and 60.13%). The FAR percentage is used to quantify the ratio of drought events that LSPEs falsely identified. The average false alarms of LSPEs are 27.93% for MSWEP, 37.04% for PERSIANN-CDR, and 37.32% for CHIRPS. High false alarms were found in the eastern Tianshan and southern Xinjiang regions. The MISS index is adopted to track drought events that are detected by gauges but missed by LSPEs. The average MISS percentage is greater than 30% for all three LSPEs. Similar to FAR, the high MISS percentages are mainly located in the eastern Tianshan region and the southern Xinjiang regions.

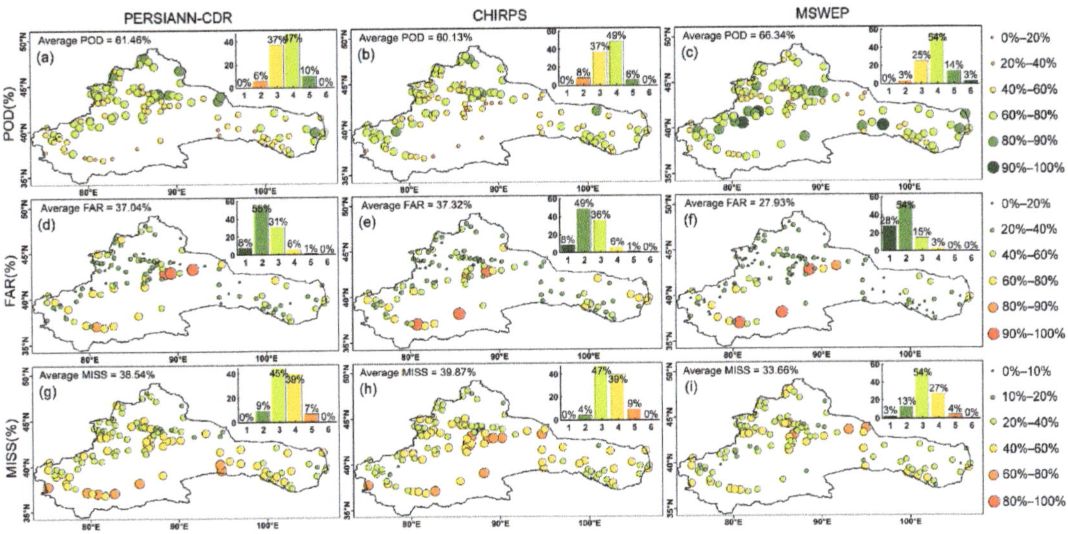

Figure 8. Spatial pattern of categorical indexes (**a–c**) POD, (**d–f**) FAR, (**g–i**) MISS.

Generally, MSWEP was found to have a higher ability in identifying drought events, with lower missing and false events based on POD, FAR and MISS percentages.

Figure 9 shows the difference in drought characteristics between LSPEs and gauges, including MDD, MDS, MDI, and MDP. The abilities to describe the duration, severity, intensity, and peak value of drought events are quantified using MDD, MDS, MDI, and MDP differences, respectively.

Three LSPEs tend to overestimate drought duration for more than half of the stations. PERSIANN-CDR has the best agreement with gauges for drought duration, with an average value of 0.25 months. About 26% of stations show an MDD overestimation for more than a month (Figure 9a). MSWEP's ability to determine the length of drought is limited. Drought events identified by MSWEP last 1 month longer in 40% of stations than gauges (Figure 9c). Similarly, all LSPEs overestimate drought severity in the majority of stations, with an average MDS overestimation of 0.31 for CHIRPS, 0.64 for MSWEP, and 0.85 for PERSIANN-CDR. The overestimation of drought severity can largely be attributed to an overestimation of drought duration (Figure 9d–f). Drought intensity is defined as the ratio of drought severity to drought duration. All LSPEs seem to have a lower drought intensity error with an average MDI difference of less than 0.05. This also suggests that drought severity overestimation is mainly caused by an overestimation of drought duration. From the bar plots, we can find that more than 60% of stations have an absolute MDI difference of less than 0.1 for three LSPEs. Among these three LSPEs, the amplitudes of overestimation for PERSIANN-CDR are relatively higher, especially for stations in the Tianshan Mountain area (Figure 9g). The drought peak value is a measurement of a drought's most severe month during one drought event. The spatial pattern of the mean drought peak value (MDP) is illustrated in Figure 9j–l. PERSIANN-CDR seems to have greatly overestimated the drought peak value of drought events (Figure 9j). This phenomenon is consistent with the results shown in Figure 9. More than 70% of stations show an overestimation of drought peak value, and 37% of stations show an overestimation of more than 0.2. However, CHIRPS tends to underestimate drought peak value in 75% of stations, with an average MDP of 0.15 (Figure 9k). MSWEP shows the best performance in estimating drought peak values. The absolute MDP difference is less than 0.1 in about half stations (Figure 9l).

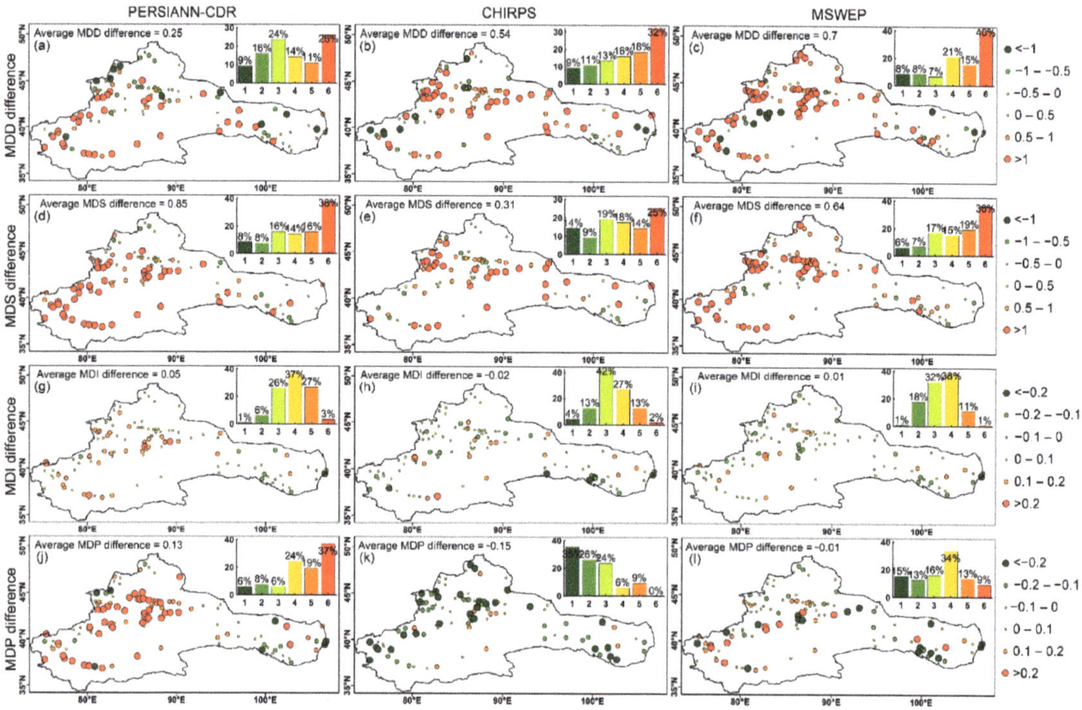

Figure 9. Spatial distribution of drought characteristic difference between LSPEs and gauge for (**a–c**) MDD, (**d–f**) MDS, (**g–i**) MDI and (**j–l**) MDP.

Generally, LSPEs tend to overestimate drought severity, which is mainly due to an overestimation of drought duration. MSWEP has a higher ability to predict drought peak values. PERSIANN-CDR tends to overestimate drought peak values, while the situation is the opposite for CHIRPS, with significant underestimation of drought peak values.

3.4. Performance of LSPE in Estimating Specific Drought Events

Drought events were identified using spatially averaged 3-month SPI values, and four drought events were selected based on various features to investigate the performance of three LSPEs during specific drought events. The specific drought characteristics of these three drought events are given in Table 3. The following is an illustration of the specific conditions of drought events, as well as a comparison of LSPEs and gauges.

The first was the most severe drought in recorded history during the study period, occurred between December 1984 and November 1986, and has been reported in some studies [31,34,64]. In September 1985, the drought event reached a peak value of 1.28 with a long duration (DD = 24 months), and high severity (DS = 11.88) and intensity (DI = 0.61). Both PERSIANN-CDR and CHIRPS overestimate the drought duration by 5 and 4 months, respectively. This is mainly due to the fact that they correctly predicted the drought's end time (November 1986), but predicted the drought's start time to be earlier. Due to their overestimation of duration, PERSIANN-CDR and CHIRPS both overestimated drought severity by 84.6% and 16%, respectively. PERSIANN-CDR also overestimated the drought intensity (25%) and its peak value (26%). MSWEP captured the start, end, and peak time accurately, with the same duration of 24 months. Despite an overestimation of drought severity (21%), it accurately captures the drought's intensity and peak value.

Table 3. Features of selected drought events. Drought characteristics based on the gauge are highlighted in bold font. The overestimated values were marked in red color while underestimated values were marked in blue color.

Event Index	Product	Start Time–End Time (yyyymm–yyyymm)	Peak Time (yyyymm)	DD (Months)	DS	DI	DP
DE1	Gauge	198412–198611	198509	24	11.88	0.61	1.28
	PERSIANN-CDR	198407–198611	198602	29	21.94	0.76	1.61
	CHIRPS	198408–198611	198602	28	13.79	0.54	1.35
	MSWEP	198412–198611	198509	24	14.36	0.60	1.35
DE2	Gauge	199105–199107	199107	3	1.83	0.61	0.66
	PERSIANN-CDR	199104–199107	199105	4	2.49	0.62	0.94
	CHIRPS	199105–199107	199106	3	2.19	0.73	0.88
	MSWEP	199104–199107	199106	4	1.99	0.50	0.67
DE3	Gauge	199701–199712	199711	12	8.03	0.67	1.24
	PERSIANN-CDR	199701–199802	199711	14	10.84	0.77	1.90
	CHIRPS	199701–199712	199711	12	9.22	0.77	1.21
	MSWEP	199701–199712	199711	12	9.71	0.81	1.50
DE4	Gauge	200606–200703	200610	10	3.43	0.34	0.86
	PERSIANN-CDR	200605–200702	200610	10	3.25	0.33	0.77
	CHIRPS	200604–200704	200702	13	5.33	0.41	0.80
	MSWEP	200605–200703	200610	11	3.61	0.32	0.84

The second drought event, which lasted from May to July 1991, was short (only 3 months) but intense, with a drought intensity of 0.61 and a peak value of 0.66. Despite its low severity (1.83), the summer drought of 1991 was reported to have a significant impact on local agriculture in Xinjiang [35]. All three LSPEs overestimated drought severity and peak value, with different magnitudes of overestimation. CHIRPS accurately captured the drought duration, while PERSIANN-CDR and MSWEP detected it as being a month longer. In estimating drought severity and peak values, PERSIANN-CDR had the worst performance with 36% and 42% overestimation, respectively. The MSWEP showed the closest drought severity and peak values with gauges.

The third drought event (January–December 1997) was characterized by its long length and high severity. The drought event had a relatively longer duration (12 months), as well as high severity (8.03) and high intensity (0.67), and has also been reported in other studies [65]. Both CHIRPS and MSWEP could accurately identify the drought start and end time, as well as its peak time. PERSIANN-CDR identified the correct start and peak times, but a later ending time, resulting in a 3-month overestimation of drought duration and a 35% overestimation of drought severity. Despite accurate drought time detection, CHIRPS and MSWEP also overestimated drought severity, intensity, and peak value. CHIRPS performed better in terms of quantifying drought characteristics during this drought event.

The fourth drought event was selected because of its low severity (3.43) and intensity (0.34), but a relatively long duration (10 months). Drought conditions were reported to have lasted more than half a year in 70% of Xinjiang stations during the drought [66]. PERSIANN-CDR showed a slight underestimation of drought severity, intensity, and peak value. This phenomenon could be related to the accurate estimation of drought duration. CHIRPS overestimated drought duration by 3 months, severity by 55%, and intensity by 21%, but drought peak values were underestimated. MSWEP showed the best performance, with a 5% overestimation of severity, a 5% underestimation of intensity, a 2% underestimation of peak values, and a 1-month longer duration.

Overall, LSPEs tend to overestimate drought severity, owing to a failure to capture the start and end of the drought time. MSWEP has better performance in estimating severity, intensity, and peak values. PERSIANN-CDR seems to have relatively poor performance, with systematic overestimation of drought severity, intensity, and peak values.

The spatial distribution of SPI on drought peak time for the four selected drought events studied is shown in Figure 10. To easily quantify the performances of LSPEs, the station ratio under different dry conditions for both LSPE and gauge was also calculated and shown in the concentric pie chart. Please note that the pie chart on the outside represents the station percentage for gauge, while the inner pie chart is the station percentage for LSPE. It is noted that only stations and corresponding grids from LSPEs were calculated in this section.

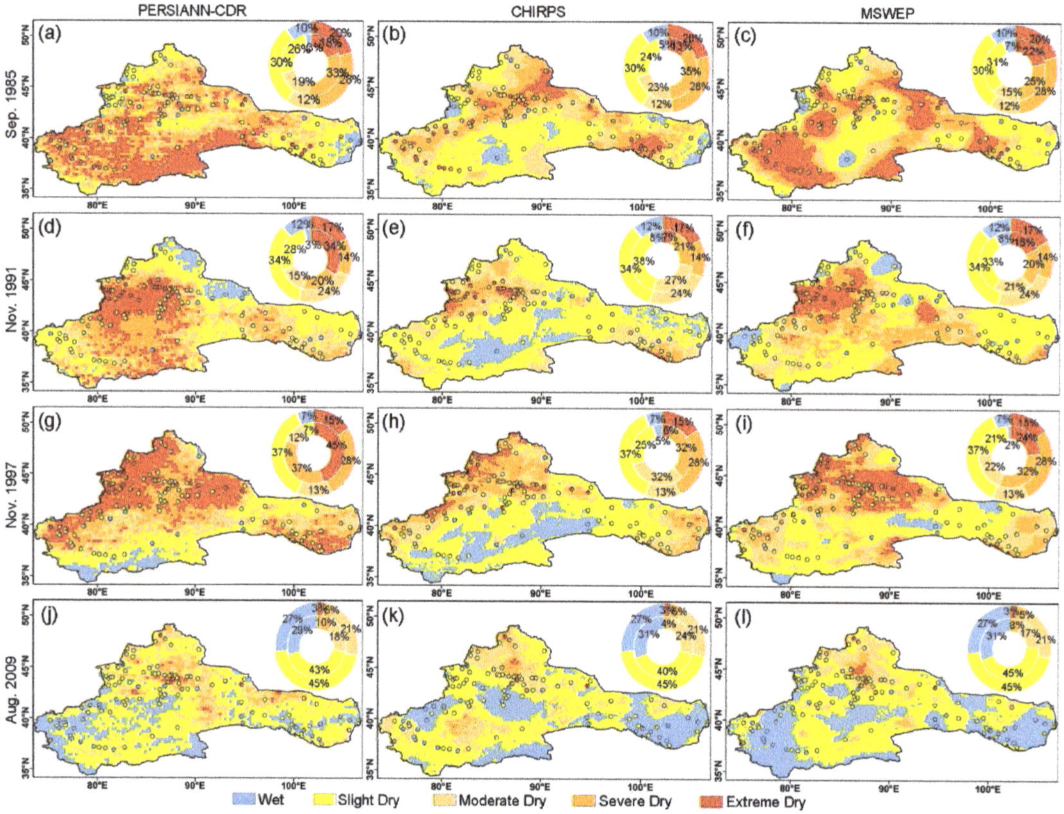

Figure 10. (**a–l**) Spatial patterns of SPI from the gauge and three satellite-based precipitation products for the peak time of four selected drought events. The outside and inside concentric pie charts show the percentage of stations and grids under different dry intensities, respectively.

For severe drought events, PERSIANN-CDR appears to catch larger drought areas, which is mainly due to the fact that some wet stations were falsely identified as experiencing drought conditions. The overestimation of extreme drought areas was revealed in November 1991 and November 1997, when 17% and 30% of stations, respectively, were incorrectly identified as being in extreme drought conditions. CHIRPS, on the other hand, tends to miss some extreme drought regions for all four selected drought events. The underestimation of extreme drought areas can be clearly identified in the drought months of September 1985 (Figure 10b), November 1991 (Figure 10e), and November 1997 (Figure 10h). MSWEP appears to have closer percentages with gauges, indicating that it is better at capturing extreme drought conditions than PERSIANN-CDR and CHIRPS. However, the dry conditions (SPI < 0) were also overestimated. During the first three severe drought events,

LSPEs performed well in capturing drought areas but missed 3% of extreme drought areas. (Figure 10j–l).

Generally, all three LSPEs were able to capture the overall drought pattern; however, there is still a lot of room for improvement when it comes to capturing drought patterns of varying intensities, especially for more severe drought events. MSWEP performs better than other LSPEs in terms of capturing drought spatial patterns.

4. Discussion

Drought index is important in drought monitoring and assessments [67,68]. Many drought indices have been developed, like SPI, Standardized Precipitation Evapotranspiration Index (SPEI) [69,70], Palmer Drought Severity Index (PDSI) [71], and so on [72]. Many studies have reviewed the characteristics of existing drought indices from different perspectives [17,72–75]. Each drought index has its own set of advantages and weaknesses. The researchers or users could select the drought index based on objective conditions and their specific needs. For example, the objective of this study is to evaluate LSPEs' performance for meteorological drought monitoring compared to the valuable ground-based rainfall observations. The SPI was selected for this study because of its multiple advantages, including ease of calculation, applicability for both wet and dry conditions, ease of comparison in different regions, and flexibility of time scales [8,54,76]. SPI is solely based on precipitation without considering other hydrological parameters like potential evapotranspiration (PET). We did not use a drought index that includes PET, such as SPEI, because PET calculation is complex and unstable, and many other parameters (e.g., wind speed, relative humidity, solar radiation, etc.) involved in PET calculation are unavailable in this region [77]. Therefore, this analysis is based on the assumption that no other parameter's effects were considered during the study period. Due to the role of the precipitation parameter in the calculation of various drought indices varying, the results may differ when another drought index is used. When studying the meteorological drought changes by considering the temperature effects, users may consider drought indices like SPEI or PDSI.

5. Conclusions

This study conducted an evaluation of three long-term satellite-based precipitation estimates in simulating precipitation and estimating drought features from 1983 to 2013 against ground-based precipitation observations. Our main findings are summarized below:

All three LSPEs show comparable performance in simulating precipitation variation, replicating dry evolution, identifying drought events, and characterizing drought features. However, we discovered that in dry months, these LSPEs tend to overestimate dry magnitude and have a relatively high false alarm rate (>27%) and missing rate (>33%). They tend to overestimate drought severity, owing to an overestimation of drought duration. LSPEs have advantages in capturing drought peak times, but they tend to overestimate drought peak values.

CHIRPS performs better at capturing precipitation patterns with high spatial resolution, but it has a weakness in estimating temporal variation; the weakness of CHIRPS in accurately capturing the temporal evolution of precipitation results in a clear lack of ability to replicate SPI variations.

PERSIANN-CDR tends to overestimate precipitation in many regions and performs poorly in simulating precipitation spatial patterns. When precipitation is overestimated, SPI values for dry months and drought characteristics (like drought severity and peak value) are overestimated. There are also obvious overestimations of the drought area in specific severe drought events.

MSWEP performs admirably when it comes to estimating precipitation spatial pattern and temporal variation. It also demonstrates marked superiority over the other two products in terms of capturing drought events, with a high probability of detection (66%)

and a lower false alarm and missing ratio. It proves a higher ability in estimating event characteristics like drought peak value, severity, and area.

The strengths and weaknesses of LSPEs identified in this study could provide useful scientific evidence for users of satellite precipitation products in data selection and error tracking. Furthermore, they could be useful for the use of satellite-based precipitation products in drought monitoring applications, especially for areas with sparse rain-gauge stations. Moreover, the results of this study may also assist algorithm developers to improve the performance of LSPE generating algorithms.

Author Contributions: Conceptualization, P.D.M.; methodology and software, H.G. and M.L.; validation, X.M. and L.Z.; formal analysis, V.N.; investigation, H.G.; data curation, H.G.; writing—original draft preparation, H.G.; writing—review and editing, H.G. and V.N.; supervision, A.B.; funding acquisition, H.G. All authors have read and agreed to the published version of the manuscript.

Funding: This research was jointly funded by the National Natural Science Foundation of China (Grant No. 42001363), the Third Xinjiang Scientific Expedition Program (Grant No. 2021XJKK140304), the New Water Resources Strategy Research Project in NJ (403-1005-YBN-FT6I-8).

Institutional Review Board Statement: Not applicable.

Informed Consent Statement: Not applicable.

Data Availability Statement: Precipitation Estimation from Remotely Sensed Information using Artificial Neural Networks–Climate Data Record (PERSIANN-CDR) was retrieved from https://climatedataguide.ucar.edu/climate-data/persiann-cdr-precipitation-estimation-remotely-sensed-information-using-artificial (accessed on 15 August 2020); Rainfall Estimates from Rain Gauge and Satellite Observations (CHIRPS) is available on the website of University of California (https://www.chc.ucsb.edu/data/chirps, accessed on 10 August 2020) and Multi-Source Weighted-Ensemble Precipitation (MSWEP) can be obtained from http://www.gloh2o.org/mswep/ (accessed on 3 February 2021).

Acknowledgments: The authors thank the anonymous reviewers for their constructive comments and suggestions that greatly improved this paper.

Conflicts of Interest: The authors declare no conflict of interest.

Appendix A

Statistics Evaluation Metrics

$$RB = \sum_{i=1}^{N}(S_i - G_i) / \sum_{i=1}^{N}(G_i) \times 100 \tag{A1}$$

$$CC = \left(\sum_{i=1}^{N}(S_i - \overline{S})(G_i - \overline{G})\right) / \left(\sqrt{\sum_{i=1}^{N}(S_i - \overline{S})^2}\sqrt{\sum_{i=1}^{N}(G_i - \overline{G})^2}\right) \tag{A2}$$

$$RMSE = \sqrt{\frac{1}{N}\sum_{i=1}^{N}(S_i - G_i)^2} \tag{A3}$$

where S and G represent LSPE and gauge, i means the sample number for evaluated variables like precipitation or SPI values.

Appendix B

Table A1. Dry/wet classification based on SPI values.

SPI Value	Category	Probability (%)
>2	Extremely wet	2.3
1.5~1.99	Severe wet	4.4
1.0~1.49	Moderately wet	9.2
0~0.99	Slightly wet	34.1
−0.99~0	Slightly dry	34.1
−1~−1.49	Moderately dry	9.2
−1.5~−2	Severe dry	4.4
<−2	Extremely dry	2.3

Appendix C

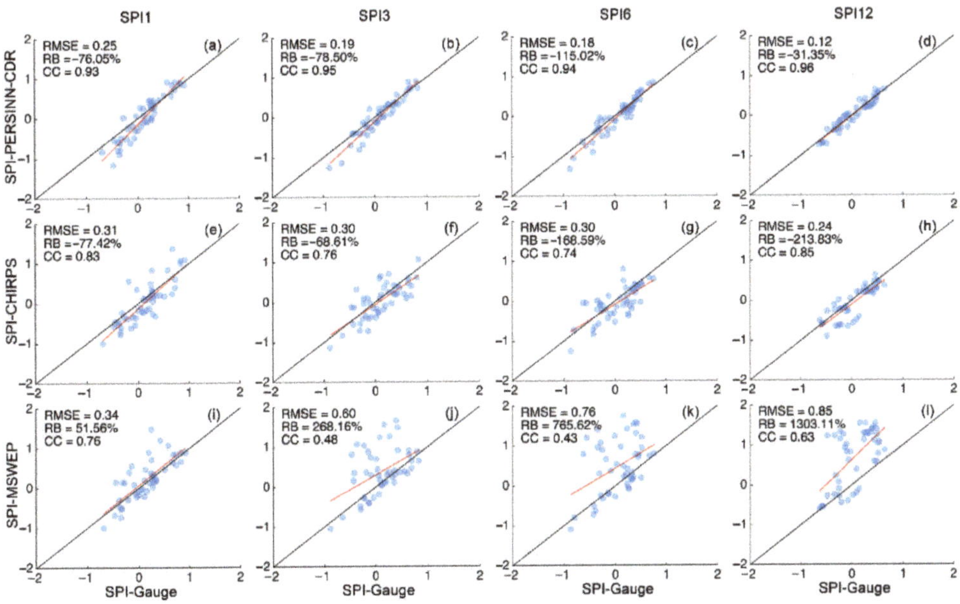

Figure A1. (a–l) Comparison of SPI between gauge observations and LSPEs from 1992 to 1995. The blue points represent pairs of SPI values between Gauge and LSPEs, while the black and red lines represent equally 1:1 and linear regression lines.

References

1. IPCC. *Managing the Risks of Extreme Events and Disasters to Advance Climate Change Adaptation*; Field, C.B., Barros, V., Stocker, T.F., Qin, D., Dokken, D.J., Ebi, K.L., Mastrandrea, M.D., Mach, K.J., Plattner, G.-K., Allen, S.K., et al., Eds.; Cambridge University Press: Cambridge, UK, 2012; 582p.
2. Field, C.B.; Barros, V.R.; Dokken, D.J.; Mach, K.J.; Mastrandrea, M.D.; Bilir, T.E.; Chatterjee, M.; Ebi, K.L.; Estrada, Y.O.; Genova, R.C. (Eds.) *Climate Change 2014: Impacts, Adaptation, and Vulnerability. Part A: Global and Sectoral Aspects. Contribution of Working Group II to the Fifth Assessment Report of the Intergovernmental Panel on Climate Change*; Cambridge University Press: Cambridge, UK, 2014; 1132p.
3. Patrick, E. Drought Characteristics and Management in Central Asia and Turkey. In *FAO Water Reports (FAO)*; Food and Agriculture Organization of the United Nations: Rome, Italy, 2017; 114p.
4. Xu, H.J.; Wang, X.P.; Zhang, X.X. Decreased Vegetation Growth in Response to Summer Drought in Central Asia from 2000 to 2012. *Int. J. Appl. Earth Obs. Geoinf.* **2016**, 52, 390–402. [CrossRef]

5. Montaseri, M.; Amirataee, B. Comprehensive Stochastic Assessment of Meteorological Drought Indices. *Int. J. Climatol.* **2017**, *37*, 998–1013. [CrossRef]
6. Kamali, B.; Abbaspour, K.C.; Lehmann, A.; Wehrli, B.; Yang, H. Identification of Spatiotemporal Patterns of Biophysical Droughts in Semi-Arid Region—A Case Study of the Karkheh River Basin in Iran. *Hydrol. Earth Syst. Sci.* **2015**, *12*, 5187–5217.
7. Liu, X.; Xiufang, Z.; Yaozhong, P.; Shuangshuang, L.; Yanxu, L.; Yuqi, M. Agricultural Drought Monitoring: Progress, Challenges, and Prospects. *J. Geogr. Sci.* **2016**, *26*, 750–767. [CrossRef]
8. McKee, T.B.; Doesken, N.J.; Kleist, J. The Relationship of Drought Frequency and Duration to Time Scales. In Proceedings of the Eighth Conference on Applied Climatology, Anaheim, CA, USA, 17–22 January 1993.
9. Svoboda, M.; Hayes, M.; Wood, D. Standardized Precipitation Index User Guide. In *Technical Report WMO—No. 1090*; Hayes, M., Wood, D., Svoboda, M., Eds.; WMO: Geneva, Switzerland, 2012; 24p.
10. Shi, L.; Feng, P.; Wang, B.; Li Liu, D.; Yu, Q. Quantifying Future Drought Change and Associated Uncertainty in Southeastern Australia with Multiple Potential Evapotranspiration Models. *J. Hydrol.* **2020**, *590*, 125394. [CrossRef]
11. Kumar, K.; Krishna, K.; Kumar, R.; Rakhecha, P.R. Comparison of Penman and Thornthwaite Methods of Estimating Potential Evapotranspiration for Indian Conditions. *Theor. Appl. Climatol.* **1987**, *38*, 140–146. [CrossRef]
12. Valipour, M. Temperature Analysis of Reference Evapotranspiration Models. *Meteorol. Appl.* **2015**, *22*, 385–394. [CrossRef]
13. Damberg, L.; AghaKouchak, A. Global Trends and Patterns of Drought from Space. *Theor. Appl. Climatol.* **2014**, *117*, 441–448. [CrossRef]
14. Thenkabail, P. *Remote Sensing of Water Resources, Disasters, and Urban Studies*; CRC Press: Boca Raton, FL, USA, 2015.
15. Suliman, A.H.A.; Awchi, T.A.; Al-Mola, M.; Shahid, S. Evaluation of Remotely Sensed Precipitation Sources for Drought Assessment in Semi-Arid Iraq. *Atmos. Res.* **2020**, *242*, 105007. [CrossRef]
16. Xu, Z.; Wu, W.; He, H.; Wu, H.; Zhou, J.; Zhang, Y.; Guo, X. Evaluating the Accuracy of Mswep V2.1 and Its Performance for Drought Monitoring over Mainland China. *Atmos. Res.* **2019**, *226*, 17–31. [CrossRef]
17. AghaKouchak, A.; Farahmand, A.; Melton, F.S.; Teixeira, J.; Anderson, M.C.; Wardlow, B.D.; Hain, C.R. Remote Sensing of Drought: Progress, Challenges and Opportunities. *Rev. Geophys.* **2015**, *53*, 452–480. [CrossRef]
18. Agutu, N.O.; Awange, J.L.; Zerihun, A.; Ndehedehe, C.E.; Kuhn, M.; Fukuda, Y. Assessing Multi-Satellite Remote Sensing, Reanalysis, and Land Surface Models' Products in Characterizing Agricultural Drought in East Africa. *Remote Sens. Environ.* **2017**, *194*, 287–302. [CrossRef]
19. Thavorntam, W.; Tantemsapya, N.; Armstrong, L. A Combination of Meteorological and Satellite-Based Drought Indices in a Better Drought Assessment and Forecasting in Northeast Thailand. *Nat. Hazards* **2015**, *77*, 1453–1474. [CrossRef]
20. AghaKouchak, A.; Nakhjiri, N. A near Real-Time Satellite-Based Global Drought Climate Data Record. *Environ. Res. Lett.* **2012**, *7*, 044037. [CrossRef]
21. Guo, H.; Chen, S.; Bao, A.; Hu, J.; Gebregiorgis, A.S.; Xianwu, X.; Zhang, X. Inter-Comparison of High-Resolution Satellite Precipitation Products over Central Asia. *Remote Sens.* **2015**, *7*, 7181–7211. [CrossRef]
22. De Jesús, A.; Breña-Naranjo, J.A.; Pedrozo-Acuña, A.; Alcocer Yamanaka, V.H. The Use of Trmm 3b42 Product for Drought Monitoring in Mexico. *Water* **2016**, *8*, 325. [CrossRef]
23. Naumann, G.; Barbosa, P.; Carrao, H.; Singleton, A.; Vogt, J. Monitoring Drought Conditions and Their Uncertainties in Africa Using Trmm Data. *J. Appl. Meteorol. Climatol.* **2012**, *51*, 1867–1874. [CrossRef]
24. Yan, N.; Wu, B.; Chang, S.; Bao, X. Evaluation of Trmm Precipitation Product for Meteorological Drought Monitoring in Hai Basin. *IOP Conf. Ser. Earth Environ. Sci.* **2014**, *17*, 12093. [CrossRef]
25. Zeng, H.W.; Li, L.J.; Li, J.Y. The Evaluation of Trmm Multisatellite Precipitation Analysis (Tmpa) in Drought Monitoring in the Lancang River Basin. *J. Geogr. Sci.* **2012**, *22*, 273–282. [CrossRef]
26. Tan, M.; Tan, K.; Chua, V.; Chan, N. Evaluation of Trmm Product for Monitoring Drought in the Kelantan River Basin, Malaysia. *Water* **2017**, *9*, 57. [CrossRef]
27. Gao, F.; Zhang, Y.; Ren, X.; Yao, Y.; Hao, Z.; Cai, W. Evaluation of Chirps and Its Application for Drought Monitoring over the Haihe River Basin, China. *Nat. Hazards* **2018**, *92*, 155–172. [CrossRef]
28. Guo, H.; Bao, A.; Liu, T.; Ndayisaba, F.; He, D.; Kurban, A.; De Maeyer, P. Meteorological Drought Analysis in the Lower Mekong Basin Using Satellite-Based Long-Term Chirps Product. *Sustainability* **2017**, *9*, 901. [CrossRef]
29. Oliveira-Júnior, J.F.; Silva Junior, C.A.; Teodoro, P.E.; Rossi, F.S.; Blanco, C.C.J.; Lima, M.; Gois, G.; Correira Filho, W.L.F.; Barros Santiago, D.; Santos Vanderley, M.H.G. Confronting Chirps Dataset and in Situ Stations in the Detection of Wet and Drought Conditions in the Brazilian Midwest. *Int. J. Climatol.* **2021**, *41*, 478–493. [CrossRef]
30. Alijanian, M.; Rakhshandehroo, G.R.; Mishra, A.; Dehghani, M. Evaluation of Remotely Sensed Precipitation Estimates Using Persiann-Cdr and Mswep for Spatio-Temporal Drought Assessment over Iran. *J. Hydrol.* **2019**, *579*, 124189. [CrossRef]
31. Guo, H.; Bao, A.; Liu, T.; Chen, S.; Ndayisaba, F. Evaluation of Persiann-Cdr for Meteorological Drought Monitoring over China. *Remote Sens.* **2016**, *8*, 379. [CrossRef]
32. Santos, C.A.G.; Neto, R.M.B.; do Nascimento, T.V.M.; da Silva, R.M.; Mishra, M.; Frade, T.G. Geospatial Drought Severity Analysis Based on Persiann-Cdr-Estimated Rainfall Data for Odisha State in India (1983–2018). *Sci. Total Environ.* **2021**, *750*, 141258. [CrossRef]
33. Bothe, O.; Fraedrich, K.; Zhu, X. Precipitation Climate of Central Asia and the Large-Scale Atmospheric Circulation. *Theor. Appl. Climatol.* **2012**, *108*, 345–354. [CrossRef]

34. Mahmood, R.; ShuangLin, L.; Babar, K. Causes of Recurring Drought Patterns in Xinjiang, China. *J. Arid. Land* **2010**, *2*, 279–285.
35. Li, Y.; Yao, N.; Sahin, S.; Appels, W.M. Spatiotemporal Variability of Four Precipitation-Based Drought Indices in Xinjiang, China. *Theor. Appl. Climatol.* **2016**, *129*, 1017–1034. [CrossRef]
36. Guo, H.; Bao, A.; Ndayisaba, F.; Liu, T.; Jiapaer, G.; El-Tantawi, A.M.; De Maeyer, P. Space-Time Characterization of Drought Events and Their Impacts on Vegetation in Central Asia. *J. Hydrol.* **2018**, *564*, 165–178. [CrossRef]
37. Guo, H.; Bao, A.; Liu, T.; Jiapaer, G.; Ndayisaba, F.; Jiang, L.; Kurban, A.; De Maeyer, P. Spatial and Temporal Characteristics of Droughts in Central Asia During 1966–2015. *Sci. Total Environ.* **2018**, *624*, 1523–1538. [CrossRef] [PubMed]
38. Wu, Y.; Bake, B.; Zhang, J.; Rasulov, H. Spatio-Temporal Patterns of Drought in North Xinjiang, China, 1961–2012 Based on Meteorological Drought Index. *J. Arid. Land* **2015**, *7*, 527–543. [CrossRef]
39. Beck, H.E.; Wood, E.F.; Pan, M.; Fisher, C.K.; Miralles, D.G.; van Dijk, A.I.J.M.; McVicar, T.R.; Adler, R.F. Mswep V2 Global 3-Hourly 0.1° Precipitation: Methodology and Quantitative Assessment. *Bull. Am. Meteorol. Soc.* **2019**, *100*, 473–500. [CrossRef]
40. Ashouri, H.; Hsu, K.L.; Sorooshian, S.; Braithwaite, D.K.; Knapp, K.R.; Cecil, L.D.; Nelson, B.R.; Prat, O.P. Persiann-Cdr Daily Precipitation Climate Data Record from Multisatellite Observations for Hydrological and Climate Studies. *Bull. Am. Meteorol. Soc.* **2015**, *96*, 69–83. [CrossRef]
41. Funk, C.; Peterson, P.; Landsfeld, M.; Pedreros, D.; Verdin, J.; Shukla, S.; Husak, G.; Rowland, J.; Harrison, L.; Hoell, A.; et al. The Climate Hazards Infrared Precipitation with Stations—A New Environmental Record for Monitoring Extremes. *Sci. Data* **2015**, *2*, 50066. [CrossRef]
42. Shen, Y.; Xiong, A. Validation and Comparison of a New Gauge-Based Precipitation Analysis over Mainland China. *Int. J. Climatol.* **2016**, *36*, 252–265. [CrossRef]
43. Shen, Y.; Xiong, A.Y.; Wang, Y.; Xie, P.P. Performance of High-Resolution Satellite Precipitation Products over China. *J. Geophys. Res. Atmos.* **2010**, *115*, D02114. [CrossRef]
44. Beck, H.E.; van Dijk, A.I.J.M.; Levizzani, V.; Schellekens, J.; Miralles, D.G.; Martens, B.; de Roo, A. Mswep: 3-Hourly 0.25 Global Gridded Precipitation (1979–2015) by Merging Gauge, Satellite, and Reanalysis Data. *Hydrol. Earth Syst. Sci.* **2017**, *21*, 589–615. [CrossRef]
45. Wang, R.; Chen, J.; Wang, X. Comparison of Imerg Level-3 and Tmpa 3b42v7 in Estimating Typhoon-Related Heavy Rain. *Water* **2017**, *9*, 276. [CrossRef]
46. Cai, Y.; Jin, C.; Wang, A.; Guan, D.; Wu, J.; Yuan, F.; Xu, L. Comprehensive Precipitation Evaluation of Trmm 3b42 with Dense Rain Gauge Networks in a Mid-Latitude Basin, Northeast, China. *Theor. Appl. Climatol.* **2015**, *126*, 659–671. [CrossRef]
47. Hofstra, N.; New, M.; McSweeney, C. The Influence of Interpolation and Station Network Density on the Distributions and Trends of Climate Variables in Gridded Daily Data. *Clim. Dyn.* **2012**, *35*, 841–858. [CrossRef]
48. Mair, A.; Fares, A. Comparison of Rainfall Interpolation Methods in a Mountainous Region of a Tropical Island. *J. Hydrol. Eng.* **2011**, *16*, 371–383. [CrossRef]
49. Pradhan, R.K.; Markonis, Y.; Vargas Godoy, M.R.; Villalba-Pradas, A.; Andreadis, K.M.; Nikolopoulos, E.I.; Papalexiou, S.M.; Rahim, A.; Tapiador, F.J.; Hanel, M. Review of Gpm Imerg Performance: A Global Perspective. *Remote Sens. Environ.* **2022**, *268*, 12754. [CrossRef]
50. Edwards, D.C. *Characteristics of 20th Century Drought in the United States at Multiple Time Scales*; Air Force Institute of Technology: Wright-Patterson AFB, OH, USA, 1997.
51. Santos, J.F.; Pulido-Calvo, I.; Portela, M.M. Spatial and Temporal Variability of Droughts in Portugal. *Water Resour. Res.* **2010**, *46*, W03503. [CrossRef]
52. Belayneh, A.; Adamowski, J. Standard Precipitation Index Drought Forecasting Using Neural Networks, Wavelet Neural Networks, and Support Vector Regression. *Appl. Comput. Intell. Soft Comput.* **2012**, *2012*, 1–13. [CrossRef]
53. Huang, S.Z.; Li, P.; Huang, Q.; Leng, G.Y.; Hou, B.B.; Ma, L. The Propagation from Meteorological to Hydrological Drought and Its Potential Influence Factors. *J. Hydrol.* **2017**, *547*, 184–195. [CrossRef]
54. Mpelasoka, F.; Hennessy, K.; Jones, R.; Bates, B. Comparison of Suitable Drought Indices for Climate Change Impacts Assessment over Australia Towards Resource Management. *Int. J. Climatol.* **2008**, *28*, 1283–1292. [CrossRef]
55. Yevjevich, V.M. *An Objective Approach to Definition and Investigations of Continental Hydrologic Droughts*; Hydrology Papers; Colorado State University: Fort Collins, CO, USA, 1967; Volume 23, No. 23.
56. Wilks, D.S. *Statistical Methods in the Atmospheric Sciences*, 3rd ed.; Academic Press: Cambridge, MA, USA, 2011; Volume 100.
57. Guo, H.; Bao, A.; Ndayisaba, F.; Liu, T.; Kurban, A.; De Maeyer, P. Systematical Evaluation of Satellite Precipitation Estimates over Central Asia Using an Improved Error-Component Procedure. *J. Geophys. Res. Atmos.* **2017**, *122*, 10906–10927. [CrossRef]
58. Gosset, M.; Viarre, J.; Quantin, G.; Alcoba, M. Evaluation of Several Rainfall Products Used for Hydrological Applications over West Africa Using Two High-Resolution Gauge Networks. *Q. J. R. Meteorol. Soc.* **2013**, *139*, 923–940. [CrossRef]
59. AghaKouchak, A.; Mehran, A. Extended Contingency Table: Performance Metrics for Satellite Observations and Climate Model Simulations. *Water Resour. Res.* **2013**, *49*, 7144–7149. [CrossRef]
60. Sadeghi, M.; Nguyen, P.; Naeini, M.R.; Hsu, K.; Braithwaite, D.; Sorooshian, S. Persiann-Ccs-Cdr, a 3-Hourly 0.04° Global Precipitation Climate Data Record for Heavy Precipitation Studies. *Sci. Data* **2021**, *8*, 157. [CrossRef] [PubMed]
61. Alijanian, M.; Rakhshandehroo, G.R.; Mishra, A.K.; Dehghani, M. Evaluation of Satellite Rainfall Climatology Using Cmorph, Persiann-Cdr, Persiann, Trmm, Mswep over Iran. *Int. J. Climatol.* **2017**, *37*, 4896–4914. [CrossRef]

62. Brasil Neto, R.M.; Guimarães Santos, C.A.; da Costa Silva, J.F.C.B.; da Silva, R.M.; dos Santos, C.A.C.; Mishra, M. Evaluation of the Trmm Product for Monitoring Drought over Paraíba State, Northeastern Brazil: A Trend Analysis. *Sci. Rep.* **2021**, *11*, 1097. [CrossRef] [PubMed]
63. Zhong, R.; Chen, X.; Lai, C.; Wang, Z.; Lian, Y.; Yu, H.; Wu, X. Drought Monitoring Utility of Satellite-Based Precipitation Products across Mainland China. *J. Hydrol.* **2019**, *568*, 343–359. [CrossRef]
64. Yao, J.; Tuoliewubieke, D.; Chen, J.; Huo, W.; Hu, W. Identification of Drought Events and Correlations with Large-Scale Ocean-Atmospheric Patterns of Variability: A Case Study in Xinjiang, China. *Atmosphere* **2019**, *10*, 94. [CrossRef]
65. Yu, M.; Li, Q.; Hayes, M.J.; Svoboda, M.D.; Heim, R.R. Are Droughts Becoming More Frequent or Severe in China Based on the Standardized Precipitation Evapotranspiration Index: 1951–2010? *Int. J. Climatol.* **2014**, *34*, 545–558. [CrossRef]
66. Yao, J.; Zhao, Y.; Yu, X. Spatial-Temporal Variation and Impacts of Drought in Xinjiang (Northwest China) During 1961–2015. *PeerJ* **2018**, *6*, e4926. [CrossRef]
67. Lei, T.; Wu, J.; Li, X.; Geng, G.; Shao, C.; Zhou, H.; Wang, Q.; Liu, L. A New Framework for Evaluating the Impacts of Drought on Net Primary Productivity of Grassland. *Sci. Total Environ.* **2015**, *536*, 161–172. [CrossRef]
68. Ayantobo, O.O.; Li, Y.; Song, S.; Yao, N. Spatial Comparability of Drought Characteristics and Related Return Periods in Mainland China over 1961–2013. *J. Hydrol.* **2017**, *550*, 549–567. [CrossRef]
69. Beguería, S.; Vicente-Serrano, S.M.; Reig, F.; Latorre, B. Standardized Precipitation Evapotranspiration Index (Spei) Revisited: Parameter Fitting, Evapotranspiration Models, Tools, Datasets and Drought Monitoring. *Int. J. Climatol.* **2014**, *34*, 3001–3023. [CrossRef]
70. Vicente-Serrano, S.M.; Begueria, S.; Lopez-Moreno, J.I. A Multiscalar Drought Index Sensitive to Global Warming: The Standardized Precipitation Evapotranspiration Index. *J. Clim.* **2010**, *23*, 696–718. [CrossRef]
71. Palmer, W.C. *Meteorological Drought*; U.S. Department of Commerce, Weather Bureau: Washington, DC, USA, 1965; Volume 30.
72. Svoboda, M.; Fuchs, B. *Handbook of Drought Indicators and Indices*; World Meteorological Organization: Geneva, Switzerland, 2016; pp. 1–44.
73. Bachmair, S.; Stahl, K.; Collins, K.; Hannaford, J.; Acreman, M.; Svoboda, M.; Knutson, C.; Helm Smith, K.; Wall, N.; Fuchs, B.; et al. Drought Indicators Revisited: The Need for a Wider Consideration of Environment and Society. *Wiley Interdiscip. Rev. Water* **2016**, *3*, 516–536. [CrossRef]
74. Pedro-Monzonís, M.; Solera, A.; Ferrer, J.; Estrela, T.; Paredes-Arquiola, J. A Review of Water Scarcity and Drought Indexes in Water Resources Planning and Management. *J. Hydrol.* **2015**, *527*, 482–493. [CrossRef]
75. Hao, Z.C.; Singh, V.P. Drought Characterization from a Multivariate Perspective: A Review. *J. Hydrol.* **2015**, *527*, 668–678. [CrossRef]
76. Ashraf, M.; Routray, J.K. Spatio-Temporal Characteristics of Precipitation and Drought in Balochistan Province, Pakistan. *Nat. Hazards* **2015**, *77*, 229–254. [CrossRef]
77. Ukkola, A.; De Kauwe, M.G.; Roderick, M.L.; Abramowitz, G.; Pitman, A.J. Robust Future Changes in Meteorological Drought in Cmip6 Projections Despite Uncertainty in Precipitation. *Geophys. Res. Lett.* **2020**, *47*, e2020GL087820. [CrossRef]

remote sensing

MDPI

Article

Interannual Variability of Water Level in Two Largest Lakes of Europe

Andrey G. Kostianoy [1,2,3,*], Sergey A. Lebedev [3,4,5], Evgeniia A. Kostianaia [1,3] and Yaan A. Prokofiev [5]

1 P.P. Shirshov Institute of Oceanology, Russian Academy of Sciences, Nakhimovsky pr. 36, 117997 Moscow, Russia; evgeniia.kostianaia@ocean.ru
2 Laboratory of Integrated Research of Water Resources, S.Yu. Witte Moscow University, Second Kozhukhovsky pr. 12, Build. 1, 115432 Moscow, Russia
3 Geophysical Center, Russian Academy of Sciences, Molodezhnaya Str. 3, 119296 Moscow, Russia; lebedev@wdcb.ru
4 Laboratory of Geoinformatics and Rational Nature Management, Maykop State Technological University, Pervomayskaya Str. 191, 385000 Maykop, Russia
5 Department of Higher Mathematics N1, National Research University "MIET", Shokina Place 1, Zelenograd, 124498 Moscow, Russia; t1007821@org.miet.ru
* Correspondence: kostianoy@ocean.ru

Abstract: Regional climate change affects the state of inland water bodies and their water balance, which is determined by a number of hydrometeorological and hydrogeological factors. An integral characteristic of changes in the water balance is the behavior of the level of lakes and reservoirs, which not only largely determines the physical and ecological state of water bodies, but also significantly affects the coastal infrastructure and socio-economic development of the region. This paper investigates the interannual variability of the level of the Ladoga and Onega lakes, the largest lakes in Europe located in the northwest of Russia, according to satellite altimetry data for 1993–2020. For this purpose, we used three specialized altimetry databases: DAHITI, G-REALM, and HYDROWEB. Water level data from these altimetry databases were compared with in-situ records at water level gauge stations. Information on air temperature (1945–2019) and precipitation (1966–2019) acquired at three meteostations located at Ladoga and Onega lakes was used to investigate interannual trends in the regional climate change. Finally, we discuss the potential impact of the lake level rise and regional climate warming on the infrastructure and operability of railways in this region.

Keywords: Lake Ladoga; Lake Onega; water level; air temperature; precipitation; climate change; satellite altimetry; railway infrastructure; meteostations; water level gauge stations

1. Introduction

The state of lakes, water reservoirs, and inland seas is an important indicator of ongoing regional climate change, including extreme climate events. Changes in climatic or long-term weather conditions affect the state and variability of natural parameters of inland water bodies: water level, water temperature, water salinity, vertical stratification of waters, currents, frontal zones, upwellings, eddies and gyres, ice cover and ice thickness, water turbidity, ecosystems, biodiversity, vegetation, algal bloom, etc. All over the world, many lakes are under the threat of drying up and even disappearing [1–14].

Lakes are included in the list of 54 essential climate variables (ECVs) of the Global Climate Observing System (GCOS), which make a decisive contribution to the characterization of the Earth's climate [15]. Water level, lake area, lake surface temperature, ice area and thickness, and lake water color are key parameters of the WMO Global Terrestrial Network of Lakes [16], which focuses on biodiversity, climate change, and sustainable development. The most important physical characteristic of the state of lakes is the water level, which is measured in-situ at level gauges and remotely using satellite altimetry [17–21]. The water level in the lakes must be measured daily and with an accuracy of at least 3 cm [15].

The subarctic zone of Russia is particularly vulnerable to climate change, as the permafrost zone contains infrastructure worth hundreds of billions of dollars. At the same time, according to observational data since the mid-1970s, the average temperatures here have been growing 2.5 times faster than the rest of the planet. If the frozen stratum thaws, then due to the significant ice content in them the average subsidence of the soil can be 10 m or more. The hydrological regime of numerous small rivers and lakes can change, as well as groundwater level and flux, river runoff, floods, etc. The railway infrastructure in the subarctic regions is operated in extremely difficult engineering-geological and landscape-climatic conditions, being continuously exposed to various external influences, leading to deformations of the track and artificial structures [22]. Among them are coastal abrasion, mudflows, floods, water and thermal erosion, landslides and debris, karst sinkholes, suffusion subsidence, thermokarst, solifluction (mass wasting), bursting deformations, frost heaving, avalanches, etc. [22,23].

Island permafrost zones (or sporadic permafrost zones, SPZ) are located on the Kola Peninsula [22,23], situated northward of the White Sea (Figure 1). The infrastructure of Russian railways in this region is especially vulnerable to the negative factors of regional climate change, since the main problem of the Volkhovstroy–Murmansk section is the fact that out of 1320 km of its length, more than 340 km are single-track sections, which limits its throughput and makes it vulnerable. In this regard, the study of climate change in the Murmansk Region, the Republic of Karelia, and the Arkhangelsk Region is extremely important for the Russian railways (RZhD). This study is a continuation of the research started in [22,24–27].

Figure 1. Map of Ladoga and Onega lakes with the route of the White Sea–Baltic Sea shipping canal (shown by the red line) (Courtesy: Norman Einstein, https://commons.wikimedia.org/wiki/File:White_Sea_Canal_map.png. (Accessed on 22 November 2021)).

Ladoga and Onega lakes are the two largest lakes in Europe, which are located in the northwest of Russia (Figure 1), based on their surface, volume, and watershed area (Table 1). There is a dense network of Russian railways in the area of these lakes, which surrounds Ladoga Lake from all sides and goes along the western coast of Lake Onega (Figure 2). Both lakes have been the focus of research and monitoring since the first expeditions led by Russian Captain-Lieutenant D. Selyaninov in 1763–1765 and Colonel A.P. Andreev in 1858–1866 [28–32]. Since 1956, the Laboratory of Limnology (now the Institute of Limnology of the Russian Academy of Sciences) and dozens of research institutes of the Russian Academy of Sciences, Russian Hydrometeorological Service, and universities began carrying out complex limnological studies (hydrological, hydrochemical, and hydrobiological) of the Ladoga and Onega lakes [29–32].

Table 1. Physical characteristics of the five largest lakes in Europe.

Parameter	Ladoga (Russia)	Onega (Russia)	Vänern (Sweden)	Saimaa (Finland)	Peipus (Estonia, Russia)
Watershed area (km^2)	276,000	66,284	46,800	61,054	47,800
Max length (km)	219	245	140	190	152
Max width (km)	138	91.6	65	85	47
Surface of the area (km^2)	17,870	9720	5650	4380	3555
Average depth (m)	5	30	27	17	7.1
Max depth (m)	230	127	106	82	15.3
Volume (km^3)	838	285	153	36	25

Water balance, lake level, and regional climate change are among the main directions of scientific research and monitoring of the Ladoga and Onega lakes [29–35]. Traditionally, this research was done based on the data provided by meteostations located along the coasts of the lakes and in the surrounding regions. The first studies of these lakes (water level and ice cover) using satellite altimetry were carried out in [36–40]. According to the results obtained by Lebedev et al. [38,39], the temporal variability of the levels of the lakes Ilmen′, Ladoga, and Pskovsko-Chudskoe (Peipus or Peipsi) is characterized by a wave with a period of four–five years, and the level of Lake Onega is characterized by a wave with a period of 15 years. In the period from 1993 to 2011, the level of Lake Ilmen′ rose at a rate of 1.17 ± 0.95 cm/year, Lake Ladoga at 0.24 ± 0.10 cm/year, Lake Peipus at 1.39 ± 0.18 cm/year, and Lake Onega at 0.18 ± 0.09 cm/year.

The aim of this paper is to investigate interannual variability of the level of Ladoga and Onega lakes using altimeters installed in different years on the satellites Topex/Poseidon, Jason-1, Jason-2, Jason-3, GFO-1, ENVISAT, SARAL, and Sentinel-3A for the period from 1993 to 2020. For this purpose, we used three specialized altimetry databases: DAHITI, G-REALM, and HYDROWEB. Water level data from these altimetry databases were compared with in-situ records at water level gauge stations. Information on air temperature (1945–2019) and precipitation (1966–2019) acquired at three meteostations located at Ladoga and Onega lakes was used to investigate interannual trends in the regional climate change. Finally, we discuss the potential impact of the lake level rise and regional climate warming on the infrastructure and operability of railways in this region.T

Figure 2. Network of Russian railways around Ladoga and Onega lakes (red lines) (modified from http://www.200stran.ru/maps_group26_item2118.html (Accessed on 22 November 2021)).

2. Study Area

Lake Ladoga (Figure 1) is the largest lake in Europe, with a surface area of 17,870 km^2, a volume of 838 km^3, and a maximum depth of 230 m. The length of the lake from south to north is 219 km, the greatest width is 138 km, and the average height above sea level is 4.84 m. The main flowing rivers are the Svir, Volkhov, Vuoksa, Syas', Naziya, and Morye (35 rivers in total); the outflowing river is the Neva, which flows into the Gulf of Finland. The catchment area is 276,000 km^2. Ladoga is located in the Republic of Karelia and the Leningrad Region and belongs to the Baltic Sea Basin [29–32].

The average air temperature in the area of Lake Ladoga is +3.2 °C. The average temperature of the coldest month (February) is −8.8 °C, the warmest (July) +16.3 °C. Average annual precipitation is 475 mm. The smallest monthly amount of precipitation falls in February–March (24 mm), the highest in September (58 mm). Approximately 85% (3820 mm) of the incoming part of the water balance gives the inflow of river water, 13% (610 mm) atmospheric precipitation, and 2% (90 mm) the inflow of groundwater. About 92% (4170 mm) of the expenditure part of the water balance goes to the Neva runoff and 8% (350 mm) to evaporation from the water surface. Seasonal and interannual variability of the water balance leads to significant variability in the level of Lake Ladoga [41].

Lake Onega is located in close proximity to Lake Ladoga (Figure 1), but the interannual variability of its level is absolutely different from the variability of the Ladoga level [41]. Lake Onega is the second largest lake after Lake Ladoga in Europe, with a surface area of 9720 km^2, a volume of 285 km^3, and a maximum depth of 127 m. About fifty rivers run to Lake Onega, and only the Svir flows out of the lake. The catchment area is 66,284 km^2.

Lake Onega is located in the Republic of Karelia, Leningrad, and Vologda regions of Russia and belongs to the Baltic Sea Basin [31,34,35].

Approximately 74% of the incoming part of the lake's water balance (15.6 km^3/year) belongs to the river runoff. More than 1000 watercourses flow into the lake, of which 52 rivers are more than 10 km long and eight rivers are more than 100 km long. The largest of them are the Vodla and the Suna. About 25% of the incoming part of the water balance belongs to precipitation and about 1% to groundwater. About 84% of the expenditure part of the water balance belongs to the runoff from the lake along with the Svir River (on average 17.6 km^3/year) and 16% to evaporation from the water surface. Besides, the flow from the lake is regulated by the Verkhne-Svirskaya hydroelectric power station (HPP). The hydrological regime of Lake Onega is characterized by a spring rise of water which lasts 1.5–2 months, with annual fluctuations in the water level up to 0.9–1 m. The highest water levels in the lake are observed in June–August and the lowest in March–April [41].

3. Data and Methods

Satellite altimetry data have been successfully used to study the hydrological regime of lakes and rivers in South America, Africa, and Siberia [42–45]; water levels in the lower reaches of the Volga River [46]; the cascade of the Volga water reservoirs [47–50]; Lake Peipus located at the border between Russia and Estonia [38,39,41,51,52]; and Lake Ilmen' [38,39,51] located not far from Lake Ladoga; and other inland water bodies [53]. For an overview of satellite altimetry applications for inland water monitoring, see [18,54–56].

In our recent publication [41] we analyzed the level of the Ladoga, Onega, Peipus, and Ilmen' lakes based on the satellite altimetry data from the Topex/Poseidon (August 1992–January 2006), Jason-1 (December 2001–July 2013), Jason-2 (June 2008–October 2019), and Jason-3 (January 2016–present), whose ground tracks pass over these lakes (Figure 3). This continuity in the satellite measurements allowed for obtaining a continuous series of data on the lake level along the same tracks from September 1992 to the present with a time step of 10 days. In the abovementioned study, the observation period was selected from 1 January 1993 to 31 December 2020 [41]. Data were obtained from the HYDROWEB database of the French Space Agency (CNES) and the Laboratory for the Study of Geophysics and Satellite Oceanography (LEGOS) [37].

In the present research, we used three specialized databases, DAHITI, G-REALM, and HYDROWEB—all focused on the water level of lakes, water reservoirs, and rivers—which contain data from almost all altimetry satellites operated between October 1992 and December 2020 (Table 2). These databases were used to investigate interannual variability of the water level in Lakes Ladoga and Onega, to compare the databases between themselves, and to compare the databases with in-situ records at water level gauge stations in both lakes.

The Database for Hydrological Time Series of Inland Waters (DAHITI) was developed by the Deutsches Geodätisches Forschungsinstitut der Technischen Universität München, Germany (DGFI-TUM) in 2013. DAHITI provides a variety of hydrological information on 5507 lakes, reservoirs, rivers, and wetlands derived from satellite data over all continents from multi-mission satellite altimetry and optical remote sensing imagery [57]. G-REALM (www.pecad.fas.usda.gov/cropexplorer, accessed on 6 December 2021) was developed by the United States Department of Agriculture (USDA) and is focused on lake and reservoir level variations using the TOPEX/Poseidon, Jason-1, Jason-2, and Envisat satellites. It produces and distributes archived and NRT water level parameters on a network of more than 200 lakes worldwide [6,45]. Hydroweb (https://hydroweb.theia-land.fr, accessed on 6 December 2021) was developed at LEGOS (France). It produces and delivers river level data from satellite altimetry on more than 12,548 virtual stations spread over the largest river basins worldwide. It also provides lake level, surface extent, and volume variations for 310 lakes. Some of the products are delivered in the near real-time (NRT) mode [18,37].

Daily lake level data from water level gauge stations Petrokrepost, Novaya Sviritsa, Syas'skye Ryadki at Lake Ladoga, and Voznesenye, Petrozavodsk, Kondopoga at Lake Onega for the period from 1 January 2001 to 31 December 2020 were used for verification of three satellite altimetry databases (Figure 3). Data on daily water level values were taken from the Information System on Water Resources and Water Management of the Russian River Basins at the Center of Register and Cadastre of the Russian Federation (http://gis.vodinfo.ru, accessed on 22 November 2021). Geographic coordinates and "zero" levels of these gauge stations are listed in Table 3.

Figure 3. *Cont.*.

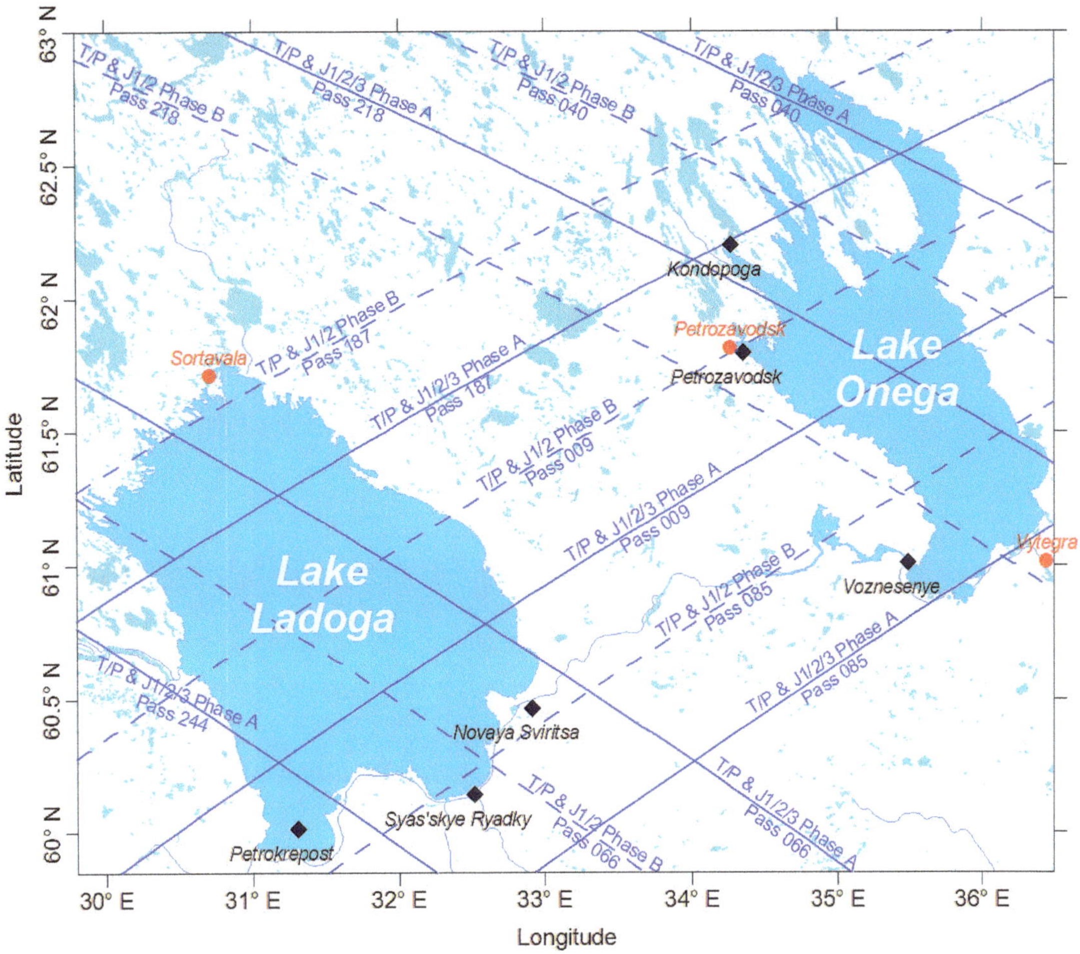

Figure 3. Watershed area of the Neva River (upper panel). Network of ground tracks of altimetry satellites TOPEX/Poseidon (T/P), Jason-1 (J1), Jason-2 (J2), and Jason-3 (J3) (lower panel). Locations of water level gauge stations and meteostations are shown on the map by black and red dots, respectively.

We have to note here that in the past two decades there has been a development of alternative in-situ methods of measurements of the water level in water reservoirs based on the geomatic methods and time-lapse photography, which is a very accurate and low-cost methodology [58,59]. These new methods can be applied in the case of a lack of water level gauge stations, expensive infrastructure, administrative barriers, etc. Satellite altimetry, remote, indirect, and different in-situ methods of water level measurements have their own advantages and disadvantages. For example, remote sensing has a global coverage and unified system of measurements for the World Ocean, inland seas, lakes, water reservoirs, and rivers, but it has a set of known disadvantages related to processing of data (a set of different corrections and retracking should be applied) and measurements in the coastal zone and in the narrow water bodies. Moreover, measurements are done along the fixed ground tracks which may cross the water body too close to the shoreline, and repetition of measurements is of 10 days or more. On the other hand, in-situ measurements allow to measure the water level with a frequency starting from minutes with much higher accuracy, but their location at coasts leads to the impact of specific local currents, winds, waves, orography effects, etc. Gauge stations also cannot be installed at remotely located water bodies with a lack of coastal infrastructure. There are many examples of this in deserts of Central Asia or in the mountains of South America. We can conclude briefly that every method has its own advantages and disadvantages which can be specific for every water body, and only a combination of different methods can provide the best result.

Table 2. A list of satellites and their track numbers which cross Ladoga and Onega lakes included in the DAHITI, G-REALM, and HYDROWEB databases.

Lake	Satellite	Track Number
Database for Hydrological Time Series of Inland Waters (DAHITI)		
Ladoga	TOPEX/Poseidon, Jason-1/2/3	066, 187
Onega	TOPEX/Poseidon, Jason-1/2/3	009, 218
Global Reservoirs and Lakes Monitor (G-REALM)		
Ladoga	TOPEX/Poseidon, Jason-1/2/3	066
Onega	TOPEX/Poseidon, Jason-1/2/3	218
HYDROWEB		
Ladoga	Topex/Poseidon	009, 066, 187, 244
	Jason-1	009, 066, 187
	Jason-2/3	009, 066, 187
	GFO-1	016, 139, 225, 332, 418
	ENVISAT	272, 358, 369, 455, 730, 816, 902, 913, 999
	SARAL	272, 358, 455, 730, 816, 902, 913, 999
	Sentinel-3A	100, 111, 214, 328, 756, 767
Onega	Topex/Poseidon	009, 218
	Jason-1/2/3	009, 040, 187, 218
	GFO-1	016, 081, 102, 483
	ENVISAT	100, 186, 197, 558, 644, 655
	SARAL	100, 197, 558, 644, 655
	Sentinel-3A	442, 556, 567, 670, 681

Table 3. A list of water level gauge stations at Ladoga and Onega lakes.

Lake	Water Level Gauge Station	"Zero" Level of the Gauge Station, m ASL *	Latitude N	Longitude E
Ladoga	Petrokrepost	0.0	59.95	31.03
	Novaya Sviritsa	0.0	60.47	32.9
	Syas'skye Ryadki	0.0	60.15	32.5
Onega	Voznesenye	31.8	61.36	35.35
	Petrozavodsk	31.8	61.73	34.33
	Kondopoga	31.8	62.61	34.43

* ASL is absolute sea level.

Regional climate change is a critical factor for changes in the hydrological regime, water balance, and water level of the Ladoga and Onega lakes [29–35]. To clarify the rates of warming of the atmosphere and precipitation namely at Ladoga and Onega lakes, we calculated interannual trends in air temperature and precipitation recorded at three coastal meteostations (Figure 3, Table 4). To reveal general cyclicality in the interannual variability of the water level, air temperature, and precipitation over Ladoga and Onega lakes, we performed wavelet analysis (the Ricker wavelet transform) of these data after preliminary exclusion the linear trends from these time series.

Table 4. A list of meteostations at Ladoga and Onega lakes.

Lake	Meteostation (WMO N)	Elevation, m ASL	Latitude N	Longitude E
Ladoga	Sortavala (N 22802)	+19.0	61.72	30.72
Onega	Petrozavodsk (N22820)	+110.0	61.82	34.27
	Vytegra (N 22837)	+56.0	61.02	36.45

We used data on monthly average air temperature for the meteostation Sortavala (WMO Station N 22802) at Lake Ladoga for 1945–2019, and at Lake Onega for meteostations Petrozavodsk (22820) for 1949–2019 and Vytegra (22837) for 1945–2019 to calculate interannual trends in air temperature change (Figure 3, Table 4). Monthly averaged sum of precipitation for the meteostation Sortavala (WMO station N 22802) at Lake Ladoga for 1966–2019, and at Lake Onega for the meteostations Petrozavodsk (22820) for 1966–2019 and Vytegra (22837) for 1966–2019 was used to calculate interannual trends in atmospheric precipitation.

The standard method of least squares, which is a common approach in regression analysis, was used to estimate linear trends in temporal variability of the water level, air temperature, and precipitation on Lakes Ladoga and Onega, as well as relationships between satellite altimetry databases and in-situ records at water level gauge stations.

Seasonality was not removed before estimating the linear trend because the seasonal cycles in the water level in these lakes are not stable. Moreover, for Lake Onega a seasonal cycle is modulated by the runoff through the hydropower station which is determined by the needs in the amount of electricity production. In such specific cases, removal of the "standard" seasonal trend leads to introduction of artificial errors in the time series.

To show the relationship between altimetry and in-situ datasets we drew scatterplots, which is a common procedure for such kinds of tasks. Correlation coefficient R and coefficient of determination R^2 were used to show the correlational relationships. The values of Nash-Sutcliffe model efficiency coefficient (NSE) as an additional characteristic for relationships between all three databases and in-situ records at gauge stations, and Percent bias (PBIAS), which is a characteristic of the systematic error, were calculated.

4. Results

4.1. Lake Ladoga

Figure 4 shows seasonal and interannual variability of the Lake Ladoga level from October 1992 to December 2020 derived from the satellite altimetry data. The gaps in the satellite altimetry databases during wintertime when the lakes were covered with ice were filled by data derived from the appropriate regressions obtained by calibration/validation of remote sensing data by in-situ measurements (see Section 4.3 Validation and calibration of altimetry data). The lowest level is observed in winter, the highest in summer, which corresponds to the seasonal variability of precipitation in the drainage basin. The range of seasonal fluctuations varies within 0.5–1.6 m, with the greatest values of the amplitude of level fluctuations observed during sharp interannual changes in the water balance. Over the entire observation period from October 1992 to December 2020, there were three maxima of the water level in the lake, which exceeded the mark of 5.5 m based on the HYDROWEB data. This occurred on 16 June 1995 (5.56 m), 17 June 2005 (5.59 m), and 15 May 2018 (5.74 m). DAHITI data, which give the highest water level values, show two additional maxima in June 2010 of about 5.8 m and June 2013 of 5.6 m (Figure 4). During the same time, based on the HYDROWEB data there were two minima of the water level which were below the 3.6 m mark—on 17 January 2003 (3.46 m) and 7 March 2015 (3.6 m).

The maximum range of lake level fluctuations for 28 years was 2.28 m as recorded in the HYDROWEB database.

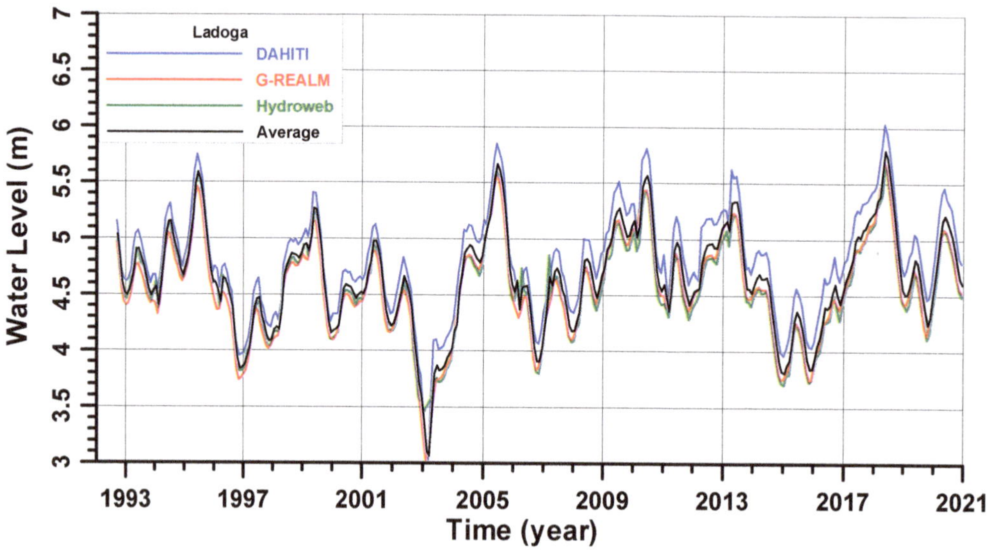

Figure 4. Interannual and seasonal variability of the Lake Ladoga level from October 1992 to December 2020 based on the DAHITI (blue line), G-REALM (red line), and HYDROWEB (green line) databases. The black line shows average value calculated from all three databases.

According to Figure 4, the DAHITI data show larger values than those from G-REALM and HYDROWEB, which are close to each other. To show the difference between all three databases, we calculated an averaged (arithmetic mean) value of the water level based on all three databases (see the black line in Figure 4) and calculated the difference between data from every database and their average value (Figure 5). Thus, the DAHITI data are 0.1–0.25 m larger than the average value with a steady increase from the beginning of the 1990s to the end of 2020. The G-REALM and HYDROWEB data are lower than the average and are better correlated with each other. In interannual variability, the G-REALM data seem to be much more stable (around −0.1 m) with no drift, while the HYDROWEB data show a negative drift from −0.02 m in the beginning of the 1990s to −0.12 m by the end of 2020 (Figure 5).

It looks like both maximum and minimum levels occur with a cycle of 10–13 years. A characteristic feature of the interannual variability of the Lake Ladoga level is a sharp rise in the level by about 2 m to the maximum levels in two–three years after the lowest values, and the subsequent rather sharp drop by about 1.5–2 m in about two years to the next minimum values (Figure 4). For example, after the interim maximum of 1995 there was an irregular decrease of the water level until the minimum of 2003. Then, a sharp rise of the water level was observed until 2005, followed by a subsequent irregular decrease of the water level by 2015. Then, we observe again a sharp rise of the water level until 2018, and again followed by a general decrease of the water level. If the observed periodicity (regularity) continues, then in the next five–six years, we can expect a further decrease in the level of Lake Ladoga by less than one meter relative to the level observed at the end of December 2020.

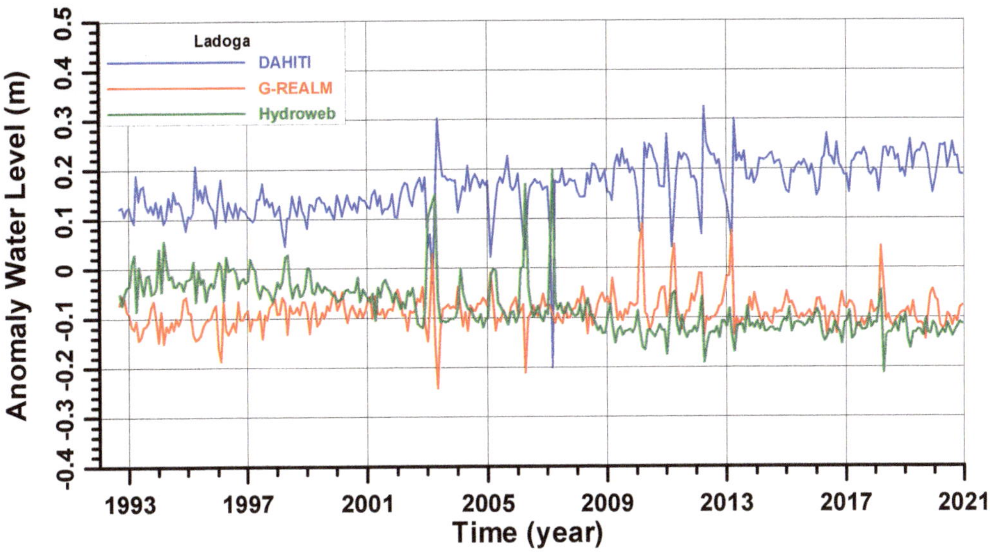

Figure 5. Interannual and seasonal variability of water level difference between the average value and the values from the DAHITI (blue line), G-REALM (red line), and HYDROWEB (green line) databases for Lake Ladoga from October 1992 to December 2020.

We summarized the main characteristics of the Lake Ladoga level derived from all three altimetry databases in Table 5. The mean lake level lies in the range between 4.55 m and 4.81 m ASL and the standard deviation (SD) of data is of 0.2 m. The main differences between the DAHITI, G-REALM, and HYDROWEB databases arise in the determination of the minimum and maximum absolute values, as well as in the amplitude of lake level variability. The difference between the databases in the extreme states of the lake level reaches almost 0.5 m. This leads to a large difference in the observed amplitude of lake level variability, being the highest for DAHITI (3.15 m) and lowest for HYDROWEB (2.28 m) (Table 5). There is a notable difference in the linear trends of the Lake Ladoga level change between three databases for the period from October 1992 to December 2020 (Table 4). DAHITI gives +1.17 cm/year, G-REALM +0.81 cm/year, and HYDROWEB +0.27 cm/year. For the averaged values of the lake level, it will be +0.75 cm/year.

Table 5. The main characteristics of the Lake Ladoga level derived from the DAHITI, G-REALM, and HYDROWEB databases.

Parameter	DAHITI	G-REALM	HYDROWEB	Average
Number of measurements	1810	1073	1536	1473
Mean lake level (m ASL)	4.81	4.55	4.62	4.64
Standard deviation (m)	0.21	0.19	0.19	0.21
Minimum (m ASL)	3.01	2.99	3.46	3.07
Maximum (m ASL)	6.18	5.70	5.74	5.79
Amplitude (m)	3.15	2.71	2.28	2.73
Linear trend (cm/year)	+1.17	+0.81	+0.27	+0.75

4.2. Lake Onega

Figure 6 shows the seasonal and interannual variability of the Lake Onega level from October 1992 to December 2020 derived from the satellite altimetry data. The lowest level was observed in spring, the highest in summer. The range of seasonal fluctuations varied within 0.4–1.0 m, with the greatest values of the amplitude of level fluctuations observed during sharp interannual changes in the water balance. Since the flow from the lake is regulated by the Verkhne-Svirskaya HPP (the maximum discharge is 47 km^3/year), the interannual variability of the Lake Onega level does not correspond to the neighboring Lake Ladoga (see Figure 4). According to the HYDROWEB data, for the observation period from October 1992 to December 2020, there were two maxima of the water level, which exceeded the mark of 34.1 m–on 20 June 1995 (34.22 m) and on 14 December 2008 (34.14 m). During the same period, there were two minima of the water level in the lake, which were below 32.8 m–on 8 March 2011 (32.76 m) and on 17 March 2019 (32.73 m). The maximum range of lake level fluctuations for 28 years was 1.49 m (Figure 6).

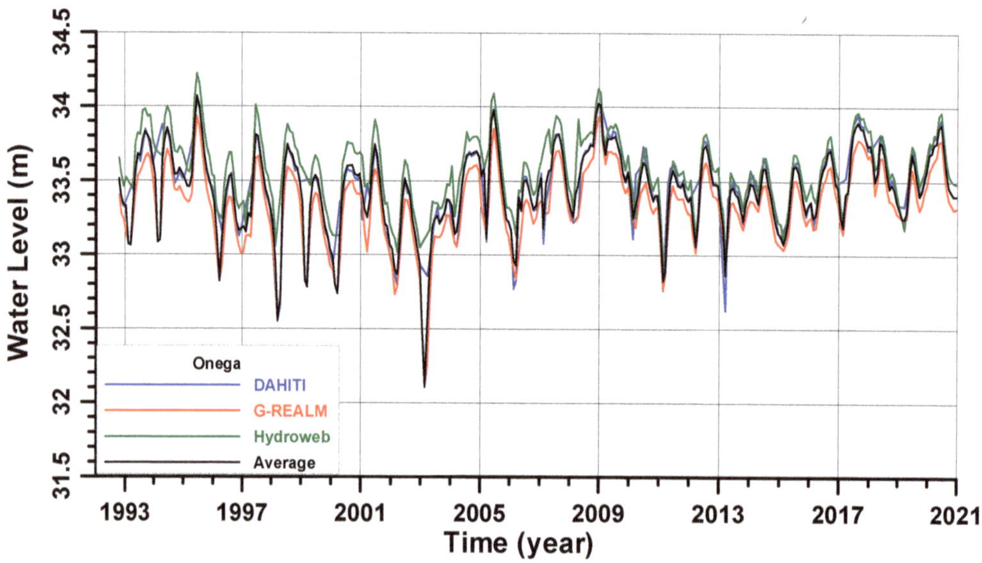

Figure 6. Interannual and seasonal variability of the Lake Onega level from October 1992 to December 2020 based on the DAHITI (blue line), G-REALM (red line), and HYDROWEB (green line) databases. The black line shows average value calculated from all three databases.

According to Figure 7, the HYDROWEB data show larger values than those from DAHITI, which corresponds well to the average values from all three databases, and G-REALM, which show lower values. To show the difference between all three databases, we calculated an averaged value of the water level based on all three databases (see the black line in Figure 6) and calculated the difference between data from every database and their average value (Figure 7). On average, the HYDROWEB data are up to 0.1–0.2 m larger than the average value with a steady decrease from the beginning of the 1990s to the end of 2020. The DAHITI data fluctuate around zero value of the anomaly. G-REALM data seem to be a mirror of the HYDROWEB data in terms of the anomalies because they show lower values than the average. In the interannual variability, the G-REALM data differences drift from -0.1 to -0.2 m in the beginning of the 1990s to -0.1 m by the end of 2020 (Figure 7). It looks like there are no evident cycles in the interannual variability of the Lake Onega level, which is explained by the regulation of its water balance by the Verkhne-Svirskaya HPP.

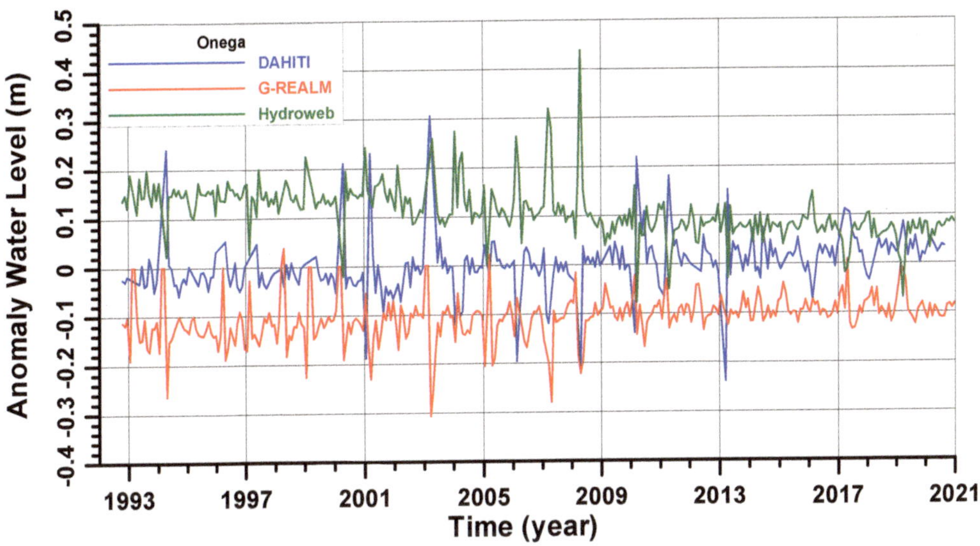

Figure 7. Interannual and seasonal variability of the water level difference between the average value and the values from the DAHITI (blue line), G-REALM (red line), and HYDROWEB (green line) databases for Lake Onega from October 1992 to December 2020.

We summarized the main characteristics of the Lake Onega level derived from all three altimetry databases in Table 6. The mean lake level lies in the range between 33.35 m and 33.58 m ASL, and the dispersion of data is of 0.06 m. The main differences between the DAHITI, G-REALM, and HYDROWEB databases arise in the determination of the minimum and maximum absolute values, as well as in the amplitude of lake level variability. The difference between the databases in the extreme states of the lake level reaches 0.65 m for a minimum level and 0.27 m for a maximum level of the lake. This highest amplitude of the lake level change was recorded for G-REALM (1.87 m) and the lowest one for HYDROWEB (1.49 m) (Table 6). There is a notable difference in the linear trends of the Lake Onega level change among three databases for the period from October 1992 to December 2020 (Table 6). DAHITI gives +0.41 cm/year, G-REALM +0.62 cm/year, and HYDROWEB gives a negative trend of −0.27 cm/year. The averaged values of the lake level will be +0.47 cm/year, which is 37% less than for Lake Ladoga. Again, this can be explained by the regulation of its water level by the Verkhne-Svirskaya HPP.

Table 6. The main characteristics of the Lake Onega level derived from the DAHITI, G-REALM, and HYDROWEB databases.

Parameter	DAHITI	G-REALM	HYDROWEB	Average
Number of measurements	1160	1031	1250	1147
Mean lake level (m ASL)	33.50	33.35	33.58	33.44
Standard deviation (m)	0.06	0.07	0.05	0.08
Minimum (m ASL)	32.52	32.08	32.73	32.11
Maximum (m ASL)	34.10	33.95	34.22	34.07
Amplitude (m)	1.58	1.87	1.49	1.96
Linear trend (cm/year)	+0.41	+0.62	−0.27	+0.47

4.3. Validation and Calibration of Altimetry Data

To verify satellite altimetry data, daily data from three gauge stations at Lake Ladoga and three gauge stations at Lake Onega were used for the period from 1 January 2001 to 31 December 2020. Data on daily water level values were taken from the Information System on Water Resources and Water Management of the Russian River Basins at the Center of Register and Cadastre of the Russian Federation (http://gis.vodinfo.ru, accessed on 22 November 2021). Examples of temporal variability of the water level at the Petrokrepost (Lake Ladoga) and Voznesenye (Lake Onega) gauge stations for the period from 1 January 2015 to 31 December 2019 are shown in Figure 8. To show the discrepancy between in-situ data and satellite altimetry products, as well as the difference between all three databases, we plotted over the appropriate data from the DAHITI, G-REALM, and HYDROWEB databases.

Figure 8a clearly shows that the data from the DAHITI database for Lake Ladoga exceed the in-situ data from the Petrokrepost gauge station by an average of 0.4 m. The maximum deviation was observed in autumn 2015 at 0.8 m. Data from the G-REALM and HYDROWEB databases exceed in-situ measurements by 0.3 m and 0.15 m, respectively. Maximum deviation of 0.5 m was also observed in autumn 2015. From October 2016, the average difference between the G-REALM and HYDROWEB databases and in-situ records was of 0.15 and 0.10 m, respectively (Figure 8a). Therefore, both databases fit better in the in-situ measurements.

For Lake Onega, the DAHITI data exceed the in-situ measurements at the Voznesenye water level gauge station by an average of 0.3 m (Figure 8b). The maximum deviation of 0.57 m was observed in the spring of 2019. The data from the G-REALM and HYDROWEB databases on average exceed the in-situ measurements by 0.15 m and 0.3 m, respectively. The maximum deviations of 0.4 m and 0.7 m, respectively, were also observed in the spring of 2019. In general, for Lake Ladoga, the DAHITI and HYDROWEB data are close to each other, but the best fit with the in-situ records is G-REALM (Figure 8b).

It is interesting that all three databases show different behavior for the Ladoga and Onega lakes which are located close to each other. For example, for Lake Ladoga DAHITI shows larger values than those from G-REALM and HYDROWEB, and for Lake Onega HYDROWEB shows larger values than those from DAHITI and G-REALM (see Figures 5, 7 and 8). But the abovementioned analysis could not answer the question: Which database better describes real variations of the water level in the lakes? To validate these databases, we built scatterplots and calculated linear regressions and correlations between the water level derived from the DAHITI, G-REALM, and HYDROWEB databases and all six in-situ water level gauge stations located at both lakes for the period from 1 January 2001 to 31 December 2020 (Tables 7 and 8). Figure 9 shows two examples of this analysis for Lake Ladoga (Petrokrepost) and Lake Onega (Voznesenye).

For Lake Ladoga, for the Petrokrepost gauge station we have the correlation coefficient R in the range from 0.965 (DAHITI) to 0.976 (G-REALM), for Syas'skye Ryadki from 0.929 (HYDROWEB) to 0.984 (G-REALM), and for Novaya Sviritsa from 0.919 (HYDROWEB) to 0.927 (DAHITI) (Table 7). For every gauge station we have high, but different correlations for every database. The observed discrepancy can be explained by geographic peculiarities (coastline, depth, prevailed winds, etc.) at the gauge station locations. Thus, for Petrokrepost the best database is G-REALM with R = 0.976, for Syas'skye Ryadki G-REALM with R = 0.984, and for Novaya Sviritsa DAHITI with R = 0.927 (Table 7). Therefore, it looks like G-REALM better describes the Lake Ladoga level variability.

For all gauge stations at Lake Onega, the G-REALM database shows the highest correlation coefficient R: for Kondopoga (0.924), Petrozavodsk (0.928), and Voznesenye (0.937). HYDROWEB shows second rank correlations (R = 0.906–0.917) for all gauge stations. DAHITI shows the lowest R (0.893–0.896) for all gauge stations (Table 8). Thus, the best database for Lake Onega is also G-REALM.

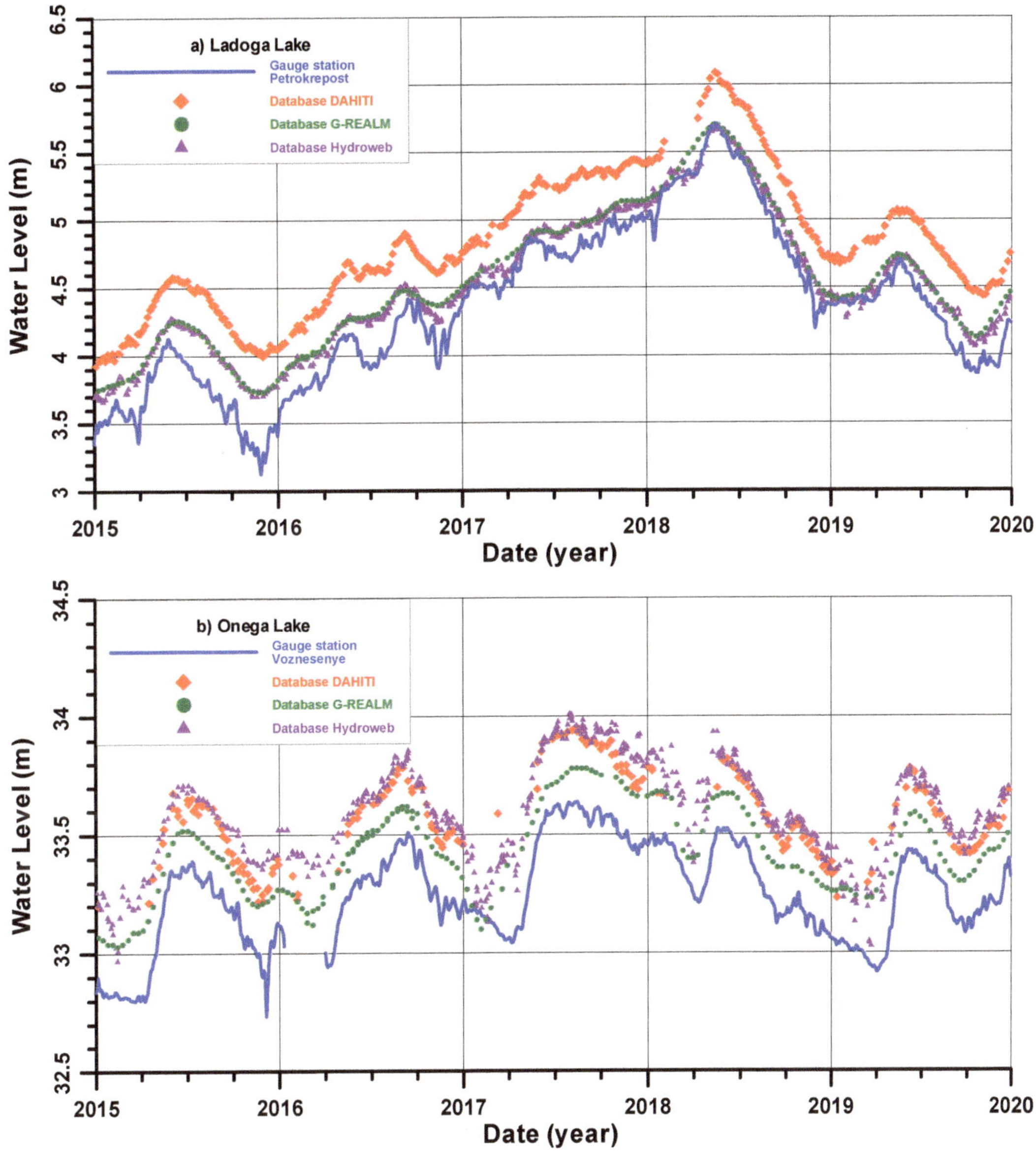

Figure 8. Variability in the level of Lake Ladoga (**a**) and Lake Onega (**b**) according to the water level gauges stations Petrokrepost (**a**) and Voznesenye (**b**) (shown by the blue line), and data from the DAHITI (red diamonds), G-REALM (green circles), and HYDROWEB (purple triangles) databases for the period from 1 January 2015 to 31 December 2019.

Table 7. The main correlation characteristics between the Lake Ladoga water level derived from the DAHITI, G-REALM, and HYDROWEB altimetry databases, and in-situ water level gauge stations for the period from 1 January 2001 to 31 December 2020. Values in bold show the largest correlation coefficient.

Water Level Gauge Station	Parameter	DAHITI	G-REALM	HYDROWEB
Petrokrepost	Correlation coefficient, R	0.966	**0.976**	0.971
	Arithmetical mean, AM (m)	4.40	4.41	4.47
	Standard deviation, SD (m)	0.44	0.43	0.41
	Coefficients in a linear regression equation (y = ax + b), where Y is water level (m) derived from altimetry data and X is water level (m) derived from in-situ gauge stations.	a = +1.103 b = −0.97	a = +1.135 b = −0.80	a = +1.138 b = −0.79
	Coefficient of determination, R^2	0.933	0.953	0.942
	Nash–Sutcliffe model efficiency coefficient (NSE)	0.933	0.953	0.942
	Percent bias (PBIAS)	0.47%	0.32%	0.36%
Syas'skye Ryadki	R	0.978	**0.984**	0.929
	AM (m)	4.62	4.62	4.69
	SD (m)	0.37	0.36	0.38
	Coefficients in a linear regression	a = +0.938 b = +0.08	a = +0.969 b = +0.21	a = +1.009 b = +0.04
	R^2	0.957	0.967	0.863
	NSE	0.957	0.967	0.863
	PBIAS	0.20%	0.15%	0.71%
Novaya Sviritsa	R	**0.927**	0.924	0.919
	AM (m)	4.80	4.81	4.87
	SD (m)	0.36	0.35	0.36
	Coefficients in a linear regression	a = +0.907 b = +0.41	a = +0.950 b = +0.47	a = +0.957 b = +0.46
	R^2	0.860	0.853	0.844
	NSE	0.8560	0.853	0.844
	PBIAS	0.65%	0.69%	0.71%

Table 8. The main correlation characteristics between the Lake Onega water level derived from the DAHITI, G-REALM, and HYDROWEB altimetry databases, and in-situ water level gauge stations for the period from 1 January 2001 to 31 December 2020. Values in bold show the largest correlation coefficient.

Water Level Gauge Station	Parameter	DAHITI	G-REALM	HYDROWEB
Kondopoga	R	0.893	**0.924**	0.907
	AM (m)	33.22	33.18	33.21
	SD (m)	0.18	0.2	0.18
	Coefficients in a linear regression	a = +0.778 b = +7.15	a = +0.856 b = +4.63	a = +0.845 b = +4.83

Table 8. *Cont*

Water Level Gauge Station	Parameter	DAHITI	G-REALM	HYDROWEB
	R^2	0.797	0.855	0.822
	NSE	0.797	0.855	0.822
	PBIAS	0.03%	0.03%	0.02%
Petrozavodsk	R	0.896	0.928	0.917
	AM (m)	33.22	33.19	33.22
	SD (m)	0.18	0.19	0.18
	Coefficients in a linear regression	a = +0.776 b = +7.22	a = +0.850 b = +4.82	a = +0.844 b = +4.89
	R^2	0.804	0.862	0.841
	NSE	0.804	0.862	0.841
	PBIAS	0.03%	0.02%	0.02%
Voznesenye	R	0.895	0.937	0.906
	AM (m)	33.23	33.21	33.24
	SD (m)	0.18	0.2	0.18
	Coefficients in a linear regression	a = +0.806 b = +6.24	a = +0.892 b = +3.46	a = +0.859 b = +4.40
	R^2	0.8	0.877	0.821
	NSE	0.8	0.877	0.821
	PBIAS	0.03%	0.02%	0.03%

There are several reasons to explain the different correlations among the three databases for every gauge station. First, DAHITI and G-REALM use TOPEX/Poseidon, Jason-1/2/3 satellites, while HYDROWEB additionally uses GFO-1, Envisat, SARAL, and Sentinel-3A (see Table 2). The NASA-built nadir pointing Radar Altimeter using C band (5.3 GHz) and Ku band (13.6 GHz) for measuring height above sea surface was operated on TOPEX/Poseidon. The same type of radar altimeters Poseidon-2 was installed onboard Jason-1, Poseidon-3 onboard Jason-2, and Poseidon-3B onboard Jason-3. Slightly different types of altimeters were installed onboard GFO-1, Envisat, SARAL, and Sentinel-3A satellites. Second, ground tracks for TOPEX/Poseidon and Jason-1/2/3 are the same, but they differ from ground tracks of other satellites. Thus, the distance between the ground tracks and gauge station as well as geographic orientation of ground tracks is different for different sets of satellites. Third, DAHITI, G-REALM, and HYDROWEB use different algorithms for processing altimetry data, in particular, corrections for wet and dry troposphere, ionosphere, solid Earth tide, Pole tide, range bias, and geoid. Finally, retracking of satellite altimetry data is done with the use of different algorithms.

It is evident that the specific location of gauge stations in Ladoga and Onega lakes may have an impact on the validation and calibration of individual altimeters as well as of altimetry databases. For instance, three gauge stations are located along the southeastern coast of Lake Ladoga. Moreover, the Petrokrepost gauge station is located in the entrance to the Neva River, Syas'skye Ryadki is located in the mouth of the Syas' River which runs to the Volkhov Bay of Lake Ladoga, and Novaya Sviritsa is located in the mouth of the Svir' River which runs to Lake Ladoga (Figure 3). The Kondopoga, Petrozavodsk, and Voznesenye gauge stations at Lake Onega are located at the western shore but in the narrow and long bays of the lake (Figure 3). The location of the gauge stations in the mouths of small rivers and in the bays may cause problems in correct validation and calibration of satellite altimetry data for these water bodies. In this respect, they are not representative for satellite altimetry, but there is no choice.

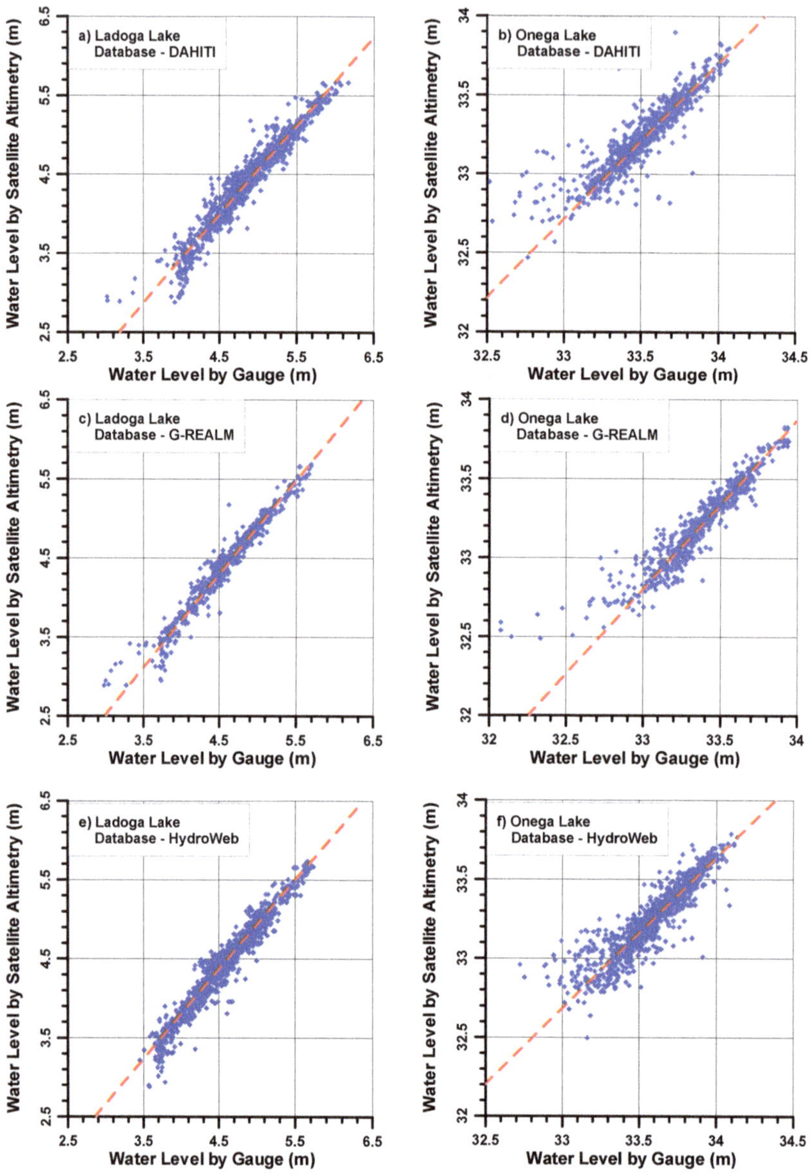

Figure 9. Scatterplots between water levels derived from the DAHITI (**a–b**), G-REALM (**c–d**), and HYDROWEB (**e–f**) altimetry databases and in-situ water level gauge stations for Lake Ladoga (Petrokrepost) and Lake Onega (Voznesenye) from 1 January 2001 to 31 December 2020.

The DAHITI, G-REALM, and HYDROWEB databases can be additionally calibrated against these three gauge stations at Lake Ladoga and three gauge stations at Lake Onega based on the linear regression equations presented in Tables 7 and 8. Moreover, in wintertime and in the presence of ice, the abovementioned equations can be used to fill the gaps in water level data in the satellite altimetry databases based on daily data records on water level at gauge stations.

5. Discussion

Regional climate change is a critical factor for changes in the hydrological regime and water balance of Ladoga and Onega lakes [29–35]. Since the early 2000s, several papers have been published on regional climate change in the study region, mainly in the drainage basin of the White Sea and Karelia. However, in most works the study period ended by 2000, therefore critical climate change in the first two decades of the 21st century was not reflected [27].

For example, in [60–62] it was shown that for the White Sea region since 1985 a period began which was characterized by positive anomalies of the eastern transport of air masses, while the frequency of occurrence of the western forms of transport was close to the multiyear average. Negative air temperature anomalies in the late 1980s changed to positive ones up to 2.0–2.5 °C, and atmospheric pressure anomalies changed their sign from positive to negative. A decrease in atmospheric pressure is explained by an increase in western transports, an increase in cyclonic activity, and an increase in air temperature is explained by the influx of warm air masses from the Atlantic. It is interesting to note that from 1930 to 2000, a weak negative trend in the air temperature was observed at almost all meteorological stations in the region [62]. The spectrum of fluctuations in surface air temperature in Arkhangelsk for 176 years revealed dominant fluctuations for a period of 4–5 and 11–13 years, which indicates the influence of large-scale climatic processes on the formation of long-term variability of hydrometeorological characteristics of the White Sea Region [62]. It is expected that the air temperature in the region by the middle of the 21st century will rise by 2.5–2.7 °C [62].

The temperature regime of the territory of Karelia in 1950–2011 is considered in [63,64]. It was shown that since 1989 only positive air temperature anomalies of the order of 1–2 °C have been observed, compared with the average value for 1961–1990. The greatest warming was observed in winter and in March from 0.3 °C to 0.6 °C/10 years. An increase in air temperature obviously leads to an increase in soil temperature. Analysis of observational data using soil temperature sensors at depths from 20 to 320 cm showed that during the 1990s and 2000s, the average annual soil temperature at various depths up to 320 cm in Karelia exceeded the climatic norm by 1.0–1.5 °C in the southern regions and by 0.5–1.0 °C in the northern regions of Karelia. It should be noted that in southern Karelia from December to mid-April the climatic norm of soil temperature was between −1.5 and −2.0 °C at a depth of 20 cm, however, during 1991–2011 soil freezing was not noted, while in the central and northern regions of Karelia it was still observed. At other depths the mean monthly temperature values were positive throughout the year. In [65], according to satellite data for 2002–2019, it was shown that the rate of growth of soil temperature in some areas of the Kola Peninsula can reach +0.1 °C/year.

In [66,67], to analyze the interannual variability of the thermal and ice regime of the lakes of Karelia, data on air temperature from adjacent meteorological stations were used. Analysis of these series for 1953–2011 showed an increasing trend in air temperature of 0.2–0.3 °C/10 years, and in 1976–2014 of 0.57–0.67 °C/10 years in the area between Lake Ladoga and the White Sea, and at the Valaam weather station (Lake Ladoga) up to +1.12 °C/10 years in July.

In [68], it was shown that for the Kola Peninsula air temperature from 1878 to 2013 increased at an average rate of 1 °C/100 years, while the growth rate was noticeably increasing in 1980–2009 by +0.55 °C/10 years. In 2000–2009, the maximum values of the average surface air temperature anomalies were observed in winter (+1.8 °C), and the minimum in summer (+0.75 °C). The authors pay attention to the significant heterogeneity of temperature changes on the Kola Peninsula. For example, on the Murmansk coast facing the Barents Sea, warming was noticeable and statistically significant, while on the Tersk and Kandalaksha shores of the White Sea the trend values were significantly less and practically insignificant. This testifies to the mosaic nature of microclimatic conditions, which for geomorphological processes in small basins of mountain rivers are more important than regional trends.

In [69,70], for the White Sea drainage basin it was shown that, based on the data of meteorological stations, the average air temperature values for 1991–2017 increased by 0.8–1.2 °C compared to the standard climatic period of 1961–1990. The observational data indicate an almost synchronous nature of the variability of the mean annual air temperature over the entire catchment area of the White Sea. As for different seasons of the year, the greatest increase in temperature was recorded for the winter months, especially for January (the average values for 1991–2017 exceeded the climatic norm by 1.7–2.5 °C).

In [71], for the drainage basin of Lake Onega it was shown that in the second half of the 21st century air temperature will increase by an average of 4.53 °C compared to the period 1951–2000, and the most significant warming is expected in winter. It is expected that the increase in average January temperatures in 2001–2050 will be of 5.6 °C, and in 2051–2100 of 8.6 °C.

Recently, Serykh et al. [27] investigated changes in near surface air temperature and vorticity of the wind speed field of the White Sea and the territory of the Murmansk and Arkhangelsk regions and the Republic of Karelia. They analyzed the monthly average NCEP/NCAR reanalysis data for the period 1950–2020. The average air temperature growth estimated using a linear trend was +0.24 °C/10 years. Against the background of this linear growth, significant interdecadal changes in air temperature were observed. The following periods are highlighted: strengthening of climate continentality (1950–1976), a more maritime climate (1977–1998), and the rapid growth of air temperature (1999–2020). The transition from a period of increasing climate continentality to a period of a more maritime climate is associated with an increase in the influence of the North Atlantic on the study region. A hypothesis has been put forward that the period of rapid growth of air temperature is caused by the transition of the climatic system of the western part of the Russian Arctic into a new phase state.

This study shows [27] that in the White Sea Region in the last two decades (1999–2020) there was a significant change in the regional climate, which was expressed in the warming of this region from +0.9 to + 1.5 °C compared to previous years (1977–1998), in a sharp increase of air temperature (+0.4–+ 1.0°C over 10 years), and to a shift of the +2 °C isotherm by 550 km northward up to the southern part of the White Sea and to the complete disappearance of average negative temperatures. The latter fact is extremely important, since it means a phase transition which will lead to thawing of permafrost soils throughout the territory of the Karelia, Murmansk, and Arkhangelsk regions. The thawing of permafrost soils and a significant increase in average temperatures will lead to a change in the water balance of numerous rivers and lakes in the study region, to an increase in geomorphological processes such as water-snow flows (a type of mudflows) and landslides [68] and may have a negative impact on the transport infrastructure [72], including the infrastructure and operability of Russian railways in the study region.

To clarify warming and precipitation rates namely at Ladoga and Onega lakes, we calculated interannual trends in air temperature for 1945–2019 and the sum of precipitation for 1966–2019 recorded at meteostations Sortavala (Lake Ladoga, WMO N 22802), Petrozavodsk (Lake Onega, WMO N 22820), and Vytegra (Lake Onega, WMO N 22837) (Table 9). Indeed, during the past 75 years there has been a warming at all meteostations, with an average rate in the range of +0.029–0.031 °C/year (Table 9). Since 1945, the average air temperature increased by about 2.5 °C from 2.0–2.5 °C to about 5.0 °C in 2019. We observed that since 2000 the monthly average air temperature during summer has reached 20 °C more often than in the previous period, but a general increase in summer temperatures is not significant. On the other hand, winter temperatures after 1987 have gotten warmer and warmer, and during the past 30 years have always been higher than −15 °C. This means that the winter period has shortened, which results in a decrease of time periods with negative temperatures, snowfalls, and ice cover, as well as a change of the hydrological regime of the watershed area of the lakes and water-air heat and mass transfer over Ladoga and Onega lakes.

Table 9. Characteristics of linear regression trends in air temperature for 1945–2019 and precipitation for 1966–2019 at meteostations Sortavala (Lake Ladoga, WMO N 22802), Petrozavodsk (Lake Onega, WMO N 22820), and Vytegra (Lake Onega, WMO N 22837).

WMO Meteostation Number	Station	Lake	Time Period	Y = aX + b		Coefficient of Determination, R^2
				a	b	
		Air Temperature Trend (°C/year)				
22802	Sortavala	Ladoga	1945–2019	+0.029	−54.47	0.0046
22820	Petrozavodsk	Onega	1949–2019	+0.030	−57.41	0.0042
22837	Vytegra	Onega	1945–2019	+0.031	−57.72	0.0044
		Sum of Precipitation Trend (mm/month/year)				
22802	Sortavala	Ladoga	1966–2019	+0.227	−399.40	0.0119
22820	Petrozavodsk	Onega	1966–2019	+0.086	−122.79	0.0020
22837	Vytegra	Onega	1966–2019	+0.114	−170.54	0.0030

Analysis of meteostations' records shows a notable increase in precipitation with an average rate from +0.086 to +0.227 mm/month/year (Table 9) for the past 54 years depending on the meteostation location. This means that monthly mean precipitation increased from 4.6 to 12.3 mm/month or by 10–25% from 1966 to 2019, which is a significant value. We also observed that during the past 25 years heavy rains seem to occur more often because precipitation over 150 mm/month, which is triple the yearly average value, is registered more often. All these factors directly impact the water regime on the watershed area of the lakes, the water balance, and the level of Ladoga and Onega lakes.

To reveal general cyclicality in the interannual variability of the water level, air temperature, and precipitation over Ladoga and Onega lakes, we performed wavelet analysis of these data after preliminary exclusion the linear trends from these time series. As for the water level, we used averaged values obtained from the DAHITI, G-REALM, and HYDROWEB databases for 1993–2019. For air temperature and precipitation, we used records from the meteostations Sortavala at Lake Ladoga and Vytegra for Lake Onega for 1993–2019 to have the same time interval with altimetry data. The Ricker wavelet (or Mexican-hat wavelet) and Morlet wavelet are the most common tools for hydrometeorological applications [73–77].

The Ricker wavelet transform of the time series of the monthly average water level shows a powerful oscillation with a period of 12.5 ± 1.9 years for Lake Ladoga (Figure 10a) and 11.7 ± 2.4 for Lake Onega (Figure 11a). In some time intervals, weak fluctuations in the water level of Lake Ladoga were shown with a period of 2.98 ± 0.43 years (Figure 10a), and Lake Onega with a period of 2.96 ± 0.41 years (Figure 11a). The Ricker wavelet transform of the time series of the monthly average air temperature and precipitation at the Sortavala weather station (Lake Ladoga) (Figure 10b,c) and Vytegra weather station (Lake Onega) (Figure 11b,c) showed fluctuations with periods of two years and from 8 to 21 years; and from 5–7 to 18 years, respectively.

Thus, the wavelet analysis showed cyclicity in the water level on periods of 3 and 12 years, in air temperature on periods of 2 years and from 8 to 21 years, and in precipitation on periods from 5–7 to 18 years. Filatov et al. [34] showed oscillations in the water level and air temperature at Ladoga and Onega lakes for 1900–2014 with periods of 2, 5–7, and 30 years. The discrepancy can be explained by a much longer time period (115 years) used by Filatov et al. [34], as only 27 years were used in our analysis. It can also be explained by a hypothesis that has been put forward by Serykh et al. [27] that the period of rapid growth of air temperature (1999–2000) was caused by the transition of the climatic system of the western part of the Russian Arctic into a new phase state. Our period of investigation of 1993–2019 fits this new state of regional climate with new or not-yet-established oscillations. Finally, concerning the water level, we used satellite altimetry data and Filatov et al. [34]

used in-situ data. As was already mentioned, these are a bit different data measured at different places (in the middle of the lake in case of altimetry and in river mouths or narrow bays for gauge stations) and with different frequency (daily at gauge stations and every 10 days in case of altimetry).

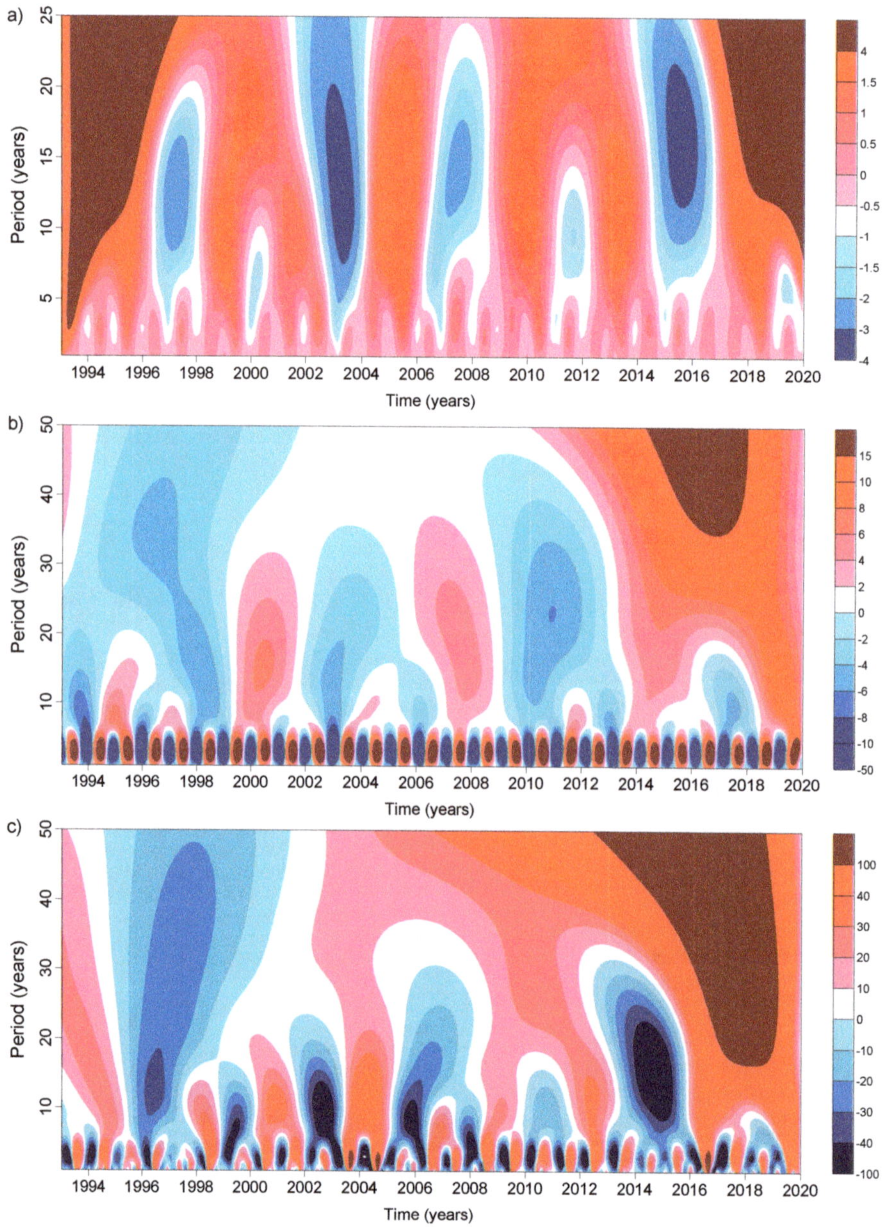

Figure 10. The wavelet transform of the monthly average water level derived from satellite altimetry (**a**), air temperature, (**b**) and precipitation (**c**) at the Sortavala meteostation (Lake Ladoga) for 1993–2019.

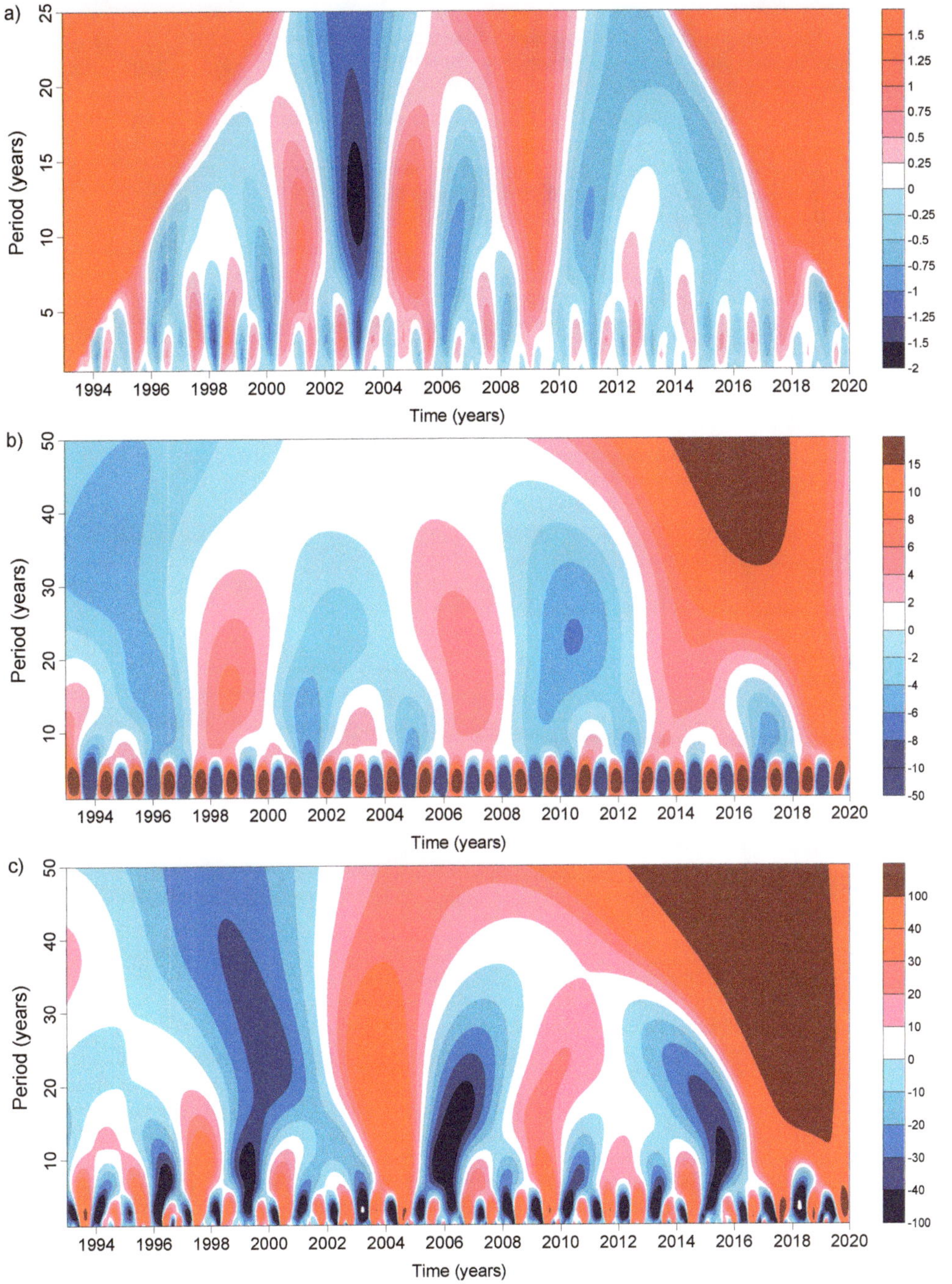

Figure 11. The wavelet transform of the monthly average water level derived from satellite altimetry (**a**), air temperature, (**b**) and precipitation (**c**) at the Vytegra meteostation (Lake Onega) for 1993–2019.

6. Conclusions

The conducted study of the interannual variability of the water level of Lake Ladoga for more than 28 years (October 1992–December 2020) showed that the change in the amplitude of interannual fluctuations of the level is in the range between 2.28 m (HYDROWEB) and 3.15 m (DAHITI). The linear interannual trend varies from +0.27 cm/year (HYDROWEB) to +1.17 cm/year (DAHITI). For Lake Onega the amplitude of water level variations changes from 1.49 m (HYDROWEB) to 1.87 m (G-REALM). The linear interannual trend varies from −0.27 cm/year (HYDROWEB) to +0.62 cm/year (G-REALM).

Bogdanov et al. [33] showed interannual variability of the Lake Ladoga level at the Valaam gauge station for 1859–2001 with a general decrease of the water level with a rate of −4.24 mm/year. Unfortunately, the periodicity in water level variations was not established. It is interesting that this general water level decrease corresponds exactly to the Earth crust rise of +3.3 mm/year measured in 1999–2009 at the Valaam and +4.2 mm/year at the Sortavala meteostations at Lake Ladoga with the help of high-resolution GPS receivers [78].

It seems that a long-term decreasing trend in water level variability has changed to an increasing one at the turn of the 20th and 21st centuries, simultaneously with a sharp change in the regional climate. The water level rise in both lakes seems to be related to a significant increase in precipitation over the lakes by 10–25% from 1966 to 2019. An increase in air temperature may have changed the hydrological regime of both lakes affecting snow/rain, ice cover, groundwater, and runoff of numerous small rivers. This may have led to an additional increase in the incoming part of water balance and, as a result, to a steady increase in the water level.

Such significant interannual fluctuations in the level of both lakes, as well as high trends in the water level rise, taking into account the expected changes in the regional climate (increase in air and soil temperature, and precipitation, change in snow and ice cover, hydrological regime and water balance), can lead to flooding of roads and railways and the corresponding infrastructure (bridges, embankments, power lines, etc.) located in the vicinity of Ladoga and Onega lakes [22,23,41].

Thawing of permafrost soils and a significant increase in average temperatures will lead to a change in the water balance of numerous rivers and lakes in the study region, to an increase in geomorphological processes such as water-snow flows (a type of mudflows) and landslides [68], and may have a negative impact on the transport infrastructure [72], including the infrastructure and operability of Russian railways in the study region.

In [72], it was shown that for the territory of the Kola Peninsula, which is characterized by high-temperature permafrost of a rare-island nature, infrastructure, whose base is formed by piles 5 m deep or less, is exposed to very high risks of decreasing functionality, down to the zero level. For example, with relatively small warming up to +1 °C, piles 5 m deep, as a rule, still provide functionality $U = 0.65$–0.85 ($U = 1$ corresponds to the maximum level of functionality for a basic, unchanged climate), which is considered as an average level of a climatic risk in relation to the object of transport infrastructure. Warming to +2 °C often already leads to a decrease in functionality to $U < 0.5$ (in the presence of soils with low humidity) and warming to + 3 °C for objects of this group can be considered catastrophic; functionality is reduced to the level $U = 0$–0.35, which corresponds to an unacceptably high level of climatic risks. An extremely important conclusion follows from the study: Objects on pile foundations with a depth of less than 6 m are under an increased risk even when the temperature is up to +2 °C, therefore, engineering protection of these objects should be carried out at a rate that outstrips regional climate warming [72].

Google Earth is a handy tool to search for critical areas in terms of potential impact of the water level rise in Ladoga and Onega lakes on the Russian railway infrastructure. Google Earth allows for checking the distance between the shoreline and railway tracks, as well as checking the height of the railway track over the water level. If both parameters are minimal, we consider this location to be a critical one. In this way, we checked the whole length of railways which encircle Lake Ladoga and a part of railways which passes along Lake Onega on its western shore (Figure 2). In the area of Lake Ladoga, one should pay attention to the northeastern coast of the lake near the settlements of Tuloksa, Vidlitsa, Salmi, Pitkyaranta, Sortavala, and Lake Karkumlampi. In the area of Lake Onega, such critical locations are situated in the northwestern part of its coast, near the settlements of Shuya, Yanishpole, Kondopoga, and Lake Kedr. Since the flow from Lake Onega is regulated by the Verkhne-Svirskaya HPP, its level can be maintained at a safe level with the help of this hydraulic structure. It is thanks to the hydroelectric power station that the range of interannual fluctuations of Lake Onega is only 1.49–1.87 m, while for Lake Ladoga this is between 2.28 m and 3.15 m.

It is necessary to conduct further studies of the level of Ladoga and Onega lakes, and other lakes in Karelia, to identify trends and cyclicality of its variability and assess its relationship with change in air and soil temperature, precipitation, and snow cover. Special attention should be paid to further validation and calibration of the DAHITI, G-REALM, and HYDROWEB satellite altimetry databases because they differ significantly among each other and from in-situ records at the local water level gauge stations. The obtained linear regressions between altimetry and in-situ data could help in recalculation of water level datasets specific for Ladoga and Onega lakes, as well as for filling the gaps in data related to ice cover during wintertime. All these efforts will make it possible to forecast sustainable development, management, and safe functioning of the railway transport infrastructure in the Karelia and the Murmansk region.

Author Contributions: Conceptualization, all authors; methodology, A.G.K. and S.A.L.; software, S.A.L. and Y.A.P.; validation, S.A.L. and Y.A.P.; formal analysis, A.G.K. and S.A.L.; investigation, all authors; data curation, S.A.L. and Y.A.P.; writing—original draft preparation, A.G.K. and E.A.K.; writing—review and editing, A.G.K. and E.A.K.; visualization, S.A.L. and Y.A.P.; supervision, A.G.K.; funding acquisition, A.G.K. and S.A.L. All authors have read and agreed to the published version of the manuscript.

Funding: The research was supported by the Russian Science Foundation (RNF) Grant N 21-77-30010 (2021–2024) "System analysis of geophysical process dynamics in the Russian Arctic and their impact on the development and operation of the railway infrastructure".

Institutional Review Board Statement: Not applicable.

Informed Consent Statement: Not applicable.

Data Availability Statement: All freely available data are mentioned in section on Data and Methods.

Acknowledgments: The authors would like to thank Remote Sensing, MDPI and Geophysical Center, and Russian Academy of Sciences for coverage of the APC.

Conflicts of Interest: The authors declare no conflict of interest.

References

1. Kostianoy, A.G.; Zavialov, P.O.; Lebedev, S.A. What do we know about dead, dying and endangered lakes and seas? In *Dying and Dead Seas. Climatic versus Anthropic Causes*; Nihoul, J.C.J., Zavialov, P.O., Micklin, P.P., Eds.; NATO ARW/ASI Series; Kluwer Acad. Publ.: Dordrecht, The Netherlands, 2004; pp. 1–48.
2. Kostianoy, A.G. Dead and Dying Seas. In *Encyclopedia of Water Science*; Taylor & Francis: Abingdon, UK, 2007; pp. 1–3. [CrossRef]
3. Adrn, R.; O'Reilly, C.M.; Zagarese, H.; Baines, S.B.; Hessen, D.O.; Keller, W.; Livingstone, D.M.; Sommaruga, R.; Straile, D.; VanDonk, E.; et al. Lakes as sentinels of climate change. *Limnol. Oceanogr.* **2009**, *54*, 2283–2297. [CrossRef] [PubMed]
4. Schindler, D.W. Lakes as sentinels and integrators for the effects of climate change on watersheds, airsheds, and landscapes. *Limnol. Oceanogr.* **2009**, *54*, 2349–2358. [CrossRef]
5. Williamson, C.E.; Saros, J.E.; Vincent, W.F.; Smol, J.P. Lakes and reservoirs as sentinels, integrators, and regulators of climate change. *Limnol. Oceanogr.* **2009**, *54*, 2273–2282. [CrossRef]
6. Birkett, C.; Reynolds, C.; Beckley, B.; Doorn, B. From research to operations: The USDA global reservoir and lake monitor. In *Coastal Altimetry*; Springer: Berlin/Heidelberg, Germany, 2011; pp. 19–50.
7. Kostianoy, A.G. Seas: Dead and Dying. In *Encyclopedia of Water Science*, 2nd ed.; Taylor & Francis: Abingdon, UK, 2011; pp. 1029–1031. [CrossRef]
8. Kostianoy, A.G. Inland seas and lakes: Central Asia case study. In *Encyclopedia of Environmental Management*; Jorgensen, S.E., Ed.; Taylor & Francis: New York, NY, USA, 2013; Volume II, pp. 1460–1467.

9. Williamson, C.E.; Brentrup, J.A.; Zhang, J.; Renwick, W.H.; Hargreaves, B.R.; Knoll, L.B.; Overholt, E.P.; Rose, K.C. Lakes as sensors in the landscape: Optical metrics as scalable sentinel responses to climate change. *Limnol. Oceanogr.* **2014**, *59*, 840–850. [CrossRef]
10. Crétaux, J.-F.; Biancamaria, S.; Arsen, A.; Bergé-Nguyen, M.; Becker, M. Global surveys of reservoirs and lakes from satellites and regional application to the Syrdarya river basin. *Environ. Res. Lett.* **2015**, *10*, 015002. [CrossRef]
11. Sulistioadi, Y.B.; Tseng, K.-H.; Shum, C.K.; Hidayat, H.; Sumaryono, M.; Suhardiman, A.; Setiawan, F.; Sunarso, S. Satellite radar altimetry for monitoring small rivers and lakes in Indonesia. *Hydrol. Earth Syst. Sci.* **2015**, *19*, 341–359. [CrossRef]
12. Kostianoy, A.G. Inland Seas and Lakes: Central Asia Case Study. In *Environmental Management Handbook, Second Edition, Volume 4: Managing Water Resources and Hydrological Systems*; Fath, B.D., Jorgensen, S.E., Eds.; Taylor and Francis: Abingdon, UK; CRC Press: Boca Raton, FL, USA, 2020; pp. 337–348. [CrossRef]
13. Tortini, R.; Noujdina, N.; Yeo, S.; Ricko, M.; Birkett, C.M.; Khandelwal, A.; Kumar, V.; Marlier, M.E.; Lettenmaier, D.P. Satellite-based remote sensing data set of global surface water storage change from 1992 to 2018. *Earth Syst. Sci. Data* **2020**, *12*, 1141–1151. [CrossRef]
14. Pham-Duc, B.; Sylvestre, F.; Papa, F.; Frappart, F.; Bouchez, C.; Crétaux, J.F. The Lake Chad hydrology under current climate change. *Sci. Rep.* **2020**, *10*, 5498. [CrossRef]
15. GOSIC. 2021. Available online: https://www.ncdc.noaa.gov/gosic (accessed on 14 November 2021).
16. GTN-L. 2021. Available online: https://community.wmo.int/governance/commission-membership/executive-council/executive-council-members/global-climate-observing-system-gcos/gcos-steering-committee/terrestrial-observation-panel-climate/global-terrestrial-network-lakes (accessed on 14 November 2021).
17. Frappart, F.; Calmant, S.; Cauhopé, M.; Seyler, F.; Cazenave, A. Preliminary results of ENVISAT RA-2-derived water levels validation over the Amazon basin. *Remote Sens. Environ.* **2006**, *100*, 252–264. [CrossRef]
18. Crétaux, J.-F.; Nielsen, K.; Frappart, F.; Papa, F.; Calmant, S.; Benveniste, J. Hydrological applications of satellite altimetry: Rivers, lakes, man-made reservoirs, inundated areas. In *Satellite Altimetry over Oceans and Land Surfaces*; Earth Observation of Global Changes; Stammer, D., Cazenave, A., Eds.; CRC Press: Boca Raton, FL, USA, 2017; pp. 459–504.
19. Shu, S.; Liu, H.; Beck, R.A.; Frappart, F.; Korhonen, J.; Lan, M.; Xu, M.; Yang, B.; Huang, Y. Evaluation of historic and operational satellite radar altimetry missions for constructing consistent long-term lake water level records. *Hydrol. Earth Syst. Sci.* **2021**, *25*, 1643–1670. [CrossRef]
20. Bergé-Nguyen, M.; Cretaux, J.F.; Calmant, S.; Fleury, S.; Satylkanov, R.; Chontoev, D.; Bonnefond, P. Mapping mean lake surface from satellite altimetry and GPS kinematic surveys. *Adv. Sp. Res.* **2021**, *67*, 985–1001. [CrossRef]
21. Frappart, F.; Blarel, F.; Fayad, I.; Bergé-Nguyen, M.; Crétaux, J.-F.; Shu, S.; Schregenberger, J.; Baghdadi, N. Evaluation of the Performances of Radar and Lidar Altimetry Missions for Water Level Retrievals in Mountainous Environment: The Case of the Swiss Lakes. *Remote Sens.* **2021**, *13*, 2196. [CrossRef]
22. Kostianaia, E.A.; Kostianoy, A.G.; Scheglov, M.A.; Karelov, A.I.; Vasileisky, A.S. Impact of regional climate change on the infrastructure and operability of railway transport. *Transp. Telecommun.* **2021**, *22*, 183–195. [CrossRef]
23. *World Experience in the Construction, Maintenance and Repair of Railway Infrastructure Facilities in Permafrost Conditions*; Center for Scientific and Technical Information and Libraries: Moscow, Russia, 2017; 126p. (In Russian)
24. Serykh, I.V.; Kostianoy, A.G. Seasonal and interannual variability of the Barents Sea temperature. *Ecol. Montenegrina* **2019**, *25*, 1–13. [CrossRef]
25. Serykh, I.V.; Kostianoy, A.G. About the climatic changes in the temperature of the Baltic Sea. *Fundam. Appl. Hydrophys.* **2019**, *12*, 5–12. (In Russian) [CrossRef]
26. Serykh, I.V.; Tolstikov, A.V. *On the Reasons for the Long-Term Variability of the Near-Surface Air Temperature over the White Sea*; Moscow University Bulletin; Series 5: Geography; Moscow University: Moscow, Russia, 2020; Volume 4, pp. 83–95. (In Russian)
27. Serykh, I.V.; Kostianoy, A.G.; Lebedev, S.A.; Kostianaia, E.A. On the transition of temperature regime of the White Sea Region to a new phase state. *Fundam. Appl. Hydrophys.* 2022, *in press* (In Russian)
28. Andreev, A.P. *Lake Ladoga*; Part 1–2; Publishing House of Maritime Department in the Main Admiralty: St. Petersburg, Russia, 1875; Part I, 263p, Part II, 135p. (In Russian)
29. Filatov, N.N. (Ed.) *Ladoga Lake*; Petrozavodsk, Karelian Research Center, Russian Academy of Sciences: Petrozavodsk, Russia, 2000; 490p. (In Russian)
30. Rumyantsev, V.A.; Drabkova, V.G. *Ladoga Lake: Past, Present and Future*; Publ. House "Nauka": St. Petersburg, Russia, 2002; 327p. (In Russian)
31. Rukhovets, L.; Filatov, N. (Eds.) *Ladoga and Onego—Great European Lakes*; Observations and Modelling; Springer: Berlin/Heidelberg, Germany; Praxis Publishing Ltd.: Chichester, UK, 2010; 315p.
32. Rumyantsev, V.A.; Kondratyev, S.A. *Ladoga*; Publ. House of Institute of Limnology of Russian Academy of Sciences: St.-Petersburg, Russia, 2013; 568p. (In Russian)
33. Bogdanov, V.I.; Kravchenkova, T.G.; Malova, T.I.; Marinich, M.A. Changes in the level of Lake Ladoga according to observations of 1859–2001 on Valaam. *Doklady Akademii Nauk.* **2002**, *386*, 672–675. (In Russian)
34. Filatov, N.N.; Vyruchalkina, T.Y.; Diansky, N.A.; Nazarova, L.E.; Sinyukovich, V.N. Intracental variability of the level of the largest lakes in Russia. *Doklady Akademii Nauk.* **2016**, *467*, 589–593. (In Russian)

35. Filatov, N.; Baklagin, V.; Efremova, T.; Nazarova, L.; Palshin, N. Climate change impacts on the watersheds of Lakes Onego and Ladoga from remote sensing and in situ data. *Inland Waters* **2019**, *9*, 130–141. [CrossRef]
36. Kouraev, A.V.; Shimaraev, M.N.; Buharizin, P.I.; Naumenko, M.A.; Cretaux, J.F.; Mognard, N.; Legresy, B.; Remy, F. Ice and Snow Cover of Continental Water Bodies from Simultaneous Radar Altimetry and Radiometry Observations. *Surv. Geophys.* **2008**, *29*, 271–295. [CrossRef]
37. Cretaux, J.-F.; Jelinski, V.; Calmant, S.; Kouraev, A.; Vuglinski, V.; Bergé-Nguyen, M.; Gennero, M.-C.; Nino, F.; Abarco Del Rio, R.; Cazenave, A.; et al. SOLS, a lake database to monitor in near real time water level storage variations from remote sensing data. *Adv. Space Res.* **2011**, *47*, 1497–1507. [CrossRef]
38. Lebedev, S.A.; Troitskaya, Y.I.; Rybushkina, G.V.; Dobrovolsky, M.N. Satellite altimetry of large lakes of the Baltic Basin. In Proceedings of the 2014 IEEE/OES Baltic International Symposium (BALTIC), Tallinn, Estonia, 27–29 May 2014; IEEE: Piscataway, NJ, USA, 2014; pp. 1–5. [CrossRef]
39. Lebedev, S.A.; Troitskaya, Y.I.; Rybushkina, G.V.; Dobrovolsky, M.N. Interannual variability of the lake levels in northwest Russia based on satellite altimetry. In *Complex Interfaces under Change: Sea—River—Groundwater—Lake, Proceedings of the HP2/HP3, IAHS-IAPSO-IASPEI Assembly, Gothenburg, Sweden, 22–26 July 2013*; IAHS Publ.: Oxfordshire, UK, 2015; pp. 91–97. [CrossRef]
40. Karetnikov, S.G.; Naumenko, M.A.; Guzivatiy, V.V.; Shimaraev, M.N.; Kouraev, A.V. Concordance of interannual variability in ice regime of Baikal and lake Ladoga. *Geogr. Nat. Resour.* **2016**, *37*, 135–143. [CrossRef]
41. Kostianoy, A.G.; Kostianaia, E.A.; Lebedev, S.A.; Serykh, I.V.; Prokofiev, Y.A. Interannual variability in the level of lakes in northwestern Russia. In Proceedings of the VI International Scientific and Practical Conference "Fundamental and Applied Aspects of Geology, Geophysics and Geoecology Using Modern Information Technologies", Maykop State Technological University, Maykop, Russia, 17–21 May 2021; Part 1. pp. 158–167. (In Russian)
42. Campos, I.O.; Mercier, F.; Maheu, C.; Cochonneau, G.; Kosuth, P.; Blitzkow, D.; Cazenave, A. Temporal variations of river basin waters from Topex/Poseidon satellite altimetry. Application to the Amazon basin. *Earth Planet. Sci.* **2001**, *333*, 633–643. [CrossRef]
43. Birkett, C.M.; Mertes, L.A.K.; Dunne, T.; Costa, M.H.; Jasinski, M.J. Surface water dynamics in the Amazon Basin: Application of satellite radar altimetry. *J. Geophys. Res.* **2002**, *107*, 8059. [CrossRef]
44. Kouraev, A.V.; Zakharova, E.A.; Samain, O.; Mognard, N.M.; Cazenave, A. Ob' river discharge from TOPEX/Poseidon satellite altimetry (1992–2002). *Remote Sens. Environ.* **2004**, *93*, 238–245. [CrossRef]
45. Birkett, C.M.; Beckley, B. Investigating the performance of the Jason-2/OSTM radar altimeter over lakes and reservoirs. *Mar. Geodesy* **2010**, *33*, 204–238. [CrossRef]
46. Lebedev, S.A.; Kostianoy, A.G. *Satellite Altimetry of the Caspian Sea*; Publ. House "Sea": Moscow, Russia, 2005; 366p. (In Russian)
47. Rybushkina, G.V.; Troitskaya, Y.I.; Soustova, I.A.; Balandina, G.N.; Lebedev, S.A.; Kostianoy, A.G. Satellite altimetry of inland water bodies. *Tr. GOIN (Proc. Russ. State Oceanogr. Inst.)* **2011**, *213*, 179–192. (In Russian)
48. Troitskaya, Y.I.; Rybushkina, G.V.; Soustova, I.A.; Balandina, G.N.; Lebedev, S.A.; Kostianoy, A.G.; Panyutin, A.A.; Filina, L.V. Satellite altimetry of inland water bodies. *Water Resour.* **2012**, *39*, 184–199. [CrossRef]
49. Troitskaya, Y.; Rybushkina, G.; Soustova, I.; Balandina, G.; Lebedev, S.; Kostianoy, A. Adaptive retracking of Jason-1 altimetry data for inland waters: The example of the Gorky Reservoir. *Int. J. Remote Sens.* **2012**, *33*, 7559–7578. [CrossRef]
50. Troitskaya, Y.; Rybushkina, G.; Soustova, I.; Lebedev, S. Adaptive Retracking of Jason-1, 2 Satellite Altimetry Data for the Volga River Reservoirs. *IEEE J. Sel. Top. Appl. Earth Obs. Remote Sens.* **2014**, *7*, 1603–1608. [CrossRef]
51. Kostianoy, A.G.; Lebedev, S.A.; Kazmina, M.V.; Tsepelev VYu Varlashina, V.M.; Neschadimova, T.G. *Satellite Monitoring of Transboundary Waters of Russia and Estonia*; Publ. House "Signal": Moscow, Russia, 2012; 16p, (In Russian and English)
52. Liibusk, A.; Kall, T.; Rikka, S.; Uiboupin, R.; Suursaar, Ü.; Tseng, K.-H. Validation of Copernicus sea level altimetry products in the Baltic Sea and Estonian Lakes. *Remote Sens.* **2021**, *12*, 4062. [CrossRef]
53. Berry, P.A.M.; Garlick, J.D.; Freeman, J.A.; Mathers, E.L. Global inland water monitoring from multimission altimetry. *Geophys. Res. Lett.* **2005**, *32*, L16401. [CrossRef]
54. Calmant, S.; Seyler, F. Continental surface waters from satellite altimetry. *C. R. Geosci.* **2006**, *338*, 1113–1122. [CrossRef]
55. Cretaux, J.-F.; Calmant, S.; Abarca del Rio, R.; Kouraev, A.; Berge-Nguyen, M.; Maisongrande, P. Lakes studies from satellite altimetry. In *Coastal Altimetry*; Vignudelli, S., Kostianoy, A.G., Cipollini, P., Benveniste, J., Eds.; Springer: Berlin/Heidelberg, Germany, 2011; pp. 509–534. [CrossRef]
56. International Altimetry Team (358 Co-Authors). Altimetry for the Future: Building on 25 Years of Progress. *Adv. Space Res.* **2021**, *68*, 319–363. [CrossRef]
57. Schwatke, C.; Dettmering, D.; Bosch, W.; Seitz, F. DAHITI—An innovative approach for estimating water level time series over inland waters using multi-mission satellite altimetry. *Hydrol. Earth Syst. Sci.* **2015**, *19*, 4345–4364. [CrossRef]
58. Molina, J.L.; Rodríguez-Gonzálvez, P.; Molina, M.C.; González-Aguilera, D.; Espejo, F. Geomatic methods at the service of water resource smodelling. *J. Hydrol.* **2014**, *509*, 105–162. [CrossRef]
59. García-López, S.; Ruiz-Ortiz, V.; Muñoz-Pérez, J.J. Time-lapse photography for monitoring reservoir leakages (Montejaque dam, Andalusia, southern Spain). *Hydrol. Res.* **2018**, *49*, 281–290. [CrossRef]
60. Filatov, N.N. (Ed.) *Climate of Karelia: Variability and Impact on Water Bodies and Catchments*; Karelian Research Center, Russian Academy of Sciences: Petrozavodsk, Russia, 2004; 224p. (In Russian)
61. Filatov, N.; Johannessen, O.M.; Pozdnyakov, D.; Bobylev, L.P.; Pettersson, L. *White Sea: Its Marine Environment and Ecosystem Dynamics Influenced by Global Change*; Springer-Praxis: Chichester, UK, 2005; 530p.

62. Filatov, N.N.; Terzhevik, A.Y. (Eds.) *The White Sea and Its Catchment Area under the Influence of Climatic and Anthropogenic Factors;* Institute of Northern Water Problems, Karelian Research Center, Russian Academy of Sciences: Petrozavodsk, Russia, 2007; 349p. (In Russian)
63. Nazarova, L.E. Variability of mean long-term values of air temperature in Karelia. *Izv. Russ. Geogr. Soc.* **2014**, *46*, 27–33. (In Russian)
64. Nazarova, L.E. Climate of the Republic of Karelia (Russia): Air temperature, variability and changes. *Geopolit. Ecogeodyn. Reg.* **2014**, *10*, 746–749. (In Russian)
65. Tronin, A.A.; Gorny, V.I.; Kiselev, A.V.; Kritsuk, S.G.; Latypov, I.S. Satellite mapping of the thermal response of northern European ecosystems to climate change. In Proceedings of the All-Russian Scientific Conference with International Participation "Ecology and Space" to the Centenary of Academician of the Russian Academy of Sciences K.Ya. Kondratyev, St. Petersburg, Russia, 25–26 May 2020; pp. 72–78. (In Russian)
66. Efremova, T.V.; Pal'shin, N.I. Influence of climate change on the thermal and ice regime of the lakes of Karelia according to the data of instrumental observations. In *Ecological Problems of Northern Regions and Ways to Solve Them, Proceedings of the VI All-Russian Scientific Conference with International Participation Devoted to 120 Years Anniversary of G.M. Kreps and 110 Years An-niversary of O.I. Semenov-Tyan-Shanskiy, Apatity, Russia, 10–14 October 2016;* Institute of North Industrial Ecology Problems of the North of the Kola Science Center of the Russian Academy of Sciences: Apatity, Russia, 2016; pp. 189–193. (In Russian)
67. Efremova, T.V.; Pal'shin, N.I.; Belashev, B.Z. Water temperature of different types of lakes in Karelia under conditions of climate change (according to the data of instrumental measurements in 1953–2011). *Water Resour.* **2016**, *43*, 228–238. (In Russian) [CrossRef]
68. Romanenko, F.A.; Shilovtseva, O.A. *Geomorphological Processes in the Mountains of the Kola Peninsula and Climate Change;* Vestnik Moscow University, Series 5 Geography; Moscow University: Moscow, Russia, 2016; Volume 6, pp. 78–86. (In Russian)
69. Nazarova, L.E. Modern climatic conditions of the White Sea catchment area. *Izv. Russ. Geogr. Soc.* **2017**, *149*, 16–24. (In Russian)
70. Filatov, N.N.; Nazarova, L.E.; Druzhinin, P.V. Influence of climatic and anthropogenic factors on the state of the "White Sea—Catchment" system. *Trans. Karelian Res. Cent. Russ. Acad. Sci.* **2019**, *9*, 30–50. (In Russian)
71. Kondratyev, S.A.; Nazarova, L.E.; Shmakova, M.V.; Bovykin, I.V. *Assessment of Runoff Changes in the Drainage Basin of Lake Onega as a Result of Possible Changes in Climatic Parameters (According to Mathematical Modeling);* Institute of Limnology RAS, Institute of Water Problems of the North, Karelian Research Center, Russian Academy of Sciences: St. Petersburg, Russia, 2007; 30p. (In Russian)
72. Yakubovich, A.N.; Yakubovich, I.A. Forecasting the impact of climatic changes on the functionality of the transport infrastructure of the permafrost zone in Russia. *Intellect. Innov. Investig.* **2019**, *1*, 104–110. (In Russian)
73. Ricker, N. Further developments in the wavelet theory of seismogram structure. *Bull. Seismol. Soc. Am.* **1943**, *33*, 197–228. [CrossRef]
74. Ricker, N. Wavelet functions and their polynomials. *Geophysics* **1944**, *9*, 314–323. [CrossRef]
75. Grossmann, A.; Morlet, J. Decomposition of Hardy functions into square integrable wavelets of constant shape. *SIAM J. Math. Anal.* **1984**, *15*, 723–736. [CrossRef]
76. Astaf'eva, N.M. Wavelet analysis: Basic theory and some applications. *Physics-Uspekhi* **1996**, *39*, 1085–1108. [CrossRef]
77. Mallat, S. *Wavelet Tour of Signal Processing: The Sparse Way*, 3rd ed.; Academic Press: Cambridge, MA, USA; Elsevier: Amsterdam, The Netherlands, 2009; 824p. [CrossRef]
78. Galaganov, O.N.; Gorshkov, V.L.; Guseva, T.V.; Rosenberg, N.K.; Perederin, V.P.; Scherbakova, N.V. Recent crustal motion of Ladoga-Onega Region revealed from satellite and ground measurements. *Curr. Probl. Remote Sens. Earth Space* **2011**, *8*, 130–136. (In Russian)

remote sensing

MDPI

Article

Satellite Remote Sensing of Water Quality Variation in a Semi-Enclosed Bay (Yueqing Bay) under Strong Anthropogenic Impact

Bozhong Zhu [1,2], Yan Bai [1,2,*], Zhao Zhang [3], Xianqiang He [1,2,4], Zhihong Wang [4], Shugang Zhang [3] and Qian Dai [5]

1 Southern Marine Science and Engineering Guangdong Laboratory (Guangzhou), Guangzhou 511458, China; bozhongzhu@sio.org.cn (B.Z.); hexianqiang@sio.org.cn (X.H.)
2 State Key Laboratory of Satellite Ocean Environment Dynamics, Second Institute of Oceanography, Ministry of Natural Resources, Hangzhou 310012, China
3 Wenzhou Marine Environment Inspection Center Station, State Oceanic Administration, Wenzhou 325000, China; zzhao@ecs.mnr.gov.cn (Z.Z.); zhangsg@ecs.mnr.gov.cn (S.Z.)
4 School of Oceanography, Shanghai Jiao Tong University, Shanghai 200030, China; wangzh@sio.org.cn
5 Zhejiang Marine Fisheries Research Institute, Zhoushan 316021, China; huoyingcomcn@163.com
* Correspondence: baiyan@sio.org.cn

Abstract: The semi-enclosed bays impacted by heavy anthropogenic activities have weak water exchange and purification capacities. Most of the sea bays have suffered severe eutrophication, water quality deterioration, ecosystem degradation and other problems. Although many countries and local governments have carried out corresponding environmental protection actions, the evaluation of their effectiveness still requires monitoring technology and data support for long-term water environment change. In this study, we take Yueqing Bay, the fourth largest bay in China, as a case to study the satellite-based water quality monitoring and variation analysis. We established a nutrient retrieval model for Yueqing Bay to produce a long-term series of nutrient concentration products in Yueqing Bay from 2013 to 2020, based on Landsat remote sensing images and long-term observation data, combined with support vector machine learning and water temperature and satellite spectra as input parameters, and then we analyzed its spatiotemporal variations and driving factors. In general, nutrient concentrations in the western part of the bay were higher than those in the eastern part. Levels of dissolved inorganic nitrogen (DIN) were lower in summer than in spring and winter, and reactive phosphate (PO_4-P) levels were lower in summer and higher in autumn. In terms of natural factors, physical effects (e.g., seasonal variations in flow field) and biological effects (e.g., seasonal differences in the intensity of plankton photosynthesis) were the main causes of seasonal differences in nutrient concentration in Yueqing Bay. Nutrient concentration generally increased from 2013 to 2015 but decreased slightly after 2015. Over the past decade, the economy and industry of Yueqing Bay basin have developed rapidly. Wastewater resulting from anthropogenic production and consumption was transported via streams into Yueqing Bay, leading to the continuous increase in nutrient concentrations (the variation rates: $a_{DIN} > 0$, $a_{PO_4-P} > 0$), which directly or indirectly caused high nutrient concentrations in some areas of the bay (e.g., Southwest Shoal at the mouth of Yueqing Bay). After 2015, the various ecological remediation policies adopted by cities around Yueqing Bay have mitigated, to some extent, the increasing nutrient concentration trends (the variation rates: $a_{DIN} < 0$, $a_{PO_4-P} < 0$), but not significantly ($P > 0.1$). The environmental restoration of Yueqing Bay also requires continuous and long-term ecological protection and restoration work to be effective. This research can provide a reference for ecological environment monitoring and remote sensing data application for similar semi-enclosed bays, and support the sustainable development of the bay.

Keywords: nutrient; satellite remote sensing; support vector machine; long time series; driving factor; Yueqing Bay

1. Introduction

With the rapid development of the coastal economy and continuous urbanization, over 60% of the world's population live within 200 km of a coastline [1], and more than 40% of China's population are centered in coastal areas [2]. As such, the impact of anthropogenic activity on the regional environment has also increased, natural resources are exploited on an ever-larger scale, and natural coastal areas and marine ecosystems are under pressure from intensive economic development and coastal zone projects [3–5]. At present, ~41% of sea areas worldwide, especially estuaries and bays, have been seriously disturbed by anthropogenic activities [3]. As a result of ever-increasing urbanization, continuous mariculture scale-up, and rapid development of the ocean economy, land-derived pollutants (e.g., nitrogen and phosphorus) are continuously entering into waters, resulting in water quality deterioration and eutrophication in local areas. In particular, as typical ecologically sensitive areas heavily disturbed by land–sea interactions and anthropogenic activities, semi-enclosed and enclosed bays exhibit weaker water exchange and purification capacities, and thus are highly vulnerable environments [6–8]. As reported by the Ministry of Ecology and Environment of the People's Republic of China, 9.4% of nearshore seawaters in China are worse than grade IV in quality, with particularly high levels of dissolved inorganic nitrogen (DIN) and reactive phosphate (PO_4-P) [9]. Therefore, to monitor variations in water quality of nearshore seas and bays, it is essential to conduct dynamic monitoring of nitrogen and phosphorus elements in water and investigate their driving factors, and then implement corresponding measures and policies to control and improve water quality status.

The accumulation of observation data provides an important basis for regional water quality monitoring. Based on measured DIN, PO_4-P, and chlorophyll data from Chesapeake Bay, Harding et al. reported that DIN concentration doubled from 1945 to 1980, leading to an increase in chlorophyll concentration and serious ecosystem damage, but DIN concentration and eutrophication decreased somewhat from 1981 to 2012 due to the control of freshwater influx and DIN concentration [10]. Based on annual and monthly field measurement data of dissolved inorganic carbon (DIC), dissolved organic carbon (DOC), chlorophyll, and nutrients, Atsushi et al. investigated long-term trends in water quality status in Tokyo Bay, and showed that total input of land-derived pollutants has decreased to some extent due to gradual improvement in coastal wastewater treatment measures in recent years [11,12]. Based on field measurement data at monitoring stations, hydrological characteristics, and status of land-derived sewage, Cao et al. investigated nearshore marine pollution in Hangzhou Bay and found that serious nutrient pollution was primarily affected by topography, hydrological environment, and industrial development in the bay area [13]. The addition of land-derived nutrients, eutrophication, and other issues have led to water quality deterioration in bays heavily affected by anthropogenic activities, prompting governments to take corresponding measures, including strict control of the total amount of pollutants discharged into the sea, strengthening the pollutants treatment, and restricting the human activities in the bay [14,15], as well as protecting the natural shoreline, restoring the ecological environment of mudflats and wetlands, promoting the ecological functions of the bay, etc. [15–17]. To evaluate remediation performance and assess and modify existing management measures, continuous monitoring data are still needed, while the large-scale and long-term simultaneous observation of water quality data in major bays is generally insufficient. Satellite remote sensing technology has the advantages of long time series, large-scale simultaneous observation, and historical backtracking, which can overcome the shortcomings of in situ surveys and enable the observation of spatiotemporal variations in water quality.

Water quality parameters with optical activity, such as chlorophyll, colored dissolved organic matter, total suspended matters and turbidity, can be retrieved by the satellite monitoring spectrum [18–24] and are widely used for the satellite-based monitoring of water quality [25–28]. However, nutrients are not directly linked to the spectral signal and the variation mechanism is extremely complex, making nutrient retrieval a difficult task [29]. Although some studies had attempted to retrieve nutrients by optically active variables [30–32], these algorithms were extremely regionally applicable and depended heavily on field measurement data, making them difficult to apply to other regions. Nevertheless, remote sensing is still an effective tool for water quality monitoring, with satellites such as the Landsat series and MODIS having been applied quite successfully to this aspect [33–36]. Therefore, the establishment of nutrient retrieval models based on field measured data can help to clarify the spatiotemporal change patterns of water quality in a region and explore the influencing factors, thereby providing theoretical and scientific support for water and ecological environment governance decisions.

Yueqing Bay is a typical semi-enclosed tidal bay and the fourth largest in area in China. Its tidal range, tidal influx, water exchange capacity, and other hydrological properties are representative of semi-enclosed tidal bays, to some extent. Moreover, Yueqing Bay has complex water exchange processes, including exchange between incoming offshore waters and internal bay water and between bay mouth waters and diluted waters from the Ou River. From the outlet to the bay head, internal waters vary considerably in water exchange capacity, as it takes less than 5 days for 90% of the waters near the outlet to be replaced by offshore waters, compared to 40 days for 90% of the waters in the whole bay to be replaced [37,38]. In addition, more than 30 streams crisscross along Yueqing Bay, and the bay mouth is located on the north side of the Ou River estuary, one of the eight major river systems in Zhejiang, China. Hence, given these natural geographic conditions, Yueqing Bay is greatly affected by land area change. In recent decades, cities along Yueqing Bay have witnessed rapid economic development, dense population growth, and surging demand for urban land for industries and agriculture. Human demand for ocean space development has increased significantly, and since the 1950s, reclaimed tidal flats along Yueqing Bay have increased to 165.1 km^2, accounting for 38% of total tidal flat area and 18% of total sea area in the bay [39]. Along with accelerated urbanization, intensified anthropogenic activities, and increased domestic sewage and industrial and agricultural wastewater discharge, total input of land-derived pollutants into the bay has increased, further affecting water quality. Under highly intensive anthropogenic activities, the environment along Yueqing Bay has experienced rapid and dramatic deterioration. Since 2006, local governments have undertaken various measures for marine environment pollution remediation and controlled the use of phosphorus and land-derived pollutants. Based on nutrient data observed in May and August from 2000 to 2014, Guo et al. studied interannual variations in nutrient salt concentrations in the fishing waters of Yueqing Bay and found that concentrations increased continuously from 2002 to 2006 due to high river runoff and industrial and agricultural wastewater and domestic sewage, but that the rising trend slowed slightly after 2006 due to the implementation of national policies and measures to control pollutant discharge [40]. Given the intense anthropogenic activities, in-depth study of variations in environmental status and the main driving factors of Yueqing Bay could offer effective reference for environmental remediation of this and other semi-enclosed bays worldwide and may be of importance to the rapidly developing global bay economy.

Based on water quality data of Yueqing Bay collected from 2014 to 2020 (provided by Wenzhou Environmental Monitoring Center) and Landsat-8 remote sensing data, we built a nutrient concentration retrieval model for Yueqing Bay using support vector machine (SVM) learning with water temperature, red-light waveband, near-infrared waveband, and short infrared waveband as input parameters, and then, spatiotemporal variations in nutrient concentration in the bay retrieved from remote sensing data (2013–2020) were studied along with influencing factors, such as hydrodynamic and environmental information, river basin-related management policies, and anthropogenic activities. This study can provide an effective reference for the remote sensing dynamic monitoring of water quality and provide technical support for the management of environmental protection and remediation of similar bays.

2. Materials and Methods

2.1. Study Area

Located along the southern coast of Zhejiang, China (as shown in Figure 1a), on the north side of the Ou River estuary (27°59′09″–28°24′26″N, 120°57′55″–121°17′09″E), Yueqing Bay is the fourth largest bay in China and the largest semi-enclosed bay in Zhejiang, China, with a total sea area of 463.6 km^2 and total shoreline length of 42 km. The bay is 10 km wide, on average, with a 21 km wide outlet and 4.5 km wide narrowest section. Average water depth is 10 m and mean tidal inflow/outflow is ~20×10^8 m^3. Tidal waves propagate from the East China Sea to Yueqing Bay, and except for Haishan and Jiangxia, which experience irregular semidiurnal tides, the other areas of the bay are under regular semidiurnal tides [41]. More than 30 streams of varying size drain into the bay, including the Qing River, Bai Creek, Ling Creek, and Jiangxia Creek [42], and mean annual runoff of the river basin is 10.3×10^8 m^3 [41]. The western part of Yueqing Bay belongs to Yueqing, Wenzhou, Zhejiang, China, while the northern and eastern parts belong to Wenling, Yuhuan, Taizhou, Zhejiang, China, respectively (as shown in Figure 1b). The Yueqing Bay river basin is densely populated, and the three largest cities along the bay have a total population of about 3 million [43]. Surging activities, such as aquatic breeding and sea reclamation, have significantly increased economic revenue and development [43].

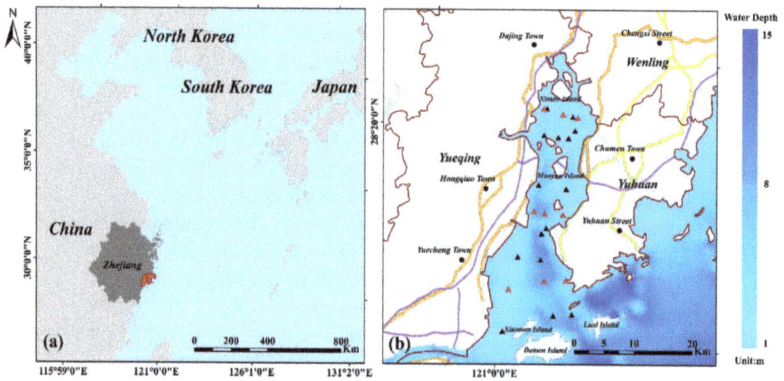

Figure 1. Location of study area (**a**) and cities along Yueqing Bay: Yueqing, Wenling, and Yuhuan (**b**). (Red triangles in (**b**) are nine fixed survey stations for each survey and black triangles are non-fixed survey stations; geographical map is sourced from China National Geographic Information Public Service Platform (https://www.tianditu.gov.cn/, accessed on 21 October 2021)).

Situated in the subtropical monsoon climate zone, Yueqing Bay is subject to various environmental factors, such as topography and marine climate, resulting in a land-bay microclimate with good heat, moisture, and temperature, resulting in a mean annual air temperature of 17–17.5 °C and mean annual rainfall of 1191.7–1506.8 mm [44]. Yueqing Bay contains 54,527.9 acres of tidal flats, and over 20 species of commercial fish, 58 species of shellfish, and 60 species of crustaceans, and is thus a natural farm of shellfish breeding. Furthermore, with a mean tidal range of more than 4 m (maximum of 8.53 m) [41], Yueqing Bay has considerable tidal energy, which is useful for tidal power plants, e.g., the Jiangxia pilot tidal power plant has an installed capacity of up to 550,000 kW [39,42].

Our study area included Yueqing Bay and its three main coastal cities, namely, Yueqing, Wenling, and Yuhuan. We investigated the spatiotemporal variational patterns in nutrient concentration in the waters from 2013 to 2020 as well as the main driving factors.

2.2. Data Source

2.2.1. Satellite Remote Sensing Data

Land satellite Landsat-8 Operational Land Imager (OLI) high-spatial-resolution (30 m) multi-spectral remote sensing data (released by United States Geological Survey (USGS)) were used for analysis. Since its launch in 2013, data from the OLI has been widely applied in long-term water quality monitoring for inland rivers, lakes, nearshore seas, and bays. The Landsat-8 OLI remote sensing data were downloaded from the USGS website (https://earthexplorer.usgs.gov/, accessed on 30 August 2021), and remote sensing images of the study area (strip no.: 118040) during 2013–2020 with sky cover less than 30% were selected. In total, 63 Landsat Level-1 remote sensing scenes met the above criteria (Table 1), including more than 10 remote sensing images per season and more than five remote sensing images per year covering most months, seasons, and years. To obtain the remote sensing reflectance product, we performed radiometric calibration and atmospheric correction of the data using ENVI v5.3 (Exelis Visual Information Solutions (Boulder, CO, USA)), with atmospheric correction conducted using the FLAASH algorithm in ENVI v5.3 [45].

Table 1. Valid Landsat-8 remote sensing images from January 2013 to December 2020 (green cells and numbers represent presence of data and quantity of valid data, respectively, while white cells represent absence of data).

	2013	2014	2015	2016	2017	2018	2019	2020	Total
1		2	1	1		1	1		6
2			1	1	1				3
3			1	1	1				3
4			2		1		1	1	5
5	1		1	1					3
6	1	1	1				1		4
7	1			2	2	2	1		8
8	2		1	2	1	1	1	1	9
9	1		1			1	1	1	5
10	1				2	1	1		5
11	1	1	1				1		4
12	2	1				2	1	2	8
Total	10	5	10	8	8	8	9	5	63

2.2.2. Field Measurement Data

For this study, environmental water quality survey data of Yueqing Bay were provided by Wenzhou Environmental Monitoring Center. During 2014–2020, monthly voyage observations were performed, and in situ sampling and observation and data acquisition were conducted at nine fixed stations and several non-fixed stations. Monitoring parameters included DIN (nitrite–nitrogen, nitrate–nitrogen, and ammonia–nitrogen), PO_4-P, water temperature, chlorophyll, dissolved oxygen, salinity, and pH. Both sampling and testing were performed in accordance with the Code of Practice for Marine Monitoring Technology [46]. During 2014–2020, the in situ sampling period and statistical sampling size were relatively stable and more than 45 pairs of observed DIN and PO_4-P data per month were retained, representing primary characteristics of nutrients in Yueqing Bay.

2.3. Analytical Methods and Related Supplementary Data

2.3.1. Matching Satellite and Field Measurement Data

In this study, satellite data were matched with field measurement data. Spatially, 1×1 pixels (30 m $\times$ 30 m) closest to a survey station in latitude and longitude were matched; temporally, satellite data differing from field measurement data in observation time by less than ± 3 days were matched, which was an optimal choice to obtain enough representative data in the shortest possible time interval. We could not obtain enough matchup data for the satellite algorithm development if we set up a shorter time interval, due to the long revisiting period of 16 days for Landsat and the limited number of field measurements. To extract reflectance and other information, the "ExtractMultiValuesToPoints" feature of ArcGIS v10.7 (Environmental Systems Research Institute, Inc. (Redlands, CA, USA)) was used to generate a vector file based on latitude and longitude information of a data observation station, with corresponding remote sensing reflectance information then extracted.

Using the above matching approach for data during 2014–2020, 80 matched DIN data pairs and 73 matched PO_4-P data pairs were generated. We randomly sampled 65% of the data as a training dataset for algorithm construction, while the remaining 35% of data served as a validation dataset for algorithm accuracy assessment. In general, the matched data encompassed and reflected spatiotemporal variability of the field measurement data, and thus could be used to construct and assess the nutrient retrieval algorithm for waters of Yueqing Bay over a long time series.

2.3.2. SVM Algorithm and Statistical Parameters of Model Assessment

Machine learning-based regression algorithms, e.g., random forest, artificial neural network, and SVM algorithms, have been widely applied in water quality research [47–49]. In this study, based on matched data pairs, we used the SVM algorithm to construct a nutrient retrieval algorithm for waters of Yueqing Bay. This was mainly due to the smaller data size, as the random forest and artificial neural network algorithms are prone to overfitting. In contrast, the SVM algorithm attempts to map training data points to high-dimensional eigenspace based on kernel function, resulting in many decision-making boundaries (hyperplanes) in eigenspace that can separate different classes of data points. In addition, the SVM algorithm does not involve probability measures or the law of large numbers; moreover, the purpose of SVM optimization is to minimize structuring risk rather than empirical risk, thereby reducing the risk of overfitting [50]. Regarding prediction, the training process generated an SVM model with multiple input parameters for automatic optimization. Based on the trained SVM model, we predicted dependent variables (e.g., DIN and PO_4-P) by inputting new training parameters (e.g., remote sensing reflectance).

During analysis, SVM algorithm construction, parameter tuning, simulation, assessment, and prediction were performed using Matlab R2020b (MathWorks (Natick, MA, USA)).

The fitting effect during algorithm construction was expressed by coefficients of determination (R^2) and significance level (*P*) resulting from statistical analysis. In addition, to assess the accuracy of the model-predicted results quantitatively, the following statistical parameters were used: (1) bias, i.e., general bias information of the predicted result; (2) Mean Absolute Error (MAE), i.e., mean value of absolute error between predicted and true values; and (3) Root Mean Squared Error (RMSE), i.e., square root of the ratio of squared bias between observed and true values to number of observations. Results were statistically analyzed using Microsoft Excel 2016 (Microsoft (Redmond, Washington, DC, USA)):

$$Bias = (DIN(PO_4 - P)^i_{RS} - DIN\left(PO_4 - P)^i_{in\text{-}situ}\right) / DIN(PO_4 - P)^i_{in\text{-}situ} \quad (1)$$

$$MAE = mean(abs(DIN(PO_4 - P)^i_{RS} - DIN(PO_4 - P)^i_{in\text{-}situ}) / DIN(PO_4 - P)^i_{in\text{-}situ})) \quad (2)$$

$$RMSE = (sum((DIN(PO_4 - P)^i_{RS} - DIN(PO_4 - P)^i_{in\text{-}situ})^2 / (N-1))^{0.5} \quad (3)$$

where $DIN(PO_4 - P)^i_{RS}$ and $DIN(PO_4 - P)^i_{in\text{-}situ}$ are the predicted and observed results, respectively, and N is the number of data pairs, $i \in [1, N]$.

2.3.3. Hydrodynamic Model for Yueqing Bay

The Finite Volume Coastal Ocean Model (FVCOM) was used to simulate three-dimensional (3D) hydrodynamics of Yueqing Bay. FVCOM is a 3D model for nearshores, estuaries, and oceans based on the finite volume method [51,52]. Unstructured triangular meshes were used in the horizontal direction and the σ coordinate system was used in the vertical direction to better resolve the complex shorelines and topography of Yueqing Bay.

This model was previously validated and applied by our research team in the Changjiang River estuary [53].

The model consisted of 14,197 points in the horizontal direction and 26,959 triangular meshes, the mesh resolution of the bay was 1000 m, and the mesh resolutions of the open bounds were 4–20 km. In the model, the S-57 electronic navigational chart was used for water depth data of Yueqing Bay and the waters near Wenzhou Port. For the open bounds, the model superposed the predicted tidal levels of eight main tidal constituents, namely, M2 (12.42 h), N2 (12.66 h), S2 (12.00 h), K2 (11.97 h), K1 (23.93 h), O1 (25.82 h), P1 (24.07 h), and Q1 (26.87 h), and background flow field data from HYCOM, among which the predicted tidal levels had a temporal resolution of 1 h and the background flow field data had a temporal resolution of 1 month. On the river bounds, climatological mean discharges of the Yangtze, Qiantang, Tiao, Yong, Jiao, Ou, Feiyun, and Ao rivers, and three major mountain creek-based riverlets, flowing into Yueqing Bay were superposed. Discharge data for the Yangtze and Qiantang rivers were sourced from Bulletin of River Sediments in China for 2002–2020, while discharge data for other rivers were estimated from the TRMM 3B43 climatological mean precipitation product for 1998–2019 using the river discharge estimation formula of Liu et al. (2020) [54]. For the sea-air upper bounds, wind data were sourced from the European Centre for Medium-Range Weather Forecasts (ECMWF), with a spatial resolution of $0.125° \times 0.125°$ and a temporal resolution of 6 h.

2.3.4. Characterization of Intensity of Anthropogenic Activities in Cities along Yueqing Bay

Based on the administrative boundaries of Yueqing, Wenling, and Yuhuan along Yueqing Bay, point-of-interest (POI) data of the various cities in 2020 were acquired using the "POI retrieval" feature in Baidu Maps (http://lbsyun.baidu.com/index.php?title=webapi accessed on 30 August 2021), which are a type of vector point data containing latitude and longitude coordinates, facility name, and facility type [55]. Each vector point in space represents one facility space in real life and corresponds to commercial and human socioeconomical activities, such as traveling and dining, and governmental services. The POI data are applicable for analysis of anthropogenic activities, economic development status, and distribution from a spatial perspective [56,57].

Since many categories of facilities are closely associated with each other in functional properties, POI data are often manifested in aggregated form in geographic space, i.e., hot spots [58]. Usually, "density" is used to characterize aggregation intensity of POI data in a local area quantitatively, while kernel density estimation is a common and effective density estimation method [59] for nonparametric analysis of geospatial distribution, reflecting the effect of a geographic phenomenon attenuating with increasing spatial distance. Given an arbitrary point in space, kernel density is the sum of weights of all sample points within a circle with a given radius whose center is at the point. Such a weight is inversely proportional to the distance between the sample point and point [60,61]. Here, kernel density was calculated using the following formula:

$$f(x) = \frac{1}{nh}\sum_{i=1}^{n} K\left(\frac{x - x_1}{h}\right), \quad (4)$$

where $f(x)$ represents the kernel density function of point x; h is the bandwidth, i.e., radius of the circle; n is the number of sample points within the circle; K is the spatial weight function; and $(x - x_1)$ is the distance between the sample point and point x.

Given the above understanding, intensities of anthropogenic activities in cities along the Yueqing Bay were spatially analyzed based on POI data by means of kernel density estimation.

3. Results

3.1. Algorithm Construction and Validation

Nutrients are non-optically active substances, and their concentrations have no significant spectral features [62]. However, nutrients are to some extent related to the physical properties of water, such as water temperature, and optically active substances in water, such as suspended solids and chlorophyll-a [63–65], which are associated with spectra [66–70]. In this study, waveband information was used as a variable input to retrieve nutrient concentrations in the waters of Yueqing Bay.

Correlation analysis was carried out for temperature and waveband information. As shown in Figure 2, DIN was highly correlated with sea surface temperature (SST) ($R = -0.84$), as well as the red-light (Band 4), near-infrared (Band 5) ($R > 0.1$), and short infrared wavebands (Bands 6 and 7) ($R < -0.1$); PO_4-P was highly correlated with SST ($R = -0.70$), as well as the red-light (Band 4) and near-infrared wavebands (Band 5) ($R > 0.15$). Therefore, SST and the red-light, near-infrared, and two short infrared wavebands were selected as input parameters of the DIN retrieval model, and SST, red-light waveband, and near-infrared waveband were selected as input parameters of the PO_4-P retrieval model.

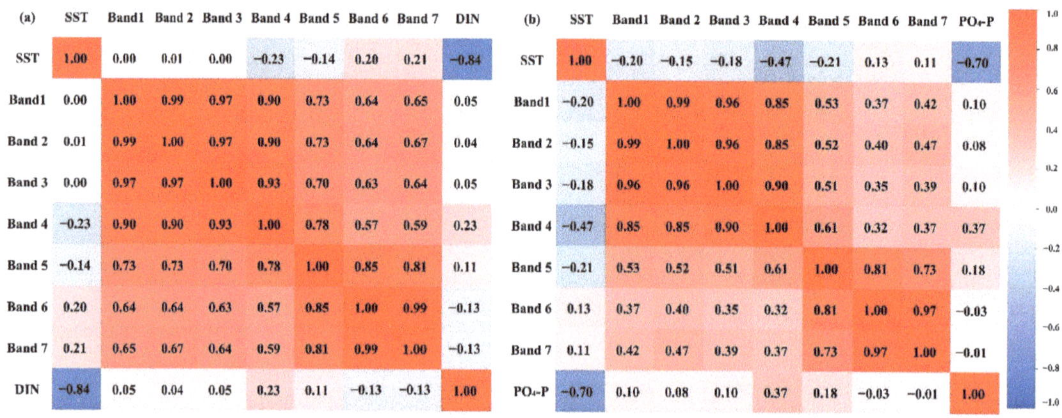

Figure 2. Correlation analysis of input variables versus output variables: (**a**) correlation between DIN and each factor; (**b**) correlation between PO_4-P and each factor.

Based on the data arrays acquired by matching, a nutrient retrieval model for Yueqing Bay was obtained by SVM. As shown in Figure 3a,b, for the training set of the DIN retrieval model, $R^2 = 0.8391$, $P < 0.001$, bias was (−0.24,0.32), MAE = 3.48%, and RMSE = 0.0550 mg/L, and for the validation set of the DIN retrieval model, $R^2 = 0.8129$, $P < 0.001$, bias was (−0.40,0.30), MAE = 4.72%, and RMSE = 0.0608 mg/L; for the training set of the PO_4-P retrieval model, $R^2 = 0.8589$, $P < 0.001$, bias was (−0.49,0.24), MAE = 0.23%, and RMSE = 0.0036 mg/L, and for the validation set of the PO_4-P retrieval model, $R^2 = 0.7606$, $P < 0.001$, bias was (−0.36, 0.60), MAE = 0.45%, and RMSE = 0.0063 mg/L.

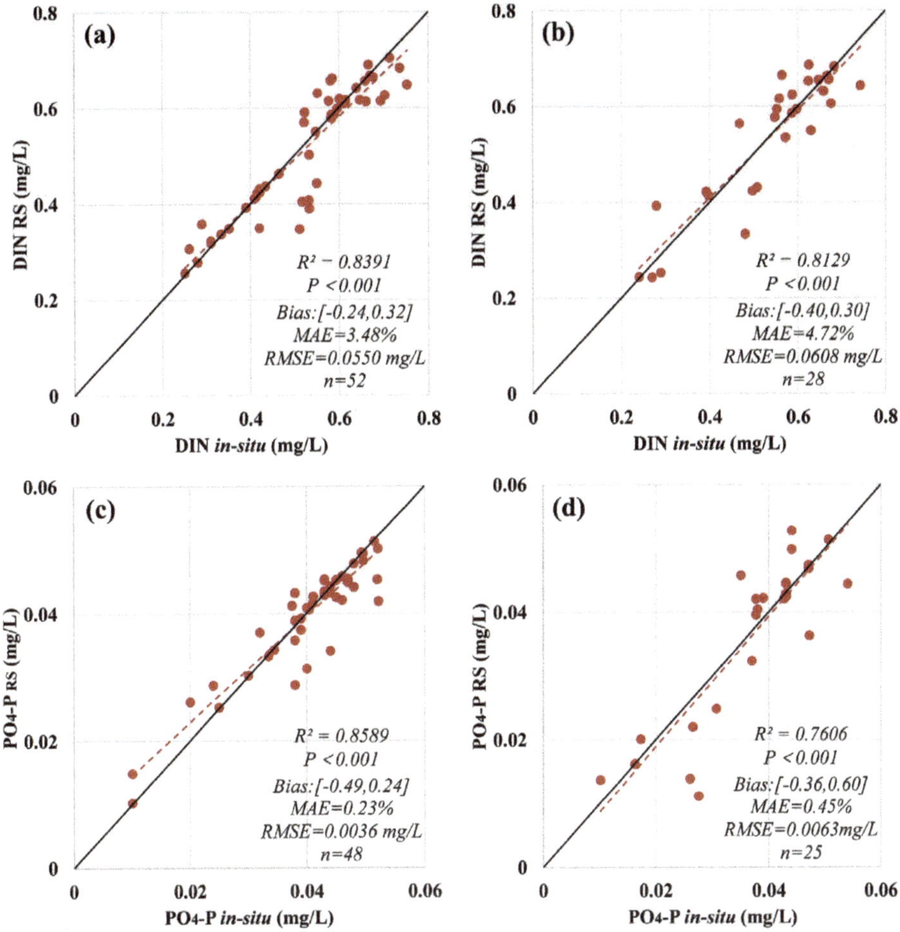

Figure 3. Model validation: (**a**) DIN training set; (**b**) DIN validation set; (**c**) PO_4-P training set; (**d**) PO_4-P validation set.

3.2. Spatiotemporal Variations in Nutrient Concentrations in Yueqing Bay

3.2.1. Spatial Distribution of Nutrient Concentrations in Yueqing Bay

From the perspective of the spatial distribution of nutrient concentration based on remote sensing retrieval, in spring, there were two high DIN regions in the narrowest area of the middle part of the bay and Southwest Shoal at the bay mouth, while DIN concentrations were more uniform in most other areas. In summer, high DIN values remained in the narrowest area of the middle part of the bay and Southwest Shoal, while at the bay head, DIN concentrations on the west side of Ximen Island were slightly higher than those on the east side. In autumn, high DIN concentrations were mainly observed in the stream on the east side of Ximen Island at the bay head, Southwest Shoal at the bay mouth, and the sea area on the north side of Damen Island, while low concentrations were mainly observed in the narrowest area of the middle part of the bay and coastal area to the southwest side of Yuhuan Island. In winter, high DIN concentrations were observed in the stream on the east side of Ximen Island at the bay head, Southwest Shoal at the bay mouth, and sea areas on the north side of Damen and Xiaomen islands, while low concentrations were observed in the narrowest area of the middle part of the bay through the sea area in the middle of the bay mouth (Figure 4a–d).

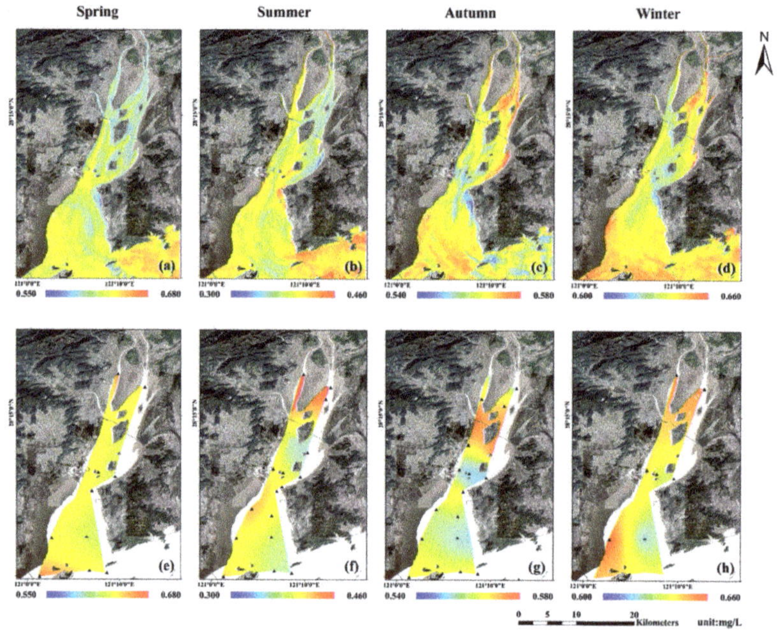

Figure 4. Seasonal spatial distribution of DIN concentrations in Yueqing Bay: (**a–d**): seasonal average of retrieval results based on remote sensing data from 2013 to 2020; (**e–h**): Kriging interpolation results based on the seasonal average of field measurement data from 2014 to 2020. (Black triangles in (**e–h**) are survey stations).

In spring, there was a high PO_4-P concentration zone in the Southwest Shoal through the sea area in the middle of the Yueqing Bay mouth, while low concentrations were observed in the sea areas near Maoyan Island and the southwest side of Yuhuan Island. In summer, high PO_4-P concentrations were observed in the Southwest Shoal, while low concentrations were observed at the bay head and sea area between Yuhuan Island and Damen Island. In autumn, a high PO_4-P concentration zone was observed in the middle part of the bay through the sea area on the north side of Damen Island, while low concentrations were mainly observed in the coastal area of the Southwest Shoal. In winter, high PO_4-P concentrations were observed in the Southwest Shoal and sea area on the north sides of Damen and Xiaomen islands, while low concentrations were observed in the stream on the west side of Ximen Island at the bay head and in the narrowest area of the middle part of the bay (Figure 5a–d).

Based on field measurement data from 2014 to 2020, the spatial distribution of nutrient concentrations in Yueqing Bay was obtained by Kriging interpolation (Figures 4e–h and 5e–h), and a comparison was made with the remote sensing retrieval results (Figures 4a–d and 5a–d). In general, the spatial distribution of the satellite-derived nutrient concentration was consistent with that of the field measured data; high-value zones were situated in the streams on both sides of Ximen Island at the bay head, narrowest area of the middle part of the bay, and Southwest Shoal at the bay mouth, while low-value zones were situated in the sea area on the west side of Yuhuan Island. Due to the limited sampling data and coverage, there were no data in the nearshore and the distribution of interpolation results might be unreasonable in the nearshore. As shown in Figure 5f, there was an obvious mutation in the concentration of PO_4-P from the middle of the bay to the bay mouth, which might be the artificial error by interpolation. The spatial distribution of satellite-derived nutrients was more complete in spatial coverage.

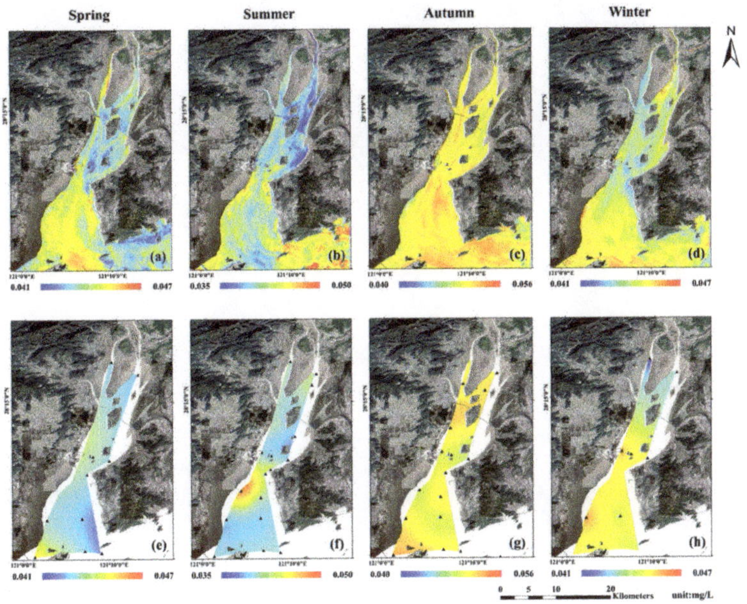

Figure 5. Seasonal spatial distribution of PO_4-P concentrations in Yueqing Bay: (**a**–**d**): seasonal average of retrieval results based on remote sensing data from 2013 to 2020; (**e**–**h**): Kriging interpolation results based on the seasonal average of field measurement data from 2014 to 2020. (Black triangles in (**e**–**h**) are survey stations).

The distribution of nutrient concentrations reported by Zeng et al. [71] in Yueqing Bay from 2006 to 2007 was basically consistent with our satellite retrieval results, that is, the high nutrient concentrations were mainly concentrated in the Southwest Shoal and the narrowest part of the middle part of the bay, and the low concentrations were mainly concentrated in the west side of Yuhuan Island. Through model simulation, Li et al. [72] found that the nutrient concentrations in the western part of the bay were higher than those in the eastern part. However, Li et al. [72] reported that high nutrient concentrations were observed in the bay head and low nutrient concentrations were observed in the bay mouth, and there were no obvious high nutrient concentrations in the narrowest area of the middle part of the bay, which was slightly different from the results of this study. The main reason would be the different input parameters of the nutrient concentration retrieval model. The water quality model constructed by Li et al. [72] mainly considered the influence of water flow and diffusion, while we used the temperature and spectral information as the main parameters to retrieve the nutrient concentrations. At the same time, on both sides, the narrowest part in the middle part of the bay might be affected by two power plants, the Huaneng Yuhuan Power Plant (put into operation in 2007) and the Zheneng Yueqing Power Plant (put into operation in 2008). Although the influence range of warm water drainage of the power plant was limited, it may still have a certain impact on the satellite-derived nutrient concentrations in the narrowest part of the middle part of the bay. Therefore, more data are needed for refined verification and analysis.

3.2.2. Seasonal Variations in Nutrient Concentrations in Yueqing Bay

Based on analysis of the remote sensing retrieval results (Table 2), seasonal DIN concentrations varied within 0.211–0.687 mg/L, with levels lowest in summer (averaging

0.3940 mg/L), highest in spring and winter (averaging 0.6189 mg/L and 0.6396 mg/L, respectively), and reaching 0.5648 mg/L in autumn. The PO_4-P concentrations varied seasonally within 0.016–0.056 mg/L, with levels lowest in summer (averaging 0.0414 mg/L), highest in autumn (averaging 0.0508 mg/L), and reaching 0.0439 mg/L and 0.0444 mg/L in spring and winter, respectively.

Table 2. Minimum, maximum, and mean nutrient concentrations retrieved by remote sensing in each season in Yueqing Bay.

Parameter	Season	Min	Max	Mean
DIN	Spring	0.5119	0.6635	0.6189
	Summer	0.2106	0.4601	0.3940
	Autumn	0.4453	0.6128	0.5648
	Winter	0.5603	0.6872	0.6396
PO_4-P	Spring	0.0374	0.0507	0.0439
	Summer	0.0158	0.0503	0.0414
	Autumn	0.0403	0.0562	0.0508
	Winter	0.0386	0.0482	0.0444

Based on the field measurement data (Figure 6) in Yueqing Bay, DIN concentrations were lower in summer (mean: 0.482 mg/L) and higher in spring and winter (means: 0.73 mg/L and 0.66 mg/L, respectively); PO_4-P concentrations were lower in summer (mean: 0.043 mg/L) and higher in autumn and winter (means: 0.047 mg/L and 0.048 mg/L, respectively). Nutrient concentrations varied greatly, with the DIN concentration ranging from 0.2 to 1.0 mg/L and PO_4-P concentration ranging from 0.01 to 0.08 mg/L.

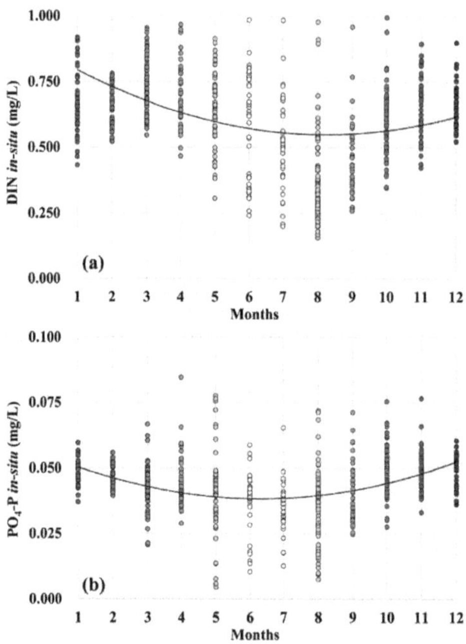

Figure 6. Observed nutrient concentrations in Yueqing Bay during 2014–2020: (**a**) monthly variations in DIN concentration; (**b**) monthly variations in PO_4-P concentration; each point represents field measurement data at one station.

3.2.3. Long Time-Series Variations in Nutrient Concentrations in Yueqing Bay

As shown in Figure 7a, the DIN concentration in Yueqing Bay exhibited a notable rising trend ($P = 0.0262$, $t = 2.5341$) during 2013–2015 and a non-significant decreasing trend during 2015–2020 ($P = 0.1233$, $t = -1.5729$). As shown in Figure 7b, the PO_4-P concentration in Yueqing Bay exhibited a non-significant increasing trend ($P = 0.3196$, $t = 1.0383$) during 2013–2015 and a non-significant decreasing trend ($P = 0.8906$, $t = -0.1383$) during 2015–2020.

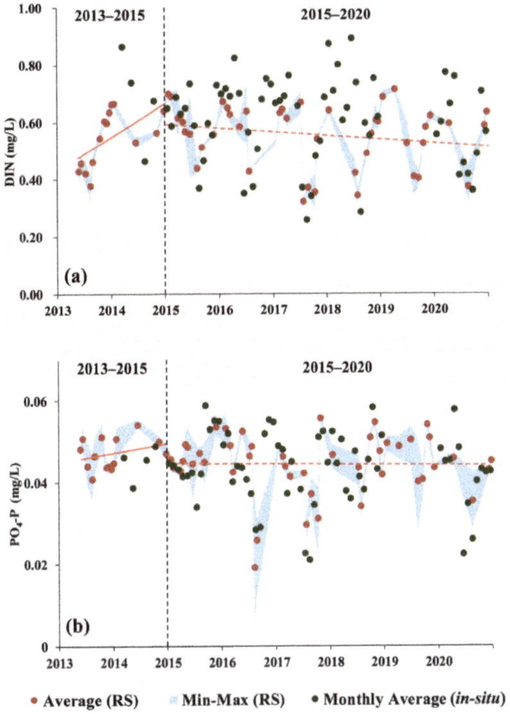

Figure 7. Interannual time-series variations in region-averaged nutrient concentrations in Yueqing Bay from 2013 to 2020: (**a**) interannual time-series variations in DIN concentrations; (**b**) interannual time-series variations in PO_4-P concentrations. (Red scatter points represent mean region-averaged nutrient concentrations in Yueqing Bay; green scatter points represent monthly means of observed nutrient concentrations; upper bound of blue zone represents maximum nutrient concentration within corresponding time interval, while lower bound represents minimum nutrient concentration within corresponding time interval; vertical black dashed lines correspond to 2015; red solid lines represent trend lines during 2013–2015; red dashed lines represent trend lines during 2015–2020).

Comparing field measurement data with the remote sensing retrieval results (Figure 7), the long time-series variation trends in remote sensing retrieval generally agreed with the observed results, but due to the acquisition time of satellite remote sensing images and other objective factors, the field measurement data of high DIN concentration could not be matched during the data matching process, resulting in the DIN retrieval results being slightly lower than the field measurement data in some periods. At the same time, abnormally low nutrient concentrations were found in Yueqing Bay in the summer of 2017.

To further elucidate the water quality trends in local regions of Yueqing Bay under strong anthropogenic impact, three 500 m × 500 m regions located at coastal stream estuaries were selected (Figure 8a), i.e., East Stream of the bay head on the east side of Ximen Island (#1), West Stream of the bay head on west side of Ximen Island and at Qing River estuary (#2), and Southwest Shoal at the bay mouth (#3), where mean, maximum, and minimum nutrient concentrations were extracted. As shown in Figure 8b–g, during 2013–2015, the increasing nutrient concentrations in these three regions were all higher than the mean nutrient concentration rate in Yueqing Bay (see slopes in Table 3). Among these results, the PO_4-P concentrations in the East Stream (#1) and West Stream of the bay head (#2) increased at the fastest rate (Figure 8c,e), $a_{PO4\text{-}P(\#1)} = 1.98 \times 10^{-5}$, $a_{PO4\text{-}P(\#2)} = 2.03 \times 10^{-5}$). After 2015, nutrient levels in the waters of the three regions decreased to variable extents. Notably, the nutrient concentrations in the East Stream (#1) decreased at a slower rate than Yueqing Bay overall (Figure 8b,e), especially DIN ($a_{DIN(\#1)} = -1.30 \times 10^{-5}$), while the decreasing nutrient concentrations in the West Stream (#2) and Southwest Shoal (#3) were higher than the mean decreasing rate in Yueqing Bay (Figure 8d–g), especially PO_4-P in Southwest Shoal (#3) ($a_{PO4\text{-}P(\#3)} = -2.90 \times 10^{-6}$). Table 3 shows the statistical parameters of the time-series regression of nutrient concentrations in the waters of various regions and Yueqing Bay.

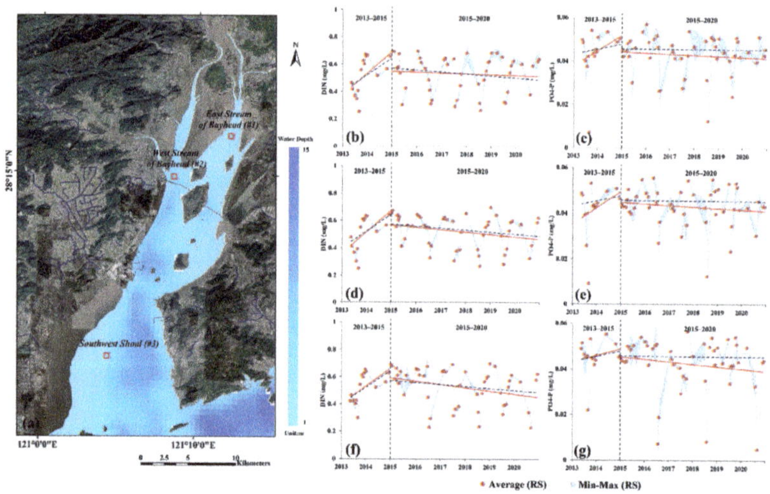

Figure 8. Long time-series variation in nutrient concentrations retrieved by remote sensing in local regions of Yueqing Bay: (**a**) spatial distribution map of data extraction regions; (**b**) variations in DIN concentration in East Stream of the bay head; (**c**) variations in PO_4-P concentration in East Stream of the bay head; (**d**) variations in DIN concentration in West Stream of the bay head; (**e**) variations in PO_4-P concentration in West Stream of the bay head; (**f**) variations in DIN concentration in Soutwest Shoal; (**g**) Variations in PO_4-P concentration in Southwest Shoal. (Red scatter points represent mean nutrient concentrations in Yueqing Bay; upper bound in blue zone represents maximum nutrient concentration within corresponding time interval, while lower bound represents minimum nutrient concentration within corresponding time interval; vertical black dashed lines correspond to 2015; red solid lines represent trend lines of nutrient concentration in various local regions; blue dashed lines represent trend lines of nutrient concentration in the whole of Yueqing Bay).

Table 3. Statistical regression analysis of nutrient concentration in local regions of Yueqing Bay.

Parameter	Time Interval	Region	Slope	Intercept	R^2	*P*	t
DIN	2013–2015	East Stream of bay head	4.30×10^{-4}	−17.4851	0.3058	**0.0325**	2.3929
		West Stream of bay head	4.30×10^{-4}	−17.3174	0.3248	**0.0266**	2.5006
		Southwest Shoal	3.50×10^{-4}	−14.2267	0.4779	**0.0285**	2.4626
		Yueqing Bay	3.20×10^{-4}	−12.97	0.3486	**0.0262**	2.5342
	2015–2020	East Stream of bay head	-1.30×10^{-5}	1.1009	0.0044	0.6505	−0.456
		West Stream of bay head	-4.80×10^{-5}	2.5658	0.072	0.0623	−1.9094
		Southwest Shoal	-6.40×10^{-5}	3.2988	0.0984	**0.0282**	−2.265
		Yueqing Bay	-4.00×10^{-5}	2.18	0.0556	0.1233	−1.573
PO_4-P	2013–2015	East Stream of bay head	1.98×10^{-5}	−0.7802	0.0925	0.2704	1.1511
		West Stream of bay head	2.03×10^{-5}	−0.804	0.1003	0.25	1.2042
		Southwest Shoal	7.91×10^{-6}	−0.2834	0.0326	0.5198	0.6616
		Yueqing Bay	6.20×10^{-6}	−0.2111	0.0824	0.3196	1.0383
	2015–2020	East Stream of bay head	-1.50×10^{-6}	0.1055	0.0118	0.4579	−0.7485
		West Stream of bay head	-1.70×10^{-6}	0.1154	0.0177	0.362	−0.9205
		Southwest Shoal	-2.90×10^{-6}	0.1663	0.0273	0.2566	−1.1484
		Yueqing Bay	-2.00×10^{-7}	0.0545	0.0004	0.8906	−0.1383

Notes: Regression equation $y = ax + b$, where a is slope and b is intercept. Values in bold indicate significance $P < 0.05$.

4. Discussion

4.1. Influences of Rivers, Hydrodynamics, and Biological Effects

Yueqing Bay is the largest semi-enclosed bay in Zhejiang and is influenced by the offshore Fujian-Zhejiang Coastal Current and Taiwan Warm Current [73,74]. Although there is no large river along the bay providing land-derived input, Yueqing Bay is on the north side of the Ou River estuary and more than 30 streams are distributed in the bay area [42]. Based on HYCOM background flow field, runoff, wind farm, and tidal data, the climatological flow field of nearshore sea areas in the middle and southern parts of Zhejiang and Yueqing Bay in summer and winter was acquired by FVCOM numerical simulation.

Based on variations in flow fields in the nearshore sea areas in summer (Figure 9a) and winter (Figure 9c), the high-temperature low-nutrient Taiwan Warm Current flowing from south to north was the dominant water mass in summer [74], while in winter, the Taiwan Warm Current weakened significantly and retreated towards the south, and the low-temperature, nutrient-rich Fujian-Zhejiang Coastal Current flowing from north to south gradually became the dominant water mass, with a flow field intensity much higher than that of the Taiwan Warm Current [73].

Waters in Yueqing Bay are affected by streams inside the bay and water from the Ou River (Figure 9). These waters are mainly manifested by bay head water flowing past the bay center and turning to the west side of Yuhuan Island, then flowing out; by waters in the Southwest Shoal flowing to the east and converging with water from the bay head; and by a small amount of water from the Ou River flowing past the north side of Damen Island and Xiaomen Island, draining likewise into the water stream from the bay head, and finally flowing out of the channel between Luxi Island and Damen Island. As seen in Figures 4a–d and 5a–d, the spatial distribution and transport direction of nutrient concentrations in Yueqing Bay remained generally consistent with the above results, indicating that river transport has a great impact on the spatial distribution of nutrients in the bay. In addition, Yueqing Bay is affected by multiple influencing factors, including inflowing streams, the Ou River, tides, and flow fields, and thus the bay mouth zone is hydrodynamically active, with frequent resuspension activity [75]. Sediment resuspension results in the redistribution of nutrients at the sediment–water interface, which can have an obvious impact on migration, transformation, and storage of nitrogen and phosphorus [76,77]. Thus, the complex water environment of Yueqing Bay is subjected to many factors.

In addition, observed chlorophyll concentration was used to characterize biomass and photosynthesis intensity and their relationship to nutrient concentration. The results show that, in summer (June to August), chlorophyll concentration increased significantly, resulting in high primary productivity, while in autumn and winter, chlorophyll concentration started to decline significantly (Figure 10a), resulting in lower primary productivity; correspondingly, in summer, nutrient concentration in Yueqing Bay decreased markedly until autumn and winter when the nutrient concentration began to increase again. Based on multi-linear curve fitting, we fit the climatological mean values of DIN, PO_4-P, and chlorophyll concentration during 2014–2020. As shown in Figure 10b, a significant inverse correlation existed between chlorophyll concentration and DIN and between chlorophyll concentration and PO_4-P ($R^2 = 0.7192$, $P < 0.005$), i.e., DIN and PO_4-P concentrations decreased with increasing chlorophyll. Thus, seasonal variation in photosynthesis intensity has a certain impact on nutrient concentration in Yueqing Bay. In spring and summer, plankton photosynthesis was intense, and high levels of DIN and PO_4-P were consumed, while in autumn and winter, plankton photosynthesis was weakened, thereby contributing to the accumulation of DIN and PO_4-P.

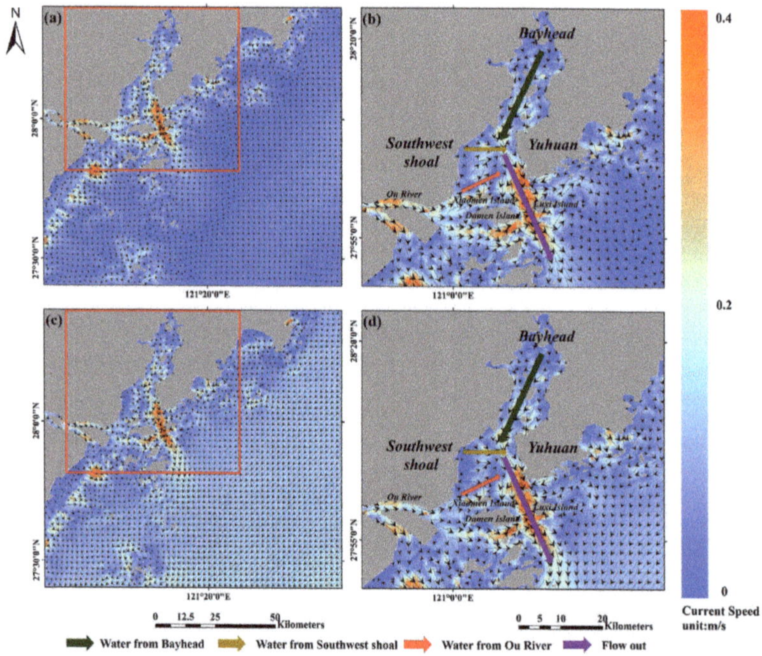

Figure 9. Distribution of flow fields in sea areas along the coasts in the middle and southern parts of Zhejiang and Yueqing Bay in summer and winter: (**a**) distribution of flow fields in sea areas of Zhejiang in summer; (**b**) distribution of flow fields in Yueqing Bay in summer; (**c**) distribution of flow fields in sea areas of Zhejiang in winter; (**d**) distribution of flow fields in Yueqing Bay in winter. (Black arrow represents flow field direction, length of black arrow represents flow field intensity; green, yellow, red, and purple arrows represent directions of water from bay head, Southwest Shoal, Ou River, and final outflowing stream, respectively).

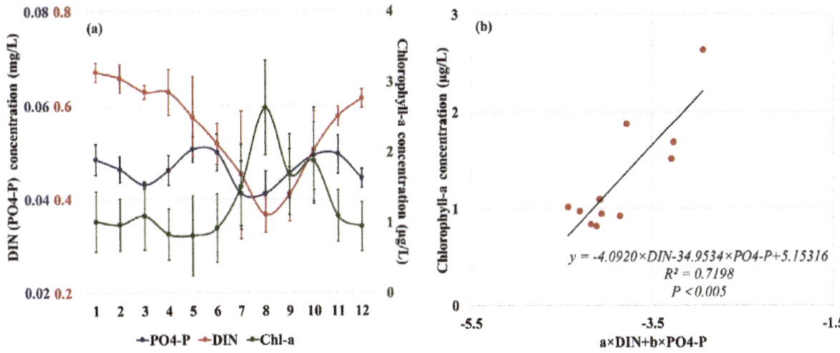

Figure 10. Inter-monthly variations (a) and correlation analysis (b) of nutrient and chlorophyll concentrations in Yueqing Bay during 2013–2020. (In (b), abscissa represents sum of DIN concentration multiplied by a constant and PO_4-P concentration multiplied by another constant after multi-linear curve fitting, which was used to characterize nutrient level in water. Ordinate represents chlorophyll concentration in water. Formula represents a relational formula for fitting nutrient and chlorophyll concentrations. As DIN and PO_4-P concentrations are multiplied by negative constants, nutrient concentrations are inversely correlated to chlorophyll concentration).

According to the available data analysis, from the perspective of natural factors, physical effects, such as the Fujian-Zhejiang Coastal Current and Taiwan Warm Current, combined with biological effects, such as seasonal variation in the intensity of plankton photosynthesis in the bay have an important impact on the seasonal variation of nutrient concentration in Yueqing Bay, but more parameters are needed for detailed comprehensive analysis.

4.2. Influences of Intensity of Anthropogenic Activities in River Basin

Changes in the intensity of anthropogenic activities inside the river basin can have an immediate impact on nutrient level in land-derived water. To investigate the impact of anthropogenic activities and stream transport on the spatial distribution of nutrient concentration in the bay, the intensity of anthropogenic activities along Yueqing Bay and spatial distribution status were analyzed by kernel density using POI data. As shown in Figure 11a, six regions along Yueqing Bay had strong anthropogenic activities, namely, Yuecheng Town (#1), Hongqiao Town (#2), and Dajing Town (#3) in Yueqing City, Chengxi Subdistrict (#4) in Wenling, and Chumen Town (#5) and Yucheng Subdistrict (#6) in Yuhuan. Based on water system distribution, elevation, and other geographic information of the bay (Figure 11b,c), streams flowing past Yuecheng Town (#1), Hongqiao Town (#2), and Dajing Town (#3) eventually drain into Yueqing Bay. Dajing Town (#3) is nearest to Yueqing Bay, but anthropogenic intensity in this region was weaker than that in the other regions (low kernel density); streams passing the Chengxi Subdistrict (#4) eventually flow into Yueqing Bay, but they are farther away. Streams flowing past Chumen Town (#5) are situated in Xuanmen Bay, but water flow is blocked by dams inside Xuanmen Bay and thus water is not continuously transported to Yueqing Bay; streams flowing past Yucheng Subdistrict (#6) eventually drain into the sea area on the east side of Yuhuan Island without having an immediate impact on the waters in Yueqing Bay. Based on the spatial distribution of nutrient concentration in Yueqing Bay (Figure 11b), high-concentration zones were situated in shoals southwest of the bay mouth (corresponding to Yuecheng Town (#1) and Hongqiao Town (#2)), streams on both sides of Ximen Island at the bay head (corresponding to Dajing Town (#3) and Chengxi Subdistrict (#4)), and coastal sea areas beyond the Xuanmen Bay dams (corresponding to Chumen Town (#5)), whereas low-concentration zones were situated in the sea area on the west side of Yuhuan Island (corresponding to the Yucheng Subdistrict (#6)). These results indicate that the intensity of coastal anthropogenic activity and transport of inflowing streams are the main driving factors contributing to the high or low concentrations of nutrients in waters in local regions inside the bay.

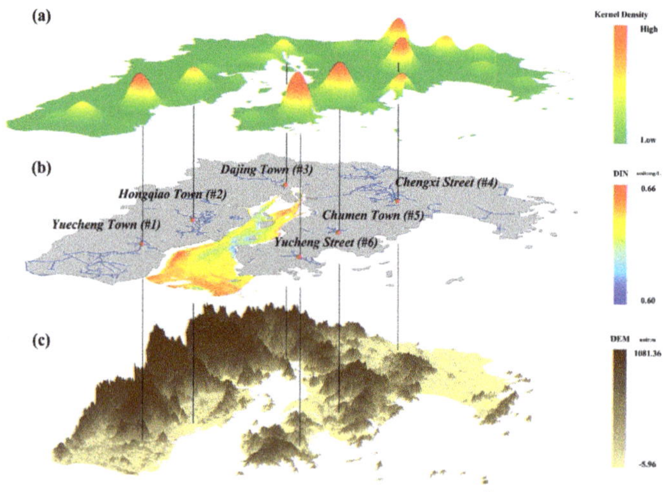

Figure 11. Spatial distribution of intensity of anthropogenic activities along Yueqing Bay and nutrient concentration inside Yueqing Bay: (**a**) spatial distribution of intensity of anthropogenic activities; (**b**) distribution of inflowing streams and nutrient concentration in Yueqing Bay; (**c**) digital elevation model (DEM), where nutrient concentration data in (**b**) are sourced from Figure 4d.

4.3. Influences of Urban Management in River Basin and Changes in Bay Governance Policy

Since its development in 1980, Yueqing Bay has also experienced many problems similar to the world-famous bay areas (e.g., Tokyo Bay), such as increased pollutant discharge and ecosystem damage [11,12]. As shown in Figure 12, compared to 1980, nutrient concentrations in Yueqing Bay have increased about 5–7-fold. From 1988 to 2005, cities along Yueqing Bay have undertaken significant land reclamation in the bay to address urban land shortages, totaling 61.81 km^2. As a continuation of the coastal reclamation plan of Xuanmen Bay in Yuhuan, Xuanmen phase II further strengthened the reclamation of land from the sea on the basis of the construction of Xuanmen dam in Xuanmen phase I, and the reclamation area reached 37.34 km^2. The implementation of Xuanmen phase I reclamation project directly blocked the exchange of water in Xuanmen Port and waters inside and outside the bay, while the large-scale reclamation of land from the sea in Xuanmen phase II further resulted in a remarkable decline in tidal influx in the bay. During 1988–2005, tidal influx from the bay center to the bay head decreased by 10%–20% and the water exchange rate declined significantly. Nutrients in the bay failed to be transported to the open sea and thus accumulated inside the bay, resulting in a marked nutrient concentration increase during this period [72]. After 2005, sea reclamation, urban areas, and anthropogenic activities increased, with an accompanying increase in the total discharge of land-derived pollutants and their transportation along inflowing streams to Yueqing Bay, resulting in further enhancement of nutrient concentrations in Yueqing Bay [40,43]. By 2018, the Yueqing Bay River basin contained a population of about 3 million and small enterprises had expanded, with industrial wastewater discharge in the bay increasing accordingly (up to 3 million tons each year). In addition, the aquatic farming industry in the bay has continued to develop. Since 2011, annual gross output of aquatic products in the three cities along Yueqing Bay has exceeded 80 million tons [43,78]; urban domestic sewage, coastal sewage outfalls, and pollutants from aquatic farming have drained continuously into Yueqing Bay.

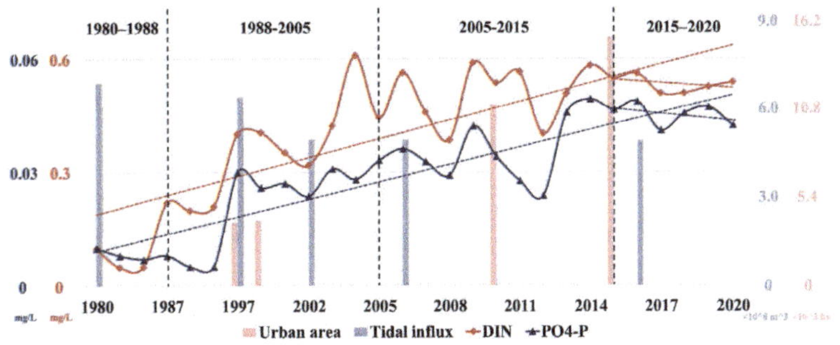

Figure 12. Variations in coastal urban development, tidal influx, and nutrient concentration in Yueqing Bay during 1980–2020. (Tidal influx and urban area data as well as nutrient concentration data of Yueqing Bay during 1980–2013 were sourced from the literature [40,43,72], nutrient concentration data of Yueqing Bay during 2013–2020 were sourced from remote-sensing-retrieved data in this study, vertical black dashed lines correspond to 1998, 2005, and 2015, respectively, and red and blue dashed lines are trend lines of DIN and PO_4-P, respectively).

As seen in Figure 13, gross output values of the primary sectors and industrial in cities along Yueqing Bay have shown a growing tendency. During 2011–2020, gross output values of both sectors increased by about RMB 90 billion, while gross output of aquatic products generally increased by ~16 million tons over the seven years, decreased somewhat after 2017, and decreased to 86.5 million tons by 2020. Most inflowing streams along Yueqing Bay run past rural towns. Many are classified as grade IV–V surface water due to increasing input from industrial wastewater, agricultural irrigation run-off, domestic sewage, and aquatic farming pollutants. Gao et al. [79] reported that most inflowing streams in Yueqing Bay have a total nitrogen and phosphorus content of 1.5–6.0 mg/L and 0.06–0.2 mg/L, respectively. At present, the output value of various industries continues to grow, the total population in the basin continues to increase, and rural towns continue to develop. Although the influence of the current pollution control plan, which highlighted the control of total nitrogen and total phosphorus emissions from rivers entering the sea and direct discharge sea pollution sources, achieved results and controlled the nutrient concentration in surface water to a certain extent, land-based pollution is still one of the main sources of pollutants in Yueqing Bay at this stage.

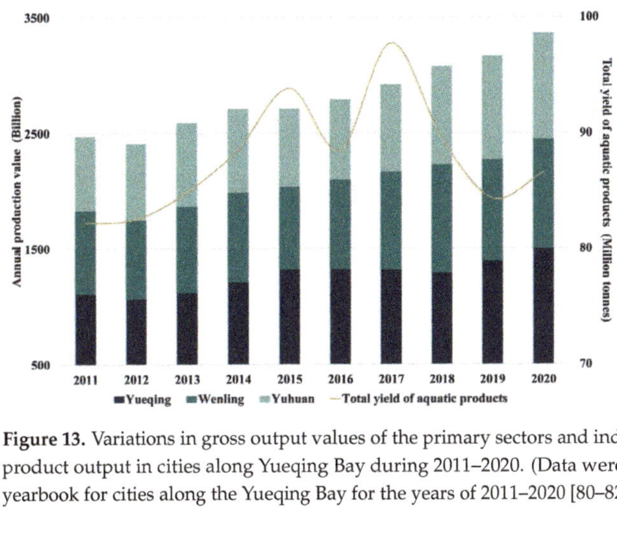

Figure 13. Variations in gross output values of the primary sectors and industrial and gross aquatic product output in cities along Yueqing Bay during 2011–2020. (Data were sourced from statistical yearbook for cities along the Yueqing Bay for the years of 2011–2020 [80–82]).

Time-series variations in nutrient concentrations in Yueqing Bay during 2013–2020 (Figure 5) showed abnormally low nutrient concentrations in the summer of 2017, which was mainly due to prolonged drought, with monthly mean rainfall lower than perennial monthly mean rainfall; indeed, only 2 days in August showed precipitation >5 mm and monthly total rainfall was only 51.2 mm in Yueqing City [82]. The drought led to a significant decline in water levels and river discharge and a lower total content of nutrient transported by streams into Yueqing Bay, resulting in abnormally low nutrient concentrations in the bay. This climatic event showed that industrial wastewater, agricultural irrigation water, domestic sewage, and aquatic farming pollutants in cities along Yueqing Bay are the main sources of nutrient pollution, and inflowing streams are the main mode of pollutant transport.

In May 2014, the "Five Water Treatment Actions" program (i.e., wastewater treatment, flood control, stagnant water drainage, water supply assurance, and water conservation) was implemented in Zhejiang for the purpose of water control, and mandatory water quality indicators have been used to control pollutant discharge and increase sustainable development [83]. In the context of this program, various ecological governance and remediation policies were proposed to address issues such as pollutant discharge control and environmental protection in cities along Yueqing Bay and to improve water quality in the bay by human intervention. In 2016, the Yueqing municipal government proposed policies to rationalize marine resource exploitation and utilization, control entry of polluting enterprises, and develop the ecological sector to control land-derived pollutants and pollutant discharge entering Yueqing Bay [84]. In 2021, the Yuhuan municipal government also proposed actions to govern and monitor pollutants discharged into the sea to improve water quality status in the bay by ecosystem remediation and pollutant control [85]. In addition to relevant pollution control and ecological remediation policies, after 2017, the development of the agricultural, forestry, husbandry, fishing, and aquatic product sectors along Yueqing Bay increased at somewhat slower rates, which had a positive impact on total input of land-derived pollutants and pollution control.

As seen in Figure 7, after 2015, the DIN and PO_4-P concentrations in Yueqing Bay decreased slightly but non-significantly ($P > 0.1$), also with no further growth noted, indicating that the ecological governance and remediation policies proposed by the cities along Yueqing Bay have had a positive impact on water quality status in Yueqing Bay.

As seen in Figure 8, for stream estuaries corresponding to regions along Yueqing Bay with intense anthropogenic activity (i.e., Yuecheng Town (#1), Hongqiao Town (#2), Dajing Town (#3), and Chengxi Subdistrict (#4) in Figure 12), nutrient concentrations changed distinctly before and after 2015. During 2013–2015, nutrient concentrations in all three regions increased at higher rates than the mean increasing rate in Yueqing Bay. In the East Stream (#1 in Figure 8) and West Stream of the bay head (#2 in Figure 8), the PO_4-P concentration increased at the fastest rate (Figure 8c,e and Table 3). After 2015, nutrient levels in the three regions declined to variable extents. In the East Stream (#1), nutrient concentrations in the water decreased at a slower rate, in particular, DIN decreased at a slightly lower rate than that in Yueqing Bay, while in the West Stream (#2 in Figure 8) and Southwest Shoal (#3 in Figure 8), nutrient concentrations decreased at a higher rate than that in Yueqing Bay, especially in the Southwest Shoal (#3 in Figure 8), where PO_4-P decreased markedly. On the west side of the bay, remediation polies to control PO_4-P were initiated relatively earlier by Yueqing city [40], resulting in significantly lower PO_4-P concentrations on the west side of Yueqing Bay. After 2015, a series of ecological governance and remediation policies were issued and implemented by local city governments along Yueqing Bay. These measures slowed the rise in nutrient concentrations in the bay and mitigated the water quality deterioration trend, especially in Yueqing City, where PO_4-P concentrations in inflowing streams were effectively controlled, resulting in a reduction in concentrations in the sea area on the west side of the bay. However, despite the mitigation effects, nutrient concentrations in Yueqing Bay remain high. For example, DIN concentrations only reach grade III seawater criteria (0.30 mg/L < DIN ≤ 0.40 mg/L) in summer, with levels higher than the upper limit specified for grade IV seawater (0.40 mg/L < DIN ≤ 0.50 mg/L) in the other seasons; furthermore, PO_4-P concentrations exceed grade IV seawater criteria (0.30 mg/L < PO_4-P ≤ 0.045 mg/L) all year [86].

The restoration of bay ecosystems is a long process; thus, continuous long-term ecological remediation efforts are essential for ecosystem restoration and regeneration. As an important semi-enclosed bay in the Kanto Area of Japan, Tokyo Bay is surrounded by several large cities, including Tokyo, Yokohama, and Kawasaki. During its development, coastal engineering, rapid urbanization, and other anthropogenic activities resulted in significant sea reclamation, pollutant discharge, ecosystem damage, and water quality deterioration in the bay [11,12]. In response, the Japanese government launched the "Tokyo Bay Restoration Plan" in 2001 to address these ecological problems, including the specific policies on wastewater discharge control, river dredging, wetland ecosystem restoration, and monitoring system establishment, etc. After years of governance and remediation, total pollutants (e.g., nitrogen and phosphorus) in Tokyo Bay have decreased, and water quality, while still poor, has not deteriorated further [87]. That is, although ecological regeneration policies have shown some success, the environment inside Tokyo Bay remains poor due to its early and high intensity development and deterioration. In comparison, Yueqing Bay developed much later, and ecological remediation has been implemented gradually over the past few years. Thus, from the perspective of the developmental and remediation history of Tokyo Bay, long-term ecological remediation efforts are still needed to improve the environment in Yueqing Bay and to eliminate or reduce the negative impacts of coast engineering and other anthropogenic activities.

5. Conclusions

This paper focused on water quality monitoring and variation analysis of a semi-enclosed bay (Yueqing Bay) in China, which is under strong anthropogenic impact. By constructing a regional remote sensing retrieval algorithm, the long time series remote sensing monitoring of water quality changes in Yueqing Bay is realized. On this basis, the impact of relevant management policies and anthropogenic activities in the basin on water quality variations in the bay can be understood, which can provide an effective reference basis for ecological environmental protection and restoration work in Yueqing Bay and other similar bays.

Based on long time-series Landsat-8 satellite remote sensing data, field measurement data, and supplementary statistical data, we constructed a model to retrieve nutrient concentrations in Yueqing Bay using SVM, and further investigated the spatiotemporal variations and main influencing mechanisms of nutrient concentrations in the bay during 2013–2020.

The nutrient retrieval results show that DIN concentration varied seasonally within 0.211–0.687 mg/L, with the lowest level in summer (averaging 0.3940 mg/L); PO_4-P concentration varied seasonally within 0.016–0.056 mg/L, with the lowest level in summer (averaging 0.0414 mg/L) and highest level in autumn (averaging 0.0508 mg/L). From a spatial distribution perspective, nutrient concentrations in Yueqing Bay were generally higher in the west and lower in the east, with high-concentration zones observed in the narrowest area of the middle part of the bay, Southwest Shoal at the bay mouth, and sea area on the north side of Damen and Xiaomen islands, and low-concentration zones observed in coastal waters on the west side of Yuhuan Island. During 2013–2015, the DIN and PO_4-P concentrations in Yueqing Bay showed an increasing tendency (DIN, $P = 0.0262$, $t = 2.5341$; PO_4-P, $P = 0.3196$, $t = 1.0383$), whereas, during 2015–2020, the DIN and PO_4-P concentrations showed a decreasing though non-significant tendency ($P > 0.1$).

The observed and remote-sensing-retrieved data were highly consistent in terms of seasonal variations, long time-series variations, and the spatial distribution of nutrient concentrations in Yueqing Bay. The field measurement data were mainly sourced from in situ surveys by marine research vessels, with high labor and material costs, limited observation stations, small spatial coverage, and low observation frequency. Based on its spatiotemporal continuity, field measurement data are less effective for investigations into the complex issues in Yueqing Bay, such as large-scale pollution prevention, dynamic monitoring of environmental pollution, long time series recording, multiple driving factors, and environmental remediation and restoration. In contrast, remote sensing technology not only shortens the data acquisition period but also improves the timeliness and spatial coverage of marine data. Hence, in comparison to field measurement data, the satellite remote sensing imagery of bays can be used to acquire environmental data with larger coverage and more continuous time series to investigate variation trends and driving factors.

Based on numerical simulation, kernel density estimation, and related literature data, seasonal variations in nutrient concentration in Yueqing Bay were mainly controlled by physical effects, including the Taiwan Warm Current and Fujian-Zhejiang Coastal Current, and biological effects, such as seasonal variations in the intensity of plankton photosynthesis in the bay. Furthermore, spatial distribution characteristics of nutrient concentration were mainly affected by synergy between the intensity of anthropogenic activities and transport by rivers along Yueqing Bay. Compared to 1980, nutrient concentrations in Yueqing Bay have increased 5–7-fold, mainly due to significantly enhanced anthropogenic activities from 1998 to 2015; in particular, during 1988–2005, large-area reclamation projects (e.g., Xuanmen phase II project) have resulted in a remarkable decline in tidal influx in the bay and a significant decrease in the water exchange rate, with nutrients thus accumulating inside the bay, rather than being transported to the open sea, resulting in a significant increase in nutrient concentration during this period. After 2005, the urban area along Yueqing Bay increased significantly, resulting in an increase in anthropogenic activities, discharge of land-derived pollutants, and transportation of pollutants along inflowing streams into Yueqing Bay, and thus caused the second rapid increase in nutrient concentrations in the bay waters [40,43]. Since 2011, the overall GDP of the three cities along Yueqing Bay has increased continuously, e.g., from 2011 to 2020, the GDP of the primary and secondary sectors increased by ~RMB 90 billion and annual aquatic production exceeded 80 million tons [43,76]. Due to this rapid development, industrial wastewater, agricultural irrigation run-off, domestic sewage, and aquatic farming pollutants increased, resulting in the gradual rise in nutrient concentrations in inflowing streams and the bay, especially the estuaries. After 2015, the main cities along Yueqing Bay implemented various restoration policies, and after 2017, the rising trend in gross aquatic product output slowed, leading to a decline (though non-significant) in the nutrient concentrations in waters inside the bay and improvement in water quality deterioration, with an overall positive impact on ecological remediation in Yueqing Bay.

Compared with Tokyo Bay, Yueqing Bay developed much later, and ecological remediation has been implemented gradually over the past few years. Thus, from the perspective of the developmental and remediation history of Tokyo Bay, long-term ecological remediation efforts are still needed to improve the environment in Yueqing Bay and to eliminate or reduce the negative impacts of coast engineering and other anthropogenic activities.

This paper demonstrates the satellite remote sensing techniques to monitor water quality variation in a bay under strong anthropogenic impact. An updated satellite algorithm is still needed in future work with more data collection and a deep mechanism understanding of the nutrient variation. The ± 3 day time interval we used in this study to guarantee enough matched data for regional algorithm development may induce errors. Moreover, satellite remote sensing only relates to the surface layer, and the detection of profile information should also be considered, although the water column was mixing in the coast and bay much more than in the open ocean under the highly dynamic hydrological condition. In general, satellite remote sensing data with a high spatial resolution have a strong ability to monitor the water quality of bays and gulfs, but they still need to be updated and combined with more field observations to provide better data support for environmental management and in-depth mechanism understanding.

Author Contributions: Conceptualization, B.Z. and Y.B.; methodology, B.Z. and X.H.; software, Z.W.; validation, B.Z.; investigation, Z.Z., S.Z. and Q.D.; data curation, Z.Z., S.Z. and Q.D.; writing—original draft preparation, B.Z.; writing—review and editing, Y.B. and X.H.; visualization, B.Z. and Z.W. All authors have read and agreed to the published version of the manuscript.

Funding: This study was supported by the National Key Research and Development Program of China (Grant #2017YFA0603003), and Key Special Project for Introduced Talents Team of Southern Marine Science and Engineering Guangdong Laboratory (Guangzhou) (GML2019ZD0602), Scientific Research Fund of the Second Institute of Oceanography, MNR (Grant #YJJC2105), the National Natural Science Foundation of China (Grants #41825014), and Zhejiang Provincial Natural Science Foundation of China (2017R52001, LR18D060001)).

Institutional Review Board Statement: Not applicable.

Informed Consent Statement: Not applicable.

Data Availability Statement: Data available in a publicly accessible repository that does not issue DOIs. Landsat series data were downloaded from https://earthexplorer.usgs.gov/ (accessed on 30 August 2021). The Point of Interest data were obtained from the Baidu Maps and can be found here: http://lbsyun.baidu.com/index.php?title=webapi (accessed on 30 August 2021). HYCOM series data were download from https://www.hycom.org/ (accessed on 30 August 2021). TRMM_3B43 Rainfall Estimate data were download from https://disc.gsfc.nasa.gov/ (accessed on 30 August 2021). ECMWF-ERA5 hourly wind field data were download from https://cds.climate.copernicus.eu/ (accessed on 30 August 2021).

Acknowledgments: We thank SOED/SIO/MNR satellite ground station, satellite data processing & sharing center, and marine satellite data online analysis platform (SatCO2) for their help with data collection and processing. We thank Wenzhou Environmental Monitoring Center for their help with field measurement data collection and processing. We thank United States Geological Survey (USGS) (https://earthexplorer.usgs.gov/ (accessed on 30 August 2021)) for providing the Landsat series satellite data and Baidu map (http://lbsyun.baidu.com/index.php?title=webapi (accessed on 30 August 2021)) for providing the Point of Interest data. We thank the National Ocean Partnership Program (NOPP) (https://nopp.org/ (accessed on 30 August 2021)) for providing the HYCOM series data. We thank the Goddard Earth Sciences Data and Information Services Center (GES DISC) (https://disc.gsfc.nasa.gov/ (accessed on 30 August 2021)) for providing the TRMM_3B43 Rainfall Estimate data. We thank the European Union's Copernicus Earth Observation Programme (https://cds.climate.copernicus.eu/ (accessed on 30 August 2021)) for providing the ECMWF-ERA5 hourly wind field data.

Conflicts of Interest: The authors declare no conflict of interest.

References

1. Creel, L. *Ripple Effects: Population and Coastal Regions*; Population Reference Bureau: Washington, DC, USA, 2003.
2. Liu, L.; Han, H. Research on countermeasures of coastal zone management. *Mar. Dev. Manag.* **2007**, *6*, 51–55.
3. Halpern, B.S.; Walbridge, S.; Selkoe, K.A.; Kappel, C.V.; Micheli, F.; D'Agrosa, C.; Bruno, J.F.; Casey, K.S.; Ebert, C.; Fox, H.E.; et al. A global map of human impact on marine ecosystems. *Science* **2008**, *319*, 948–952. [CrossRef] [PubMed]
4. Worm, B.; Hilborn, R.; Baum, J.K.; Branch, T.A.; Collie, J.S.; Costello, C.; Fogarty, M.J.; Fulton, E.A.; Hutchings, J.A.; Jennings, S.; et al. Rebuilding global fisheries. *Science* **2009**, *325*, 578–585. [CrossRef] [PubMed]
5. Zhai, T.; Wang, J.; Fang, Y.; Qin, Y.; Huang, L.; Chen, Y. Assessing ecological risks caused by human activities in rapid urbanization coastal areas: Towards an integrated approach to determining key areas of terrestrial-oceanic ecosystems preservation and restoration. *Sci. Total Environ.* **2020**, *708*, 135153. [CrossRef] [PubMed]
6. Furukawa, K.; Okada, T. Tokyo Bay: Its environmental status—Past, present, and future. In *The Environment in Asia Pacific Harbours*; Springer: Berlin/Heidelberg, Germany, 2006; pp. 15–34. [CrossRef]
7. Ribbe, J.; Wolff, J.-O.; Staneva, J.; Gräwe, U. Assessing water renewal time scales for marine environments from three-dimensional modelling: A case study for Hervey Bay, Australia. *Environ. Modell. Softw.* **2008**, *23*, 1217–1228. [CrossRef]
8. Roberts, A.D.; Prince, S.D. Effects of urban and non-urban land covers on nitrogen and phosphorus runoffs to Chesapeake Bay. *Ecol. Indic.* **2010**, *10*, 459–474. [CrossRef]
9. Ministry of Ecology Environment of the People's Republic of China. Bulletin on China's Ecological Environment in 2020. Available online: https://www.mee.gov.cn/hjzl/sthjzk/zghjzkgb/ (accessed on 10 October 2021).
10. Harding, L.W.; Gallegos, C.L.; Perry, E.S.; Miller, W.D.; Adolf, J.E.; Mallonee, M.E.; Paerl, H.W. Long-term trends of nutrients and phytoplankton in Chesapeake Bay. *Estuar. Coasts* **2016**, *39*, 664–681. [CrossRef]

11. Kubo, A.; Hashihama, F.; Kanda, J.; Horimoto-Miyazaki, N.; Ishimaru, T. Long-term variability of nutrient and dissolved organic matter concentrations in Tokyo Bay between 1989 and 2015. *Limnol. Oceanogr.* **2019**, *64*, S209–S222. [CrossRef]
12. Kubo, A.; Kanda, J. Coastal urbanization alters carbon cycling in Tokyo Bay. *Sci. Rep.* **2020**, *10*, 20413. [CrossRef]
13. Cao, F.F.; Dai, K.; Tao, Q.R.; Fan, C.Y.; He, L.L. Marine pollution status of nearshore areas of Hangzhou Bay and its governance countermeasures. *Environ. Sci. Technol.* **2020**, *43*, 60–69. [CrossRef]
14. Greening, H.; Janicki, A.; Sherwood, E.T.; Pribble, R.; Johansson, J.O.R. Ecosystem responses to long-term nutrient management in an urban estuary: Tampa Bay, Florida, USA. *Estuar. Coastal Shelf Sci.* **2014**, *151*, A1–A16. [CrossRef]
15. Liu, M.; Bi, Y.P.; Gong, Y.J.; Xi, X.H.; Lei, L.Y. Integrated restoration of ecological environments in typical bays:in a case of Jinzhou Bay and Hulushan Bay in Liaoning Province. *J. Dalian Ocean. Univ.* **2014**, *29*, 272–275. [CrossRef]
16. Hassett, B.; Palmer, M.; Bernhardt, E.; Smith, S.; Carr, J.; Hart, D. Restoring watersheds project by project: Trends in Chesapeake Bay tributary restoration. *Front. Ecol. Environ.* **2005**, *3*, 259–267. [CrossRef]
17. Perry, W. Elements of South Florida's Comprehensive Everglades Restoration Plan. *Ecotoxicology* **2004**, *13*, 185–193. [CrossRef]
18. Gitelson, A.A.; Dall'Olmo, G.; Moses, W.; Rundquist, D.C.; Barrow, T.; Fisher, T.R.; Gurlin, D.; Holz, J. A simple semi-analytical model for remote estimation of chlorophyll-a in turbid waters: Validation. *Remote Sens. Environ.* **2008**, *112*, 3582–3593. [CrossRef]
19. Brivio, P.; Giardino, C.; Zilioli, E. Determination of chlorophyll concentration changes in lake garda using an image-based radiative transfer code for Landsat TM images. *Int. J. Remote Sens.* **2001**, *22*, 487–502. [CrossRef]
20. Brivio, P.A.; Giardino, C.; Zilioli, E. Validation of satellite data for quality assurance in lake monitoring applications. *Sci. Total Environ.* **2001**, *268*, 3–18. [CrossRef]
21. Nelson, N.B.; Siegel, D.A. Chromophoric DOM in the open ocean. In *Biogeochemistry of Marine Dissolved Organic Matter*; Academic Press: San Diego, CA, USA, 2002; pp. 547–578.
22. Somvanshi, S.; Kunwar, P.; Singh, N.B.; Shukla, S.P.; Pathak, V. Integrated remote sensing and GIS approach for water quality analysis of gomti river, Uttar Pradesh. *Int. J. Environ. Sci.* **2012**, *3*, 62–74.
23. Ritchie, J.C.; Schiebe, F.R.; Mchenry, J.R. Remote sensing of suspended sediments in surface waters. *Photogramm. Eng. Remote Sens.* **1976**, *42*, 1539–1545.
24. Mohd Hasmadi, I.; Norsaliza, U. Analysis of SPOT-5 data for mapping turbidity level of river klang, peninsular malaysia. *Appl. Remote Sens. J.* **2010**, *1*, 14–18.
25. Lim, J.; Choi, M. Assessment of water quality based on Landsat 8 operational land imager associated with human activities in Korea. *Environ. Monit. Assess.* **2015**, *187*, 384. [CrossRef] [PubMed]
26. Alparslan, E.; Coskun, H.G.; Alganci, U. Water quality determination of Küçükçekmece Lake, Turkey by using multispectral satellite data. *Sci. World J.* **2009**, *9*, 1215–1229. [CrossRef] [PubMed]
27. Shahraiyni, T.H.; Schaale, M.; Fell, F.; Fischer, J.; Preusker, R.; Vatandoust, M.; Shouraki, B.S.; Tajrishy, M.; Khodaparast, H.; Tavakoli, A. Application of the active learning method for the estimation of geophysical variables in the caspian sea from satellite ocean colour observations. *Int. J. Remote Sens.* **2007**, *28*, 4677–4683. [CrossRef]
28. Zhang, C.; Han, M.I.N. Mapping chlorophyll—A concentration in Laizhou Bay using Landsat 8 oli data. In Proceedings of the 36th IAHR World Congress, The Hague, The Netherlands, 28 June–3 July 2015. [CrossRef]
29. Gholizadeh, M.H.; Melesse, A.M.; Reddi, L. A Comprehensive Review on Water Quality Parameters Estimation Using Remote Sensing Techniques. *Sensors* **2016**, *16*, 1298. [CrossRef] [PubMed]
30. Carlson, R.E. A trophic state index for lakes. *Limnol. Oceanogr.* **1977**, *22*, 361–369. [CrossRef]
31. Busse, L.B.; Simpson, J.C.; Cooper, S.D. Relationships among nutrients, algae, and land use in urbanized southern California streams. *Can. J. Fish. Aquat. Sci.* **2006**, *63*, 2621–2638. [CrossRef]
32. Uusitalo, R.; Yli-Halla, M.; Turtola, E. Suspended soil as a source of potentially bioavailable phosphorus in surface runoff waters from clay soils. *Water Res.* **2000**, *34*, 2477–2482. [CrossRef]
33. Kim, S.-I.; Kim, H.-C.; Hyun, C.-U. High resolution ocean color products estimation in Fjord of Svalbard, arctic sea using Landsat-8 oli. *Korean J. Remote Sens.* **2014**, *30*, 809–816. [CrossRef]
34. Allan, M.G.; Hamilton, D.P.; Hicks, B.J.; Brabyn, L. Landsat remote sensing of chlorophyll a concentrations in central north island lakes of new zealand. *Int. J. Remote Sens.* **2011**, *32*, 2037–2055. [CrossRef]
35. He, W.; Chen, S.; Liu, X.; Chen, J. Water quality monitoring in a slightly-polluted inland water body through remote sensing—Case study of the Guanting Reservoir in Beijing, China. *Front. Environ. Sci. Eng. China* **2008**, *2*, 163–171. [CrossRef]
36. Qi, L.; Hu, C.; Duan, H.; Barnes, B.B.; Ma, R. An EOF-based algorithm to estimate chlorophyll a concentrations in Taihu Lake from MODIS land-band measurements: Implications for near real-time applications and forecasting models. *Remote Sens.* **2014**, *6*, 10694–10715. [CrossRef]
37. Li, J. Hydraulic Properties of Environment in the Yueqing Bay. Master's Thesis, Zhejiang University, Hangzhou, China, 2010.
38. Luo, F.; Liao, G.H.; Yang, C.H.; Xu, X.H. Study on the features of water exchange in Yueqingwan Bay. *J. Mar. Sci.* **2011**, *29*, 79–88. [CrossRef]
39. China Bays Compiling Committee. *Chinese Harbours and Embayments*; Ocean Press: Beijing, China, 1993; Volume 6.
40. Guo, Y.M.; Zhu, J.P.; Gu, J.; Bao, J.J.; He, Y.N.; Liu, Q. Interannual variations of nutrient contents in fishing waters of the Yueqing Bay in May and August during 2000–2014. *Mod. Agric. Sci. Technol.* **2015**, 254–255. [CrossRef]

41. Mu, J.B.; Wen-Ren, Y.Z.; Ying, C. Impact of Tidal Flat Reclamation on Water Exchange in the Yueqing Bay. In Proceedings of the 14th National Conference on Hydrodynamics and the 28th National Symposium on Hydrodynamics: Changchun, Jilin, China, 8–10 August 2017.
42. Zhang, L.J. Environment and Resource Characteristics and Integrated Coastal Zone Management of the Yueqing Bay in Zhejiang Province. *Resour. Sci.* **2001**, *22*, 57–61. [CrossRef]
43. Yu, M.J. *Response of Landscape Pattern of the Yueqing Bay Coastal Zone to Human Disturbance*; Zhejiang Normal University: Jinhua, China, 2020.
44. Zhou, H. *Zhejiang Islands*; Higher Education Press: Beijing, China, 1998.
45. Cooley, T.; Anderson, G.P.; Felde, G.W.; Hoke, M.L.; Ratkowski, A.J.; Chetwynd, J.H.; Gardner, J.A.; Adler-Golden, S.M.; Matthew, M.W.; Berk, A.; et al. FLAASH, a MODTRAN4-based atmospheric correction algorithm, its application and validation. In Proceedings of the IEEE International Geoscience and Remote Sensing Symposium, Toronto, ON, Canada, 24–28 June 2002; pp. 1414–1418.
46. State Oceanic Administration (Now the Ministry of Natural Resources of The People's Republic of China). *Code of Practice for Marine Monitoring Technology*; State Oceanic Administration: Beijing, China, 1991.
47. Kupssinskü, L.S.; Guimaraes, T.T.; De Souza, E.M.; Zanotta, D.C.; Veronez, M.R.; Gonzaga, L., Jr.; Mauad, F.F. A method for chlorophyll-a and suspended solids prediction through remote sensing and machine learning. *Sensors* **2020**, *20*, 2125. [CrossRef] [PubMed]
48. Gómez, D.; Salvador, P.; Sanz, J.; Casanova, J.L. A new approach to monitor water quality in the Menor sea (Spain) using satellite data and machine learning methods. *Environ. Pollut.* **2021**, *286*, 117489. [CrossRef] [PubMed]
49. Wagle, N.; Acharya, T.D.; Lee, D.H. A comprehensive review on application of machine learning algorithms for water quality parameter estimation using remote sensing data. *Sens. Mater.* **2020**, *32*, 3879–3892. [CrossRef]
50. Hsu, C.-W.; Chang, C.-C.; Lin, C.-J. A Practical Guide to Support Vector Classification. 2003. Available online: https://www.ee.columbia.edu/~{}sfchang/course/svia/papers/svm-practical-guide.pdf (accessed on 10 October 2021).
51. Chen, C.; Liu, H.; Beardsley, R.C. An unstructured grid, finite-volume, three-dimensional, primitive equations ocean model: Application to coastal ocean and estuaries. *J. Atmos. Ocean. Technol.* **2003**, *20*, 159–186. [CrossRef]
52. Chen, C.; Beardsley, R.C.; Cowles, G. Finite volume coastal ocean. *Oceanography* **2006**, *19*, 78–89. [CrossRef]
53. Wang, Z.; Bai, Y.; He, X.; Tao, B.; Li, T.; Chen, X.; Wang, T.; Gong, F. Estimating particulate organic carbon flux in a highly dynamic estuary using satellite data and numerical modeling. *Remote Sens. Environ.* **2021**, *252*, 112116. [CrossRef]
54. Liu, D.; Bai, Y.; He, X.; Chen, C.-T.A.; Huang, T.-H.; Pan, D.; Chen, X.; Wang, D.; Zhang, L. Changes in riverine organic carbon input to the ocean from mainland China over the past 60 years. *Environ. Int.* **2020**, *134*, 105258. [CrossRef] [PubMed]
55. Xu, Z.N.; Gao, X.L. A novel method for identifying the boundary of urban built-up areas with POI data. *Acta Geogr. Sin.* **2016**, *71*, 928–939. [CrossRef]
56. Chi, J.; Jiao, L.M.; Dong, T.; Gu, Y.Y.; Ma, Y.L. Quantitative Identification and Visualization of Urban Functional Area Based on POI Data. *J. Geomat.* **2016**, *41*, 68–73. [CrossRef]
57. Yu, B.C.; Liu, Y.X.; Chen, G. Urban Spatial Structure of Port City in South China Sea Based on Spatial Coupling between Nighttime Light Data and POI. *J. Geo.-Inf. Sci.* **2018**, *20*, 854–861. [CrossRef]
58. Yu, W.H.; Ai, T.H.; Liu, P.C.; He, Y.K. Network Kernel Density Estimation for the Analysis of Facility POI Hotspots. *Acta Geod. Cartogr. Sin.* **2015**, *44*, 1378–1383. [CrossRef]
59. Yu, W.H.; Ai, T.H. The Visualization and Analysis of POI Features under Network Space Supported by Kernel Density Estimation. *Acta Geod. Cartogr. Sin.* **2015**, *44*, 82–90. [CrossRef]
60. Wang, Y.; He, H. *Spatial Data Analysis Methods*; Science Press: Beijing, China, 2007; pp. 66–69. ISBN 978-7-03-018966-0.
61. Yu, W.H.; Ai, T.H.; Yang, M.; Liu, J.P. Detecting "Hot Spots" of Facility POIs Based on Kernel Density Estimation and Spatial Autocorrelation Technique. *Geomat. Inf. Sci. Wuhan Univ.* **2016**, *41*, 221–227. [CrossRef]
62. Zaneveld, J.; Ronald, V. *Bulletin of the American Meteorological Society*; American Meteorological Society: Boston, MA, USA, 1995; pp. 60–63. Available online: http://www.jstor.org/stable/26231610 (accessed on 10 October 2021).
63. Brodie, J.; Schroeder, T.; Rohde, K.; Faithful, J.; Masters, B.; Dekker, A.; Brando, V.; Maughan, M. Dispersal of suspended sediments and nutrients in the Great Barrier Reef lagoon during river-discharge events: Conclusions from satellite remote sensing and concurrent flood-plume sampling. *Mar. Freshwater Res.* **2010**, *61*, 651–664. [CrossRef]
64. Huang, C.; Guo, Y.; Yang, H.; Li, Y.; Zou, J.; Zhang, M.; Lyu, H.; Zhu, A.; Huang, T. Using remote sensing to track variation in phosphorus and its interaction with chlorophyll-a and suspended sediment. *IEEE J. Sel. Top. Appl. Earth Obs. Remote Sens.* **2015**, *8*, 4171–4180. [CrossRef]
65. Huang, Y.; Fan, D.; Liu, D.; Song, L.; Ji, D.; Hui, E. Nutrient estimation by HJ-1 satellite imagery of Xiangxi bay, Three Gorges Reservoir, China. *Environ. Earth Sci.* **2016**, *75*, 633. [CrossRef]
66. Mao, Z.; Chen, J.; Pan, D.; Tao, B.; Zhu, Q. A regional remote sensing algorithm for total suspended matter in the East China Sea. *Remote Sens. Environ.* **2012**, *124*, 819–831. [CrossRef]
67. Gurlin, D.; Gitelson, A.A.; Moses, W.J. Remote estimation of chlorophyll-a concentration in turbid productive waters—Return to a simple two-band NIR-red model? *Remote Sens. Environ.* **2011**, *115*, 3479–3490. [CrossRef]

68. Gilerson, A.A.; Gitelson, A.A.; Zhou, J.; Gurlin, D.; Moses, W.; Ioannou, I.; Ahmed, S.A. Algorithms for remote estimation of chlorophyll-a in coastal and inland waters using red and near infrared bands. *Opt. Express* **2010**, *18*, 24109–24125. [CrossRef] [PubMed]
69. Knaeps, E.; Ruddick, K.G.; Doxaran, D.; Dogliotti, A.I.; Nechad, B.; Raymaekers, D.; Sterckx, S. A SWIR based algorithm to retrieve total suspended matter in extremely turbid waters. *Remote Sens. Environ.* **2015**, *168*, 66–79. [CrossRef]
70. Nechad, B.; Ruddick, K.G.; Park, Y. Calibration and validation of a generic multisensor algorithm for mapping of total suspended matter in turbid waters. *Remote Sens. Environ.* **2010**, *114*, 854–866. [CrossRef]
71. Zeng, J.N.; Pan, J.M.; Liang, C.J.; Chen, Q.Z. *Comprehensive Investigation Report on the Ecological Environment of Key Harbors in Zhejiang Province*; China Ocean Press: Beijing, China, 2011; pp. 114–118.
72. Lee, J. Numerical Simulation of DIN&DIP Distribution in Yueqing Bay and Their Response to Important Coastal Engineerings. Master's Thesis, Zhejiang University, Hangzhou, China, 2005.
73. Wang, C.; Guo, X.F.; Fang, J.; Li, Q.S. Characteristics of seasonal spatial expansion of Fujian and Zhejiang Coastal Current and their bay effects. *J. Appl. Oceanogr.* **2018**, *37*, 1–8. [CrossRef]
74. Shi, X.Y.; Li, H.S.; Wang, H.; Wang, L.S.; Zhang, C.S. Preliminary discussion about hydrochemical properties of Taiwan Warm Current in summer and their influences on harmful algal bloom-prevalent zones in East China Sea. *Oceanol. Limnol. Sin.* **2013**, *44*, 1208–1215. [CrossRef]
75. Ji, X.M.; Zhang, Y.Z.; Zhu, D.K. Study on Marine Environment and Recent Coastal Evolution of Yueqing Bay, Zhejiang Province, China. *Mar. Sci. Bull.* **2006**, *25*, 44–53. [CrossRef]
76. Qin, B.; Hu, W.; Gao, G.; Luo, L.; Zhang, J. Dynamics of sediment resuspension and the conceptual schema of nutrient release in the large shallow Lake Taihu, China. *Chin. Sci. Bull.* **2004**, *49*, 54–64. [CrossRef]
77. Wang, P.P. *Effects of Sediment Resuspension in the Yangtze River Estuary and Neighboring Sea Areas on Nutrients in Waters*; Ocean University of China: Qingdao, China, 2015.
78. Peng, X.; Qiu, J.B.; Chen, S.B.; Xie, Q.L.; Li, S.L.; Huang, X.L. The study on the vulnerability of the ecosystem in Yueqing Bay. *J. Mar. Sci.* **2009**, *27*, 111–118. [CrossRef]
79. Gao, S.Q.; Lu, Y.; Zeng, J.N.; Yao, L.K.; Gao, D.W. Characteristics of the Aquatic Environment and Causative Analysis of Eutrophication in Yueqing Bay. *Mar. Sci. Bull.* **2005**, *24*, 25–32. [CrossRef]
80. Wenling Municipal People's Government. Statistical Bulletin. Available online: http://www.wl.gov.cn/col/col1402302/index.html (accessed on 10 October 2021).
81. Yuhuan Municipal People's Government. Statistical Bulletin. Available online: http://www.yuhuan.gov.cn/col/col1229302012/index.html (accessed on 10 October 2021).
82. Yueqing Municipal People's Government. Statistical Bulletin. Available online: http://www.yueqing.gov.cn/col/col1229248995/index.html (accessed on 10 October 2021).
83. Liu, H.Z.; Shan, B.Q.; Zhang, W.Q.; Shen, Y.Y.; Shi, Z.D. Exploration of Successful Water Pollution Control Strategy in a Province with Serious Water Pollution—Achievement of Water Co-governance in Zhejiang Province and Suggestions of Future Promotion. *Environ. Prot. Sci.* **2015**, *3*, 47–52. [CrossRef]
84. Chen, X.X.; Yu, X.X.; Huang-Ye, Q.H. Yueqing City to forge modern spatial planning pattern of national land by "integrating multiple plans". *Zhejiang Land Resour.* **2020**, *201*, 30–32. [CrossRef]
85. Chu, X.D. Yuhuan to construct practice innovation bases on islands while cherishing the idea "Lucid waters and lush mountains are invaluable assets". *Zhejiang Econ.* **2021**, *688*, 72–73. [CrossRef]
86. *GB 3097-1997*; Quality Criteria for Seawater. Ministry of Ecology and Environment: Beijing, China, 1997.
87. Xiong, H.X.; Dai, M.X.; Peng, S.T.; Hu, J.B. Enlightenments of Tokyo bay revitalization action plan (2003~2012) in Janpan to Bohai bay in China. *J. Waterw. Harb.* **2020**, *41*, 119–124. [CrossRef]

remote sensing

MDPI

Article

Assessing Satellite, Land Surface Model and Reanalysis Evapotranspiration Products in the Absence of In-Situ in Central Asia

Friday Uchenna Ochege [1,2,3], Haiyang Shi [1,2,4,5], Chaofan Li [6], Xiaofei Ma [1,2], Emeka Edwin Igboeli [1,4,7] and Geping Luo [1,2,4,*]

1 State Key Laboratory of Desert and Oasis Ecology, Xinjiang Institute of Ecology and Geography, Chinese Academy of Sciences, No. 818, South Beijing Road, Urumqi 830011, China; friday@ms.xjb.ac.cn (F.U.O.); haiyang.shi@ugent.be (H.S.); mxf@ms.xjb.ac.cn (X.M.); emeka.igboeli@unn.edu.ng (E.E.I.)
2 Research Centre for Ecology and Environment of Central Asia, Chinese Academy of Sciences, Urumqi 830011, China
3 Department of Geography and Environmental Management, University of Port Harcourt, PMB 5323 Choba, East-West, Port Harcourt 500004, Nigeria
4 University of Chinese Academy of Sciences, Beijing 100039, China
5 Department of Geography, Ghent University, 9000 Ghent, Belgium
6 School of Geographic Sciences, Nanjing University of Information Science and Technology, Nanjing 210044, China; lcf@nuist.edu.cn
7 Department of Geography, University of Nigeria, Nsukka 410001, Nigeria
* Correspondence: luogp@ms.xjb.ac.cn; Tel.: +86-13899875832

Abstract: Shortfalls in regular evapotranspiration (ET) monitoring and evaluation pose a huge challenge to agricultural water resource distribution in arid Central Asia (CA). In this study, a first detailed regional assessment of GLEAM, ERA5, MERRA2, CLSM and NOAH ET products in CA was performed by systematically implementing the triple collocation (TC) method, in which about 36,936 grid cells for each ET data (within a six-triplet design) were collocated, at 0.25° and with monthly resolutions during 2003–2020. The reliability of the strategy adopted was confirmed in four arid biomes using standard evaluation metrics (R, RMSE and BIAS), and by spatiotemporal cross-validation of the six ET triplets across CA. Results show that the systematic TC method produced more robust ET product assessment metrics with reduced RMSEs compared to the initial ET product validation using in-situ, which showed weak-positive correlation and high negative bias-range (i.e., $-21.02 \leq \text{BIAS} < 16$ mm) in the four arid biomes of CA. The spatial cross-validation by TC showed that the magnitude of ET random errors significantly varies, and confirms the systematic biases with site-scale measurements. The highest ET uncertainties by CLSM (27.43%), NOAH (29.16%), MERRA2 (38.28%), ERA5 (36.75), and GLEAM (41%) were more evident in the shrubland, cropland, grassland, cropland again, and desert biomes, respectively. Moreover, error magnitudes in high altitudes (Tianshan Mountain range) are generally lower than in plain-desert areas. All ET products spatially captured ET dynamics over CA, but none simultaneously outperformed the other. These findings are invaluable in the utilization of the assessed ET products in supporting regional water resource management, particularly in CA.

Keywords: evapotranspiration; triple collocation; arid ecosystem biomes; satellite; land surface model; reanalysis

1. Introduction

Terrestrial evapotranspiration (ET) controls land surface-atmosphere interactions and is critical to the physical processes of regional hydrology [1], and agricultural water productivity [2]. ET is a key climate change indicator [3,4], however, it is difficult to calculate, and most times, it requires some thorough computational procedures, which are even more demanding in data-scarce heterogeneous regions with outstanding ecological history, e.g., Central Asia-CA [5,6]. Fortunately, remote sensing technological advancements have demonstrated the effectiveness of direct and indirect applications of sensor- and process-driven frameworks in analyzing the hydrodynamic features of various land surface fields [1,7–9], which helps to establish spatially continuous ET information for regions like CA to rely on.

Some of the globally available ET products include MOD16 (Moderate Resolution Imaging Spectroradiometer), CSIRO-PML v. 2.2.3 (Commonwealth Scientific and Industrial Research Organisation- Penman-Monteith-Leuning), ALEXI (Atmosphere-Land Exchange Inverse), GLDAS (Global Land Data Assimilation System v 2.21, example, NOAH and Catchment Land Surface Model LSM), GLEAM v. 3.5 (The Global Land Evaporation Amsterdam Model), SSEBop (Operational Simplified Surface Energy Balance), EEFLUX (Earth-Engine Flux), PT-JPL (Priestley-Taylor Jet Propulsion Laboratory), NTSG (Numerical Terradynamic Simulation Group) models [8,10–12]; and recently, the *ECOSTRESS* (ECOsystem Spaceborne Thermal Radiometer Experiment on Space Station) [13]. Reanalysis evaporation products, such as European Centre for Medium-Range Weather Forecasts Reanalysis-5 (ERA5) [14] and Modern-Era Retrospective Analysis for Research and Application Land version 2 reanalysis product (MERRA2) [15] are also extremely beneficial to a variety of hydrological applications in data-scarce regions.

The comprehensive utilization of these ET products largely depends on their global coverage, uncertainty characteristics and quality; because not all the ET models are widely available to cover complex areas or are appropriate for practical purposes. There is evidence that most satellite ET products are not suitable for arid regions due to considerable data gaps [16]. We confirmed this in a preliminary assessment of twelve ET models, but results are not shown here. Several researchers have also examined one or more of these ET models at different spatiotemporal scales with varying results. Velpuri, Senay [17] extensively evaluated two MODIS-based ET products over the conterminous United States where both products revealed average uncertainty of up to 50% and 25% at point and basin scales, respectively. Yang, Yong [18] used ChinaFlux towers to validate GLEAM in China where they found that the product produced reasonable accuracy in grassland, but revealed a systematic bias in the forest ecosystem.

Baik, Liaqat [19] assessed the uncertainties of four different ET products, including GLDAS-NOAH, GLEAM, MOD16 and MERRA over Australia using a continental grided ET data set (AWRA-L: Australian Water Resource Assessment Landscape). The study showed large uncertainties across seven different land cover classifications and four different climate zones, with correlation coefficients (R) ranging from 0.1 to 0.85. Li, Wang [20] improvised a stratification method to evaluate GLEAM, MERRA-Land, GLDAS-NOAH and EartH2Observe-En ET products. The products presented significant differences in performance, with none showing optimum results at site-scale. In East Asia, Khan, Liaqat [21] reported the physical accuracy performance of MOD16, GLEAM and GLDAS by using the extended triple collocation approach [22] and AsiaFlux measurements. They noted the underestimation of the NOAH model in grassland with corresponding overestimations in forests, paddy field ecosystems, and the improved performance of GLEAM in the cropland ecosystem. Moshir Panahi, Sadeghi Tabas [23] applied the water balance method to assess ERA5, GLEAM, and GLDAS datasets and found discrepancies in their seasonal and monthly time series over Iran. In general ET product performance largely depends on the forcing data and/or evaluation approach employed [24].

In-situ-based measurements from lysimeter, Bowen ratio, Eddy Covariance–EC, flux gradient, and large aperture scintillometer remain the most patronized methods for ET product evaluations, due to their capabilities to register almost accurate records [8]. Despite that point-based evaluations often suffer from energy balance misclosure and footprint/scale mismatch [9,25], their installation costs, system maintenance, observation heights, or limited spatial scale still pose a huge challenge. In CA, for instance, installations of in-situ-based systems require complex micrometeorological instrumentations due to terrain heterogeneity. For these same reasons, the majority of other global arid regions suffer from poor observations, which result in the scarcity of ET product assessments to support water resource management in these arid regions. Therefore, process-based methods [26,27], including those incorporating machine learning algorithms, and/or their combination with satellite-based products [9,28] have been used to evaluate and compare alternative ET estimates at various scales.

Other powerful, yet not new approaches which are rapidly evolving include the three cornered hat-TCH [29] and triple collocation-TC [22,30] methods. The TC technique is very robust [31], and has gained relevance in assessing various land surface fields, including characterizing the uncertainties associated with leaf area index (LAI), precipitation and ET [21,32,33]. Although, TC implementation experiments are mostly applied to satellite soil moisture (SM) retrievals [34,35], however, they have been largely supported by methodological advances [22,31,35] and extensive in-situ coverage of the corresponding SM variable [36]. Compared to water fluxes, poor distribution of measurement systems [37,38] pose huge limitations to their effective assessment in CA. Notwithstanding, we demonstrate the systematic implementation of the TC without in-situ components in the triplet design to achieve comparable results as with previous studies, e.g., Khan et al. [21].

In this study, the TC method which does not require high-level reference data [32,34], yet is able to characterize random error variances of land surface fields from coarse satellite data sets [31], was applied to evaluate the random error variances of five state-of-the-art ET data sets in entirely new terrain. The choice of TC method application in this study was mainly to fill the huge knowledge gap in the existing regional ET research studies aimed at characterizing the error variances and uncertainties associated with the CLSM, ERA5, GLEAM, MERRA2 and NOAH ET products, which are alternative ET estimates in arid data-scarce CA. Therefore, the main objectives are, (i) to spatially and temporarily assess the performance of the five newer/updated versions of these five ET products, by systematically implementing the TC method in CA; (ii) to compare the performances of the TC analysis and the standard evaluation metrics (SM) where in situ validation was performed; (iii) to characterize the uncertainties associated with the five ET products in four distinct arid biomes.

2. Materials and Methods

2.1. Study Area

Geographically, Central Asia (CA) is located in the heart of the Eurasian continent covering an expanse of 566×104 km^2, between longitude 46°30′ to 96°30′E and latitude 30°20′ to 50°30′N (Figure 1). CA comprises Kazakhstan, Kyrgyzstan, Tajikistan, Turkmenistan, Uzbekistan and Xinjiang in northwest China [39]. The region is characterized by large portions of deserts (~74%, e.g., the Taklamakan and Karakum deserts which are habitats to peculiar water stressed shrubs and herbs), and a small fraction of humid regions (~26%) containing croplands, broadleaf and needle leaf vegetation types [40,41]. The general vegetation type is grass. The snowcapped Mt. system in CA extends from the Altai Mt. in the north to Pamir Mt. in the south, with the Tianshan Mt. located in-between them (Figure 1). This Mt. system interacts with the dominant wind system (i.e., the westerlies) to regulate air mass circulation, control rainfall distribution and atmospheric pressure from low to mid-latitude of CA and beyond [40]. Due to water scarcity in CA, intermittent runoff rivers provide water resources at all levels to support the irrigation activities which also helps to boost the food security status of the region [6]. Largely, rainfall is low in CA (mean annuals < 250 mm), and combines with increased temperature anomalies to impact the region's local climate which further induces high vapor flux transfer from different land cover types [2]. Therefore, water imbalances remain a huge challenge in the region.

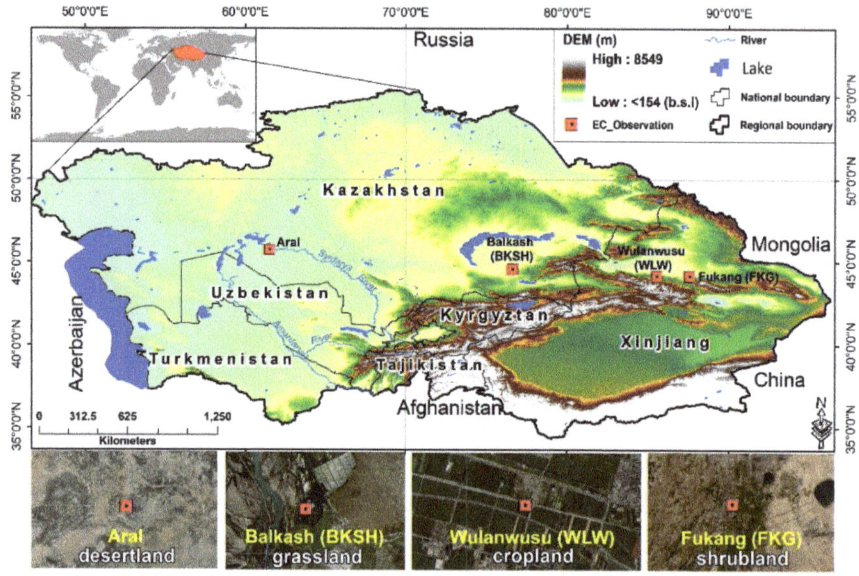

Figure 1. Digital Elevation Model–DEM of Central Asia (CA) showing the study area; DEM was derived from the Shuttle Radar Topography Mission (SRTM) elevation model, showing the snow-covered mountain ranges, lowlands, lakes and rivers. The region consists of oasis-desert ecosystems, dominated by barelands, grasslands, arableland, with patches of shrubs, forests, wetland and urban areas. The basin is well-drained by rivers Amu Darya and Syr Darya which sources are from the Tianshan Mt. system, hence, popularly referred to as the "Water Tower of Central Asia". The red dots show the location of the Eddy Covariance (EC) flux towers which served as ET validation and are directly depicted under the map.

2.2. State-of-the-Art ET Products

Five global ET data sets were comprehensively evaluated in this study (Table 1). They include Global Land Evaporation Amsterdam Model (GLEAM) satellite ET [11], two Global Land Data Assimilation System (GLDAS)- Land Surface Model (LSM) ET products, i.e., Catchment LSM [42] and NOAH LSM [10], and two reanalysis ET data sets, i.e., European Centre for Medium-Range Weather Forecasts Reanalysis-5 (ERA-5) [14] and the Modern-Era Retrospective Analysis for Research and Application Land version 2 reanalysis product (MERRA2) [15]. We assessed the GLEAM actual land evaporation dataset at a monthly time scale and at the 0.25° × 0.25° resolution from 2003 to 2020 July [11]. It was produced using a set of algorithms, such as the Priestly–Taylor formulation with explicit soil moisture stress and the Gash interception model method and forced by different satellite data [43]. The assessed CLSM's parametrization scheme maintains a temporal consistency between the ECMWF-forced (2003 to present) and the Princeton-forced (1948–2012) simulations, with the available data starting from February 2003 till date, at 0.25° spatial resolution [42]. The meteorological forcings used for simulating NOAH include air temperature, humidity, precipitation, surface pressure, downwelling and upwelling shortwave and longwave radiation, and wind speed [10]. We focused on the newer Level 4 (L4) NOAH v2.1 (0.25° × 0.25°) at the monthly temporal resolution and was evaluated during 2003–2020. We assessed the updated ERA5 with an advanced radiative transfer model based on a 10-member ensemble 4DVar [14], at 0.25° × 0.25° spatial scale from 2003 to 2020. ERA5 was updated from the ERA-Interim reanalysis [44]. Lastly, the recent MERRA2 [15] derived from energy balance at the land surface and was reinforced with the CLSM [45] was also evaluated. To align CLSM and NOAH with GLEAM, ERA5 and MERRA2, the measurements were converted from W m^{-2} to mm/m. Both ERA5 and MERRA2 were resampled to 0.25° × 0.25° to match with LSM and satellite ET data sets. A summary of these ET products is listed in Table 1.

Table 1. A summary of ET data sets that was evaluated in this study.

Product Name	Spatial Resolution	Temporal Resolution	ET Approach/ Reference	Period
CLSM L4 V2.2	0.25° × 0.25°	Daily	LSM Li, Beaudoing. [42]	February 2003–2020
ERA5	0.25° × 0.25°	Monthly	IFS Hersbach, Bell [14]	2003–2020
NOAH L4 V2.1	0.25° × 0.25°	Monthly	LSM Rodell, Houser [10]	2003–2020
MERRA2	0.625° × 0.50°	Monthly	GEOS-5 Gelaro, McCarty [15]	2003–2020
GLEAM v3.5b	0.25° × 0.25°	Monthly	Priestly–Taylor algorithm with canopy interception Martens, Gonzalez Miralles [11]	2003– July 2020

2.3. *In-Situ Measurements*

As at the time of this study, CA still does not have a single flux tower instrument in the global Flux Network (FLUXNET) system [38], which is the comprehensive platform for verifying satellite and alternative ET products [37]. Therefore, this study made use of in situ observations from four independent EC flux towers located in the crop, desert, grass and shrubland biomes in CA (red dots in Figure 1). Detailed preprocessing of the flux data sets has already been elaborated and documented [41,46]. The flux data have been extensively utilized by previous researchers, of which some are coauthors in this study [2,41,46]. The flux towers are currently managed by the Institute of Ecology and Geography, Chinese Academy of Sciences, China. Details of these flux towers are listed in Table 2.

Table 2. Summary of ground-based observation used in this study.

EC Site Name	Biome (Land Cover)	Location		Climatology (Mean/Year)		Period
		Lat.	Long.	Prec.	Temp.	
Aral Sea	Desert shrubs	45°58′N	61°05′E	140 mm	6.6 °C	April–Sept. 2012
Balkhash	Grassland	44°34′N	76°39′E	140 mm	5.7 °C	May–Oct. 2012
Fukang	Shrubland	44°17′N	87°56′E	163 mm	6.8 °C	2004–2006, 2009, 2011–2013
Wulanwusu	Cropland	44°17′N	85°49′E	210 mm	7.0 °C	2009–2013

2.4. *Evaluation Approach*

First, five state-of-the-art ET products obtained from different sources (Table 1) from February 2003 to July 2020 were bilinearly interpolated to align them to the same spatial grid of 0.25° resolution (Section 2.2). Then, the Triple Collocation (TC) method [31, 47] was applied to derive the representativeness errors in three parallel analyses of six different ET triplets (ET1, … ET6), including the independent uncertainties of each triplet component (i.e., CLSM, ERA5, GLEAM, MERRA2 and NOAH), in four arid biomes in CA (Section 2.5). Figure 2 shows the structural design of the triplets, while Figure 3 is the summary of the study's methodological framework. Due to insufficient in situ data in CA, we considered the reanalysis products (i.e., ERA5 and/or MERRA2) as a reasonable proxy in the triplet combinations; based on their practical hydrological applications [48,49] and performances in previous ET-based studies [50–52]. Figure 4 shows that the triplet design is both temporally and spatially reliable.

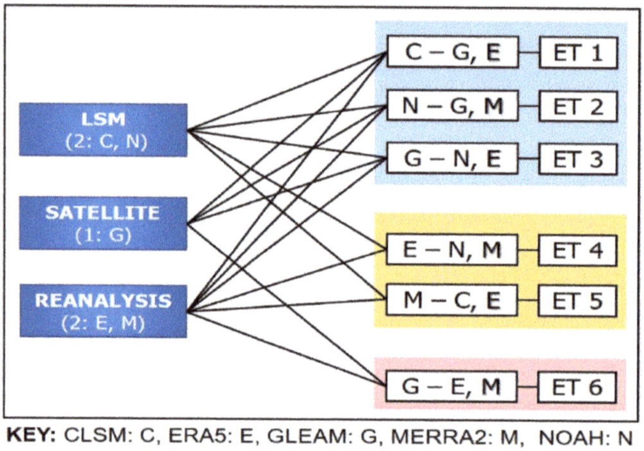

Figure 2. The Triplet design for the TC implementation. Left part of the figure shows the three ET measurement systems, including land surface model ET products (GLDAS NOAH and CLSM), satellite ET model (GLEAM), and reanalysis ET products (ERA5 and MERRA2). The triplet combinations comprise either two or three measurement systems as shown on the first column of the right panel of the figure. The blue background area shows the triplets comprising three different ET measurement systems (which is the ideal combination in TC analysis), the yellow background area consists of the triplets designed with two measurement systems (LSM and Reanalysis), and another triplet combination also comprised of two measurement systems (satellite and reanalysis) with pink background. Overall, the six triplets (ET1, … ET6) comprised of one or two reanalyses (ERA5 or MERRA2) which formed the reference component in the triplets for the TC analysis. Then the last triplet was used for validation purposes, however, all six triplets were independently evaluated to achieve a comprehensive assessment of the individual ET product in the study area.

Next, to ensure that a single product error from any one of the six triplets is stable and reliable [53], a two-step validation procedure was separately performed on the reference data sets; (i) direct linear comparison of the two references (i.e., ERA5 and MERRA2) was performed to confirm their consistency, (ii) spatial cross-validation of the triplets containing any one of the reference products (i.e., ET1 or ET2, and ET3) against a related reference-based triplet (i.e., ET4 or ET5), and then against the validation triplet (i.e., ET6). Note that GLEAM ET provides actual evaporation estimates [11,43], therefore, we combined GLEAM and the two reference data sets to give us the validation triplet–ET6. So, the spatial cross-validation was constructed as follows: ET1 vs. ET5, ET2 vs. ET4, and ET3 vs. ET6, respectively.

We also assessed the accuracy of the ET products before applying the TC method (Section 3.3). Regardless of the ET products' positions in the triplet design (Figure 2), their comprehensive spatial and temporal random error variances in the diverse arid ecosystem biomes were realized for each triplet (ET1, … ET6), as well as for each individual ET product (Sections 3.1, 3.2, 3.4 and 3.5). Similarly, the TC error structures for each product were assessed independently in the biomes using standard evaluation metrics (Section 3.6), and the relative uncertainties of each ET product were calculated using the TC-based standard error variances. Hence, allowing for a comprehensive assessment of the ET products in arid CA (Section 3.7). We defined the ET product uncertainty in this work as the RMSE expressing the variance of residual errors $\left(\Theta_{\epsilon}^{2}\right)$ of each ET product obtained from Equation (1) [32]. The study methodological framework is shown in Figure 3, while details of the procedures are presented in Section 2.5.

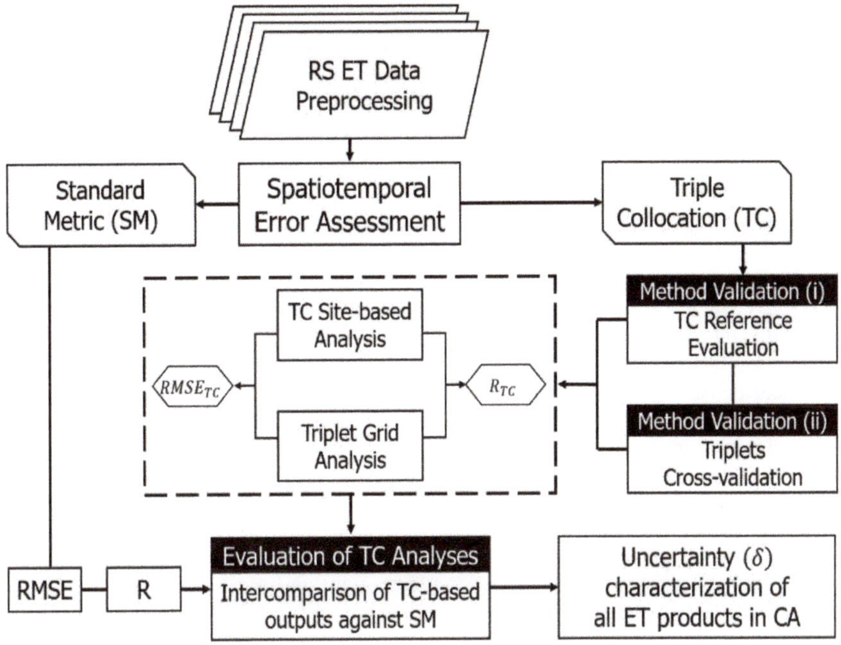

Figure 3. Methodological Framework of the study. Four major steps are involved, (i) data preparation, including ET data screening and spatial interpolation of the ET products to the same grid cells; (ii) design of triplets for the TC method, and the validation of the reference component used in the triplet combinations; (iii) the main analysis which is the applications of the triplet method (TC or TM) and the traditional model evaluation method (SM); (iv) Evaluation of the TM by comparing the TC outputs with those of the SM.

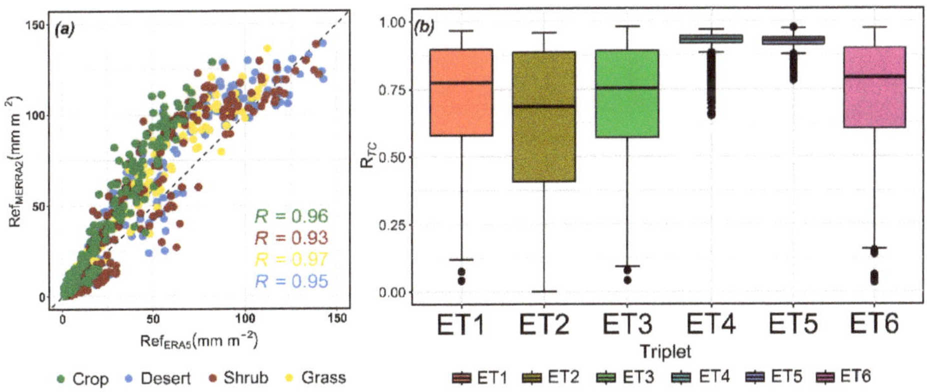

Figure 4. Linear comparison of the reference data sets used in the triplet combination. (**a**) shows the scatterplot assessing the relationship between the reference ERA5 and MERRA2 data sets in the cropland, desert, shrubland and grassland biomes in CA. (**b**) is the boxplot distribution of the 6 Triplets (i.e., ET1, … ET6) which shows that triplet combination with two reference ET products (ERA5 and MERRA2) are consistent in the triplet design (i.e., ET4 and ET5).

2.5. Characterizing Error Variances of the ET Products without In-Situ

We used the triple collocation (TC) method [47] which has the ability to estimate random error variances from point-scale observation and from coarse-resolution satellite data to characterize the ET products error variances [31]. Here, we mainly describe the direct application of TC to our ET data sets, as TC has been effectively used in numerous studies with detailed mathematical derivations already extensively documented [30,31,43,49]. Thus, we applied the TC technique which adopts a linear relationship between the ET measurement systems ($\Theta_m \rightarrow i \in [x, y, z]$, i.e., the satellite, LSM and/or reanalysis ET products of the same grid cells: 25 × 25 km^2) and their supposed true state (Θ_t, i.e., hypothetical true value) to estimate the error variances $\left(\Theta_\epsilon^2\right)$ of the ET products, as in Equation (1):

$$\Theta_m = \alpha + \beta\Theta_t + \epsilon, \tag{1}$$

where, α and β are the systematic additives and multiplicative biases, and ϵ is the random error/additive zero-mean random noise [31,43]. The TC technique (Equation (1)) does not require high-quality reference data set [54]. However, the TC application in this study was guided by the following caveats: structural errors from the collocations (ET triplets) were uncorrelated, orthogonal and independent with respect to the truth [31,55]. So, to estimate the TC error statistics, a total of six combinations (i.e., ET1, ... ET6) were realized from the five ET products based on a systematic tree selection (Figure 2), which produced several hundreds of collocated samples of the data sets, sufficient enough for the required assessment. A minimum of a hundred samples has been recommended [30,31].

The collocated data sets have the same spatial and temporal scales, and it is assumed that their error structure is Gaussian. However, where they solve for dissimilar scales (which we try to minimize during the interpolation analysis), the common variance to the lesser spatial and/or temporal scales forms part of the variances of the residual errors. These variances are representativeness errors [33,34]. Therefore, the spatiotemporally collocated triplets of the ET products were averaged by their cross-multiplied differences (Equation (2)), which is a representation of the diagonal elements of the error covariance matrix of the supposed reference data $\left(\text{i.e., } R_i = \sigma_{\epsilon_i}^2\right)$, as follows:

$$\begin{aligned} \sigma_{\epsilon_x}^2 &= \sigma_X^2 - \frac{\sigma_{XY}\sigma_{XZ}}{\sigma_{YZ}} \\ \sigma_{\epsilon_y}^2 &= \sigma_Y^2 - \frac{\sigma_{YX}\sigma_{YZ}}{\sigma_{XZ}} \\ \sigma_{\epsilon_z}^2 &= \sigma_Z^2 - \frac{\sigma_{ZX}\sigma_{ZY}}{\sigma_{XY}} \end{aligned} \tag{2}$$

where the ET triplet data variances are represented by σ_X^2, σ_Y^2 and σ_Z^2, respectively. While the random error variances of the $i \in [x, y, z]$ are represented by $\sigma_{\epsilon_x}^2$, $\sigma_{\epsilon_y}^2$ and $\sigma_{\epsilon_z}^2$, respectively, also known as the covariance of the measurements [31]. The aim of the TC method (Equation (1)) is to estimate the RMSE (hereafter, $RMSE_{TC}$) by measuring the variance of residual errors for each ET product [32]. After estimating the $RMSE_{TC}$ according to Equation (1), we calculated for the correlation coefficient–R (hereafter, R_{TC}) between each of the ET products to determine their relationship and the supposed truth, according to Equation (3).

$$R_{TC} = \sqrt{\frac{\sigma_{x,y}\sigma_{x,z}}{\sigma_X^2\sigma_{y,z}}} \tag{3}$$

where x, y, z are the independent ET data, while σ_X^2 and $\sigma_{y,z}$ are dataset variances of the dataset x and covariance between dataset x and dataset y, respectively. Derivation of R_{TC}, as adapted from the classical TC method [56] is shown by McColl, Vogelzang [22] to extend the practical application of the TC in all terrains. Next, we approximated the relative uncertainty of each ET product by expressing the variance of their residual errors (i.e., $RMSE_{TC}$) derived from Equation (1) as the ratio of standard random error $\left(\Theta_\epsilon^2\right)$ from the ET product's mean values at each of the validation site (biome), as in Equation (4) [21,32].

$$std(\epsilon_{ET}) = \mathrm{var}(\sigma_x) - \frac{\mathrm{cov}\left(\sigma_{\epsilon_x}, \sigma_{\epsilon_y}\right) \cdot \mathrm{cov}(\sigma_{\epsilon_x}, \sigma_{\epsilon_z})}{\mathrm{cov}\left(\sigma_{\epsilon_y}, \sigma_{\epsilon_z}\right)} \tag{4}$$

Due to the limited in situ observation which may require rescaling of the ET data to sufficient reference data space [31], we did not present the absolute uncertainties to avoid misinterpretation. However, as shown above, it was possible to independently derive the absolute error structures ($RMSE_{TC}$ and R_{TC}) following Equations (1) and (3), and the standard deviation $std[-]$ of random error for each of the representative triplets (ϵ_{ET}) (Equation (4)), representing the product's characteristic inconsistencies (uncertainties) with their corresponding reference [21]. Finally, the relative uncertainties were statistically analyzed according to biome type (Section 3.7).

2.6. Validation

The averages of the TC-based absolute error structures were validated spatially (continental) by cross-validation of the triplets (Section 3.2), and by using the statistical metrics of the standard evaluation method (SM) at the biomes (Section 3.6). Specifically, the point scale TC-based metrics at each of the validation sites (biomes) were compared with the standard metrics derived from the same biomes. In the grid point-by-point spatial analysis, a total of 36,936 grid cells for each ET product data were collocated for the period 2003–2020 for each time of the calculation using ERA5 or MERRA2 as the reference in the TC analysis. Then the TC-based grid cells of the ET product representativeness errors by the triplets were spatially compared against those of the validation triplet ET grid cells (i.e., ET1 vs. ET5 and ET2 vs. ET3, against ET3 vs. ET6), which revealed almost similar spatial error patterns, indicating the reliability of the triplet design used for the ET assessment in situ CA (Section 3.1).

2.7. Evaluation Metrics

Standard model performance metrics, such as the root mean square error (RMSE) and Pearson correlation coefficient (R) was used to evaluate the consistency and error biases of the ET product estimates against the in-situ measurements in the four arid ecosystem biomes (Equations (5) and (6)).

$$RMSE = \sqrt{\frac{\sum_{i=1}^{n}(x_i - y_i)^2}{n}} \tag{5}$$

$$R = \frac{\sum_{i=1}^{n}(x_i - \overline{x})(y_i - \overline{y})}{\sqrt{\sum_{i=1}^{n}(x_i - \overline{x})^2}\sqrt{\sum_{i=1}^{n}(y_i - \overline{y})^2}} \tag{6}$$

where n is the total number of observations, x is the ET product value, y is the in-situ observation value from either EC, while $\overline{x} = \frac{1}{n}\sum_{i=1}^{n} x_i \ldots n$ and $\overline{y} = \frac{1}{n}\sum_{i=1}^{n} y_i \ldots n$. Additional statistical outputs, such as the product bias and systematic errors (∞) of the ET products, obtained by calculating the standard deviation of the product from the actual mean ET in each biome are presented in Appendix A. More importantly, the TC-based method was validated by detailed statistical intercomparison with those of the standard metrics (SM) in CA (Section 3.6).

3. Results

3.1. Evaluation of the TC Method in Assessing ET Errors in CA

Figure 5 shows the intercomparison of the triplets used in assessing ET in CA. ET4 and ET5 are identical (Figure 4b). So, Figure 5 only displays the distribution of the R_{TC} performance of the triplets comprising of three separate measurement systems of ET and is regarded as the optimal combination style when using the TC method (e.g., LSM, satellite and reanalysis) [31]. These triplets are ET1, ET2 and ET3 (Figure 5). Triplet ET6 which contains only two measurement systems (i.e., satellite and reanalysis) was specifically designed for validation and cross-validation purposes. However, ET6 was also independently evaluated alongside ET4 and ET5 to achieve a comprehensive assessment of the ET products using both the three measurement systems and two measurement systems of ET (Figure 2). The detailed statistics of all the triplets are shown in Table 3. The RC_{TC} values of the main triplets ET1, ET2 and ET3 were 0.77, 0.69 and 0.75, respectively (Figure 5). Compared to other triplets previously illustrated in Figure 4, ET6 produced a very high R_{TC} value (~0.80), which most closely agree with its evaluation components (i.e., ERA5 and MERRA2). This qualifies ET6 to serve as a reliable validation triplet to ET1, ET2 and ET3, though they were all independently evaluated (Figure 5).

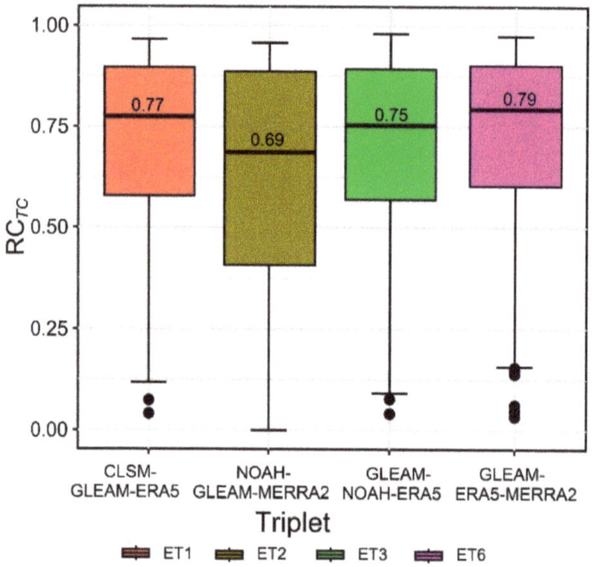

Figure 5. Distribution of validation and intercomparisons of Triplets used for the TC analysis. The upper and lower bars represent the 25th and 75th percentiles, respectively. The black horizontal line within each box is the median line with their average performance value placed on it. The interquartile range is indicated by whiskers extending from the 25th and 75th percentiles (±1.5).

Table 3. Performance statistics and intercomparison of the triplets by the TC method.

Statistics	ET1	ET2	ET3	ET4	ET5	ET6
75th Percentile	0.89	0.89	0.89	0.95	0.94	90
Median	0.77	0.69	0.75	0.93	0.92	0.79
25th Percentile	0.57	0.41	0.57	0.92	0.91	0.60
Mean	35.095	32.1975	35.47	41.66	37.38025	33.4225
SD	34.1325	31.0725	34.2125	36.91349	33.89	32.11

It is seen that triplets ET4 and ET5 outperformed others given their high median values (Figure 4b); however, ET6 showed less deviation from the mean ET compared to ET4 and ET5 (Table 3). Besides, ET4 and ET5 contain only two measurement systems. In comparison, ET1 [±0.23] and ET3 [±0.31] are significantly correlated with ET6 [±0.21], contrary to ET2 [-] based on its R_{TC} value of 0.10 ($p < 0.001$). Based on the triplets with the ideal combination measurement systems (Figure 5), it is expected that assessing the ET product characteristics in CA using ET2 might reveal a different performance than other triplets (ET1 and ET3). A cross-validation examination of the triplets was conducted to further determine the probability of the addictive impact (Section 3.2).

3.2. Evaluation of the TC Method by Triplet Cross-Validation Performance in CA

Figure 6 shows the spatial cross-validation of all the triplets. Based on the 36,936 grid sample cells generated for each triplet in the $RMSE_{TC}$ computation, triplets containing a corresponding reference component (i.e., ERA5 or MERRA2) were spatially cross-validated. In other words, the statistical matching is as follows: ET1 vs. ET5, ET2 vs. ET4, and ET3 vs. ET6 (Figure 6). In terms of $RMSE_{TC}$, the cross-validation illustrates a consistent distribution pattern of random errors, indicating that the TC approach is effective in measuring ET errors in the absence of in-situ observation in CA.

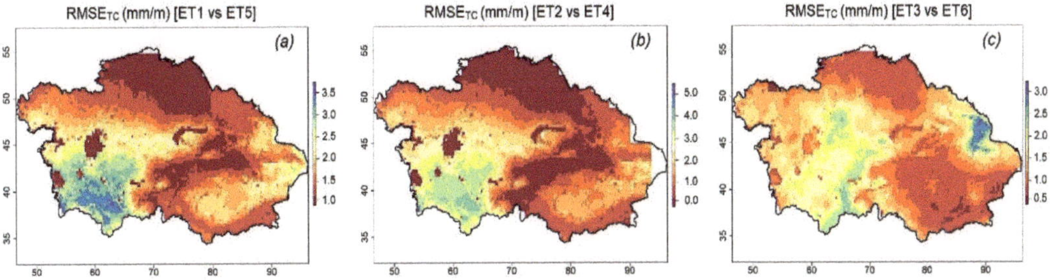

Figure 6. Spatial Cross-validation of the TC triplets. (**a–c**) represents the spatial statistical matching of ET1 vs. ET5, ET2 vs. ET4 and ET3 vs. ET6, respectively, during 2003–2020. That is, the ideal triplet combinations (ET1, ET2 and ET3) were evaluated against triplets with only two measurement systems (ET4, ET5 and ET6).

Comparatively, the plain areas of CA, which extend from the Aral Sea to the south-western parts have the highest error variabilities compared to other sections of the regions, though, at different rates (Figure 6). Figure 6c shows that the high error variabilities also extend to the northeast of CA, which includes parts of north-west Kazakhstan and north Xinjiang in China (Figure 6c). According to Figure 6a, b, this area has a low or moderate $RMSE_{TC}$. North Kazakhstan and the Tianshan Mt. ranges show the least $RMSE_{TC}$ (Figure 6a, b), whereas the rest of the CA has moderate or low $RMSE_{TC}$ (Figure 6a, b). The cross-validation results all together revealed a consistent pattern of ET error distribution in CA.

3.3. Assessment of the ET Products Using EC Flux Tower Measurements

Figure 7 present the validation scatterplots for the comparison of CLSM, ERA5, GLEAM, MERRA2, and NOAH using EC observation. The analysis was performed for the periods with available ET measurements in the desert, grassland, shrubland and cropland sites. The best performing ET models have high R values and low RMSE values, while the worst performing products have low R values and high RMSE values at the sites. Accordingly, the best performing ET models with the least RMSEs are NOAH (0.67 mm and 1.02 mm in grassland and shrubland biomes, respectively), MERRA2 (1.02 mm in cropland biome) and GLEAM (2.24 mm in the desert biome). CLSM and ERA5 showed their best performances in the desert (4.13 mm) and shrubland biomes (5.73 mm), respectively (Figure 7). The corresponding R values for these performances are 0.91 and 0.79 for NOAH in the grassland and shrubland sites, respectively; MERRA2 has an R value of 0.86 in the cropland site, while GLEAM has 0.79 in the desert site. CLSM and ERA5 have 0.80 and 0.78 in the shrubland biome, respectively. The performances varied across arid ecosystem biomes.

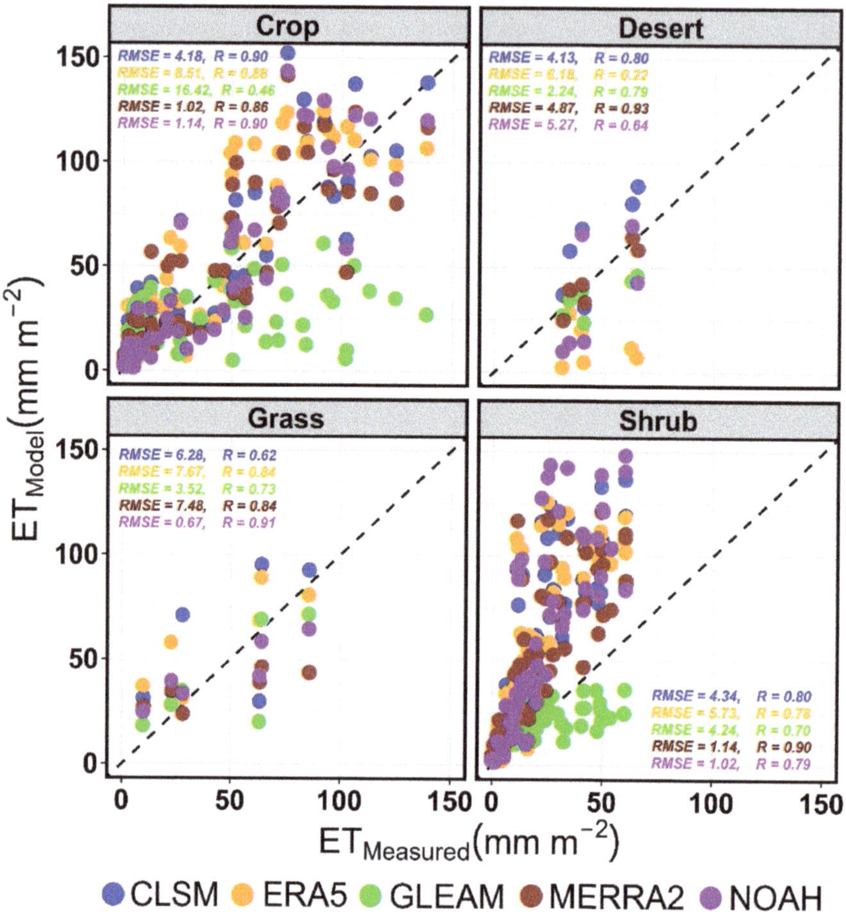

Figure 7. Analytical scatterplots of the ET products against in-situ observations and their comparative performance in desert, grassland, shrubland and cropland ecosystem biomes in CA.

Note that all the ET products show a relatively stable ET trend during the period of study, albeit with varying amplitudes in different seasons from 2003 to 2020. These might account for the differences in the ET product performance in the various biomes. However, the scatterplots show that the ET products have a weak-positive correlation with the in-situ measurements. CLSM, ERA, MERRA2 and NOAH, for example, overestimated the ET measurements in the shrubland site (Figure 7). In the cropland, GLEAM clearly underestimated ET, as did ERA5 in the desert biome (Figure 7). It is worth noting that each of the validation sites had EC flux measurements for at least one or more growing season (s). As such, validation using the in-situ observations in the corresponding months with those of the ET products provided the chance to compare the results with those derived from the TC analysis (Section 3.7).

Additional statistical analysis was performed to better understand the relationship between the ET model performance at the site scale in each biome. The clarification here is that there exist some obvious systematic errors (uncertainties) with the EC measurements, as revealed by the high ET model deviation from the actual mean ET values, resulting in higher negative biases in the biomes. The additional statistics are provided in the Appendix A (Table 1). The triplet SD provided in Table 3 differs from this result. The comparison with the TC-based metrics clearly confirms that the EC measurements have some system biases, as will be seen later in Section 3.7. In fact, the TC method exhibited less variance from the average ET values, as is evident in the SD (Table 3).

3.4. Site-Scale TC Analysis of the Triplet Components in CA

Figure 8 shows the calculated TC metrics ($RMSE_{TC}$ and R_{TC}) stratified by triplet comparison in the four validation sites in CA. Each Triplet component error can be deduced independently from the bar ranking, on the basis of higher/lower R_{TC} value and lower/higher $RMSE_{TC}$ value indicating a reasonable/poor performance of ET product in a particular biome (Figure 8). ET1 shows that CLSM and GLEAM have low $RMSE_{TC}$ in the desert (0.12 and 0.14 mm, respectively), and CLSM in the cropland with $RMSE_{TC}$ of 0. 21 mm. Conversely, ET2 showed a better performance (reduced $RMSE_{TC}$) of GLEAM ET than ET1 in the desert (0.12 mm), grassland (0.15 mm) and shrubland (0.10mm) biomes, but with high $RMSE_{TC}$ in cropland. Still based on the low $RMSE_{TC}$ values, ET2 shows that NOAH and MERRA2 ET products have good performance (0.23 mm and 0.14 mm, respectively) in the shrubland. In ET3, ERA5 showed the best performance with low $RMSE_{TC}$ in grassland (0.40 mm), shrubland (0.41 mm) and cropland (0.78 mm) with correspondingly high R_{TC} values in these biomes, respectively. NOAH only performed well in the grassland (0.70 mm) biome with the highest R_{TC} value (0.85). In general, this result clearly shows the strength and limitation of each of the triplets in determining the ET product error variance and the product's ability to retrieve/capture ET dynamics at the site-level in CA.

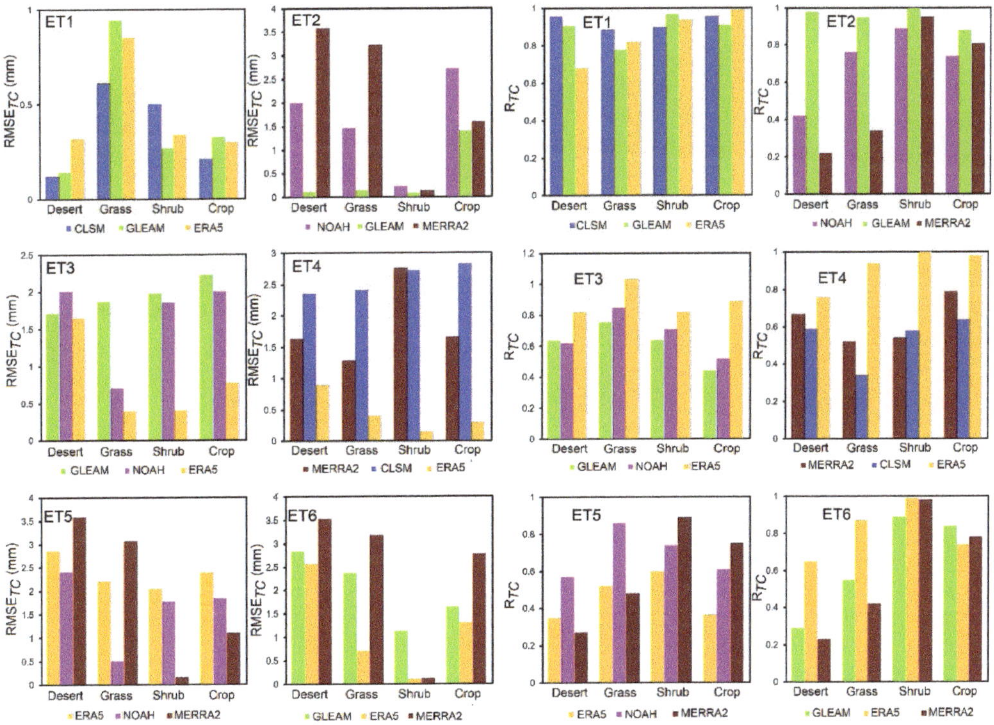

Figure 8. Site scale TC analysis in desert, grassland, shrubland and cropland ecosystem biomes in CA. Barcharts in columns 1 and 2 represents the TC-based RMSE metrics (mm) for each ET component in the triplet analysis, while those in columns 3 and 4 represents the TC-based R metrics for each ET component in the tripl analysis.

3.5. Spatiotemporal Assessment of the Triplets and Their Individual Components at Continental Scale

Figure 9 shows the ET products' spatial distribution based on TC error structures ($RMSE_{TC}$ and R_{TC}), while Figure 10 shows their component temporal statistics, illustrated with boxplots. They provide useful insights into the consistencies and discrepancies between the triplets in reproducing ET spatial and temporal patterns by each of the ET products from 2003 to 2020. The spatial trends are not entirely consistent, since all the products perform well temporally over a variety of locations (Figure 9). However, they generally conform with the spatial distribution of the validation product–ET6 (Figure 7).

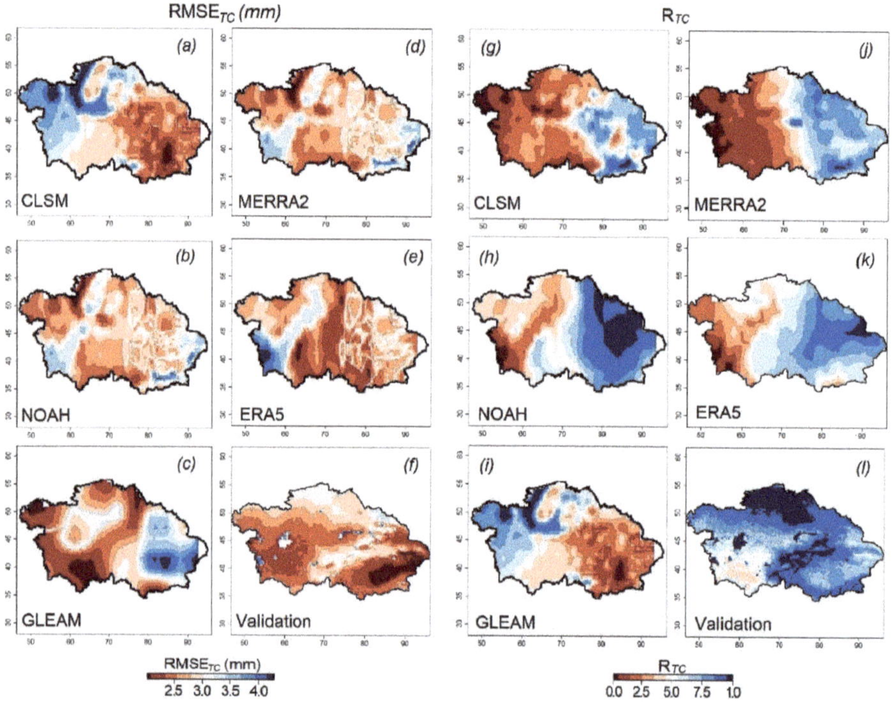

Figure 9. TC-based spatial distribution outputs for the ET products in CA during 2003–2020. Columns 1 and 2 (i.e., **(a–f)**) are the $RMSE_{TC}$ distribution pattern, and while columns 3 and 4 (i.e., **(g–l)**) are the corresponding R_{TC} distribution pattern. ET1, ET3, and ET5 used ERA5 as the triplet component, while ET2, ET4, and ET6 used MERRA2 as triplet component.

Compared to the remote sensing ET product (GLEAM), the CLSM and NOAH products showed considerably high $RMSE_{TC}$ values over the western (Kazakhstan) part of the CA and Southwest Turkmenistan, respectively (Figure 9a,b); where their R_{TC} values had almost identical scales as those of MERRA2 and ERA5, respectively (Figure 9j,k). It suggests that using ERA5 or MERRA2 as a reference in the triplets (Figure 2) produces similar distribution patterns of $RMSE_{TC}$ and R_{TC}. GLEAM exhibited low $RMSE_{TC}$ values in the west and central parts (Northwest Kazakhstan, Uzbekistan, Turkmenistan and parts of Tajikistan), which corresponds to high R_{TC} values in these regions (Figure 9c,i). In the eastern areas, GLEAM is reasonably distributed and indicates high RMS_{TC} that corresponds to low R_{TC} values (Figure 9c,i). These analyses suggest that the three measurement systems suffer different types of degradation in ET retrieval in CA. In the plain areas (western parts occupied by grasslands), the reference combination with LSM (CLSM and NOAH) produced high error variabilities (Figure 9a,b), while the TC reference combination with the remote sensing product produced low error variabilities, which was the opposite in the snowcapped mountains and dry desert lands in southern Xinjiang (Figure 9c,i).

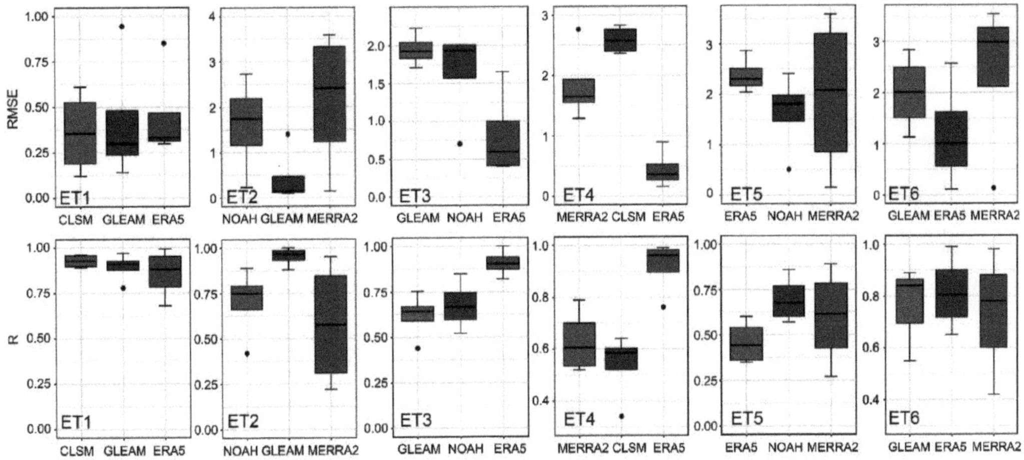

Figure 10. $RMSE_{TC}$ and R_{TC} temporal distribution for the ET products in CA during 2003–2020. *y*-axis of $RMSE_{TC}$ end at different values to improve visualization. The boxplots show the continental-scale media and interquartile range metrics for each ET product during in CA. Interpret these results by comparing the key boxplot metrics (the median, the 25th and 75th percentiles, and the interquartile range bounded by the top and bottom whiskers). Typically, ET products with low $RMSE_{TC}$ and high R_{TC} values indicate a good performance.

Figure 10 presents a detailed comparison of the calculated $RMSE_{TC}$ temporal error patterns with their corresponding R_{TC}, derived from the TC point-to-point grid analysis, using ERA5 or MERRA2 as a reference during 2003–2020. CLSM outperformed NOAH (high R_{TC} and low $RMSE_{TC}$) among the LSM parent products, as in ET1. The equivalent performance of NOAH and MERRA2 in east Xinjiang, CA, as illustrated by Figure 9b,d, may be attributable to a substantial dependency on the triplet component. This could be the reason why CLSM outperformed NOAH as shown in Figure 10. The remote sensing product (GLEAM) performed best (lowest $RMSE_{TC}$ and highest R_{TC}), as in ET2. For the reanalysis ET products, ERA5 had a better performance than MERRA2 as in ET3. These interpretations are focused on triplets with three different ET measurement systems, with each having its independent errors. However, there exist some temporal differences in the error statistics among all the five ET products. Therefore, the type of reference data utilized in the triplet combination can have a significant impact on the TC analytical performance of the ET products in CA.

3.6. Intercomparison and Evaluation of the TC Analysis against Standard Metric

Figure 11 present the intercomparison of the two evaluation methods used in this study. Figure 11a shows the RMSE values outputted by the two evaluation methods, while Figure 11b is their corresponding R values. As demonstrated in Figure 11b, the R values of both techniques are generally high. The RMSE values, on the other hand, reveal the systematic differences between the two approaches (Figure 11a). To examine the efficacy of both approaches in measuring ET accuracy in CA, an error (RMSE) threshold value of 2.50 mm was arbitrarily selected (red dotted line in Figure 11a).

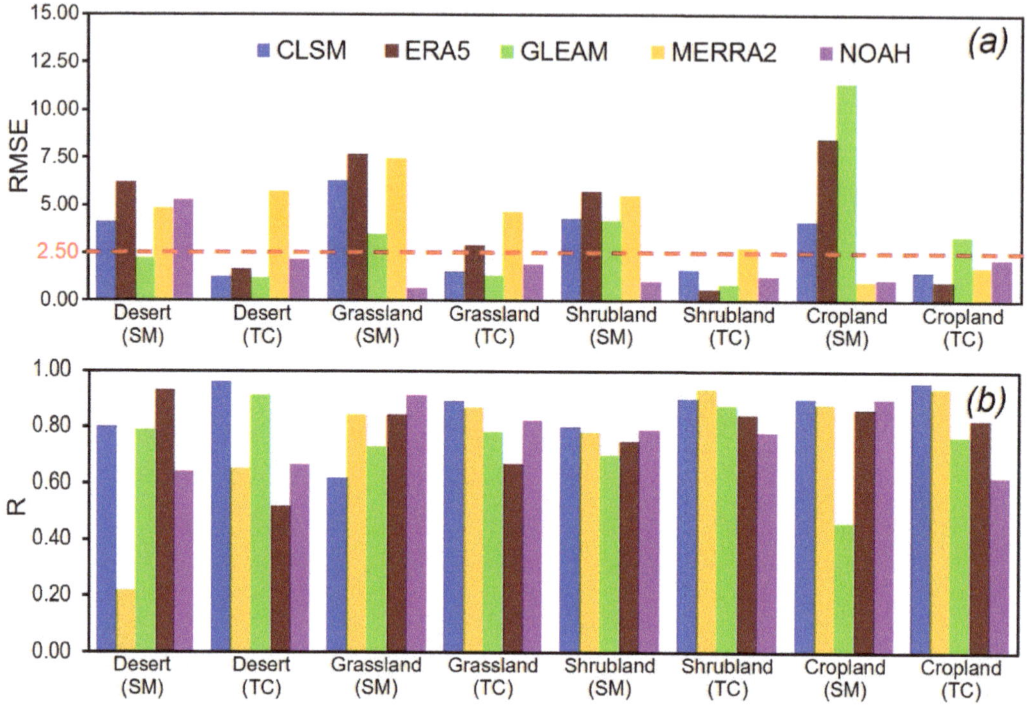

Figure 11. Intercomparison of the TC-based metrics and the standard error metrics in CA. (**a**) is the evaluation of the TC method using RMSE values obtained from the standard/traditional validation method (SM), while (**b**) is the evaluation of the TC method using R values obtained from the SM.

In general, the TC method shows that ERA5 has the least RMSE value (0.61 mm) with the highest R value (0.93) in the shrubland, while MERRA2 has the highest RMSE value (5.75 mm) and lowest R value (0.52) in the desert. In the grassland site, the best performing ET model by the SM method is NOAH, which has the lowest RMSE (0.67 mm) and the highest R value (0.91). Except for GLEAM, all the ET models in the cropland, have RMSE values < 2.2 mm according to the TC method. Whereas CLSM, ERA5 and GLEAM, all have high RMSE values of 8.51 mm, 4.18 mm and 16.42 mm, respectively, according to the SM method.

In the shrubland site, only MERRA2 exhibited a high TC RMSE (2.80 mm) value of above 2.50 mm, whereas the SM RMSE values of CLSM, ERA5, GLEAM and MERRA2 were above 2.50 mm (Figure 11a). Additionally, SM exhibited similar behavior for these four ET models in the grassland. In the desert site, GLEAM showed an RMSE value of 2.24 mm according to the SM (Figure 11a). The red dotted error margin in Figure 11a clearly shows that the TC technique produced low RMSE values in all the locations and outperformed the traditional evaluation method (SM). In Figure 12, we present a more comprehensive summary of the evaluation illustrating the degree of significance in the analyses by TC and SM methods in CA. The boldly highlighted values, with large magnitudes as indicated by the blue and red color scales, are the ET evaluation metrics with significantly better performance in the different arid ecosystem biomes (Figure 12).

Biome (Method)	R					RMSE (mm)				
	CLSM	ERA5	GLEAM	MERRA2	NOAH	CLSM	ERA5	GLEAM	MERRA2	NOAH
Desert (SM)	0.80	0.22	**0.79**	0.93	0.64	4.13	6.18	**2.24**	4.87	5.27
Desert (TC)	**0.96**	**0.65**	**0.91**	0.52	0.67	**1.24**	**1.66**	**1.20**	5.75	2.14
Grass (SM)	0.62	0.84	0.73	0.84	**0.91**	6.28	7.67	3.52	7.48	**0.67**
Grass (TC)	**0.89**	0.87	**0.78**	0.67	**0.82**	**1.51**	2.92	**1.34**	4.69	**1.89**
Shrub (SM)	0.80	0.78	0.70	0.75	**0.79**	4.34	5.73	4.24	5.55	**1.02**
Shrub (TC)	**0.90**	**0.93**	**0.88**	0.84	**0.78**	**1.62**	**0.61**	**0.87**	2.80	**1.29**
Crop (SM)	0.90	0.88	0.46	**0.86**	**0.90**	4.18	8.51	11.42	**1.02**	**1.14**
Crop (TC)	**0.96**	**0.94**	0.77	**0.83**	0.62	**1.52**	**1.02**	3.40	**1.79**	2.19

Figure 12. Inferential statistics showing the performance of the ET models by the standard metric (SM), and the tripe collocation (TC) assessment of ET without in situ component. The boldly highlighted values indicate the best performance (R and RMSE) of the ET model that are significant based on high correlation (R) and lower RMSE. Dark Blue > blue > light blue > lighter blue represents the color codes from the highest R and corresponding lower RMSE values, and vice-versa. Dark red > red > light red > lighter red represents the color codes from the lowest R and corresponding higher RMSE values, and vice-versa. Therefore, the fading blue and red scales represent degradation in the performance of the ET product in the biomes in CA. This clearly shows that the TC method can reliably assess ET with reduced uncertainty in regions without available in situ, such as CA.

3.7. Uncertainty Assessment of the ET Products Based on TC Technique

Figure 13 present the relative uncertainty of each ET model in cropland, desert, grassland and shrubland biomes. Here, we concentrate on the ET product uncertainty based on the TC standard error variances, which are expressed as the ratio of the RMSE to the mean of the ET in each biome. The highest uncertainty was associated with GLEAM in the cropland biome at 41.30%. However, it has a low uncertainty ratio in the shrubland biome, at 15%.

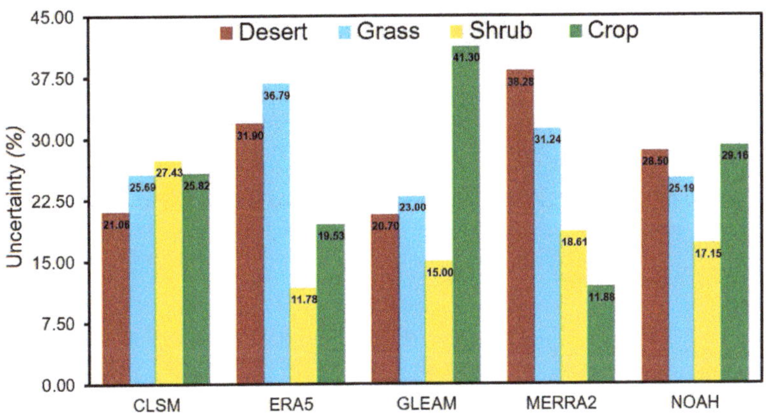

Figure 13. Uncertainty distribution of the derived from the TC analysis at the biomes in CA.

In descending order, the next products with high uncertainties are MERRA2 in the desert biome (38.28%), ERA5 in both grassland and desert biomes (36.79% and 31.90%, respectively), and MERRA2 again in the grassland biome (31.24%). NOAH depicted the highest ET uncertainties among the LSM ET products in the cropland and desert biomes, at 29.16% and 28.50%, respectively. The uncertainty ratio by CLSM ET ranges from 21.06% in the desert biome to 25.82% in the cropland biome. In general, ET products with the least uncertainty are ERA5 in the shrubland (11.78%), MERRA2 in the cropland (11.88%) and GLEAM in the shrubland (15.00%).

4. Discussion

A point-by-point grid assessment of ET dynamics using six triplets of collocated raster maps without in situ reference components was achieved in this study. Previous studies implementing TC methods in regions with sufficient in situ observations tend to include in situ component in the triplet analysis [21,49], however, LSM or reanalysis satellite reference can also be independently used and/or alongside with in situ in the TC analysis to achieve a robust assessment [34]. The implementation of six ET triplets in this study was based on a systematic triplet design, which combined three ET measurement systems (i.e., LSM, satellite, and reanalysis, hence ET1, ET2 and ET3) and Two ET measurement systems (i.e., satellite and reanalysis, hence ET4, ET5 and ET6) from 2003 to 2020 (Figure 2).

Figure 4 shows the effectiveness of the triplet design conceived for the study area. The consistency of the triplet reference components (i.e., ERA5 and MERRA2) were ascertained by their high correlation coefficients of 0.96, 0.93, 0.97 and 0.95 in the cropland, desert, grassland and shrubland biomes, respectively (Figure 4a); then, by their high matching median values of 0.93 and 0.92 as revealed by the parent triplets, as in ET4 and ET5, respectively. Figure 4b shows the performance of all the triplets, while Figure 5 shows the good performance of the triplets with the ideal combination (i.e., ET1, ET2 and ET3), and the triplet used for validation of other triplets (ET6). All the triplets were implemented independently to assess the individual ET characteristics and random error variances in the absence of a harmonized actual observation in CA. The spatial cross-validation of the triplets produced a consistent pattern of random error distribution over the study area (Figure 6).

Moreover, evaluation of the individual ET products using EC flux tower observation produced both high/low correlation coefficients and low/high RMSEs, depending on the biome type where a product was validated. Figure 7 shows, however, that the linear relationship between the ET models and in situ is not strong enough, but weakly correlated. Some products underestimated/overestimated ET in the various arid ecosystem biomes. For example, CLSM, ERA5, MERRA2 and NOAH overestimated ET in the shrubland site with an RMSE value range between 1.14–5.73 mm/m and R-value range between 0.70–0.90; whereas GLEAM underestimated ET in the shrubland and cropland biomes with R and RMSE values of 0.70 and 4.24 mm/m and 0.46 and 16.42 mm/m, respectively. Despite the underestimation of GLEAM in cropland and shrubland sites, we note that it performed reasonably well in both desert (0.79 R value and 2.24 mm RMSE) and grassland (0.73 R value and 3.52 mm RMSE) biomes.

The grassland result of this study is consistent with the report by Yang, Yong [18] that GLEAM ET was closer to the EC observation in the three grassland sites in China than in the cropland site. The result of this study exclusively pertains to the ecosystem biomes in CA, however, it suggests that GLEAM can also perform well under less water availability in drylands depending on the biome characteristics or plant functional type [57]. Previous studies found that the covariational differences between ERA5, GLEAM and GLDAS-NOAH with PET, Precipitation (P), and P plus groundwater suggest a significant amount of energy exchange during low P conditions, which is a major reason for inconsistencies and errors in these ET products [23,58]. Another explanation is that the saturation of satellite signals of the forcing drivers or model calibration strategy can greatly affect

ET product performance, especially over heterogeneous areas [24,59]. To a large extent, these issues can be resolved through the choice of forcing data used in the ET products' parametrization.

Unlike in tropical ecosystems, for instance, in East Asia, Khan, Liaqat [21] found that GLEAM and GLDAS-NOAH tend to show better correlation with in situ in cropland and tropical forest biomes than in the arid ecosystems of CA. Similarly, in Australia, Baik, Liaqat [19] reported that GLEAM and GLDAS showed better performance (R ~ 0.8) over crop and forest areas, respectively. In this study, the R-values for GLEAM and GLDAS-NOAH in situ in the cropland are 0.46 and 0.90, respectively. One major reason for ET products' general underestimation in arid biomes is the limited water resources, which is a major factor in CA, and tend to change the desert plant functional types while also constraining their effective transpiration even in the grown season [57]. The implication is that hydrological processes, for example, sublimation and snowmelt, which are common physical processes that impact water cycling and energy transfer in other sections of CA (e.g., forest and Tianshan mountain areas) can be weakened during the peak of evaporation in growing seasons [60,61].

TC analysis in this study provides in-depth intercomparisons of the ET products statistics (Figure 8). First, we noticed that all triplet components (i.e., the five ET products) exhibited different magnitudes of variability in all the biomes, which is expected. Note that the bias and gain may not be site (biome) dependent, because the capability to reproduce error variability by the ET measurement systems should ideally be performed with respect to a sufficient amount of actual observation [43], as has been demonstrated with the satellite soil moisture data sets [48,49]. This is not the case in this study (Figure 4), as we used triplet collocations without in situ but reference data (Figure 2).

In terms of physical accuracy, the point estimated R values obtained for CLSM, ERA5, GLEAM, MERRA2, and NOAH ranges from 0.89–0.96, 0.65–0.94, 0.78–0.91, 0.52–0.84, and 0.62–0.82, respectively. Whereas, Khan, Liaqat [21] obtained 0.35–0.60, 0.63–0.84 and 0.48–0.79 for MOD16, GLEAM and GLDAS-NOAH. Our TC point estimation range are slightly higher for GLEAM (0.78–0.91) and GLDAS-NOAH (0.62–0.82), which shows the difference in the application of the TC method in different ecosystems (i.e., arid and tropical biomes). However, the point-wise TC analysis predicts much better results of land surface fields (e.g., soil moisture), because it allows the gains to change from point to point, and with the ERA-Land reanalysis, better performance is guaranteed [49].

As shown in Figure 6, the spatial distribution of the cross-validation outputs shows a consistent pattern of ET error patterns over CA, but at varying magnitudes (Figure 6). It explains the spatial discrepancies by the ET models in different parts of CA (Figure 9). These impacts might have resulted from topographic and climatic constraints prevalent across CA [40]. CA has two distinct landscape types: the plain areas dominated by desert grasslands and the elevated mountain ranges with patches of vegetated zones. Regions occupied by these two physical environments are typified by different climatology that produces varying hydrological processes and energy exchanging patterns [60], which can affect the performance of the different ET products in the arid environment as evident in Figure 10.

In summary, none of the ET products performed well in all the biomes at the same time, indicating their biome-specificity (Figure 13). Different forms and causes of uncertainties [8, 29] can be attributed to the performance of these ET products in CA. Besides that, CA is poorly observed, the physical investigation of different land surface fields (notably ET) is usually compounded by the region's remoteness and complicated geography. All the ET products, particularly the LSM products (i.e., CLSM and NOAH) exhibited the best stable monthly temporal pattern with each other, but not exactly with the observation in the shrubland.

However, because the TC method does not calculate biases [43], so the additional statistical analysis revealed both the biases and the systematic errors (mean $\pm$ SD) associated with the EC flux tower measurements (Table 1). The site-scale bias range, starting from the worst performing are $-10 \leq$ bias ≤ 16, $-21 \leq$ bias ≤ 4.43, $-14.92 \leq$ bias 4.19, and -17.38 bias 2.27 in cropland, desert, grassland and shrubland biomes, respectively. As a result of the systematic errors connected with EC flux tower measurements (Table 1), describing the absolute uncertainties associated with the five ET products in the various biomes without some type of expert preconception is challenging. Therefore, we can only attribute the inherent inconsistencies displayed by the ET products in the four arid ecosystem biomes as only illustrative of their corresponding ET product biases, as demonstrated by the TC analysis [21].

5. Conclusions

This study spatially and temporarily evaluated the random error variances in five state-of-the-art ET products, including CLSM, ERA5, GLEAM, MERRA2 and NOAH, from 2003 to 2020 in CA. Both TC and EC methods were thoroughly tested under different conditions of data availability. The first major caveat was to ensure the applicability of the TC method to assess ET data sets without in situ observations in CA. The spatiotemporal collocations in the TC analysis, which was based on a systematic triplet design encompassing three and two ET measurement systems was constructed and implemented, yielding six independent ET triplets (i.e., ET1 ... ET6, Figure 2).

The correlation of the reanalysis reference (ERA5 or MERRA2) used in the triplets against in situ was very high in the cropland, desert, grassland and shrubland biomes (i.e., 0.96, 0.93, 0.97 and 0.95, respectively) (Figure 4a). The validation of the triplet types which were based on spatial and temporal intercomparison of the ET measurement systems revealed that triplets with the ideal combinations (i.e., Satellite, LSM and Reanalysis) do not necessarily outperform triplets with only two ET measurement systems (i.e., Reanalysis, LSM, and Reanalysis, or Satellite, Reanalysis and Reanalysis) when assessing ET dynamics in the absence of in situ; this is especially true for the biomes in CA. Overall, all the ET products have the capacity to capture ET dynamics at different levels of accuracy, however, none of the products simultaneously performed well in all the four arid biomes in CA.

This study provides the first evaluation metrics on the newer/updated versions of the five ET products in CA without using in situ, even though in situ evaluation was performed but with less coverage. Therefore, the study reiterates the need to extend energy flux observation towers to the remote parts of the Eurasian continent to better understand ET dynamics in the region, including its ecological divergences with other regions of the world. Besides supporting regional water resource management, the effort will greatly enhance global change and climate change research projects; especially when these uncharted biomes located in unique climate regions of the world are specifically captured in future studies.

Author Contributions: Conceptualization, data collection, validation, F.U.O. and G.L.; formal analysis, Visualization, F.U.O., H.S. and G.L. methodology, F.U.O.; Project administration, G.L.; resources, F.U.O. and E.E.I.; data curation, F.U.O., G.L., H.S. and C.L.; Software, F.U.O. and G.L.; Supervision, G.L.; writing—original draft preparation, F.U.O.; writing—review and editing, F.U.O., G.L., C.L., H.S., E.E.I. and X.M. All authors have read and agreed to the published version of the manuscript.

Funding: This research was funded by the National Natural Science Foundation of China (Grant No. U1803243), the Strategic Priority Research Program of Chinese Academy of Sciences (Grant No. XDA20060302), and the Key projects of the Natural Science Foundation of Xinjiang Autonomous Region (Grant No. 2022D01D01).

Institutional Review Board Statement: Not applicable.

Informed Consent Statement: Not applicable.

Data Availability Statement: The source of data used in this study is provided in the manuscript.

Acknowledgments: For providing the data sets used in this study at no cost, we acknowledge the NASA's Global Modeling and Assimilation Office (GMAO) for the MERRA-2 data set, the European Centre for Medium-Range Weather Forecasts for the ERA-5 data set; the Global Land Data Assimilation System for providing the two Land Surface Evapotranspiration Models (CLSM and NOAH) and the Global Land Evaporation Amsterdam Model for the satellite ET data–GLEAM. We thank the Xinjiang Institute of Ecology and Geography, CAS for providing the eddy covariance flux tower data. We thank the editors and anonymous reviewers for their constructive comments and suggestions which helped to improve the paper.

Conflicts of Interest: The authors declare no conflict of interest or state.

Appendix A

Table 1. Additional statistics of ET product evaluation using in situ measurements in the four arid biomes in Central Asia.

ET Product	Biome	R	RMSE (mm)	BIAS (mm)	SD	Product Mean (mm/m)	Systematic Errors (∞)
CLSM	Crop	0.90	4.18	−6.18	42.37	45.50	3.13
ERA5	Crop	0.88	8.51	−10.16	43.72	49.48	5.76
GLEAM	Crop	0.46	12.42	16.14	15.51	23.18	7.67
MERRA2	Crop	0.86	1.02	−6.12	39.56	45.44	5.88
NOAH	Crop	0.90	1.14	−2.83	43.16	42.14	−1.02
CLSM	Desert	0.80	4.13	−15.09	23.31	60.91	37.60
ERA5	Desert	0.22	6.18	−13.28	9.95	12.50	2.55
GLEAM	Desert	0.79	2.24	−21.02	9.00	35.81	26.81
MERRA2	Desert	0.93	4.87	−16.98	15.13	43.84	28.71
NOAH	Desert	0.64	5.27	4.43	27.39	36.40	9.01
CLSM	Grass	0.62	6.28	−13.16	31.04	59.16	28.12
ERA5	Grass	0.84	7.67	−14.92	23.31	60.91	37.60
GLEAM	Grass	0.73	3.52	2.50	24.01	40.50	16.49
MERRA2	Grass	0.84	7.48	4.19	9.00	35.81	26.81
NOAH	Grass	0.91	0.67	2.15	15.13	43.84	28.71
CLSM	Shrub	0.80	4.34	−14.41	40.42	57.68	17.26
ERA5	Shrub	0.78	5.73	−14.58	40.78	64.85	24.07
GLEAM	Shrub	0.70	4.24	2.27	8.67	19.00	10.33
MERRA2	Shrub	0.75	5.55	−15.80	34.90	54.07	19.17
NOAH	Shrub	0.79	1.02	−17.38	46.76	60.65	13.89

References

1. Chen, J.M.; Liu, J. Evolution of evapotranspiration models using thermal and shortwave remote sensing data. *Remote Sens. Environ.* **2020**, 237, 111594. [CrossRef]
2. Ochege, F.U.; Luo, G.; Obeta, M.C.; Owusu, G.; Duulatov, E.; Cao, L.; Nsengiyumva, J.B. Mapping evapotranspiration variability over a complex oasis-desert ecosystem based on automated calibration of Landsat 7 ETM+ data in SEBAL. *GISci. Remote Sens.* **2019**, 56, 1305–1332. [CrossRef]
3. Martens, B.; Waegeman, W.; Dorigo, W.A.; Verhoest, N.E.C.; Miralles, D.G. Terrestrial evaporation response to modes of climate variability. *NPJ Clim. Atmos. Sci.* **2018**, 1, 43. [CrossRef]
4. Li, S.; Wang, G.; Sun, S.; Fiifi Tawia Hagan, D.; Chen, T.; Dolman, H.; Liu, Y. Long-term changes in evapotranspiration over China and attribution to climatic drivers during 1980–2010. *J. Hydrol.* **2021**, 595, 126037. [CrossRef]
5. Lioubimtseva, E.; Henebry, G.M. Climate and environmental change in arid Central Asia: Impacts, vulnerability, and adaptations. *J. Arid Environ.* **2009**, 73, 963–977. [CrossRef]

6. Micklin, P.; Aladin, N.V.; Plotnikov, I.S. (Eds.) *The Aral Sea. The Devastation and Partial Rehabilitation of a Great Lake*; Springer: Berlin/Heidelberg, Germany, 2014. [CrossRef]
7. Lama, G.F.C.; Crimaldi, M.; Pasquino, V.; Padulano, R.; Chirico, G.B. Bulk Drag Predictions of Riparian Arundo donax Stands through UAV-Acquired Multispectral Images. *Water* **2021**, *13*, 1333. [CrossRef]
8. Wang, K.; Dickinson, R.E. A review of global terrestrial evapotranspiration: Observation, modeling, climatology, and climatic variability. *Rev. Geophys.* **2012**, *50*, 54. [CrossRef]
9. Jung, M.; Koirala, S.; Weber, U.; Ichii, K.; Gans, F.; Camps-Valls, G.; Papale, D.; Schwalm, C.; Tramontana, G.; Reichstein, M. The FLUXCOM ensemble of global land-atmosphere energy fluxes. *Sci. Data* **2019**, *6*, 74. [CrossRef]
10. Rodell, M.; Houser, P.R.; Jambor, U.; Gottschalck, J.; Mitchell, K.; Meng, C.J.; Arsenault, K.; Cosgrove, B.; Radakovich, J.; Bosilovich, M.; et al. The global land data assimilation system. *Bull. Am. Meteorol. Soc.* **2004**, *85*, 381–394. [CrossRef]
11. Martens, B.; Gonzalez Miralles, D.; Lievens, H.; Van Der Schalie, R.; De Jeu, R.A.; Fernández-Prieto, D.; Beck, H.E.; Dorigo, W.; Verhoest, N. GLEAM v3: Satellite-based land evaporation and root-zone soil moisture. *Geosci. Model Dev.* **2017**, *10*, 1903–1925. [CrossRef]
12. Zhang, K.; Kimball, J.S.; Nemani, R.R.; Running, S.W. A continuous satellite-derived global record of land surface evapotranspiration from 1983 to 2006. *Water Resour. Res.* **2010**, *46*, 21. [CrossRef]
13. Fisher, J.B.; Lee, B.; Purdy, A.J.; Halverson, G.H.; Dohlen, M.B.; Cawse-Nicholson, K.; Wang, A.; Anderson, R.G.; Aragon, B.; Arain, M.A.; et al. ECOSTRESS: NASA's Next Generation Mission to Measure Evapotranspiration from the International Space Station. *Water Resour. Res.* **2020**, *56*, e2019WR026058. [CrossRef]
14. Hersbach, H.; Bell, B.; Berrisford, P.; Hirahara, S.; Horányi, A.; Muñoz-Sabater, J.; Nicolas, J.; Peubey, C.; Radu, R.; Schepers, D.; et al. The ERA5 global reanalysis. *Q. J. R. Meteorol. Soc.* **2020**, *146*, 1999–2049. [CrossRef]
15. Gelaro, R.; McCarty, W.; Suárez, M.J.; Todling, R.; Molod, A.; Takacs, L.; Randles, C.A.; Darmenov, A.; Bosilovich, M.G.; Reichle, R.; et al. The Modern-Era Retrospective Analysis for Research and Applications, Version 2 (MERRA-2). *J. Clim.* **2017**, *30*, 5419–5454. [CrossRef]
16. Jung, H.C.; Getirana, A.; Arsenault, K.R.; Holmes, T.R.; McNally, A. Uncertainties in Evapotranspiration Estimates over West Africa. *Remote Sens.* **2019**, *11*, 892. [CrossRef]
17. Velpuri, N.M.; Senay, G.B.; Singh, R.K.; Bohms, S.; Verdin, J.P. A comprehensive evaluation of two MODIS evapotranspiration products over the conterminous United States: Using point and gridded FLUXNET and water balance ET. *Remote Sens. Environ.* **2013**, *139*, 35–49. [CrossRef]
18. Yang, X.; Yong, B.; Ren, L.; Zhang, Y.; Long, D. Multi-scale validation of GLEAM evapotranspiration products over China via ChinaFLUX ET measurements. *Int. J. Remote Sens.* **2017**, *38*, 5688–5709. [CrossRef]
19. Baik, J.; Liaqat, U.W.; Choi, M. Assessment of satellite- and reanalysis-based evapotranspiration products with two blending approaches over the complex landscapes and climates of Australia. *Agric. For. Meteorol.* **2018**, *263*, 388–398. [CrossRef]
20. Li, S.; Wang, G.; Sun, S.; Chen, H.; Bai, P.; Zhou, S.; Huang, Y.; Wang, J.; Deng, P. Assessment of Multi-Source Evapotranspiration Products over China Using Eddy Covariance Observations. *Remote Sens.* **2018**, *10*, 1692. [CrossRef]
21. Khan, M.S.; Liaqat, U.W.; Baik, J.; Choi, M. Stand-alone uncertainty characterization of GLEAM, GLDAS and MOD16 evapotranspiration products using an extended triple collocation approach. *Agric. For. Meteorol.* **2018**, *252*, 256–268. [CrossRef]
22. McColl, K.A.; Vogelzang, J.; Konings, A.G.; Entekhabi, D.; Piles, M.; Stoffelen, A. Extended triple collocation: Estimating errors and correlation coefficients with respect to an unknown target. *Geophys. Res. Lett.* **2014**, *41*, 6229–6236. [CrossRef]
23. Moshir Panahi, D.; Sadeghi Tabas, S.; Kalantari, Z.; Ferreira, C.S.S.; Zahabiyoun, B. Spatio-Temporal Assessment of Global Gridded Evapotranspiration Datasets across Iran. *Remote. Sens.* **2021**, *13*, 1816. [CrossRef]
24. Dembélé, M.; Ceperley, N.; Zwart, S.J.; Salvadore, E.; Mariethoz, G.; Schaefli, B. Potential of satellite and reanalysis evaporation datasets for hydrological modelling under various model calibration strategies. *Adv. Water Resour.* **2020**, *143*, 103667. [CrossRef]
25. Wilson, K.; Goldstein, A.; Falge, E.; Aubinet, M.; Baldocchi, D.; Berbigier, P.; Bernhofer, C.; Ceulemans, R.; Dolman, H.; Field, C.; et al. Energy balance closure at FLUXNET sites. *Agric. For. Meteorol.* **2002**, *113*, 223–243. [CrossRef]
26. Vinukollu, R.K.; Wood, E.F.; Ferguson, C.R.; Fisher, J.B. Global estimates of evapotranspiration for climate studies using multi-sensor remote sensing data: Evaluation of three process-based approaches. *Remote Sens. Environ.* **2011**, *115*, 801–823. [CrossRef]
27. Chen, J.; Chen, B.; Black, T.A.; Innes, J.L.; Wang, G.; Kiely, G.; Hirano, T.; Wohlfahrt, G. Comparison of terrestrial evapotranspiration estimates using the mass transfer and Penman-Monteith equations in land surface models. *J. Geophys. Res. Biogeosci.* **2013**, *118*, 1715–1731. [CrossRef]
28. Xu, T.; Guo, Z.; Xia, Y.; Ferreira, V.G.; Liu, S.; Wang, K.; Yao, Y.; Zhang, X.; Zhao, C. Evaluation of twelve evapotranspiration products from machine learning, remote sensing and land surface models over conterminous United States. *J. Hydrol.* **2019**, *578*, 124105. [CrossRef]
29. Long, D.; Longuevergne, L.; Scanlon, B.R. Uncertainty in evapotranspiration from land surface modeling, remote sensing, and GRACE satellites. *Water Resour. Res.* **2014**, *50*, 1131–1151. [CrossRef]
30. Scipal, K.; Dorigo, W.; de Jeu, R. Triple collocation—A new tool to determine the error structure of global soil moisture products. In Proceedings of the 2010 IEEE International Geoscience and Remote Sensing Symposium, Honolulu, HI, USA, 25–30 July 2010; pp. 4426–4429.

31. Gruber, A.; Su, C.H.; Zwieback, S.; Crow, W.; Dorigo, W.; Wagner, W. Recent advances in (soil moisture) triple collocation analysis. *Int. J. Appl. Earth Obs. Geoinf.* **2016**, *45*, 200–211. [CrossRef]
32. Fang, H.; Wei, S.; Jiang, C.; Scipal, K. Theoretical uncertainty analysis of global MODIS, CYCLOPES, and GLOBCARBON LAI products using a triple collocation method. *Remote Sens. Environ.* **2012**, *124*, 610–621. [CrossRef]
33. Roebeling, R.A.; Wolters, E.L.A.; Meirink, J.F.; Leijnse, H. Triple Collocation of Summer Precipitation Retrievals from SEVIRI over Europe with Gridded Rain Gauge and Weather Radar Data. *J. Hydrometeorol.* **2012**, *13*, 1552–1566. [CrossRef]
34. Miyaoka, K.; Gruber, A.; Ticconi, F.; Hahn, S.; Wagner, W.; Figa-Saldaña, J.; Anderson, C. Triple Collocation Analysis of Soil Moisture from Metop-A ASCAT and SMOS Against JRA-55 and ERA-Interim. *IEEE J. Sel. Top. Appl. Earth Obs. Remote Sens.* **2017**, *10*, 2274–2284. [CrossRef]
35. Gruber, A.; De Lannoy, G.; Albergel, C.; Al-Yaari, A.; Brocca, L.; Calvet, J.C.; Colliander, A.; Cosh, M.; Crow, W.; Dorigo, W.; et al. Validation practices for satellite soil moisture retrievals: What are (the) errors? *Remote Sens. Environ.* **2020**, *244*, 111806. [CrossRef]
36. Dorigo, W.A.; Wagner, W.; Hohensinn, R.; Hahn, S.; Paulik, C.; Xaver, A.; Gruber, A.; Drusch, M.; Mecklenburg, S.; van Oevelen, P.; et al. The International Soil Moisture Network: A data hosting facility for global in situ soil moisture measurements. *Hydrol. Earth Syst. Sci.* **2011**, *15*, 1675–1698. [CrossRef]
37. Baldocchi, D.; Falge, E.; Gu, L.; Olson, R.; Hollinger, D.; Running, S.; Anthoni, P.; Bernhofer, C.; Davis, K.; Evans, R. FLUXNET: A new tool to study the temporal and spatial variability of ecosystem-scale carbon dioxide, water vapor, and energy flux densities. *Bull. Am. Meteorol. Soc.* **2001**, *82*, 2415–2434. [CrossRef]
38. Falge, E.; Aubinet, M.; Bakwin, P.S.; Baldocchi, D.; Berbigier, P.; Bernhofer, C.; Black, T.A.; Ceulemans, R.; Davis, K.J.; Dolman, A.J.; et al. *FLUXNET Research Network Site Characteristics, Investigators, and Bibliography, 2016*; ORNL Distributed Active Archive Center: Oak Ridge, TN, USA, 2017. [CrossRef]
39. Li, C.F.; Zhang, C.; Luo, G.P.; Chen, X.; Maisupova, B.; Madaminov, A.A.; Han, Q.F.; Djenbaev, B.M. Carbon stock and its responses to climate change in Central Asia. *Glob. Chang. Biol.* **2015**, *21*, 1951–1967. [CrossRef]
40. Chen, S.; Hamdi, R.; Ochege, F.U.; Du, H.; Chen, X.; Yang, W.; Zhang, C. Added Value of a Dynamical Downscaling Approach for Simulating Precipitation and Temperature over Tianshan Mountains Area, Central Asia. *J. Geophys. Res. Atmos.* **2019**, *124*, 11051–11069. [CrossRef]
41. Li, L.; Luo, G.; Chen, X.; Li, Y.; Xu, G.; Xu, H.; Bai, J. Modelling Evapotranspiration in a Central Asian Desert Ecosystem. *Ecol. Model.* **2011**, *222*, 3680–3691. [CrossRef]
42. Li, B.; Beaudoing, H.; Rodell, M. GLDAS Catchment Land Surface Model L4 daily 0.25 × 0.25 degree GRACE-DA1 V2.2. In *GLDAS_CLSM025_DA1_D, V2.2 ed.*; NASA, GSFC, Eds.; Goddard Earth Sciences Data and Information Services Center (GES DISC): Greenbelt, MD, USA, 2020. [CrossRef]
43. Miralles, D.G.; De Jeu, R.A.M.; Gash, J.H.; Holmes, T.R.H.; Dolman, A.J. Magnitude and variability of land evaporation and its components at the global scale. *Hydrol. Earth Syst. Sci.* **2011**, *15*, 967–981. [CrossRef]
44. Albergel, C.; Dutra, E.; Munier, S.; Calvet, J.C.; Munoz-Sabater, J.; de Rosnay, P.; Balsamo, G. ERA-5 and ERA-Interim driven ISBA land surface model simulations: Which one performs better? *Hydrol. Earth Syst. Sci.* **2018**, *22*, 3515–3532. [CrossRef]
45. Koster, R.D.; Suarez, M.J.; Ducharne, A.; Stieglitz, M.; Kumar, P. A catchment-based approach to modeling land surface processes in a general circulation model: 1. Model structure. *J. Geophys. Res. Atmos.* **2000**, *105*, 24809–24822. [CrossRef]
46. Li, L.; Chen, X.; van der Tol, C.; Luo, G.; Su, Z. Growing season net ecosystem CO_2 exchange of two desert ecosystems with alkaline soils in Kazakhstan. *Ecol. Evol.* **2014**, *4*, 14–26. [CrossRef]
47. Scipal, K.; Holmes, T.; de Jeu, R.; Naeimi, V.; Wagner, W. A possible solution for the problem of estimating the error structure of global soil moisture data sets. *Geophys. Res. Lett.* **2008**, *35*, 4. [CrossRef]
48. Dorigo, W.A.; Scipal, K.; Parinussa, R.M.; Liu, Y.Y.; Wagner, W.; de Jeu, R.A.M.; Naeimi, V. Error characterisation of global active and passive microwave soil moisture datasets. *Hydrol. Earth Syst. Sci.* **2010**, *14*, 2605–2616. [CrossRef]
49. Pierdicca, N.; Fascetti, F.; Pulvirenti, L.; Crapolicchio, R.; Muñoz-Sabater, J. Analysis of ASCAT, SMOS, in-situ and land model soil moisture as a regionalized variable over Europe and North Africa. *Remote Sens. Environ.* **2015**, *170*, 280–289. [CrossRef]
50. Le, M.-H.; Kim, H.; Moon, H.; Zhang, R.; Lakshmi, V.; Nguyen, L.-B. Assessment of drought conditions over Vietnam using standardized precipitation evapotranspiration index, MERRA-2 re-analysis, and dynamic land cover. *J. Hydrol. Reg. Stud.* **2020**, *32*, 100767. [CrossRef]
51. Abolafia-Rosenzweig, R.; Pan, M.; Zeng, J.L.; Livneh, B. Remotely sensed ensembles of the terrestrial water budget over major global river basins: An assessment of three closure techniques. *Remote Sens. Environ.* **2021**, *252*, 112191. [CrossRef]
52. Lu, J.; Wang, G.; Chen, T.; Li, S.; Hagan, D.F.T.; Kattel, G.; Peng, J.; Jiang, T.; Su, B. A Harmonized Global Land Evaporation Dataset from Reanalysis Products Covering 1980–2017. *Earth Syst. Sci. Data Discuss.* **2021**, *2021*, 1–28. [CrossRef]
53. Chen, F.; Crow, W.T.; Colliander, A.; Cosh, M.H.; Jackson, T.J.; Bindlish, R.; Reichle, R.H.; Chan, S.K.; Bosch, D.D.; Starks, P.J.; et al. Application of Triple Collocation in Ground-Based Validation of Soil Moisture Active/Passive (SMAP) Level 2 Data Products. *IEEE J. Sel. Top. Appl. Earth Obs. Remote Sens.* **2017**, *10*, 489–502. [CrossRef]
54. Chen, F.; Crow, W.T.; Bindlish, R.; Colliander, A.; Burgin, M.S.; Asanuma, J.; Aida, K. Global-scale evaluation of SMAP, SMOS and ASCAT soil moisture products using triple collocation. *Remote Sens. Environ.* **2018**, *214*, 1–13. [CrossRef] [PubMed]
55. Yilmaz, M.T.; Crow, W.T. Evaluation of Assumptions in Soil Moisture Triple Collocation Analysis. *J. Hydrometeorol.* **2014**, *15*, 1293–1302. [CrossRef]

56. Stoffelen, A. Toward the true near-surface wind speed: Error modeling and calibration using triple collocation. *J. Geophys. Res.* **1998**, *103*, 7755–7766. [CrossRef]
57. Zhang, C.; Li, C.; Luo, G.; Chen, X. Modeling plant structure and its impacts on carbon and water cycles of the Central Asian arid ecosystem in the context of climate change. *Ecol. Model.* **2013**, *267*, 158–179. [CrossRef]
58. Khan, M.S.; Baik, J.; Choi, M. Inter-comparison of evapotranspiration datasets over heterogeneous landscapes across Australia. *Adv. Space Res.* **2020**, *66*, 533–545. [CrossRef]
59. Badgley, G.; Fisher, J.B.; Jiménez, C.; Tu, K.P.; Vinukollu, R. On Uncertainty in Global Terrestrial Evapotranspiration Estimates from Choice of Input Forcing Datasets. *J. Hydrometeorol.* **2015**, *16*, 1449–1455. [CrossRef]
60. Yang, T.; Li, Q.; Chen, X.; Hamdi, R.; De Maeyer, P.; Kurban, A.; Li, L. Improving snow simulation with more realistic vegetation parameters in a regional climate model in the Tianshan Mountains, Central Asia. *J. Hydrol.* **2020**, *590*, 125525. [CrossRef]
61. Zhang, M.; Luo, G.; Cao, X.; Hamdi, R.; Li, T.; Cai, P.; Ye, H.; He, H. Numerical Simulation of the Irrigation Effects on Surface Fluxes and Local Climate in Typical Mountain-Oasis-Desert Systems in the Central Asia Arid Area. *J. Geophys. Res. Atmos.* **2019**, *124*, 12485–12506. [CrossRef]

remote sensing

MDPI

Article

Characterizing Leaf Nutrients of Wetland Plants and Agricultural Crops with Nonparametric Approach Using Sentinel-2 Imagery Data

Mandla Dlamini [1,2,*], George Chirima [3,4], Mbulisi Sibanda [5], Elhadi Adam [1] and Timothy Dube [2]

1 School of Geography, Archaeology and Environmental Studies, University of the Witwatersrand, Johannesburg 2000, South Africa; Elhadi.Adam@wits.ac.za
2 Institute for Water Studies, Department of Earth Science, University of the Western Cape, Cape Town 7535, South Africa; tidube@uwc.ac.za
3 Geoinformation Science Division, Agricultural Research Council-Institute for Soil, Climate and Water, Pretoria 0083, South Africa; ChirimaJ@arc.agric.za
4 Centre for Geoinformation Science, Department of Geography, Geoinformatics and Meteorology, University of Pretoria, Pretoria 0001, South Africa
5 Department of Geography, Environmental Studies and Tourism, University of the Western Cape, Cape Town 7535, South Africa; msibanda@uwc.ac.za
* Correspondence: mdlamini@uwc.ac.za

Abstract: In arid environments of the world, particularly in sub-Saharan Africa and Asia, floodplain wetlands are a valuable agricultural resource. However, the water reticulation role by wetlands and crop production can negatively impact wetland plants. Knowledge on the foliar biochemical elements of wetland plants enhances understanding of the impacts of agricultural practices in wetlands. This study thus used Sentinel-2 multispectral data to predict seasonal variations in the concentrations of nine foliar biochemical elements in plant leaves of key floodplain wetland vegetation types and crops in the uMfolozi floodplain system (UFS). Nutrient concentrations in different floodplain plant species were estimated using Sentinel-2 multispectral data derived vegetation indices in concert with the random forest regression. The results showed a mean R^2 of 0.87 and 0.86 for the dry winter and wet summer seasons, respectively. However, copper, sulphur, and magnesium were poorly correlated ($R^2 \leq 0.5$) with vegetation indices during the summer season. The average % relative root mean square errors (RMSE's) for seasonal nutrient estimation accuracies for crops and wetland vegetation were 15.2 % and 26.8%, respectively. There was a significant difference in nutrient concentrations between the two plant types, (R^2 = 0.94 (crops), R^2 = 0.84 (vegetation). The red-edge position 1 (REP1) and the normalised difference vegetation index (NDVI) were the best nutrient predictors. These results demonstrate the usefulness of Sentinel-2 imagery and random forests regression in predicting seasonal, nutrient concentrations as well as the accumulation of chemicals in wetland vegetation and crops.

Keywords: crop production; multispectral data; random forests; vegetation indices; wetlands conservation

1. Introduction

Wetlands in South Africa cover about 2.9 million hectares and about 2.4% of the country's land area [1]. They are recognised as highly valuable natural resources that sustain the livelihoods of local communities by providing a wide-ranging ecosystem goods and services that include, wild fruits, vegetables, rice and water purification [2,3]. They also mediate the adverse effects of extreme weather conditions [4] by attenuating floods and slowing down the speed of water movement [5]. Despite their valued recognition, the rate of wetland degradation worldwide and in SA remains high [6,7].

In South Africa, wetlands are some of the most threatened ecosystems, with ~50% of them in a critically endangered state [1]. Although they are vital for the livelihoods of rural communities [8], their sustainability is being threatened by the increasing incidence of climate change-driven droughts and frequent rainfall failures [9]. In sub-Saharan countries including Ethiopia, Kenya, Malawi, Tanzania, and Zambia, wetland cultivation is a recognised and common practice in arid environments [10] that generates about 37% of consumed food and 55% of cash income per household [11]. In Zimbabwe for example, Nyamadzawo, et al. [12] compared wetland gardens against upland fields and observed that more harvests from wetland gardens (2–3 t ha^{-1}) compared to upland fields (1 t ha^{-1}). Despite this higher productivity, wetland ecosystems are being adversely impacted by water extraction, urbanisation, infrastructure development, pollution, poor farming practices as well as droughts and climate variability.

Given their high agricultural productivity, and the water reticulation and other ecological roles, there is a need to explore spatially explicit techniques that can be used to assess and monitor nutrient enrichment in these ecosystems. This is essential because nutrient loading and contamination by different chemicals can disrupt the ecological functioning of these vital sub-systems. A selected example of this disruption is provided by Zhu, et al. [13] who report that an increase in foliar copper content is often associated with shrinkage of mesophyll cells and, concomitant destruction of the internal structure of a plant's leaves and its physiological and structural characteristics. Although a detailed discussion of similar adverse effects is beyond the scope of this paper, it is apparent that better understanding of the foliar nutrient composition of wetland vegetation and crops is critical for informed adoption of eco-friendly management strategies. As vegetation links the physical environment and the upper levels of the food chain [14,15], these linkages need to be properly maintained because they determine the dynamics of nutrient circulation and physiological traits of plants in these systems.

Plants require at least fourteen mineral elements for adequate nutrition [16], lack of which reduces plant growth [17]. Six of these—nitrogen (N), phosphorus (P), potassium (K), calcium (Ca), magnesium (Mg), and sulphur (S)—are critically required by plants in adequate amounts [17]. Their presence at tolerable concentration levels in the plant can now be determined using remote sensing techniques [18–27]. However, considerable research still needs to be conducted on how the concentrations of these elements influence plant growth, as most of the research has tended to be narrowly focused on foliar nitrogen, phosphorus and potassium [18–27]. This bias can be explained by the dominant use of point-based sampling techniques and laboratory assaying, which are time-consuming and expensive These limitations can be overcome by tapping on the abilities of remote sensing techniques to cost-effectively provide quantitative estimates of foliar nutrient concentrations in wetland plants.

Nitrogen, phosphorus and potassium have been widely investigated using remote sensing techniques [28–30]. Although the literature indicates that these elements can be easily characterised using various techniques, other trace elements that are found in wetlands such as magnesium (Mg), copper (Cu), boron (B) and sulphur (S) are difficult to quantify. Remote sensing-based techniques provide a convenient way to overcome this limitation. This dexterity provides partial explanation of why we decided to use both in-situ and remotely sensing (RS) techniques to assess the foliar concentrations of these chemical elements and others.

The decision to couple these techniques was reasoned to be helpful because other researchers have also done this before using RS techniques and laboratory-based spectroscopy in order to cost-effectively maximise their usefulness [31–34]. Additional evidence to support this improvisation is provided by Li, et al. [32], who used hyperspectral data and laboratory analysis to characterise nitrogen and phosphorus concentrations in wetland plants. The majority of these studies have however mostly focused on grassland and dryland woody species [35]. Less effort has been exerted towards foliar nutrients like Cu, B, and S in wetland settings [36]. Although the traditional laboratory methods have been routinely used to provide the information that is required to guide wetland conservation and management [37], these methods are confounded by reliance on labour intensive, time consuming and costly field compilation and analysis of samples [38,39].

The use of hyperspectral RS data has proven to be capable of addressing these constraints and to further facilitate detailed characterisation of wetland plant nutrient concentrations [34]. Although these datasets are useful, they their usefulness is compromised by restricted temporal and spatial coverage with airborne hyperspectral sensors offering little for large scale applications compared to their space-borne equivalents because they are prohibitively expensive. This limitation causes hyperspectral data to be inflexible, inefficient, difficult to access and inappropriate for large scale mapping. To adequately monitor nutrient concentrations in wetland plants, a monitoring system that can be applied over optimum spatial and temporal scales is required. The use of freely available satellite imagery is a viable option as it enables cost-effective mapping and regular monitoring of inaccessible and extensive wetland areas [40,41]. Overall, the literature underscores the immense potentials of the medium resolution (M-res) datasets such as the WorldView-2, 3 and Sentinel-2 (S-2) in detecting variations in plant nutrient concentrations [1,42–44] compared to the costly high-resolution sensors [45].

In the last few years, S-2 is one of M-res datasets that has attracted a lot of interest by many researchers. Although launched recently in 2015, it has already proven to be highly useful for the monitoring of vegetation quality and macronutrients such as nitrogen and phosphorus [46–48]. Its high revisit period of 5 days, spatial resolutions of 10–60 m, systematic global acquisition and open access policy have made it the workhorse of local-level real-time plant nutrient monitoring. In addition to the visible and near-infrared (NIR) wavelengths, the S-2 includes three bands in the red edge region centred at 705, 740, and 775 nm, which are suitable for the characterisation of various plant traits and vegetation monitoring including macronutrients [19,26,48–50]. For instance, S-2 has been used to characterise the seasonal and spatial variations of the nitrogen and phosphorus ratio in the alpine grasslands of China at optimal accuracies of R^2 0.49 and 0.59 and root mean square errors (RMSE) of 2.27 and 3.11 for the dry and wet seasons, respectively) [50]. Specifically, S-2 has 13 spectral bands comprising four bands, three bands and six bands with spatial resolutions of 10, 20 and 60 m that are centred at; (1) 496, 560, 665 and 835 nm (2), 703, 740, 783, 865, 1610 and 2202 nm, and (3), 443, 945 and 1373 nm respectively [51]. Because S-2 data can be accessed free of charge (https://scihub.copernicus.eu/25, accessed on June 2020), it has offered substantial opportunities for the mapping and monitoring of micro and vegetation macronutrients, especially in countries with limited access to spatial data due to financial and logistical constraints. However, unfortunately most of the studies that have sought to characterise foliar nutrients using S-2 remotely sensed data have tended to selectively focus on a narrow range of the primary nutrients notably, N, P, and K. This drawback argues for the need to explore ways through which the mapping of foliar micronutrients nutrients such as Mg, Cu, B and S, can be improved.

Evaluation of the nutritional requirements by different plants is becoming increasingly necessary because the scientific community is obliged to continue providing better means of enhancing the realisation of SDG goals that have a bearing to sustainable utilisation of the finite resources at our disposal. This asseveration is supported by the emergence of a community practice that is informed by the potential realisation of benefits from systematic utilisation of the rich RS datasets at our disposal and systematic use of the techniques that science has so far provided [50,52]. Accomplishing this is possible because RS provides robust techniques with demonstrated capabilities of providing the information required for meaningful realisation of SDGs, i.e., random forests (RF) ensemble, machine learning algorithms (MLAs) [53–55]. The use of MLAs, such as RF amongst others, in estimating nitrogen is a novel advancement that involves a fusion of several spectral vegetation indices in mapping leaf nutrient variations [56].

Although there is no machine learning method that is universally appropriate for estimating vegetation quality, several studies have evaluated the performance of the RF regression model (RF) in predicting the leaf nitrogen content of wetland vegetation [36,57].

RF has been demonstrated to be one of the most robust and widely used algorithms for this purpose [58–62]. The RF ensemble algorithm has been widely used in numerous studies to estimate plant foliar nutrient variations. This is so because, apart from being able to discern subtle variations in numerous variables, RF is able to identify the complex relationships between auto-correlated descriptors [63,64]. When implemented properly, RF offers additional advantages in that regardless of the sample size of the dataset used, it has a bootstrapping mechanism that accommodates utilisation of data implemented when drawing training data points for building trees for each model [65]. It is in this regard that RF was perceived to be a suitable algorithm for estimating foliar nutrient concentrations in the floodplain wetlands of Northern KwaZulu Natal, South Africa. The underlying research question of the study was, can Sentinel-2 data in concert with RF be used to characterise the concentrations of N, K, Ca, Mg, P, S, Zn, B and Cu in the leaves of wetland natural vegetation? We attempted to answer this question by calculating Sentinel-2-derived vegetation indices for different herbaceous species in. The results showed that there is urgent need to explore techniques that can be used to provide unitary perspectives on how these and other challenges can be addressed. This investigation attempts to do this by providing a case illustration of how RF regression modelling can be used to (1) characterise the N, K, Ca, Mg, P, S, Zn, B and Cu foliar concentrations in different wetland vegetation species b) determining the key wavelengths that are important predictors in ascertaining the biochemical leaf foliar variations in these chemicals and (3) characterising seasonal variations in wetland plant leaf nutrient content.

2. Materials and Methods

2.1. Study Area

The uMfolozi floodplain (UFS) system is located in St Lucia, a town that is situated at 28°22′S and 32°25′E in Mtubatuba Local Municipality, South Africa. The uMfolozi River consists of two main tributaries, the Black uMfolozi; that rises at around 1500 m asl. in the north and the White uMfolozi that rises to an altitude of 1620 m asl. these two tributaries converge on the sea Around 50 km west of the mouth of the uMfolozi River. Because the catchment falls within the uMfolozi–Hluhluwe Nature Reserve, the bulk of the catchment comprises natural vegetation cover. Grasslands are about 60% of natural vegetation, with 21% and being classified as thicket and bush respectively. In less than a quarter of the uMfolozi catchment, natural vegetation comes under the dominant agriculture and commercial forestry land uses. The UF floodplain is predominantly used for sugar cane cultivation (65%), and the remaining designated as protected area under the iSimangaliso Park (previously known as St. Lucia Wetlands Park).

The uMfolozi River catchment drains a portion of 11,068 km^2 of northern KwaZulu-Natal on Southern Africa's eastern seaboard. The river's surface geological layout consists of Lebombo rhyolite rock in the east and, Zululand/Maputaland rocks calcarenite, calcareous, limestone, and conglomerate formations in the north and south of the Indian Ocean respectively [66]. About 80% of the rainfall/precipitation occurs in the summer months, peaking between November and April [67]. Mean annual catchment precipitation ranges between 1288 mm in the coastal town of St. Lucia and 667 mm/a in the mid-upper catchment of the uMfolozi Game Reserve with ~914 mm/a occurring in the upper catchment in Nongoma. Mean annual evapotranspiration potential is normally more than double that of precipitation, with average amounts approximating 1800 mm/a [68]. Figure 1 shows the geographic location of the study area in KwaZulu Natal province, South Africa.

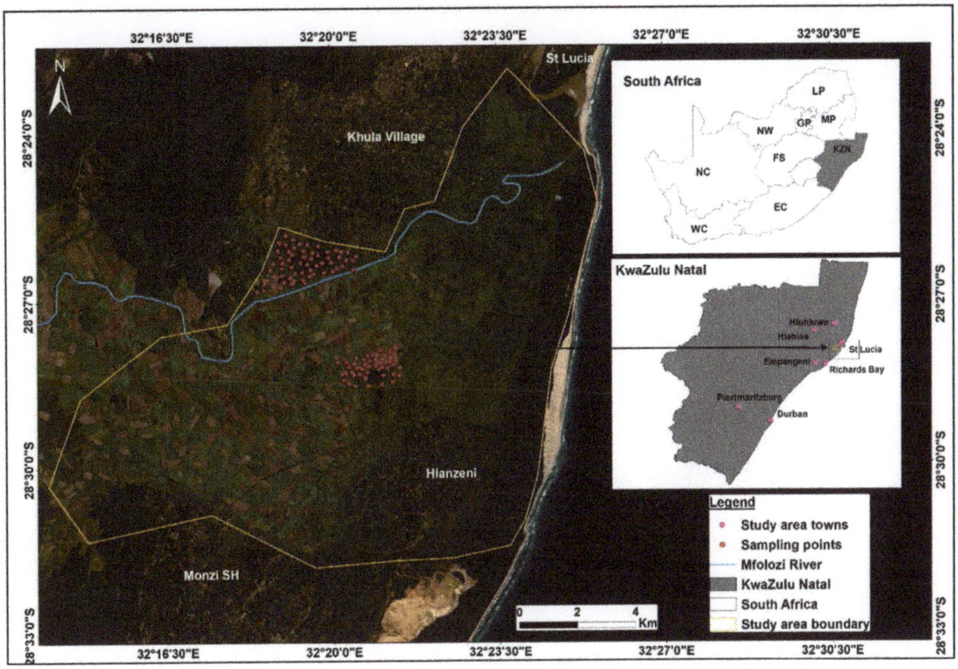

Figure 1. Geographic location of the study area in KwaZulu Natal province, South Africa.

2.2. Field Data Collection

2.2.1. Plant Leaf Sampling

The datasets comprised leaf samples of major crops (*Musa acuminate*, *Ipomoea batatas*, sugarcane and *Colocasia esculenta*) and dominant wetland vegetation species (*Phragmites australis*, *Cyperus papyrus* and *Cynodon dactylon*). Sampling points were systematically generated to cover the study area's footprint and leaves likewise selected following a sampling procedure that was designed to provide representative coverage of the sampling universe. Leaf samples were collected on the 12 March 2017 (wet/summer season) and 22 July 2017 (dry/winter season) respectively. During field investigation, the geographic positions of identifiable herbaceous species were recorded in spreadsheet inventories for further analysis and detailed verification of their geographic positions and species type

A Garmin Montana 650 global positioning system (GPS) with a rated positional accuracy of $\pm$3 m, was used to measure the location of each plant where leaves were harvested. This was followed by randomised sampling of leaves at different crown levels (top, middle, and bottom) to avoid bias. Overall, a total of 76 leaf samples (38 for crops and 38 for vegetation) were collected in summer and 85 samples (40 for crops and 45 for vegetation) collected in winter and ~150 g of leaf material collected from each crop/plant. Leaf samples were then packed in labelled ziplock plastic bags and stored in a cooler box to preserve them during transportation to the laboratory. In the laboratory, the samples were oven-dried at 70 °C for at least 24 h and milled to particle sizes of <0.5 mm.

2.2.2. Chemical Analyses

In this study, six major elements required in large amounts by plants (nitrogen, phosphorus, potassium, calcium, magnesium and sulphur) [17] and three micronutrient elements required as trace amounts (boron, zinc and copper) [69] were chosen. The dry oxidation (Dumas) method was used to determine nitrogen [67] by igniting each leaf sample in oxygen at 950 °C to produce carbon dioxide, nitrogen gas and oxides of nitrogen. These gases were passed through silvered cobalt oxide and a column of copper at 650 °C to reduce the nitrogen oxides to nitrogen gas by removing excess O_2. After removal of water vapour and CO_2, the N_2 gas was finally separated from other gases using gas chromatography, based on a helium carrier gas and detection by a thermal conductivity detector. The instrument was calibrated against the pure compound of known composition following standard procedures that been recommended for this purpose. The compound chosen for the calibration standard is phenylalanine which contains 8.48% N.

An aliquot of the digest solution was used for the inductively coupled plasma optical emission spectrometric (ICP-OES) instrument (Agilent Technologies, United States, North America) for the determination of K, Ca, Mg, P, S, Zn, B and Cu [70]. This is an Agilent 725 (700 series) simultaneous instrument which determines all the elements and wavelengths simultaneously. Thus, several of the elements may be determined at more than one wavelength allowing confirmation of the values with no increase in analysis time or consumption of digest solution. However, S was determined separately from the other elements after purging the optics of the instrument with Ar gas. This is due to the low wavelengths (<190 nm) used for detecting S and the problems caused by oxygen in the air at these very low wavelengths if the system is not purged. Each element was measured at one or two appropriate emission wavelengths that were chosen for their high sensitivity and lack of spectral interferences.

2.2.3. Image Compilation and Pre-Processing

Table 1 describes the characteristics of the S-2 images that were used and the numbers of sample sites from which leaf samples were collected during the wet (12 March 2017) and dry (22 July 2017) seasons.

Table 1. Temporal sequencing, identity details, cloud cover percentages of Sentinel-2 images that were used and field samples by scene ID.

Season	Image Acquisition Date	Sample Collection Date	Scene ID	Cloud Cover (%)	No of Samples
Wet	27 March 2017	12 March 2017	S2A_MSIL1C_20170327T074231_N0204_R049_T36JVP	0.4	76
Dry	18 July 2017	22 July 2017	S2A_MSIL1C_20170718T075211_N0205_R092_T36JVP	0	85

Source: https://www.sentinel-hub.com/develop/documentation/eo_products/Sentinel2EOproducts/ (accessed on 25 June 2020).

The S-2 images were preferred because of their optimum spatial and spectral resolutions for leaf-nutrient characterisation and the availability of footprint coverages of the study area. The difference in the number of days between collection dates of samples and image acquisition dates for wet and dry periods were restricted to 15 and 4, respectively (Table 1) in order to enhance the acquisition of similar leaf nutrient concentrations by using samples at the same phenological stages. The images were atmospherically corrected to surface reflectance using the Sen2Cor plug-in tool provided in the Sentinel Application Platform (SNAP) toolbox, as illustrated in literature [71–73]. S-2 MSI has 13 spectral bands four at 10 m (blue (band 2), green (band 3), red (band 4), and NIR 1 (band 8)), six at 20 m (RE1 to 3 (band 5, 6 and 7), NIR-2 (band 8A), SWIR 1 (band 11) and 2 (band 12), and three at 60 m spatial resolution. Except for bands 1, 9 and 10, all S-2 MSI 10 and 20mbands were used in mapping foliar nutrients in this study. The 10 m bands were all resampled to a spatial resolution of 20 m before estimating the foliar nutrients. However, bands 1, 9, and 10 which have a spatial resolution of 60 m were excluded in this study because they are not suitable for vegetation applications. The images were classified in ArcGIS 10.6 by using 70% of the data that was collected during field investigation for signature compilation with the remaining 30% being reserved for classification accuracy assessment. Vegetation indices were then computed and used with the spectral bands to estimate leaf nutrient concentrations in wetland vegetation and crops growing by using several red edge based indices (Table 2).

Table 2. Vegetation indices used in this study. The Sentinel-2 band IDs are used in the formulas (as per Section 2.2.1).

Name	Equation	References
Normalised Difference Vegetation Index (NDVI)	(NIR − R)/(NIR + R)	[74]
Simple Ratio (SR)	NIR/R	[74]
Simple Ratio 1 (SR1)	Re2/R	[74]
Simple Ratio 2 (SR2)	Re3/R	[74]
Simple Ratio 3 (SR3)	R3/R2	[75]
Red Edge Position 1 (REP1)	(R1 − R)/2	[19,76]
Red Edge 2 Position 2 (REP2)	(R2 − R1)/2	[19,76]
Red Edge 3 Position 3 (REP3)	(R3 − R2)/2	[19,76]
Chlorophyll Red Edge (CIred-edge)	(Re3/R2) − 1	[19]
Chlorophyll Green (CI green)	(R3/green) − 1	[77,78]
The MERIS Terrestrial Chlorophyll Index (MTCI 1)	(NIR − R2/R705)/(R2 − R)	[79]
Modified NDVI (nDVI)	$(B\gamma i) - (B\gamma j)/(B\gamma i) + (B\gamma j)$	[46]
Modified Simple Ratio (sR)	$(B\gamma i)/(B\gamma j)$	[46]

NB: $B\gamma i$ and $B\gamma j$ are different Sentinel 2 MSI spectral bands.

Although a large and growing body of the literature has illustrated that vegetation indices outperform general wavebands in estimating vegetation attributes [80–82], in this study we combined the vegetation indices with spectral bands considering that very few of the aforementioned studies were conducted in wetlands. Vegetation indices were used in this study because of their robustness as illustrated in the literature [50,56,57,80,83–85]. They derive their robustness from two or more wavebands. These bands are often from two different regions of the electromagnetic spectrum. Their optimal performances have also been observed to be capable of circumventing the effects of the atmospheric noise, view/sun angle soil background, topographic effects and sensitivity to vegetation spectral and temporal attributes [86]. This and other considerations explain why Sentinel-2 wavebands were used in this investigation. The processes that were used for this purpose are summarised in Figure 2 which shows how satellite image data and field data were collected and combined in RF regression modelling.

2.3. Estimation of Nutrient Concentrations Using Random Forests Regression

RF was performed in the R statistical package to estimate the concentrations of nutrients in crops and wetland vegetation. RF is a blended model that is characterised by an enormous number of trees [59,87]. The model works by repeatedly splitting each tree (remotely sensed data in this study) into increasingly homogenous subsets at each node to produce a series of terminal nodes. In this study, the training of the regression model was based on 70% of the field data and new estimations determined by sensing the input down the tree and taking the means of the response variables (nutrient contents) and the remaining 30% was used for accuracy assessment as explained earlier.

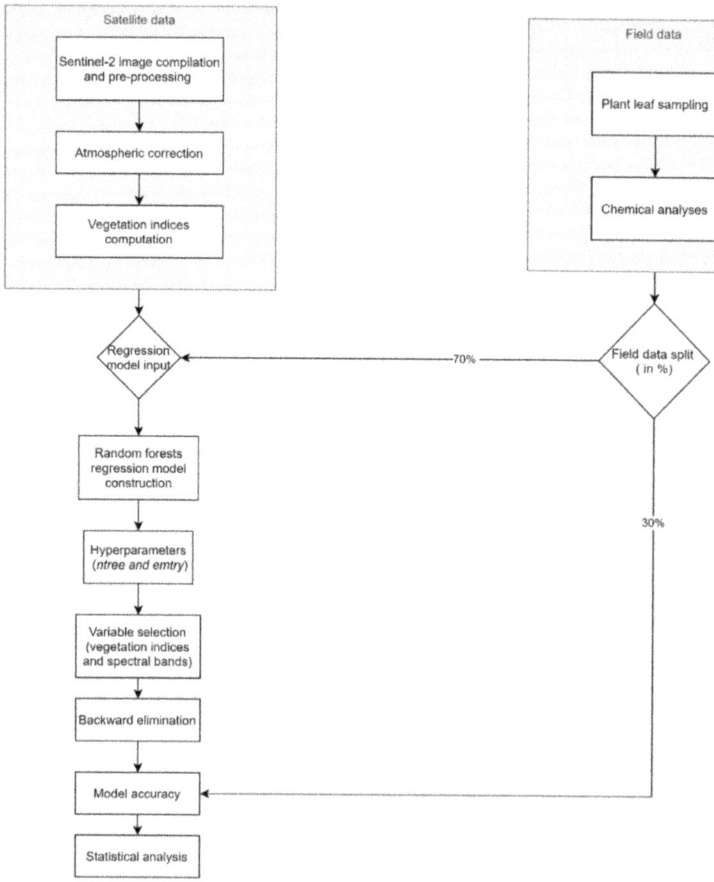

Figure 2. The flow chart summarises the overall methods for the study.

Each regression tree in the RF algorithm is built using a subset of training samples that are independently selected by replacement of the original samples [88]. A subset of a few variables was randomly selected to determine the split in order to increase the robustness of the model by increasing diversity amongst trees and avoiding overfitting the model [88] and the RF predictor was finally constructed by taking the average of overall trees. The samples that are not utilised to grow the tree are referred to as Out of the Bag (OOB) data [59] which the algorithm uses to estimate accuracy by using the difference in the mean square errors to compute the OOB error estimate [71]. The explanatory power of each variable is determined by a Gini coefficient which measures the total decrease in node impurity (weighted by the probability of reaching that node) averaged across all trees. In this study, two hyperparameters were used to tune the models, that is the number of trees (*ntree*) and the number of variables randomly sampled as candidates at each split (*mtry*). It has to be pointed out that the adjustment of these hyperparameters did not significantly change the results hence these were held constant for various models considering the variability in the number of samples in stages 1 to 3 illustrated in Table 3.

Table 3. Experiments were performed using S-2 MSI data to estimate macro and micronutrient concentrations in crops and wetland vegetation.

Analysis	Data Manipulation	Spectral Data Used
1	Separate plant functional types and seasons	Bands and vegetation band indices separately
2	Pooled based on plant functional types and both seasons	Bands and vegetation indices separately
3	Pooled based on nutrient types across all plant types	Vegetation indices

Variable selection of the most important model parameters needed for accurate nutrient estimation was accomplished by implementing the backward feature elimination method [89,90]. The variable importance in RF determines various measures such as the importance of variables based on the Gini coefficient and permutation coefficient. The variation method is considered superior to other approaches because it uses OOB assessments [59]. RF assesses the variable importance of different factors by using the mean decrease in accuracy. Increased mean variability indicates greater importance for that particular variable, while low mean values indicate a lower influence in the model. The method works by generating all the variables of the input predictor and gradually removing input predictor variables with the least relative effect. Table 3 describes the experiments that were performed with S-2 MSI data to estimate nutrient concentrations in crops and wetland vegetation. More details on the RF regression ensemble are provided elsewhere [50,59,91,92].

To estimate nutrient concentrations, data analysis was done in three analysis stages (stage 1, 2, 3). Model input variables varied in each analytical phase. For analysis I, nutrient concentrations were estimated using Sentinel 2 MSI bands only. These were bands 2, 3, 4, 5, 6, 7, 8, 8A, 11 and 12, whereas, additional vegetation indices were computed and used as a stand-alone dataset to estimate nutrient concentrations. The following vegetation indices were computed; chlorophyll green (Cl.green), chlorophyll red edge (Cl.green), red edge position (REP 1, 2 and 3), simple ratio (SR 1, 2 and 3), the NDVI index, MERIS terrestrial chlorophyll index 1 (MTCI.1), modified normalised vegetation indices (nDVI (i.e., nDVI_$B_{\gamma i_}$ $B_{\gamma j}$)), as well as modified simple ratio vegetation indices (sR (i.e., sR_$B_{\gamma i_}$ $B_{\gamma j}$, where $B\gamma i$ and $B\gamma j$ are different Sentinel 2 MSI spectral bands)). The selected indices were computed based on all possible Sentinel 2 band combinations (10 spectral bands). Then S-2 MSI bands were then used as standalone model input variables except for the wet season (March) and the dry (July) seasons (Tables 4 and 5). This was undertaken to determine the ability of Sentinel-2 MSI data in detecting the seasonal differences in leaf nutrient concentrations. For stage 2 analysis, we detected and characterised the year-round nutrient concentrations using pooled seasonal data which was categorised into two datasets (wetland vegetation and crops). In addition, spectral bands and vegetation indices were used separately to estimate the foliar nutrients (Table 6). For the third analysis, to assess the robustness of S-2 MSI data in detecting and characterising nutrient concentrations across all seasons (Figure 3), wet and dry season datasets were pooled into one dataset. Specifically, vegetation indices were used as standalone model input variables (Figure 3). All estimation models were evaluated by using the explained mean squared residual (MSR) variance (Var Expl (%), root mean square error (RMSE) and, RMSE %.

Table 4. Dry and wet season nutrient concentration estimation accuracies for crops.

		Raw Spectral Bands					Vegetation Indices				
Period	**Element**	**MSR**	**Var Expl %**	**R^2**	**RMSE**	**RMSE %**	**MSR**	**Var Expl %**	**R^2**	**RMSE**	**RMSE %**
Summer	N%	0.36	62	0.64	0.65	32	0.20	89	0.97	0.51	23
	K (mg/kg)	34,589,753.00	65	0.75	11,898.33	34	10,233,470.00	86	0.96	7357.83	28
	Ca (mg/kg)	66,890,049.00	29	0.73	5889.75	52	26,130,783.00	57	0.95	3924.56	12
	Mg (mg/kg)	5,751,843.00	35	0.74	3452.00	30	3,335,425.06	89	0.98	1969.82	32
	P (mg/kg)	157,738.20	66	0.74	768.78	31	786,589.96	90	0.94	689.53	26
	S (mg/kg)	218,191.40	54	0.69	1172.17	34	176,031.10	85	0.95	966.12	29
	Zn (mg/kg)	43.85	76	0.6	10.97	32	30.84	87	0.96	8.56	26
	B (mg/kg)	484.50	52	0.59	23.13	42	325.52	65	0.88	13.98	29
	Cu(mg/kg)	16.56	54	0.71	12.36	32	12.99	66	0.94	5.86	26
Winter	N%	0.33	73	0.56	0.69	33	0.15	83	0.71	0.82	27
	K (mg/kg)	20,789,123.00	71	0.71	6512.12	32	14,545,775.00	82	0.87	39,794.20	27
	Ca (mg/kg)	2,189,691.00	64	0.72	2289.59	36	1,876,120.00	87	0.81	1582.82	28
	Mg (mg/kg)	307,861.79	60	0.61	989.28	32	259,985.90	86	0.83	672.58	25
	P (mg/kg)	92,876.18	68	0.59	548.96	31	69,886.90	84	0.79	391.86	22
	S (mg/kg)	258,951.21	53	0.58	681.59	30	200,100.10	75	0.79	398.30	20
	Zn (mg/kg)	28.80	66	0.62	9.99	32	20.74	73	0.75	5.99	24
	B (mg/kg)	43.94	69	0.56	10.93	37	30.96	80	0.85	5.16	30
	Cu (mg/kg)	2.99	70	0.67	3.98	30	1.28	82	0.76	2.90	25

Explanation: MSR = mean squared residual, Var Expl % = variance explained percentage, R^2 = R-squared, RMSE % = root mean square error (percentage).

Table 5. Dry and wet season nutrient concentration variations accuracies for wetland vegetation.

		Raw Spectral Bands					Vegetation Indices				
Period	**Element**	**MSR**	**Var Expl %**	**R^2**	**RMSE**	**RMSE %**	**MSR**	**Var Expl %**	**R^2**	**RMSE**	**RMSE %**
Summer	N%	0.16	74	0.65	0.88	29	0.145685	69	0.82	0.575231	25
	K (mg/kg)	18,186,459.00	51	0.51	3849.73	40	13,815,824	67	0.88	2989.125	20
	Ca (mg/kg)	59,128,582.00	49	0.62	1789.77	29	4,812,124	55	0.85	1500.132	27
	Mg (mg/kg)	698,784.20	54	0.57	1123.90	35	598,798.9	64	0.93	980.012	30
	P (mg/kg)	558,987.10	51	0.53	594.48	29	498,689.99	66	0.73	512.354	24
	S (mg/kg)	8,289,574.00	13	0.36	1289.20	33	791,210.01	10	0.27	981.012	31
	Zn (mg/kg)	41.87	53	0.64	19.49	28	37.394	68.9	0.80	14.0581	22
	B (mg/kg)	98.58	58	0.57	14.55	38	89.7865	60	0.89	10.01	35
	Cu(mg/kg)	4.18	69	0.63	9.28	30	2.001087	67	0.83	6.12351	24
Winter	N%	0.19	57	0.66	1.00	29	0.12986	87	0.89	0.701	23
	K (mg/kg)	42,894,179.00	63	0.61	6188.08	29	38,215,694	80	0.97	5010.56	22
	Ca (mg/kg)	9,989,987.00	53	0.56	1684.49	29	8,614,811	73	0.98	1498.356	25
	Mg (mg/kg)	4,958,737.00	52	69.00	991.88	38	4,015,420	59	0.98	865.2113	30
	P (mg/kg)	29,879,879.00	39	0.63	497.99	24	2,132,142	45	0.90	390.12	15
	S (mg/kg)	69,688.79	67	0.67	798.39	32	65,211.32	88	0.92	701.124	26
	Zn (mg/kg)	32.99	68	0.56	7.94	26	25.096	89	0.94	5.998879	23
	B (mg/kg)	199.93	62	0.68	6.88	29	150.32135	68	0.95	4.98794	25
	Cu (mg/kg)	7.02	53	0.56	3.97	28	4.325612	74	0.95	1.9789	21

Table 6. Comparisons of nutrient estimation accuracies (RMSE %) between the summer and winter seasons in crops and vegetation, respectively.

		RMSE%	
Plants	**Statistics**	**Summer**	**Winter**
Crops	Rank Sum	138	52
	p-value (α = 0.05)	0.001	
Vegetation	Rank Sum	189.2	69.2
	p-value (α = 0.05)	0.012	

The RMSE% and the R^2 were scaled between 0 and 100. To compute the RMSE%, all RMSEs from each model were normalised using the mean of each variable and then expressed as percentages [93,94]. The RMSE% has been widely used in the literature to compare different variable estimations [94–96] hence it was adopted and used in this study. The accuracies (RMSE % and in some instances with R^2) of the training datasets were presented and used to conduct the Mann–Whitney U and the Student's T tests. The Mann–Whitney U independent samples test was then used to test whether there were significant differences at α = 0.05 between the estimation accuracies (R^2 and RMSE%) derived during the summer in relation to those derived from the winter crops and wetland vegetation, respectively (Table 6). The Mann–Whitney U Test was used following the data's significant deviation from the normal distribution at α = 0.05 based on the Kolmogorov–Smirnov Test. Similarly, a Student's t-test of independent samples was then used to assess whether the estimation accuracies derived from crops were significantly different from those derived from wetland vegetation presented in (Table 7) at α = 0.05 (Table 8). The Student's t-test was used because the data did not significantly deviate from the normal distribution based on the Kolmogorov–Smirnov test. The raster calculator tool in ArcGIS 10.6 was used to map the spatial and temporal distributions of nutrients by utilizing the RF regression model outputs and essential variables (NDVI, REP1, and band 7) to characterize the studied nutrients. Figure 2. summarises the methods that were used to determine the accuracies of these calculations.

Table 7. Seasonal nutrient estimation accuracies for crops and vegetation.

Plant Type	Nutrient	MSR	Var Expl		RMSE	
			%	R^2	RMSE	%
Wetland vegetation	N	0.186243	69.97	0.94	0.48	18
	K	65,985,615	84.82	0.83	7235.78	29
	Ca	1,869,486	81.95	0.88	2610.2	30
	Mg	247,612.98	80.83	0.75	2166.48	38
	P	202,121.7	77.85	0.78	1231.25	29
	S	3,981,612.2	39.44	0.84	310.41	12
	Zn	11.58776	88.22	0.88	4.01	14
	B	119.012	78.48	0.79	9.12	43
	Cu	0.8844	89.2	0.88	2.12	29
Crops	N	0.18799	88.51	0.94	0.301	12
	K	13,131,613	90.12	0.96	2911.59	14
	Ca	15,532,853	59.75	0.95	1433.21	20
	Mg	149,814	68.7	0.95	511.51	13
	P	99,539.7	85.62	0.95	221.1	11
	S	14,912.3	87	0.95	221.72	12
	Zn	18.13452	90.04	0.96	2.51	12
	B	123.89745	81.42	0.98	6.01	25
	Cu	4.1765523	88.61	0.88	1.7	18

Table 8. Descriptive statistics for R^2 values and RMSE% for vegetation and crop types using pooled data.

Accuracy Unit	Statistics	Vegetation	Crops
R^2	Mean	0.84	0.94
	Maximum	0.88	0.98
	Minimum	0.75	0.88
	Median	0.84	0.95
	St deviation	0.06	0.027
	p-value		0.02
RMSE %	Mean	15	27
	Maximum	25	43
	Minimum	11	12
	Median	29	29
	St deviation	13	29
	p-value		0.001

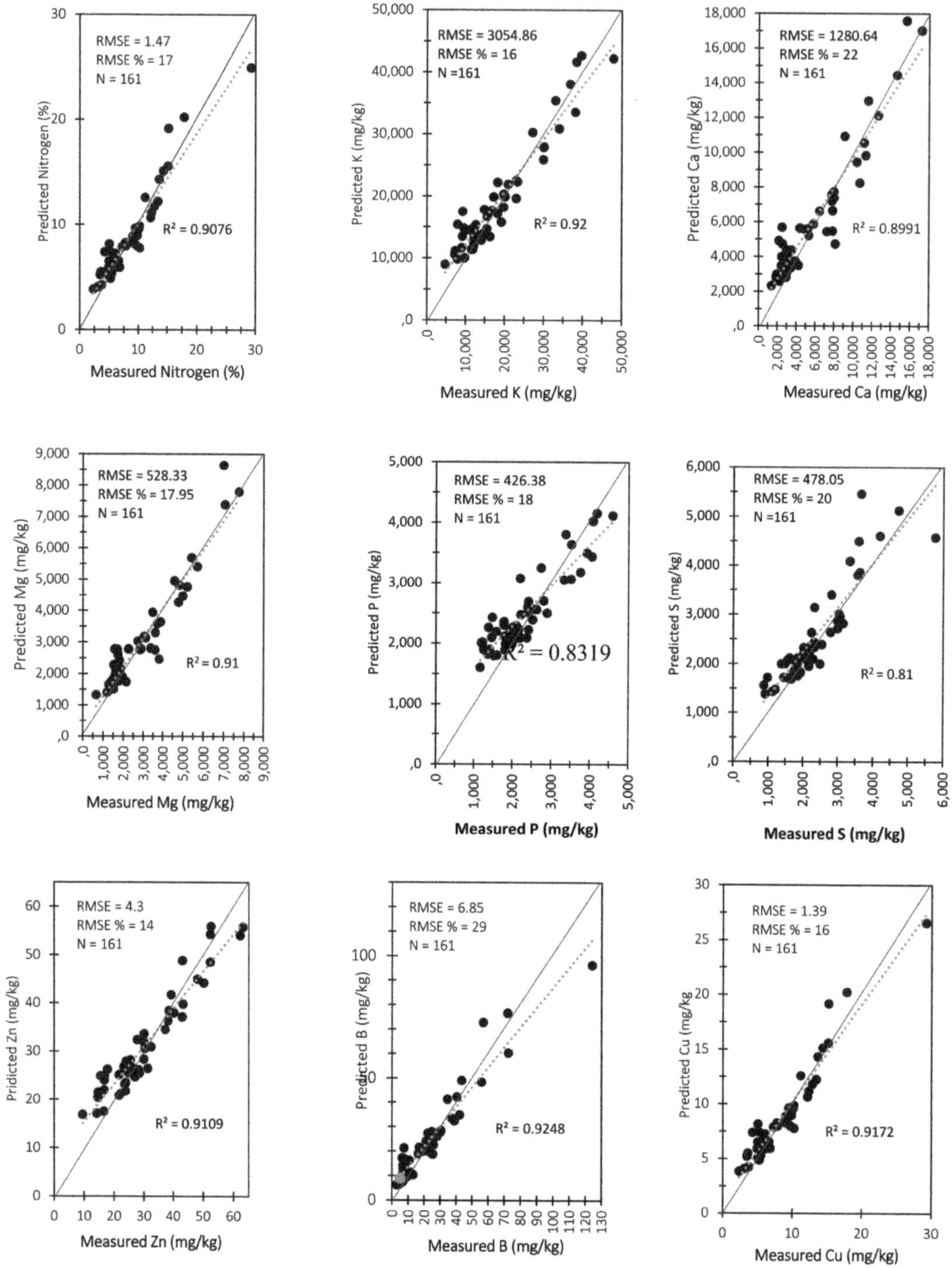

Figure 3. Accuracies derived in estimating N, K, Ca, Mg, P, S, Zn, B, Cu concentrations using testing data sets. Dotted lines in the graph illustrate the linear regression line.

3. Results

3.1. Comparison of Single Bands and Vegetation Indices in Estimating Different Nutrients in Summer and Winter Seasons

Tables 4 and 5 shows the seasonal accuracy levels (RMSE, R^2, and RMSE %) for different nutrients that were investigated based on the raw bands and vegetation indices. The RMSE, which depicts the standard deviation of residuals, shows that the linear regression model based on vegetation indices performed better than raw bands. Using raw spectral bands, all of the RMSE values for both crops and vegetation in summer and winter were greater than of the vegetation indices (Tables 4 and 5). As a result, the results of vegetation indices were used for further analysis of this study because of their lower prediction errors (RMSE values).

For both vegetation and crops, high R^2 and low RMSE% were observed for all nutrients in winter (Tables 4 and 5). For instance, Calcium (Ca) estimation in crops based on VIs exhibited an R^2 of 0.73 and RMSE of 5889.75 (mg/kg) which was optimally estimated to an R^2 of 0.95 and a RMSE of 39284.56 (mg/kg) (RMSE % = 12) based on vegetation indices in summer (Table 4). Meanwhile in winter Ca exhibited a R^2 = 0.72 and a RMSE = 2289.59 mg/kg (RMSE % = 36) based on bands only. These accuracies improved to an R^2 of 0.81, RMSE of 1582.82 mg/kg (RMSE % = 28). A similar trend was observed when estimating nutrients in wetland vegetation (Table 5). Specifically, poor accuracies of an R^2 = 0.73, RMSE = 5889.75 mg/kg (RMSE % = 52) were attained when estimating Ca in wetland vegetation using only spectral bands during the summer. These accuracies improved when vegetation indices were used with an R^2 of 0.95, RMSE of 3924.56 mg/kg and an RMSE % of 12. A similar trend was also observed in estimating Ca during the winter season. A R^2 = 0.56, RMSE = 1684.49 mg/kg and a RMSE% = 29. An improvement in the accuracies was realised in estimating Ca in using bands only. Again, an improvement was observed when estimating Ca using the vegetation indices (R^2 = 0.98, RMSE = 1498.356 mg/kg and a RMSE% = 25) during winter. The summer–winter RMSE percentages were higher in vegetation compared to crops and the differences in average values for both were significant (Table 6).

Tables 6–8 show seasonal mean comparisons of R^2 and RMSE % nutrient estimations and seasonal nutrient estimation accuracies for crops and vegetation Table 8 shows R^2 values and RMSE% for vegetation and crop types using pooled data.

3.2. Dry and Wet Season Crop and Vegetation Nutrient Estimation Using Pooled Data

In addition, the annual estimations of nutrient concentrations presented in Table 7 were used to calculate the statistics provided in Table 8. A higher mean R^2 accuracy for nutrients across plant types was observed for vegetation (0.94) when compared to a mean R^2 of 0.84 for crops (Table 8). The differences in the mean R^2 values were significant. A higher percentage of RMSE (27) was observed for crops when compared to what was observed for vegetation (15). The difference between these vegetation types and crop RMSE values was significant. In all cases, measured and estimated nutrient concentrations were strongly correlated with R^2 ranging from 0.98 for boron to 0.75 for magnesium. Although boron and magnesium showed a good correlation, both had high RMSE percentages.

Figure 2 is a composite summary of the performance of the statistical models that were used to relate foliar nutrients measured from the laboratory and data from S-2 images. Figure 3 shows the best fits for the model (in terms of R^2 and RMSE %) arranged by performance. As it can be noted, the studied nutrients' estimation performances were strong and very consistent, with R^2 values ranging from 0.81 to 0.93. The nutrients, on the other hand, showed the opposite in terms of variability based on the RMSE%. However, it's worth noting that lower RMSE values suggest a better model estimation fit. The RMSE percentages for different nutrients ranged from 14% to 29%. Zinc, potassium, sulphur, copper, and nitrogen were found to have low RMSE percentages ranging from 14% to 17%. The RMSE of calcium, boron, and magnesium were all high, with boron having the highest of 29 per cent. The overall findings revealed that several nutrients, such as calcium, boron, and magnesium, are poorly estimated, as evidenced by the RMSE per cent values.

3.3. Variable Importance Selection

The RF model was able to rank the variables by their estimation capabilities. Figure 3 shows important variables derived in estimating N, K, Ca, Mg, P, S, Zn, B and Cu. REP1 and NDVI were the most frequent vegetation indices that yielded optimal models of estimating foliar nutrients as indicated by their high % IncMSE in Figure 4.

Figure 5 illustrates the estimated spatial distributions of sulphur, potassium, magnesium, copper, nitrogen, phosphorus, zinc and boron. The lowest values correspond to areas with crop farming (dark red for waterbodies), higher values (dark green) represent the most developed vegetation (forest and grassland). It can be observed that nutrients have high concentrations in the eastern section of the study area, except for magnesium that showed high concentrations in the sugarcane infested area.

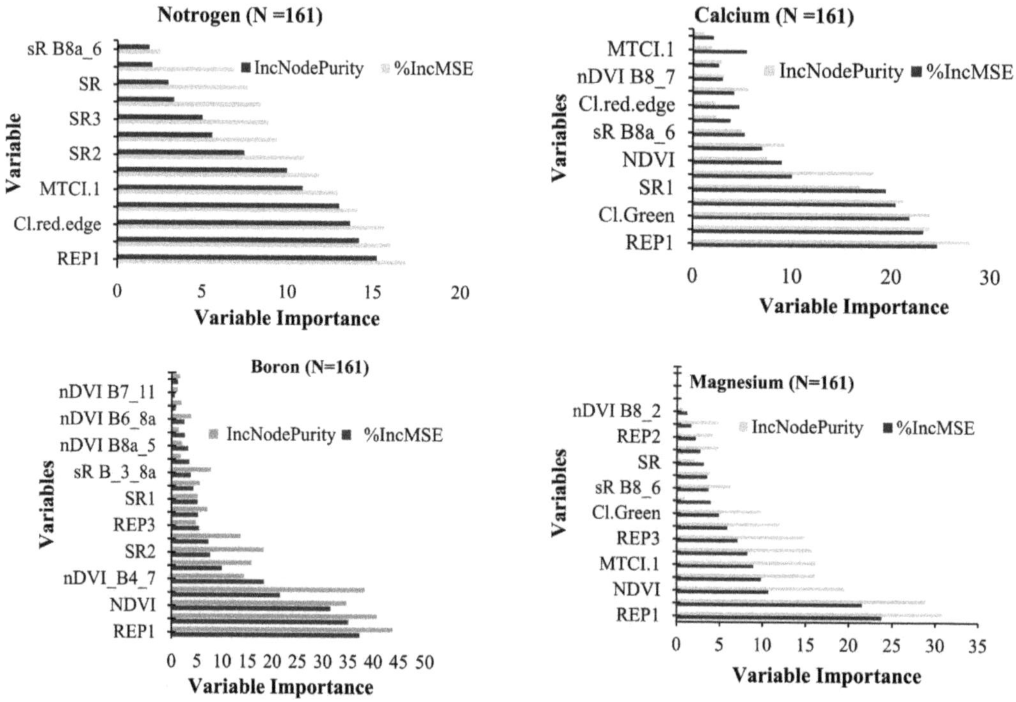

Figure 4. *Cont.*

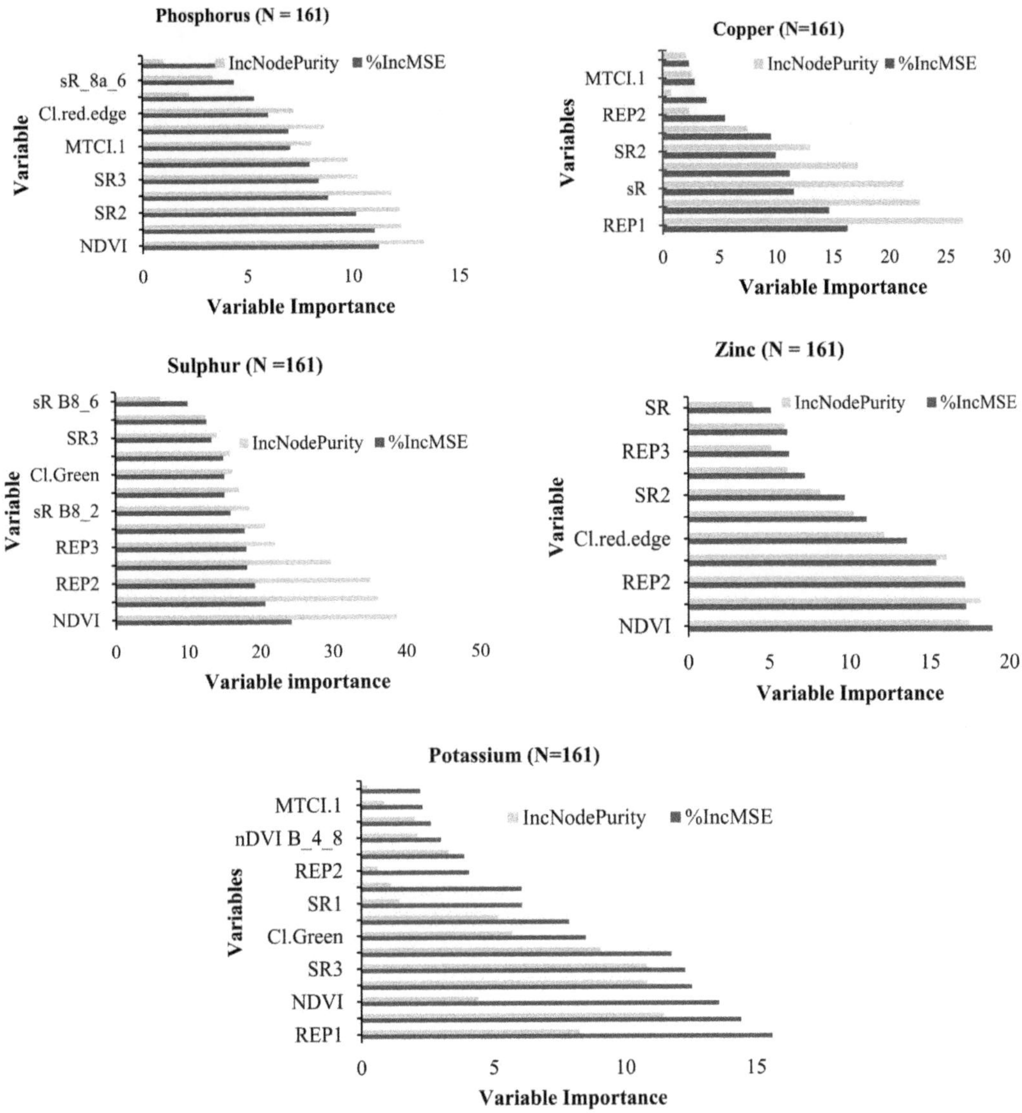

Figure 4. Importance of variables that were used to estimate N, Ca, K, Mg, P, Cu, Zn, B and, S by using S-2 MIS derived vegetation indices data. NB *, %IncMSE is Mean decrease accuracy, IncNodePurity is mean decrease Gini, REP; 1, 2 and 3 is red edge position 1, 2 and 3, chlorophyll green (Cl. Green), simple ratio (SR; 1, 2 and 3), MTCI.1 is MERIS Terrestrial Chlorophyll Index 1, NDVI is normalised difference vegetation index, nDVI (nDVI_$B_{\gamma i}$_ $B_{\gamma j}$) is modified normalised vegetation difference index and sR (sR_$B_{\gamma i}$_ $B_{\gamma j}$) _ is modified sample ratio where $B_{\gamma i}$_ $B_{\gamma j}$ are any Sentinel 2 MSI bands excluding bands 1, 9 and 10.

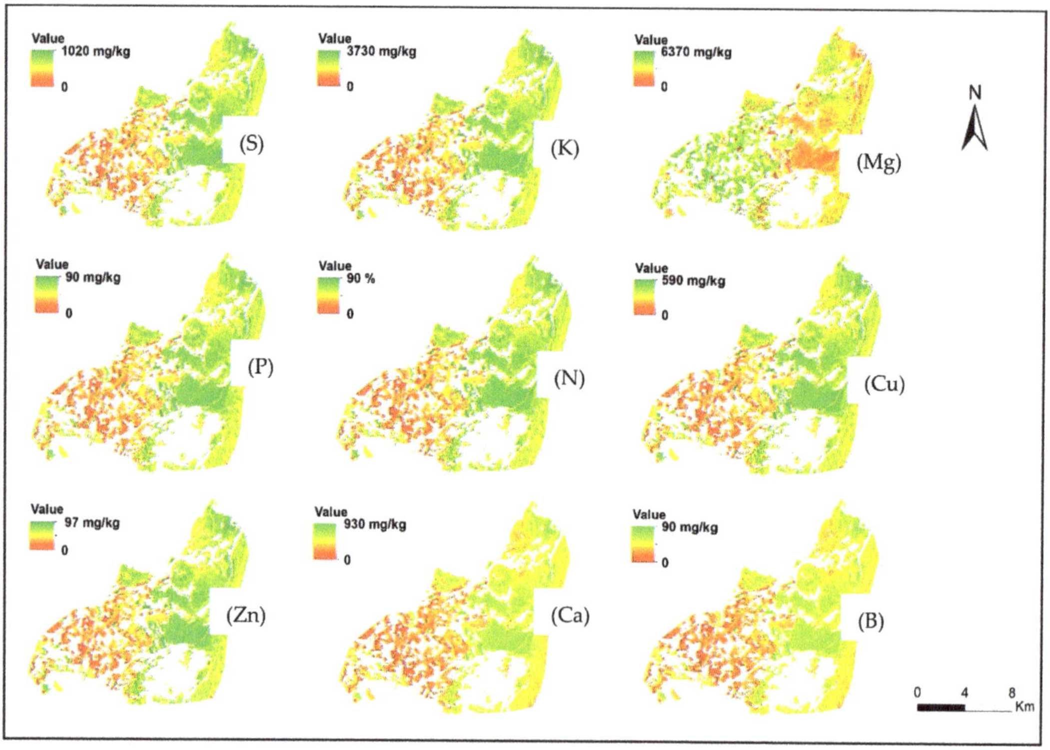

Figure 5. Spatial distribution of nutrients: S, K, Mg, Cu, N, P, Zn, Ca and B.

4. Discussion

In this study, we used the RF algorithm and S-2 on both wetlands vegetation and crops and across seasons to estimate the concentrations of N, K, Ca, Mg, P, S, Zn, B and Cu. The results show that: (a) the RF model using S-2 can estimate foliar concentration across several nutrients and (b) S-2 can estimate nutrients across plant types and seasons. Derived S-2 vegetation indices such as NDVI and bands 2 and 7 performed well in estimating nutrients in crops and vegetation. Seasonal characterisation of nutrients was also successful which could be attributed to the variability of photosynthetic pigments such as chlorophyll [36]. However, the RF model performed poorly in estimating magnesium, and sulphur in the summer season. It also performed poorly in estimating calcium, magnesium, phosphorus and boron in wetland vegetation across the seasons as demonstrated by high RMSE %s. This can be attributed to easy leaching and increased mobility of these nutrients that might have caused the decrease of their concentration during the high rainfall period hence weak correlation with S-2. Phenology is also an important factor in this result because most vegetation indices such as NDVI, especially red edge-based indices, rely on the vigour and greenness of the vegetation. Osco, et al. [97] also found nutrients like Mg, S, P, K and Ca presented inferior performances compared to nutrients such as nitrogen, zinc etc. Another contribution of this work is that it was possible to identify wavelengths and spectral regions that contributed most to nutrient estimation.

A combination of vegetation indices and spectral bands was found to be robust when compared to the raw spectral bands. As outlined above, the infusion of vegetation spectral indices proved to be important in the evaluation of the most studied nutrients [98–100]. Different indices and bands that were key to estimating nutrient concentrations include the red edge position, NDVI and band 2 (Figure 2). This observation implies that different nutrients can be estimated by these different variables. In this study, the key variables that were important in estimating most of the nutrients including nitrogen, potassium, calcium, boron and copper were the red edge position 1 computed from the NIR and red edge position bands followed by NDVI for estimating P and S. Similar findings were observed in Bush Buck Ridge, Mpumalanga, South African savannah grass where canopy nitrogen was correlated to NIR spectral region [101]. In similar findings were also found in North American forests where canopy nitrogen was correlated to both the NIR spectral region as well as NIR-based vegetation indices including NDVI [102,103].

The sensitivity of red edge bands to potassium, calcium, copper and nitrogen was not surprising. Studies by other researchers confirm the strong correlation between nitrogen concentrations and red edge bands [18,51,104,105]. The red edge is considered as the surrogate measure of vegetation chlorophyll content [51,106,107]. Therefore, in this study, an expectation of the magnesium concentration to be strongly estimated in the red edge bands was not questionable. Magnesium is located in the central of the chlorophyll molecule and it is regarded as the activator of some enzymes in plants [108]. The results by [109,110] found the NIR region to be the best in estimating magnesium concentrations. NDVI computed from red and NIR bands showed to be the key variable in estimating phosphorus and sulphur concentrations. This is similar to the findings of Lisboa, et al. [111], who found NDVI to be a useful tool in estimating nitrogen and phosphorus concentrations in sugarcane crops. Several studies have used these indices (NIR spectral region and NDVI) to study heavy metals in plants [112–114].

4.1. Estimation of Nutrients Using the RF Model

This study has shown the utility of the RF model using S-2 in estimating concentrations of N, K, Ca, Mg, P, S, Cu, Zn and B. The technique yielded high coefficients of determination ranging between 0.75 and 0.98. The technique also exhibited low RMSE % for most nutrients except for magnesium (38%), boron (43%) and calcium (30%) which emerged as difficult to detect by using the RF model and S-2. The usefulness of RF regression and remotely sensed images is demonstrated by its ability to estimate sugarcane leaf nitrogen levels from Hyperion images [104]. Multiple studies have demonstrated that RF models often perform remarkably well in different fields of scientific research including the estimation of nitrogen content [115–118]. In this study, fusing vegetation indices in the NIR spectral region and NIR-based vegetation indices (NDVI) with RF proved to be a suitable approach for estimating the studied nutrients. RF also proved to be suitable in relation to NDVI, REP1 and band 7 in developing a relationship between nutrient variations and land use land cover types (Figure 4) except for magnesium, which exhibited high concentrations in sugarcane farms where the land use land cover-nutrient effect variation was consistent. This is attributed to the antagonistic effects of (Ca and Mg) and K in sugarcane, where soils with high Ca and Mg can lower leaf K and vice versa.

4.2. Crops and Vegetation Nutrients Seasonal Estimations

Most nutrients exhibited significant relationships ($R^2 < 0.7$) between measured and estimated concentrations across seasons and plant types with few that showed a weak relationship ($R^2 > 0.5$), i.e., calcium, magnesium, phosphorus and boron). This implies that S-2 has the potential in estimating concentrations of selected chemical elements across the seasons and plant types. Generally, with the R^2 accuracies in estimating foliar nutrient concentrations, there were no significant differences within nutrients from crops and vegetation between the summer and winter seasons.

Similar findings by Gama, et al. [119] confirmed a weak relationship between leaf reflectance and concentrations of phosphorus, potassium and calcium. Poor estimations of magnesium, copper and sulphur in summer were observed which implies difficulties in estimating such nutrients in a wet season. Magnesium, copper and sulphur are micronutrients that are highly affected by the processes of reduction and oxidation (redox) in concert with the shifting of water levels which determines their seasonal concentration levels in wetland soils and water. Their poor detection in this study could therefore be explained by the seasonal variations in the amount of available water which regulates the redox processes thereby and their concentration levels in soil, water different plant systems [120–122]. For instance, a related study [121] also concluded that seasonally occurring processes such as redox and seasonal shifts in water abundance regulate the type and amount of trace elements that are available in the water.

Moreover, high the poor performance revealed by the model's high RMSE % suggesting across the seasons was also observed in calcium, magnesium, phosphorus and boron in wetland vegetation. Season-specific analyses showed that dry season models performed better than their wet season counterparts, as shown by their respective coefficient determinants of 0.87 in winter and 0.86 (Tables 4 and 5). This might be attributed to the higher reflectance in the visible spectrum (bands 2, 4 and 4) in the dry season compared to the wet season. Plants are generally less green in dry seasons due to lower moisture content and less chlorophyll which leads to lower energy absorption and greater reflection. Contrary to these observations, however, Ramoelo, et al. [123]; Cho, et al. [124] and Skidmore, et al. [125] found models such as RF to perform better in the wet season compared to the dry season.

4.3. Implications of Remote Sensing on Wetland Plant Nutrients

Over the past years, several studies [126,127] have used remote sensing and chemical analyses in estimating foliar nutrient concentrations in plants. However, these studies mostly concentrated on seasonal estimations of nitrogen in grasses. This study has attempted to build on these initiatives by broadening the application of these techniques to the estimation of different nutrients. This initiative is helpful because it offers opportunities for enhanced understanding of vegetation health which has been previously regarded as a complex research area [128]. Although limited in geographical reach, our findings show that S-2 imagery could be a significant additional source of valuable information on seasonal variations in plant nutrient content.

This study has demonstrated that RF using S-2 can be useful for monitoring and estimating various plant nutrient quantities and the quality of floodplain vegetation. S-2 images yielded significant relationships between nutrient content in the NIR spectral region and NIR-based vegetation indices (NDVI). S-2 also provided an opportunity to seasonally characterize the studied nutrients. The findings of this study present a great opportunity and technique for mapping and monitoring nutrient enrichment in wetlands beyond the characterization of plant macronutrients (NPK) across both crops and natural vegetation species. This is a step towards a time-efficient and affordable technique for mapping and monitoring wetlands from local to regional scales based on S-2's optimal spatial resolution and swath widths.

A method of performing chemical analysis in plants in the laboratory is spatially limited for many reasons. Our approach returned high coefficients of determination for most of the nutrients in crops and vegetation. By implication, remotely sensed data make up for shortcomings of the traditional methods with the advantages of real-time and rapid observations. As a result, this technique would be a useful alternative for estimating plant health conditions and nutritional status under different environmental settings.

5. Conclusions

This study demonstrated immense potentials for Sentinel-2 data to be used for estimating leaf nitrogen, potassium, calcium, magnesium, phosphorus, sulphur, zinc, boron and copper concentrations. Different vegetation indices and bands including the red edge position, NDVI and band 2 can be important in estimating different nutrients across the seasons even though foliar estimation using Sentinel-2 did not show a strong relationship in estimating leaf magnesium, copper and sulphur in a wet season. The major conclusion of this work is that the RF model was able to provide estimated nutrients with reasonable accuracies. The study, therefore, recommends that researchers elsewhere can use the methodology provided in this article as a tool in assessing and monitoring the quality of crops/vegetation in floodplain wetlands and the use of at least two scenes (satellite images) per season to achieve representative estimates over these periods. Considering that there is no consensus with regards to the performance of different machine learning techniques in characterising foliar chemical elements such as leaf nitrogen, potassium, calcium, magnesium, phosphorus, sulphur, zinc, boron and copper, future studies need to compare and assess their performance.

Author Contributions: Literature search, M.D.; figures, M.D. and M.S.; study design, M.D.; data collection, M.D. and G.C.; data analysis, M.D., M.S. and T.D.; data interpretation, M.D. and M.S.; writing, M.D. and G.C.; co-supervised, G.C.; provided guidance to the PhD study, G.C. and E.A.; guidance on the literature searches, G.C.; guidance of the study design, G.C., M.S. and T.D.; guidance and participation in results interpretation, G.C., M.S. and T.D.; supervision, G.C. and E.A.; guidance on the conceptualization of the manuscript, G.C. and E.A. All authors have read and agreed to the published version of the manuscript.

Funding: The Department of Higher Education, Science and Technology funded the study, while the Agricultural Research Council funded the analysis of plant samples.

Institutional Review Board Statement: Not applicable.

Informed Consent Statement: Not applicable.

Data Availability Statement: The data belongs to Witwatersrand University and will be kept by the authors. The data will be availed upon request for research purposes only, and the disclosure shall include the following: the title of the research or paper to be used for the stated data; the details of the organization and the supervisory body or persons concerned; the assurance that the research would not yield any commercial benefit. Our contacts are detailed in this paper to request the data.

Acknowledgments: The authors would like to thank local communities and Isimangaliso Wetland Park for allowing us to access their land during field investigation. We are also grateful to the Agricultural Research Council laboratory for analyzing our plant samples.

Conflicts of Interest: The authors declare no conflict of interest. There was no position for the funders in the study design; data collection, publication or interpretation; the writing of the manuscript or the decision to report the results of the study.

References

1. Driver, A.; Sink, K.J.; Nel, J.L.; Holness, S.; Van Niekerk, L.; Daniels, F.; Jonas, Z.; Majiedt, P.A.; Harris, L.; Maze, K. *National Biodiversity Assessment 2011: An Assessment of South Africa's Biodiversity and Ecosystems*; SANBI: Pretoria, South Africa, 2012.
2. Kotze, D.; Marneweck, G.; Batchelor, A.; Lindley, D.; Collins, N. *WET-EcoServices: A Technique for Rapidly Assessing Ecosystem Services Supplied by Wetlands*; SANBI: Pretoria, South Africa, 2009.
3. Steven, D.D.; Lowrance, R. Agricultural conservation practices and wetland ecosystem services in the wetland-rich Piedmont-Coastal Plain region. *Ecol. Appl.* **2011**, *21*, S3–S17. [CrossRef]
4. IPCC. *Intergovernmental Panel on Climate Change (IPCC)*; Fifth Assessment Report, Synthesis for Policy Makers; IPCC: Geneva, Switzerland, 2014.
5. Renaud, F.G.; Sudmeier-Rieux, K.; Estrella, M. *The Role of Ecosystems in Disaster Risk Reduction*; United Nations University Press: Tokyo, Japan, 2013.
6. Gardner, R.C.; Barchiesi, S.; Beltrame, C.; Finlayson, C.M.; Galewski, T.; Harrison, I.; Paganini, M.; Perennou, C.; Pritchard, D.E.; Rosenqvist, A.; et al. *State of the World's Wetlands and Their Services to People: A Compilation of Recent Analyses*; Ramsar Convention Secretariat: Gland, Switzerland, 2015; p. 19.
7. Davidson, N.; Fluet-Chouinard, E.; Finlayson, C. Global extent and distribution of wetlands: Trends and issues. *Mar. Freshw. Res.* **2018**, *69*, 620–627. [CrossRef]
8. Houghton, R.A.; Hackler, J.L.; Lawrence, K.T. The US carbon budget: Contributions from land-use change. *Science* **1999**, *285*, 574–578. [CrossRef] [PubMed]
9. Dai, A. Increasing drought under global warming in observations and models. *Nat. Clim. Chang.* **2013**, *3*, 52–58. [CrossRef]
10. Nyirenda, R. Gendered Access to Wetland Gardens (Dimba) in Northern Malawi. Master's Thesis, West Virginia University, Morgantown, WV, USA, 2020.
11. Msusa, H. Sustainable utilization of wetlands for food security: A case study of the Simulemba traditional Authority in the Kasungu District of Malawi. *J. Dev. Sustain. Agric.* **2011**, *6*, 86–100.

12. Nyamadzawo, G.; Wuta, M.; Nyamangara, J.; Nyamugafata, P.; Chirinda, N. Optimizing dambo (seasonal wetland) cultivation for climate change adaptation and sustainable crop production in the smallholder farming areas of Zimbabwe. *Int. J. Agric. Sustain.* **2015**, *13*, 23–39. [CrossRef]
13. Zhu, Y.; Qu, Y.; Liu, S.; Chen, S. A reflectance spectra model for copper-stressed leaves: Advances in the PROSPECT model through addition of the specific absorption coefficients of the copper ion. *Int. J. Remote Sens.* **2014**, *35*, 1356–1373. [CrossRef]
14. Verkleij, J.A.; Golan-Goldhirsh, A.; Antosiewisz, D.M.; Schwitzguébel, J.-P.; Schröder, P. Dualities in plant tolerance to pollutants and their uptake and translocation to the upper plant parts. *Environ. Exp. Bot.* **2009**, *67*, 10–22. [CrossRef]
15. Peralta-Videa, J.R.; Lopez, M.L.; Narayan, M.; Saupe, G.; Gardea-Torresdey, J. The biochemistry of environmental heavy metal uptake by plants: Implications for the food chain. *Int. J. Biochem. Cell Biol.* **2009**, *41*, 1665–1677. [CrossRef]
16. Mengel, K.; Kirkby, E.A.; Mengal, K.; Appel, T.; Kosegarton, H. *Principles of Plant Nutrition*, 5th ed.; Kluwer Academic Publishers: Dordrecht, The Netherlands, 2001.
17. White, P.; Brown, P. Plant nutrition for sustainable development and global health. *Ann. Bot.* **2010**, *105*, 1073–1080. [CrossRef]
18. Kanke, Y.; Raun, W.; Solie, J.; Stone, M.; Taylor, R. Red edge as a potential index for detecting differences in plant nitrogen status in winter wheat. *J. Plant Nutr.* **2012**, *35*, 1526–1541. [CrossRef]
19. Clevers, J.G.; Gitelson, A.A. Remote estimation of crop and grass chlorophyll and nitrogen content using red-edge bands on Sentinel-2 and-3. *Int. J. Appl. Earth Obs. Geoinf.* **2013**, *23*, 344–351. [CrossRef]
20. Özyiğit, Y.; Bilgen, M. Use of spectral reflectance values for determining nitrogen, phosphorus, and potassium contents of rangeland plants. *J. Agric. Sci. Technol.* **2013**, *15*, 1537–1545.
21. Mutanga, O.; Adam, E.; Adjorlolo, C.; Abdel-Rahman, E.M. Evaluating the robustness of models developed from field spectral data in predicting African grass foliar nitrogen concentration using WorldView-2 image as an independent test dataset. *Int. J. Appl. Earth Obs. Geoinf.* **2015**, *34*, 178–187. [CrossRef]
22. Adjorlolo, C.; Mutanga, O.; Cho, M.A. Estimation of canopy nitrogen concentration across C3 and C4 grasslands using WorldView-2 multispectral data. *IEEE J. Sel. Top. Appl. Earth Obs. Remote Sens.* **2014**, *7*, 4385–4392. [CrossRef]
23. Ramoelo, A.; Cho, M.A. Explaining leaf nitrogen distribution in a semi-arid environment predicted on Sentinel-2 imagery using a field spectroscopy derived model. *Remote Sens.* **2018**, *10*, 269. [CrossRef]
24. Mutanga, O.; Adam, E.; Cho, M.A. High density biomass estimation for wetland vegetation using WorldView-2 imagery and random forest regression algorithm. *Int. J. Appl. Earth Obs. Geoinf.* **2012**, *18*, 399–406. [CrossRef]
25. Li, F.; Wang, L.; Liu, J.; Wang, Y.; Chang, Q. Evaluation of leaf N concentration in winter wheat based on discrete wavelet transform analysis. *Remote Sens.* **2019**, *11*, 1331. [CrossRef]
26. Delloye, C.; Weiss, M.; Defourny, P. Retrieval of the canopy chlorophyll content from Sentinel-2 spectral bands to estimate nitrogen uptake in intensive winter wheat cropping systems. *Remote Sens. Environ.* **2018**, *216*, 245–261. [CrossRef]
27. Ye, X.; Abe, S.; Zhang, S. Estimation and mapping of nitrogen content in apple trees at leaf and canopy levels using hyperspectral imaging. *Precis. Agric.* **2020**, *21*, 198–225. [CrossRef]
28. Mahajan, G.; Pandey, R.; Sahoo, R.; Gupta, V.; Datta, S.; Kumar, D. Monitoring nitrogen, phosphorus and sulphur in hybrid rice (*Oryza sativa* L.) using hyperspectral remote sensing. *Precis. Agric.* **2017**, *18*, 736–761. [CrossRef]
29. Özyiğit, Y.; Bilgen, M. Estimation of Nitrogen Levels by Remote Sensing Method in Alfalfa (*Medicago sativa* L.). *Tarim Ve Doga Derg.* **2018**, *21*, 902.
30. Siqueira, R.; Longchamps, L.; Dahal, S.; Khosla, R. Use of fluorescence sensing to detect nitrogen and potassium variability in maize. *Remote Sens.* **2020**, *12*, 1752. [CrossRef]
31. Zengeya, F.M.; Mutanga, O.; Murwira, A. Linking remotely sensed forage quality estimates from WorldView-2 multispectral data with cattle distribution in a savanna landscape. *Int. J. Appl. Earth Obs. Geoinf.* **2013**, *21*, 513–524. [CrossRef]
32. Li, W.; Dou, Z.; Cui, L.; Wang, R.; Zhao, Z.; Cui, S.; Lei, Y.; Li, J.; Zhao, X.; Zhai, X. Suitability of hyperspectral data for monitoring nitrogen and phosphorus content in constructed wetlands. *Remote Sens. Lett.* **2020**, *11*, 495–504. [CrossRef]
33. George, R.; Padalia, H.; Sinha, S.; Kumar, A.S. Evaluating sensitivity of hyperspectral indices for estimating mangrove chlorophyll in Middle Andaman Island, India. *Environ. Monit. Assess.* **2019**, *191*, 1–14. [CrossRef]
34. Luo, J.; Ma, R.; Feng, H.; Li, X. Estimating the total nitrogen concentration of reed canopy with hyperspectral measurements considering a non-uniform vertical nitrogen distribution. *Remote Sens.* **2016**, *8*, 789. [CrossRef]
35. Van Asselen, S.; Verburg, P.H.; Vermaat, J.E.; Janse, J.H. Drivers of wetland conversion: A global meta-analysis. *PLoS ONE* **2013**, *8*, 81292. [CrossRef] [PubMed]
36. Ramoelo, A.; Cho, M.A.; Mathieu, R.; Madonsela, S.; Van De Kerchove, R.; Kaszta, Z.; Wolff, E. Monitoring grass nutrients and biomass as indicators of rangeland quality and quantity using random forest modelling and WorldView-2 data. *Int. J. Appl. Earth Obs. Geoinf.* **2015**, *43*, 43–54. [CrossRef]
37. Barker, A.V.; Pilbeam, D.J. *Handbook of Plant Nutrition*; CRC Press: Boca Raton, FL, USA, 2015.
38. Chhabra, A.; Manjunath, K.; Panigrahy, S. Non-point source pollution in Indian agriculture: Estimation of nitrogen losses from rice crop using remote sensing and GIS. *Int. J. Appl. Earth Obs. Geoinf.* **2010**, *12*, 190–200. [CrossRef]
39. Gilbertson, J.K.; Kemp, J.; Van Niekerk, A. Effect of pan-sharpening multi-temporal Landsat 8 imagery for crop type differentiation using different classification techniques. *Comput. Electron. Agric.* **2017**, *134*, 151–159. [CrossRef]
40. Lee, T.-M.; Yeh, H.-C. Applying remote sensing techniques to monitor shifting wetland vegetation: A case study of Danshui River estuary mangrove communities, Taiwan. *Ecol. Eng.* **2009**, *35*, 487–496. [CrossRef]

41. Held, A.; Ticehurst, C.; Lymburner, L.; Williams, N. High resolution mapping of tropical mangrove ecosystems using hyperspectral and radar remote sensing. *Int. J. Remote Sens.* **2003**, *24*, 2739–2759. [CrossRef]
42. Wood, A.; Dixon, A.; McCartney, M. People-centred wetland management. In *Wetland Management and Sustainable Livelihoods in Africa*; Routledge: Oxon, UK, 2013; pp. 23–64.
43. Ollis, D.; Snaddon, K.; Job, N.; Mbona, N. *Classification System for Wetlands and Other Aquatic Ecosystems in South Africa*; SANBI: Pretoria, South Africa, 2013.
44. Fennessy, M.; Mack, J.; Rokosch, A.; Knapp, M.; Micacchion, M. *Biogeochemical and Hydrological Investigations of Natural and Mitigation Wetlands*; Ohio EPA Technical Report WET/2004-5; Ohio EPA: Columbus, OH, USA, 2004.
45. Dube, T.; Mutanga, O.; Ismail, R. Quantifying aboveground biomass in African environments: A review of the trade-offs between sensor estimation accuracy and costs. *Trop. Ecol.* **2016**, *57*, 393–405.
46. Sibanda, M.; Mutanga, O.; Rouget, M. Examining the potential of Sentinel-2 MSI spectral resolution in quantifying above ground biomass across different fertilizer treatments. *ISPRS J. Photogramm. Remote Sens.* **2015**, *110*, 55–65. [CrossRef]
47. Otunga, C.; Odindi, J.; Mutanga, O.; Adjorlolo, C. Evaluating the potential of the red edge channel for C3 (*Festuca* spp.) grass discrimination using Sentinel-2 and Rapid Eye satellite image data. *Geocarto Int.* **2019**, *34*, 1123–1143. [CrossRef]
48. Chemura, A.; Mutanga, O.; Odindi, J.; Kutywayo, D. Mapping spatial variability of foliar nitrogen in coffee (*Coffea arabica* L.) plantations with multispectral Sentinel-2 MSI data. *ISPRS J. Photogramm. Remote Sens.* **2018**, *138*, 1–11. [CrossRef]
49. Clevers, J.G.; Kooistra, L.; van den Brande, M.M. Using Sentinel-2 Data for Retrieving LAI and Leaf and Canopy Chlorophyll Content of a Potato Crop. *Remote Sens.* **2017**, *9*, 405. [CrossRef]
50. Gao, J.; Liu, J.; Liang, T.; Hou, M.; Ge, J.; Feng, Q.; Wu, C.; Li, W. Mapping the Forage Nitrogen-Phosphorus Ratio Based on Sentinel-2 MSI Data and a Random Forest Algorithm in an Alpine Grassland Ecosystem of the Tibetan Plateau. *Remote Sens.* **2020**, *12*, 2929. [CrossRef]
51. Frampton, W.J.; Dash, J.; Watmough, G.; Milton, E.J. Evaluating the capabilities of Sentinel-2 for quantitative estimation of biophysical variables in vegetation. *ISPRS J. Photogramm. Remote Sens.* **2013**, *82*, 83–92. [CrossRef]
52. Hernández-Clemente, R.; Hornero, A.; Mottus, M.; Penuelas, J.; González-Dugo, V.; Jiménez, J.; Suárez, L.; Alonso, L.; Zarco-Tejada, P.J. Early diagnosis of vegetation health from high-resolution hyperspectral and thermal imagery: Lessons learned from empirical relationships and radiative transfer modelling. *Curr. For. Rep.* **2019**, *5*, 169–183. [CrossRef]
53. Pham, T.D.; Yokoya, N.; Bui, D.T.; Yoshino, K.; Friess, D.A. Remote sensing approaches for monitoring mangrove species, structure, and biomass: Opportunities and challenges. *Remote Sens.* **2019**, *11*, 230. [CrossRef]
54. Wu, L.; Zhu, X.; Lawes, R.; Dunkerley, D.; Zhang, H. Comparison of machine learning algorithms for classification of LiDAR points for characterization of canola canopy structure. *Int. J. Remote Sens.* **2019**, *40*, 5973–5991. [CrossRef]
55. Wolanin, A.; Camps-Valls, G.; Gómez-Chova, L.; Mateo-García, G.; van der Tol, C.; Zhang, Y.; Guanter, L. Estimating crop primary productivity with Sentinel-2 and Landsat 8 using machine learning methods trained with radiative transfer simulations. *Remote Sens. Environ.* **2019**, *225*, 441–457. [CrossRef]
56. Prado Osco, L.; Marques Ramos, A.P.; Roberto Pereira, D.; Moriya, É.A.S.; Nobuhiro Imai, N.; Takashi Matsubara, E.; Estrabis, N.; de Souza, M.; Marcato Junior, J.; Gonçalves, W.N. Predicting Canopy Nitrogen Content in Citrus-Trees Using Random Forest Algorithm Associated to Spectral Vegetation Indices from UAV-Imagery. *Remote Sens.* **2019**, *11*, 2925. [CrossRef]
57. Liang, L.; Di, L.; Huang, T.; Wang, J.; Lin, L.; Wang, L.; Yang, M. Estimation of leaf nitrogen content in wheat using new hyperspectral indices and a random forest regression algorithm. *Remote Sens.* **2018**, *10*, 1940. [CrossRef]
58. Polikar, R. Ensemble based systems in decision making. *IEEE Circuits Syst. Mag.* **2006**, *6*, 21–45. [CrossRef]
59. Breiman, L. Random forests. *Mach. Learn.* **2001**, *45*, 5–32. [CrossRef]
60. Burai, P.; Deák, B.; Valkó, O.; Tomor, T. Classification of herbaceous vegetation using airborne hyperspectral imagery. *Remote Sens.* **2015**, *7*, 2046–2066. [CrossRef]
61. Doktor, D.; Lausch, A.; Spengler, D.; Thurner, M. Extraction of plant physiological status from hyperspectral signatures using machine learning methods. *Remote Sens.* **2014**, *6*, 12247–12274. [CrossRef]
62. Ghimire, B.; Rogan, J.; Miller, J. Contextual land-cover classification: Incorporating spatial dependence in land-cover classification models using random forests and the Getis statistic. *Remote Sens. Lett.* **2010**, *1*, 45–54. [CrossRef]
63. Grimm, R.; Behrens, T.; Märker, M.; Elsenbeer, H. Soil organic carbon concentrations and stocks on Barro Colorado Island—Digital soil mapping using Random Forests analysis. *Geoderma* **2008**, *146*, 102–113. [CrossRef]
64. Prasad, A.M.; Iverson, L.R.; Liaw, A. Newer classification and regression tree techniques: Bagging and random forests for ecological prediction. *Ecosystems* **2006**, *9*, 181–199. [CrossRef]
65. Adelabu, S.; Mutanga, O.; Adam, E.; Cho, M.A. Exploiting machine learning algorithms for tree species classification in a semiarid woodland using RapidEye image. *J. Appl. Remote Sens.* **2013**, *7*, 073480. [CrossRef]
66. Garden, S. Wetland Geomorphology and Floodplain Dynamics on the Hydrologically Variable Mfolozi River, KwaZulu-Natal, South Africa. Ph.D. Thesis, University of KwaZulu-Natal, Durban, South Africa, 2008.
67. Morgenthal, T.; Kellner, K.; Van Rensburg, L.; Newby, T.; Van der Merwe, J. Vegetation and habitat types of the Umkhanyakude Node. *S. Afr. J. Bot.* **2006**, *72*, 1–10. [CrossRef]
68. Schulze, R.E. *South African Atlas of Agrohydrology and Climatology*; Water Research Commission: Pretoria, South Africa, 1997.
69. Öztürk, Ö.; Ceyhan, E.; Önder, M.; Harmankaya, M.; Hamurcu, M.; Gezgin, S. Micronutrient contents in leaves of sunflower cultivars grown with different boron doses. *Helia* **2010**, *33*, 215–220. [CrossRef]

70. Gardner, R.C.; Finlayson, M. *Global Wetland Outlook: State of the World's Wetlands and their Services to People*; Ramsar Convention: Gland, Switzerland, 2018.
71. Muller-Wilm, U.; Louis, J.; Richter, R.; Gascon, F.; Niezette, M. Sentinel-2 level 2A prototype processor: Architecture, algorithms and first results. In Proceedings of the ESA Living Planet Symposium, Edinburgh, UK, 9–13 September 2013; pp. 9–13.
72. Farahmand, N.; Sadeghi, V. Estimating Soil Salinity in the Dried Lake Bed of Urmia Lake Using Optical Sentinel-2 Images and Nonlinear Regression Models. *J. Indian Soc. Remote Sens.* **2020**, *48*, 1–13. [CrossRef]
73. Van Passel, J.; De Keersmaecker, W.; Somers, B. Monitoring woody cover dynamics in tropical dry forest ecosystems using sentinel-2 satellite imagery. *Remote Sens.* **2020**, *12*, 1276. [CrossRef]
74. Jordan, C.F. Derivation of leaf-area index from quality of light on the forest floor. *Ecology* **1969**, *50*, 663–666. [CrossRef]
75. Gitelson, A.; Merzlyak, M.N. Spectral reflectance changes associated with autumn senescence of *Aesculus hippocastanum* L. and Acer platanoides L. leaves. Spectral features and relation to chlorophyll estimation. *J. Plant Physiol.* **1994**, *143*, 286–292. [CrossRef]
76. Barnes, E.; Clarke, T.; Richards, S.; Colaizzi, P.; Haberland, J.; Kostrzewski, M.; Waller, P.; Choi, C.; Riley, E.; Thompson, T. Coincident detection of crop water stress, nitrogen status and canopy density using ground based multispectral data. In Proceedings of the Fifth International Conference on Precision Agriculture, Bloomington, MN, USA, 16–19 July 2000.
77. Gitelson, A.A.; Gritz, Y.; Merzlyak, M.N. Relationships between leaf chlorophyll content and spectral reflectance and algorithms for non-destructive chlorophyll assessment in higher plant leaves. *J. Plant Physiol.* **2003**, *160*, 271–282. [CrossRef]
78. Gitelson, A.A.; Keydan, G.P.; Merzlyak, M.N. Three-band model for noninvasive estimation of chlorophyll, carotenoids, and anthocyanin contents in higher plant leaves. *Geophys. Res. Lett.* **2006**, *33*. [CrossRef]
79. Dash, J.; Curran, P.J.; Tallis, M.J.; Llewellyn, G.M.; Taylor, G.; Snoeij, P. Validating the MERIS Terrestrial Chlorophyll Index (MTCI) with ground chlorophyll content data at MERIS spatial resolution. *Remote Sens.* **2010**, *31*, 5513–5532. [CrossRef]
80. Xue, J.; Su, B. Significant Remote Sensing Vegetation Indices: A Review of Developments and Applications. *J. Sens.* **2017**, *2017*, 1353691. [CrossRef]
81. Aklilu Tesfaye, A.; Gessesse Awoke, B. Evaluation of the saturation property of vegetation indices derived from sentinel-2 in mixed crop-forest ecosystem. *Spat. Inf. Res.* **2021**, *29*, 109–121. [CrossRef]
82. Zhao, D.; Huang, L.; Li, J.; Qi, J. A comparative analysis of broadband and narrowband derived vegetation indices in predicting LAI and CCD of a cotton canopy. *ISPRS J. Photogramm. Remote Sens.* **2007**, *62*, 25–33. [CrossRef]
83. Mahajan, G.R.; Das, B.; Murgaokar, D.; Herrmann, I.; Berger, K.; Sahoo, R.N.; Patel, K.; Desai, A.; Morajkar, S.; Kulkarni, R.M. Monitoring the Foliar Nutrients Status of Mango Using Spectroscopy-Based Spectral Indices and PLSR-Combined Machine Learning Models. *Remote Sens.* **2021**, *13*, 641. [CrossRef]
84. Pullanagari, R.R.; Kereszturi, G.; Yule, I. Integrating airborne hyperspectral, topographic, and soil data for estimating pasture quality using recursive feature elimination with random forest regression. *Remote Sens.* **2018**, *10*, 1117. [CrossRef]
85. Yang, H.; Li, F.; Wang, W.; Yu, K. Estimating Above-Ground Biomass of Potato Using Random Forest and Optimized Hyperspectral Indices. *Remote Sens.* **2021**, *13*, 2339.
86. Thenkabail, P.S.; Lyon, J.G.; Huete, A. *Hyperspectral Remote Sensing of Vegetation*; CRC Press: Boca Raton, FL, USA, 2011.
87. Ishwaran, H. The effect of splitting on random forests. *Mach. Learn.* **2015**, *99*, 75–118. [CrossRef]
88. Ok, A.O.; Akar, O.; Gungor, O. Evaluation of random forest method for agricultural crop classification. *Eur. J. Remote Sens.* **2012**, *45*, 421–432. [CrossRef]
89. Sandri, M.; Zuccolotto, P. Variable selection using random forests. In *Data Analysis, Classification and the Forward Search*; Springer: Berlin, Germany, 2006; pp. 263–270.
90. Hapfelmeier, A.; Ulm, K. A new variable selection approach using random forests. *Comput. Stat. Data Anal.* **2013**, *60*, 50–69. [CrossRef]
91. Grömping, U. Variable importance assessment in regression: Linear regression versus random forest. *Am. Stat.* **2009**, *63*, 308–319. [CrossRef]
92. Otgonbayar, M.; Atzberger, C.; Chambers, J.; Damdinsuren, A. Mapping pasture biomass in Mongolia using Partial Least Squares, Random Forest regression and Landsat 8 imagery. *Int. J. Remote Sens.* **2019**, *40*, 3204–3226. [CrossRef]
93. Shcherbakov, M.V.; Brebels, A.; Shcherbakova, N.L.; Tyukov, A.P.; Janovsky, T.A.; Kamaev, V.A.e. A survey of forecast error measures. *World Appl. Sci. J.* **2013**, *24*, 171–176.
94. Richter, K.; Hank, T.B.; Mauser, W.; Atzberger, C. Derivation of biophysical variables from Earth observation data: Validation and statistical measures. *J. Appl. Remote Sens.* **2012**, *6*, 063557. [CrossRef]
95. Karl, J.W.; McCord, S.E.; Hadley, B.C. A comparison of cover calculation techniques for relating point-intercept vegetation sampling to remote sensing imagery. *Ecol. Indic.* **2017**, *73*, 156–165. [CrossRef]
96. Chen, J.; Zhu, W.; Tian, Y.Q.; Yu, Q.; Zheng, Y.; Huang, L. Remote estimation of colored dissolved organic matter and chlorophyll-a in Lake Huron using Sentinel-2 measurements. *J. Appl. Remote Sens.* **2017**, *11*, 036007. [CrossRef]
97. Osco, L.P.; Ramos, A.P.M.; Faita Pinheiro, M.M.; Moriya, É.A.S.; Imai, N.N.; Estrabis, N.; Ianczyk, F.; Araújo, F.F.d.; Liesenberg, V.; Jorge, L.A.d.C. A Machine Learning Framework to Predict Nutrient Content in Valencia-Orange Leaf Hyperspectral Measurements. *Remote Sens.* **2020**, *12*, 906. [CrossRef]
98. Shah, S.H.; Angel, Y.; Houborg, R.; Ali, S.; McCabe, M.F. A random forest machine learning approach for the retrieval of leaf chlorophyll content in wheat. *Remote Sens.* **2019**, *11*, 920. [CrossRef]

99. Huang, S.; Miao, Y.; Yuan, F.; Gnyp, M.L.; Yao, Y.; Cao, Q.; Wang, H.; Lenz-Wiedemann, V.I.; Bareth, G. Potential of RapidEye and WorldView-2 satellite data for improving rice nitrogen status monitoring at different growth stages. *Remote Sens.* **2017**, *9*, 227. [CrossRef]
100. Kalacska, M.; Lalonde, M.; Moore, T. Estimation of foliar chlorophyll and nitrogen content in an ombrotrophic bog from hyperspectral data: Scaling from leaf to image. *Remote Sens. Environ.* **2015**, *169*, 270–279. [CrossRef]
101. Munyati, C.; Balzter, H.; Economon, E. Correlating Sentinel-2 MSI-derived vegetation indices with in-situ reflectance and tissue macronutrients in savannah grass. *Int. J. Remote Sens.* **2020**, *41*, 3820–3844. [CrossRef]
102. Lepine, L.C.; Ollinger, S.V.; Ouimette, A.P.; Martin, M.E. Examining spectral reflectance features related to foliar nitrogen in forests: Implications for broad-scale nitrogen mapping. *Remote Sens. Environ.* **2016**, *173*, 174–186. [CrossRef]
103. Wang, Z.; Wang, T.; Darvishzadeh, R.; Skidmore, A.K.; Jones, S.; Suarez, L.; Woodgate, W.; Heiden, U.; Heurich, M.; Hearne, J. Vegetation indices for mapping canopy foliar nitrogen in a mixed temperate forest. *Remote Sens.* **2016**, *8*, 491. [CrossRef]
104. Abdel-Rahman, E.M.; Ahmed, F.B.; Ismail, R. Random forest regression and spectral band selection for estimating sugarcane leaf nitrogen concentration using EO-1 Hyperion hyperspectral data. *Int. J. Remote Sens.* **2013**, *34*, 712–728. [CrossRef]
105. Ramoelo, A.; Cho, M.; Mathieu, R.; Skidmore, A.K. Potential of Sentinel-2 spectral configuration to assess rangeland quality. *J. Appl. Remote Sens.* **2015**, *9*, 094096. [CrossRef]
106. Raper, T.; Varco, J. Canopy-scale wavelength and vegetative index sensitivities to cotton growth parameters and nitrogen status. *Precis. Agric.* **2015**, *16*, 62–76. [CrossRef]
107. Schlemmer, M.; Gitelson, A.; Schepers, J.; Ferguson, R.; Peng, Y.; Shanahan, J.; Rundquist, D. Remote estimation of nitrogen and chlorophyll contents in maize at leaf and canopy levels. *Int. J. Appl. Earth Obs. Geoinf.* **2013**, *25*, 47–54. [CrossRef]
108. Marschner, H. *Mineral Nutrition of Higher Plants*, 2nd ed.; Academic Press: Cambirdge, MA, USA, 1995.
109. Ferwerda, J.G.; Skidmore, A.K. Can nutrient status of four woody plant species be predicted using field spectrometry? *ISPRS J. Photogramm. Remote Sens.* **2007**, *62*, 406–414. [CrossRef]
110. Mutanga, O.; Skidmore, A.K.; Prins, H. Predicting in situ pasture quality in the Kruger National Park, South Africa, using continuum-removed absorption features. *Remote Sens. Environ.* **2004**, *89*, 393–408. [CrossRef]
111. Pinheiro Lisboa, I.; Melo Damian, J.; Roberto Cherubin, M.; Silva Barros, P.P.; Ricardo Fiorio, P.; Cerri, C.C.; Eduardo Pellegrino Cerri, C. Prediction of sugarcane yield based on NDVI and concentration of leaf-tissue nutrients in fields managed with straw removal. *Agronomy* **2018**, *8*, 196. [CrossRef]
112. Van Deventer, H.; Cho, M.A. Assessing leaf spectral properties of Phragmites australis impacted by acid mine drainage. *S. Afr. J. Sci.* **2014**, *110*, 1–12. [CrossRef]
113. Martinez, N.; Sharp, J.; Kuhne, W.; Johnson, T.; Stafford, C.; Duff, M. Assessing the use of reflectance spectroscopy in determining CsCl stress in the model species Arabidopsis thaliana. *Int. J. Remote Sens.* **2015**, *36*, 5887–5915. [CrossRef]
114. Rathod, P.H.; Brackhage, C.; Van der Meer, F.D.; Müller, I.; Noomen, M.F.; Rossiter, D.G.; Dudel, G.E. Spectral changes in the leaves of barley plant due to phytoremediation of metals—Results from a pot study. *Eur. J. Remote Sens.* **2015**, *48*, 283–302. [CrossRef]
115. Freeman, E.A.; Moisen, G.G.; Frescino, T.S. Evaluating effectiveness of down-sampling for stratified designs and unbalanced prevalence in Random Forest models of tree species distributions in Nevada. *Ecol. Model.* **2012**, *233*, 1–10. [CrossRef]
116. Freeman, E.A.; Moisen, G.G.; Coulston, J.W.; Wilson, B.T. Random forests and stochastic gradient boosting for predicting tree canopy cover: Comparing tuning processes and model performance. *Can. J. For. Res.* **2016**, *46*, 323–339. [CrossRef]
117. Cutler, D.R.; Edwards Jr, T.C.; Beard, K.H.; Cutler, A.; Hess, K.T.; Gibson, J.; Lawler, J.J. Random forests for classification in ecology. *Ecology* **2007**, *88*, 2783–2792. [CrossRef] [PubMed]
118. Verikas, A.; Gelzinis, A.; Bacauskiene, M. Mining data with random forests: A survey and results of new tests. *Pattern Recognit.* **2011**, *44*, 330–349. [CrossRef]
119. Gama, M.J.; Cho, M.A.; Chirwa, P.; Masemola, C. Estimating mineral content of indigenous browse species using laboratory spectroscopy and Sentinel-2 imagery. *Int. J. Appl. Earth Obs. Geoinf.* **2019**, *75*, 141–150. [CrossRef]
120. Lu, Q.; Bai, J.; Gao, Z.; Wang, J.; Zhao, Q. Effects of Water Level and Salinity on Total Sulfur Contents in Salt Marsh Soils of the Yellow River Delta, China. *Wetlands* **2016**, *36*, 137–143. [CrossRef]
121. Olivie-Lauquet, G.; Gruau, G.; Dia, A.; Riou, C.; Jaffrezic, A.; Henin, O. Release of trace elements in wetlands: Role of seasonal variability. *Water Res.* **2001**, *35*, 943–952. [CrossRef]
122. Ribeiro, B.T.; Nascimento, D.C.; Curi, N.; Guilherme, L.R.G.; Costa, E.T.d.S.; Lopes, G.; Carneiro, J.P. Assessment of Trace Element Contents in Soils and Water from Cerrado Wetlands, Triângulo Mineiro Region. *Rev. Bras. de Ciência do Solo* **2019**, *43*, 0180059. [CrossRef]
123. Ramoelo, A.; Skidmore, A.K.; Cho, M.A.; Schlerf, M.; Mathieu, R.; Heitkönig, I.M. Regional estimation of savanna grass nitrogen using the red-edge band of the spaceborne RapidEye sensor. *Int. J. Appl. Earth Obs. Geoinf.* **2012**, *19*, 151–162. [CrossRef]
124. Cho, M.A.; Ramoelo, A.; Debba, P.; Mutanga, O.; Mathieu, R.; Van Deventer, H.; Ndlovu, N. Assessing the effects of subtropical forest fragmentation on leaf nitrogen distribution using remote sensing data. *Landsc. Ecol.* **2013**, *28*, 1479–1491. [CrossRef]
125. Skidmore, A.K.; Ferwerda, J.G.; Mutanga, O.; Van Wieren, S.E.; Peel, M.; Grant, R.C.; Prins, H.H.; Balcik, F.B.; Venus, V. Forage quality of savannas—Simultaneously mapping foliar protein and polyphenols for trees and grass using hyperspectral imagery. *Remote Sens. Environ.* **2010**, *114*, 64–72. [CrossRef]

126. Oliveira, L.F.R.d.; Santana, R.C.; Oliveira, M.L.R.d. Nondestructive estimation of leaf nutrient concentrations in Eucalyptus plantations. *Cerne* **2019**, *25*, 184–194. [CrossRef]
127. Mahajan, G.; Sahoo, R.; Pandey, R.; Gupta, V.; Kumar, D. Using hyperspectral remote sensing techniques to monitor nitrogen, phosphorus, sulphur and potassium in wheat (*Triticum aestivum* L.). *Precis. Agric.* **2014**, *15*, 499–522. [CrossRef]
128. Mutanga, O.; Dube, T.; Ahmed, F. Progress in remote sensing: Vegetation monitoring in South Africa. *S. Afr. Geogr. J.* **2016**, *98*, 461–471. [CrossRef]

remote sensing

MDPI

Article

Estimating Evapotranspiration of Mediterranean Oak Savanna at Multiple Temporal and Spatial Resolutions. Implications for Water Resources Management

Elisabet Carpintero [1], Martha C. Anderson [2], Ana Andreu [1], Christopher Hain [3], Feng Gao [2], William P. Kustas [2] and María P. González-Dugo [1,*]

1 Andalusian Institute for Research and Training in Agriculture and Fischeries, IFAPA, Centro Alameda del Obispo, Avd. M. Pidal s/n, 14071 Córdoba, Spain; elisabet.carpintero@juntadeandalucia.es (E.C.); anandreumz@gmail.com (A.A.)
2 Hydrology and Remote Sensing Laboratory, USDA ARS, Bldg. 007, BARC West, Beltsville, MD 20705, USA; Martha.Anderson@ARS.USDA.GOV (M.C.A.); Feng.Gao@ARS.USDA.GOV (F.G.); Bill.Kustas@ARS.USDA.GOV (W.P.K.)
3 NASA Marshall Space Flight Center, Huntsville, AL 35805, USA; christopher.hain@nasa.gov
* Correspondence: mariap.gonzalez.d@juntadeandalucia.es; Tel.: +34-671-532-700

Abstract: Mediterranean oak savanna is composed of a mixture of scattered oak trees, crops, pasture, and shrubs. It is the most widespread agroforestry landscape in Europe, and its conservation faces multiple threats including water scarcity, which has been exacerbated by global warming and greater climate variability. Evapotranspiration (ET) can be used as a proxy of the vegetation water status and response to water shortage conditions, providing relevant information about the ecosystem stability and its hydrological dynamics. This study evaluates a framework to estimate ET at multiple spatial and temporal scales and applies it to the monitoring of the oak savanna vegetation water consumption for the years 2013–2015. We used a remote sensing-based energy balance model (ALEXI/DisALEXI approach), and the STARFM data fusion technique to provide daily ET estimates at 30 m resolution. The results showed that modeled energy balance components compared well to ground measurements collected by an eddy covariance system, with root mean square error (RMSE) values ranging between 0.60 and 2.18 MJ m^{-2} d^{-1}, depending on the sensor dataset (MODIS or Landsat) and the flux. The daily 30 m ET series generated by STARFM presented an RMSE value of 0.67 mm d^{-1}, which yielded a slight improvement compared to using MODIS resolution or more simple interpolation approaches with Landsat. However, the major advantage of the high spatio-temporal resolution was found in the analysis of ET dynamics over different vegetation patches that shape the landscape structure and create different microclimates. Fine-scale ET maps (30 m, daily) provide key information difficult to detect at a coarser spatial resolution over heterogeneous landscapes and may assist management decisions at the field and farm scale.

Keywords: evapotranspiration (ET) monitoring; *dehesa*; surface energy balance; spatio-temporal resolution; management practices

1. Introduction

In water-controlled ecosystems with limited water resources, soil moisture dynamics play a central role in the existence and spatial distribution of the different vegetation functional types [1,2]. The discontinuities in the functioning of these ecosystems are related to an alternation in the dry and wet periods [3]. A feedback relationship is observed in these environments, where the vegetation water consumption strongly conditions the hydrological balance of the system, while plants are impacted by water stress situations.

Higher variability in climate patterns coupled with an increase in water demand due to anthropogenic factors is expected to cause a considerable reduction in the quality and quantity of freshwater resources in the Mediterranean basin, intensifying recurrent water scarcity problems [4,5]. Climate forecasting models suggest that these regions will be subject to hotter and drier summers, with a higher occurrence of heatwaves and an increase in the frequency and severity of droughts [6,7]. Aridity and desertification processes may lead to the abandonment of farmland and grazing areas and therefore, to land-use changes [8–10]. In this scenario of increased variability, accurate modeling of hydrological processes and their effects on the vegetation could assist decision-making when implementing government policies at different levels (local, regional, national, etc.), in order to mitigate global warming impact and reduce ecosystems' vulnerability.

Evapotranspiration (ET), or the water consumed by the canopy and evaporated from the soil, is a key process in semi-arid landscapes. ET functions as a proxy that reveals the vegetation water status, responding to situations of water scarcity. Moreover, regular monitoring of water consumption provides information about the system's stability and its hydrological dynamics. The use of thermal infrared (TIR) observations from satellite sensors in surface energy balance models to derive ET—as latent heat flux, the available energy for the ET process—has been widely validated over homogeneous crops and sparse vegetation covers [11–15], and more heterogeneous landscapes [16,17].

Some energy balance-based modeling approaches, such as the TSEB (Two Source Energy Balance) [18] and SEBS (Surface Energy Balance System) [19] models, have been applied with good results over savanna-type landscapes, at different spatial and time scales, with monitoring purposes [20–23]. The TSEB model proposes a separate estimation of the surface energy fluxes over the soil and the vegetation, based on the disaggregation of the radiometric surface temperature measured by the thermal sensors between these surface components. For large-area modeling, there are approaches for implementing the TSEB that have the advantage of not requiring any local observations as input, including the ALEXI model (Atmosphere-Land Exchange Inverse) [24,25] and the associated flux disaggregation scheme named DisALEXI [26].

The European oak savanna landscape (known as *dehesa* in Spain and *montado* in Portugal) is the most widespread agroforestry system in the continent, occupying more than three million hectares [27] in the Iberian Peninsula. It has a canopy structure composed of a mosaic of scattered oak trees (mostly *Quercus Ilex* L. and *Quercus Suber* L.), crops, pasture, and shrubs. *Dehesas* provide multiple socio-economic uses (livestock, agriculture, hunting, etc.) with an essential role in the economy of rural areas [28]. They also are biodiversity reservoirs, listed in the EU directive as a habitat with community-wide interest [29]. Despite its importance, in recent decades numerous threats are endangering the *dehesa* conservation, such as the lack of natural regeneration of tree species [30], profitability problems, or soil degradation mainly caused by the intensification of agricultural and livestock activities [27,31]. Climate change is worsening this fragile environment.

The multiple vegetation layers that compose this typical Mediterranean ecosystem vary in phenology, physiology, and function, each one contributing in a different way to the turbulent exchanges and the radiative transfer budget [32], with impact on the local microclimate and hydrology. For instance, most tree species are evergreen and tolerate high irradiance while limiting carbon assimilation either by stomatal closure or a decrease of photosynthetic capacity, which are adaptation mechanisms to cope with the water scarcity conditions [33,34]. Meanwhile, the underlayer of annual species emerges after the first rainfalls in autumn and dries up during the summer.

Although the *dehesa* can be considered a homogeneous system on a regional scale, differences in tree/grass/crops/scrub and bare soil cover fractions can be observed at finer scales. The combined differential functioning and characteristics of the different patches of vegetation affect the biogeochemical flux dynamics, resulting in a high spatial and temporal variability and creating distinct intra-ecosystem microclimates [35,36]. This structure plays an important role in *dehesas'* resilience, making the system an efficient convector of sensible heat and keeping the canopy surface temperature within the survival range [32].

Within this context, the monitoring of ET with high resolution in both time and space can help to assess landscape structure and to perform a disaggregated evaluation of areas mostly covered by grasslands, and areas with higher tree or bush coverage. The analysis of the spatial and temporal variations in the hydrological conditions controlling the production of pastures and acorns, the two primary sources of livestock feed in the *dehesa*, can provide information critical for adjusting management practices at the farm scale, with an important impact on the ecosystem conservation and profitability.

Space agencies and companies are making major efforts to distribute high-resolution products, but the current satellite programs present limitations in offering thermal information with high spatial and temporal resolution (e.g., 30 m/daily) simultaneously. An additional problem is the presence of clouds that often result in longer periods without useful images. Multi-scale remotely sensed data fusion techniques are a viable alternative to improve the spatio-temporal resolution of ET estimates derived from thermal-based modeling. The Spatial and Temporal Adaptive Reflectance Fusion Model, STARFM, was developed by Gao et al. [37] to initially integrate surface reflectance data from multiple sensors [38,39], but more recently it has been applied to fuse ET retrievals with satisfactory results in homogeneous agricultural and forested/mosaic areas [16,17,40–43]. These studies demonstrated that data fusion modeling improved the accuracy in the estimation of the water consumed by the vegetation, identifying small areas with deficits or excessive irrigation.

This research monitors the vegetation water use in the oak-grass savanna landscape of a small Mediterranean watershed of Southern Spain at different spatial and temporal scales. The specific objectives of the study were:

(i) To evaluate the utility of a surface energy balance model (ALEXI/DisALEXI) and the STARFM data fusion technique, using multiple remote sensing platforms (Landsat 7/8 and MODIS), to estimate high-resolution ET in time and space over the complex canopy structure of Mediterranean oak savannas.

(ii) To analyze the opportunities offered by this high-resolution product to provide information that is useful to improve the water and vegetation management of this agroforestry system at a field scale. To do that, we evaluated the water use patterns of the herbaceous stratum and other small heterogeneous vegetation patches typical of the *dehesa* (e.g., scrubs, humid areas, creek shore), which shape the landscape structure and reflect the existence of different micro-ecosystems and climates. Finally, the cumulative monthly ET generated by the different approaches with different spatial resolutions (1 km and 30 m) was quantified over the same vegetation patches.

2. Materials and Methods

2.1. Description of the Study Area and Experimental Site

The study was conducted over the *Martin Gonzalo* watershed (48.4 km^2), part of the larger Guadalquivir River basin located in Southern Spain (Figure 1a,c). The elevation range varies from 760 m in the north to 280 m at the outlet of the watershed, where there is a dam. The continental Mediterranean climate of this area is highly seasonal, with moderately cold winters alternating with hot and long dry summers. The rainfall presents intra- and inter-year variability, with an annual average of 895 mm (from 1990 to 2015) concentrated during spring and fall.

The landscape is mainly occupied by homogeneous *dehesa*, along with some conifer forests and olive groves. *Dehesa* is a Mediterranean oak savanna, whose canopy structure is composed of sparse clumped trees and a grassland or crops understory. In this ecosystem, extensive livestock production fed with acorns and grass is the main economic activity, but they also provide other uses, such as cork, cereal production, hunting, mushroom harvesting, and beekeeping. They offer as well multiple environmental services, such as biodiversity hotspots, water provisioning, CO_2 fixation and high diversity of habitats [28,44].

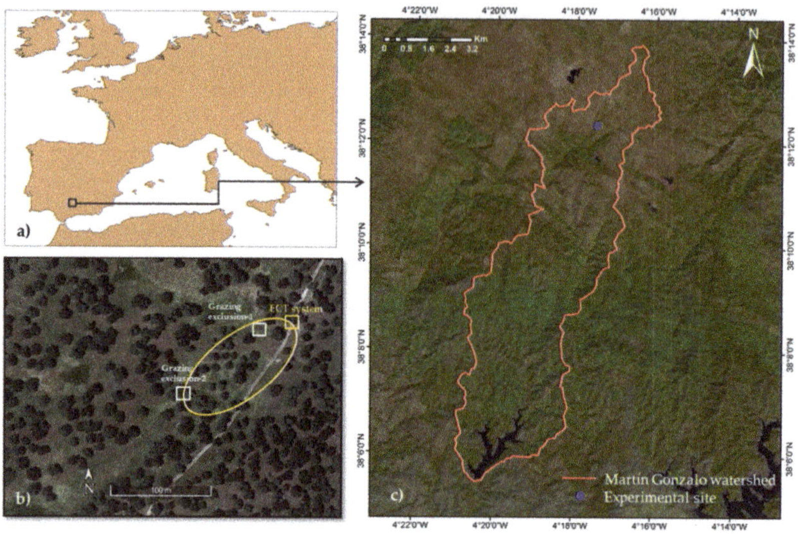

Figure 1. (**a**,**c**) Location of the study area and experimental site; (**b**) Aerial photograph of validation site showing the distribution of the three fields discussed in the text. ECT: eddy covariance tower. The orange ellipse represents the predominant fetch of the ECT.

Ground validation measurements were taken at an experimental site (*Santa Clotilde*, 38°12′ N, 4°17′ W, 736 m a.s.l, Figure 1c), located in a *dehesa* farm within the studied watershed. The setup included an eddy covariance tower (ECT) over the combined tree + grassland system and two grazing exclusion enclosures (over open grassland and under an oak tree respectively) to take into account the heterogeneity of the area (Figure 1b). All energy balance components: the turbulent fluxes of sensible heat (H) and latent heat (LE), net radiation (Rn) and the heat flux transport across the surface soil (G), were measured continuously.

The ECT system was installed on a tower at 18 m above ground level in April 2012, registering the ecosystem response as a whole (Figure 2a). The system included a 3D sonic anemometer (model CSAT3, Campbell Scientific Inc. Mention of trade names or commercial products in this publication is solely for the purpose of providing specific information and does not imply recommendation or endorsement by the U.S. Department of Agriculture) that measures horizontal and vertical fluctuations of temperature and wind speed, and a hygrometer (model KH20, Campbell Scientific Inc.) to estimate water vapor fluctuations, the latter being replaced in 2015 by an open Path CO_2/H_2O Gas Analyzer LICOR-7500, taking simultaneous measurements of carbon dioxide and water vapor in turbulent air structures (Figure 2b). These eddy covariance measurements were recorded with a frequency of 10 Hz and corrected for density effects due to heat and water vapor transfer [45].

A relative humidity and air temperature probe (model HMP155, Vaisala) and a net radiometer (model NR-Lite, Kipp&Zonen, Delft, Netherlands) to obtain the net radiation were also installed at the tower. In 2015, the net radiometer was replaced by a four-component-net-radiation sensor (model NR01, Hukseflux Thermal Sensors, Delft, Netherlands) (Figure 2b). More information on the experimental site and the equipment can be found in Andreu et al. [20] and Carpintero et al. [46].

The quality of the energy balance fluxes measured with the ECT was tested by Andreu et al. [20] for the 2012 summer season, resulting in an average closure balance (Rn − G = LE + H) of 86% of available energy. For the period 2014–2015, the closure balance was 91%. For the comparison with modeled LE, the latent heat fluxes obtained by forcing the closure of the energy balance with the residual method was preferred to direct

LE measurement. This method assumes that the sensible heat, H, is correctly measured and LE is obtained by solving the surface energy balance equation. Daytime-integrated energy fluxes were calculated by averaging half-hourly values focusing on daytime fluxes and not considering overnight fluxes.

Figure 2. (**a**) Eddy covariance tower (ECT) over the combined tree + grassland system; (**b**) equipment installed on the tower; (**c**) Grazing exclusion enclosure over open grassland.

2.2. Modeling Framework

Figure 3 shows the flowchart of the modeling framework applied for the period 2013–2015. The process starts with a global ET product (5 km, daily) developed using the ALEXI model based on MODIS day-night temperature differences [47]. In the second step, the ALEXI ET fluxes are disaggregated to higher resolutions over the study area, using the flux disaggregation scheme DisALEXI applied to both MODIS (1 km, daily) and Landsat 7/8 (60–100 m, 16 days) images. Finally, both types of ET maps, the infrequent Landsat at 30 m resolution and the daily images at MODIS-scale generated by DisALEXI, are combined using the data fusion technique STARFM to provide ET estimates with both fine spatial (30 m) and temporal (daily) resolution.

2.2.1. ALEXI/DisALEXI Model

The Atmosphere-Land Exchange Inverse (ALEXI) model [24,25] and the associated flux disaggregation technique (DisALEXI) [26] are based on the Two Source Energy Balance (TSEB) model.

The TSEB model partitions the surface blackbody radiance into the soil/substrate (T_S) and vegetated canopy (T_C) blackbody temperatures, being weighted by the cover fraction of each component at the sensor view angle, $f(\varphi)$, and resulting in:

$$T_{RAD}(\varphi) \approx \left(f(\varphi) T_c^4 + [1 - f(\varphi)] T_s^4 \right)^{\frac{1}{4}} \tag{1}$$

The surface energy balance is solved for the whole soil-canopy-atmosphere system, and for each individual component, with Equations (2) and (3):

$$Rn_s = H_s + LE_s + G \tag{2}$$

$$Rn_c = H_c + LE_c \quad (3)$$

where the subscripts "*s*" and "*c*" represent fluxes from the soil and canopy, respectively. Norman et al. [48] present a description of the original model, and further improvements can be found in Kustas et al. [18].

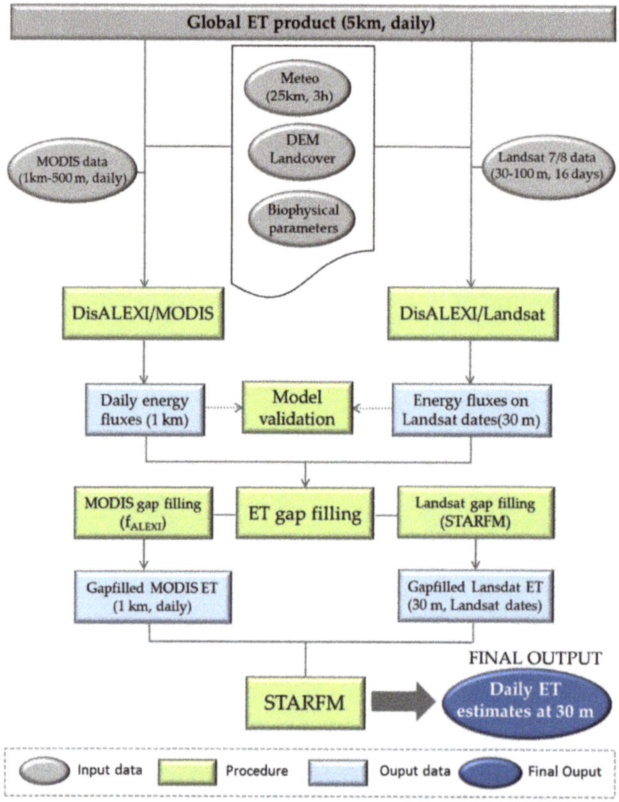

Figure 3. Flowchart of modeling framework applied to generate high resolution ET estimations.

The ALEXI model was designed to reduce the use of ancillary meteorological data while maintaining a physically realistic representation of land-atmosphere exchange over a wide range of vegetation coverages. ALEXI implements the TSEB in a time-differential mode, applying this scheme two times during the morning, using surface radiometric temperature data generally provided by geostationary satellites. In this mode, the sensibility of the model to absolute temperature biases is reduced [25]. In ALEXI, TSEB is coupled with an atmospheric boundary layer model to internally simulate the effect of land-atmosphere feedback on the near-surface air temperature [49]. One limitation of this procedure is the dependence on geostationary datasets, with different calibrations and temporal extensions. For this reason, in this research, a global ET product developed using the non-geostationary MODIS sensor day-night temperature differences has been used. As a result, ALEXI generates surface energy fluxes at the continental scale, but at the coarse spatial resolution, of several km.

The DisALEXI scheme runs the TSEB using higher resolution thermal data from polar-orbiting satellites, such as MODIS or Landsat, for mapping finer spatial resolution fluxes. It iteratively adjusts the air temperature boundary conditions such that the disaggregated daily ET flux field reaggregates up to the coarse-resolution ALEXI regional baseline. This spatial disaggregation technique facilitates consistent flux evaluations at local to continental scales [49]. This ALEXI/DisALEXI framework has been widely evaluated across the United States and Europe [49–51].

2.2.2. Remote Sensing Data Fusion Method

The STARFM model [37] is a remotely sensed data fusion algorithm that allows an improvement of the temporal representation of ET variations between clear Landsat dates. It was originally designed to blend surface reflectance data, but the fusion of ET maps has been successfully conducted over a number of land covers by Cammalleri et al. [40,41] over rain-fed and irrigated agricultural areas in the Midwestern United States and over irrigated crops in Texas, by Semmens et al. [42] and Knipper et al. [17] over California vineyards, by Anderson et al. [16] over the California Delta and by Yang et al. [43–52] over forested landscapes. It combines information at a high-temporal frequency from MODIS and high-spatial-resolution from Landsat. STARFM compares one or two Landsat/MODIS image pairs acquired on the same day to predict maps at Landsat spatial scale on other MODIS dates.

Following the weighting function (Equation (4)), STARFM model predicts ET for the central pixel of a selected moving window at generic date (t_0):

$$L\left(x_{\frac{w}{2}}, y_{\frac{w}{2}}, t_0\right) = \sum_{i=1}^{w} \sum_{j=1}^{w} \sum_{k=1}^{n} W_{ijk}\left[M\left(x_i, y_j, t_0\right) + L\left(x_i, y_j, t_k\right) - M\left(x_i, y_j, t_k\right)\right] \quad (4)$$

where w is the searching window size; ($x_{w/2}$, $y_{w/2}$) is the central pixel of this moving window; n is the number of Landsat and MODIS pairs used (in this case only one pair) and W is the weighting factor. The weighting function (W) integrates the spatial differences between the Landsat (L) and MODIS (M) images on the acquisition date t_k and also the temporal differences between MODIS images from observed and predicted dates, t_k and t_0 respectively. The prediction for the central pixel only uses spectrally (or ET) similar pixels within the searching window. Further details about the STARFM model are provided by Gao et al. [37].

2.2.3. ET Data Gap Filling

Prior to the fusion algorithm, MODIS and Landsat ET maps produced by DisALEXI were preprocessed to fill the gaps created by clouds or instrument issues (e.g., the failure of scan-line corrector in Landsat-7 ETM+). The gaps in MODIS images were filled to obtain a full daily coverage at 1 km resolution. To accomplish this, the ratio between MODIS and ALEXI ET was computed on MODIS days (f_{ALEXI} = ET MODIS/ET ALEXI), then smoothed and filled using the Savitsky–Golay method described by Sun et al. [53]. The smoothed and filled f_{ALEXI} time series maps were then multiplied by daily ALEXI ET to obtain daily MODIS ET (Figure 3). The technique used to ensure optimal spatial coverage in Landsat maps employing the STARFM model was described by Yang et al. [43] (Figure 3). It combines the gapped ET image with a prediction from STARFM on the target date using a nearby MODIS-Landsat date pair.

2.2.4. Simple ET Interpolation Methods

In parallel to the application of the STARFM framework to obtain a daily 30 m ET series, daily ET values obtained with a simpler data interpolation method were also generated for comparison purposes. It used the potential ET and MODIS ET as scaling fluxes. The objective was to evaluate the advantages of using the STARFM technique with respect to these simpler and less demanding methods. In this interpolation approach, the ratio between the Landsat ET and a daily scaling flux (potential and MODIS ET) was computed on clear dates. These ratios (FPET = Landsat ET/Potential ET and FMOD = Landsat ET/MODIS ET) were then linearly interpolated in time and fused in the daily ET estimation.

2.3. Model Input Datasets

2.3.1. Landsat Data

This work employed a set of 82 Pre-collection and Collection 1 scenes (path 200/row 33) from the Landsat 7 and 8 satellites, acquired from January 2013 to December 2015. Scenes with high cloud coverage were excluded. The Landsat surface reflectance climate data record (SR CDR) (http://espa.cr.usgs.gov/, accessed on 2 September 2017) (distributed atmospherically corrected) was used to compute the leaf area index (LAI) and albedo parameters. A simple approach proposed by Liang et al. [54] was applied to calculate albedo at a 30 m resolution using six surface reflectance bands. To estimate LAI at 30 m on Landsat overpass dates, the regression tree approach developed by Gao et al. [55] was used, which was trained with samples from the MODIS LAI products and Landsat reflectance. This method has been successfully applied to annual and woody crops [42,55]. Landsat thermal band data were corrected for atmospheric and surface emissivity effects using the atmospheric radiative transfer model MODTRAN [56], and subsequently sharpened to the shortwave bands' spatial resolution (30 m) using the Data Mining Sharpener (DMS) method [57].

2.3.2. MODIS Data

MODIS images from Collection 5 (at 1 km spatial resolution) and Collection 6 (500 m) with variable temporal resolution were also used as inputs in the DisALEXI application, including data of land surface temperature (LST; MOD11_L2), geolocation (MOD03), and LAI (MCD15A3). The radiometric temperature with a daily frequency was derived from the 5-min swath LST product, MOD11_L2. The 4-day LAI product, integrated into the TIMESAT algorithm [58], allowed for creating a smoothed and filled daily LAI time series.

As with other DisALEXI application inputs, the global albedo product from Boston University (ftp://rsftp.eeos.umb.edu/data02/Gapfilled/, accessed on 20 September 2017), in addition to the yearly and global MODIS product of land cover (MCD12Q1, 500 m) were included. The albedo data is a global gap-filled snow-free albedo product, which is temporally smoothed and composed over a global grid.

2.3.3. Meteorological Input Data and Vegetation Properties

The energy balance-based modeling framework requires several regional meteorological datasets: a series of surface atmospheric pressure, wind speed at 30 m, and air temperature and specific humidity at 2 m. These were provided every 3 h at a 25 km spatial resolution by the Climate Forecast System Reanalysis (CFSR) [59]. Hourly insolation data at a spatial resolution of 25 km were also obtained from the CFSR. As part of the regional information, a global digital elevation model at a 30 m resolution from ASTER (a product of METI and NASA) and the Andalusian land cover map 2010 were also employed.

A set of biophysical parameters associated with the vegetation in Mediterranean savanna ecosystems were used. The vegetation ground cover fraction, clumping factor and canopy height were calculated following Andreu et al. [20].

2.3.4. Input Data Filtering

In order to isolate errors caused by poor input data quality, the insolation and LST data were filtered for fluxes comparison using ground information. It is known that the quality of insolation data is the primary driver of the total energy available for the ET process [16], and TSEB is mostly driven by thermal information. Both variables were compared with the data measured at the experimental site, showing a mean relative error (mean absolute error/observed mean value; RE) of 9.4% in insolation and 14.5% in surface temperature. The days in which the RE value was higher than 25% in insolation data or higher than 50% in surface temperature were eliminated from the flux comparison analysis. Remotely sensed LST was compared with the half-hourly infrared temperature value corresponding to the time of the satellite overpass. In this case, given the variations between both values and the mismatch representativeness (IR thermometer is mostly measuring canopy—Figure 2b), the threshold considered for filtering the data is less restrictive than for radiation.

2.4. Global Remotely Sensed ET Product

The 8-day total ET provided by a product at a 500 m pixel resolution (MOD16A2, Collection 6) was used for comparison purposes. This data collection uses Mu et al.'s [60] improved algorithm. It is based on the logic of the Penman-Monteith equation which includes inputs of daily meteorological reanalysis data along with MODIS data, such as albedo, LAI, fraction of photosynthetically active radiation (FPAR) and land cover.

3. Results

3.1. Evaluation of Surface Energy Fluxes at the Flux Tower Site

The frequency distribution of the wind direction at the ECT for 2015 showed the southwest as the predominant fetch. A *footprint* analysis was performed by estimating the contribution areas to daily fluxes for 132 days of 2015 as described by Hsieh et al. [61]. An extensive description of the methodology can be found in Kustas et al. [62]. Over the period analyzed, it is observed that 78% of the system ECT fluxes were collected from an area within 130 m upwind (Figure 1b), with the maximum contribution to the measured energy fluxes at approximately 33 m upwind. According to these results, average values of a 3 × 3 grid cell (with 30 × 30 m size-cell and located to the southwest of the ECT) of the modeled fluxes were evaluated.

The relationship between the daily energy fluxes (daytime integrated fluxes) observed at the ECT and modeled by the DisALEXI application is plotted in Figure 4, using MODIS (Figure 4a) and Landsat images (Figure 4b). The comparison was made on days in which both in-situ measurements and input information for the model were available, with a larger set of usable field data corresponding to 2014 and 2015. In addition, 22 days were filtered from the comparison with DisALEXI/MODIS and 5 days for DisALEXI/Landsat, due to the lack of input data quality (for further information about the input data filtering, see Section 2.3.4). A total of 313 days were used in the comparison for MODIS data and 44 images for Landsat. In all three years, there was a general shortage of clear images and/or field measurements during the last part of the winter and early spring. This is the rainy season for this region and the absence of available data hindered incorporating enough observations representative of this period in the statistical analysis. Statistical metrics, such as mean absolute error (MAE), root mean square error (RMSE), mean bias error (MBE) and coefficient of determination (R^2) are presented for all the daily energy fluxes generated by DisALEXI application for the period 2013–2015 (Table 1).

Table 1. Statistical metrics of daily energy fluxes generated by DisALEXI application with MODIS and Landsat images for 2013–2015.

	Flux	$\bar{O}$ (MJ m^{-2} d^{-1})	MAE (MJ m^{-2} d^{-1})	RMSE (MJ m^{-2} d^{-1})	MBE (MJ m^{-2} d^{-1})	R^2
MODIS	Rn	12.93	0.91	1.15	0.11	0.95
	G	1.59	0.56	0.69	0.35	0.52
	H	7.34	1.64	1.99	−0.60	0.69
	LE	4.01	1.54	1.87	0.37	0.69
Landsat	Rn	12.90	0.91	1.15	−0.33	0.94
	G	1.52	0.46	0.60	0.32	0.60
	H	7.52	1.56	2.18	−1.16	0.68
	LE	3.86	1.46	1.85	0.50	0.56

$\bar{O}$ = mean observed flux; MAE = mean absolute error; RMSE = root mean square error; MBE = mean bias error; R^2 = coefficient of determination.

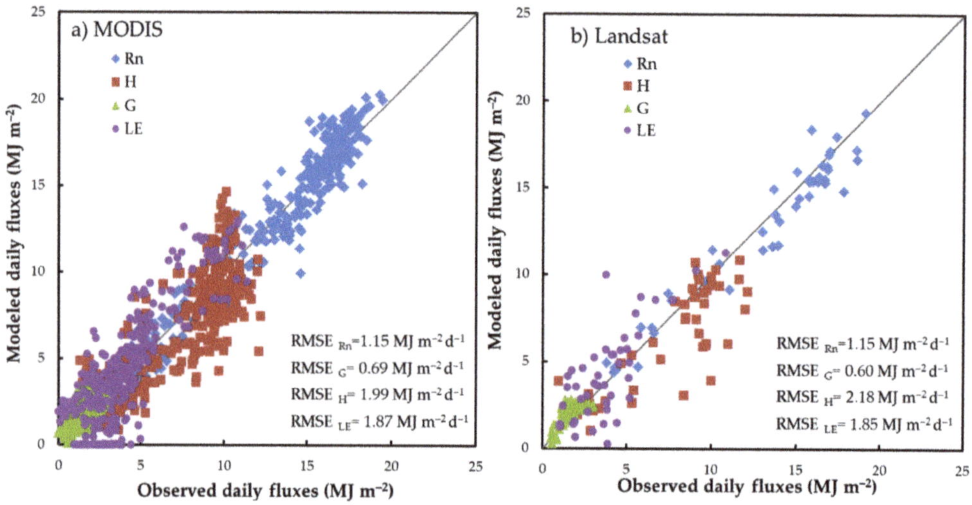

Figure 4. Comparison between observed and modeled daytime integrated energy fluxes obtained from DisALEXI procedure using (**a**) MODIS and (**b**) Landsat images for the years 2013–2015.

Modeled Rn generally agreed with in-situ measurements (RMSE value of 1.15 MJ m^{-2} d^{-1} and R^2 = 0.94–0.95). Rn behavior was different depending on the satellite sensor, slightly overestimated with MODIS data and underestimated when using Landsat images. Insolation (solar incoming radiation) inputs for both applications were the same, but with different albedo and fractional cover values to derive the energy available for the photosynthesis process. The soil heat flux estimations presented a good agreement (RMSE of 0.69 for MODIS and 0.60 MJ m^{-2} d^{-1} for Landsat), worthy of mention in this ecosystem with bare soil patches and high temperatures, where G can use up to 20% of the available energy.

RMSE and MAE errors observed with Landsat images were equal or slightly lower than those generated with MODIS for Rn, G and LE, possibly due to the better footprint representation of the higher resolution input data. The RMSE value for H was equal to 1.99 MJ m^{-2} d^{-1} when using MODIS data, and 2.18 MJ m^{-2} d^{-1} with Landsat. A general underestimation of the H flux can be observed in Figure 4, corroborated by the negative MBE values obtained with both sensors. Modeled LE fluxes had RMSE values of 1.87 MJ m^{-2} d^{-1} with MODIS, and 1.85 MJ m^{-2} d^{-1} with Landsat (in mass flux units, 0.76 mm d^{-1}), with an overestimation of the modeled fluxes (MBE of 0.37 and 0.50 MJ m^{-2} d^{-1} or 0.15 and 0.20 mm d^{-1}, respectively), except for low LE values. On the other hand, the values of the coefficient of determination were in the range 0.56–0.69 using Landsat and MODIS for both convective fluxes (H and LE).

3.2. *Analysis of ET Time Series from DisALEXI and STARFM*

Figure 5 depicts the ET time series generated by DisALEXI (MODIS—1 km and Landsat—30 m) and STARFM approaches for the years 2013–2015, along with the closure-corrected ET measurements acquired by the *dehesa* ECT system and the rainfall observations. It can be observed in Figure 5 that in this ecosystem (with ~25% fraction tree cover) the annual ET curve is bimodal, with two distinct peaks of different sizes. The largest peak occurs during the spring, reaching maximum values of around 4–5 mm d^{-1}. A second and smaller peak appears in autumn, with maximum values of around 2–2.5 mm d^{-1} in 2013–2014 and 1.5 mm d^{-1} in 2015. This pattern is closely linked to the distribution of the annual rainfall throughout the year, defining the general pattern of soil water availability, and the vegetation growth dynamics, as well as by the energy available for evaporation.

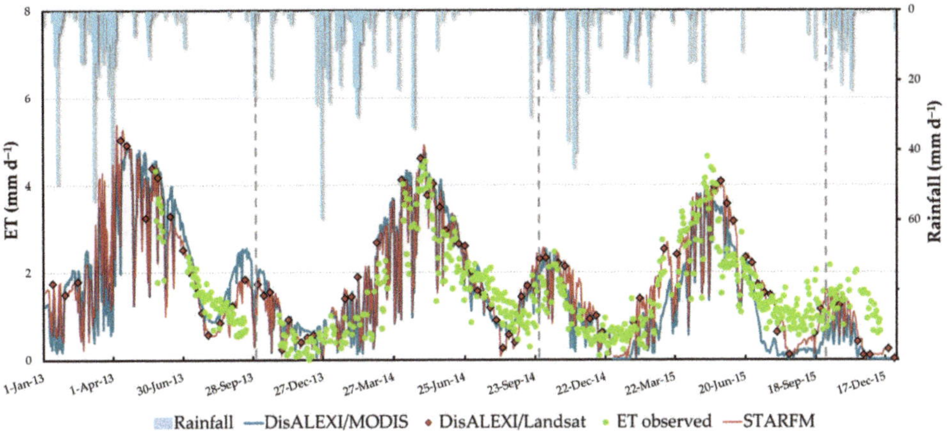

Figure 5. *ET* time series over the *dehesa* ecosystem from DisALEXI/MODIS, DisALEXI/Landsat, STARFM approach and observed data with closure correction for 2013, 2014 and 2015. Blue bars illustrate rainfall events and gray vertical lines indicate the starting of a hydrological year.

The modeled ET reasonably reproduced the temporal dynamics of consumptive water use observed by the flux tower instruments. In general terms, the maximum peaks reached in the spring seasons, as well as the gradual decrease of ET during the end of the spring and summer seasons (periods with higher data available) were accurately reproduced by the STARFM and DisALEXI approaches (Figure 5).

The annual ET modeled by STARFM was equal to 600 mm in the hydrological year 2013/2014 (1 October to 30 September), and 578 mm in the year 2014/2015, with rainfall values of 704 mm and 511 mm for each hydrological year, respectively. These results reflect the evolving vegetation water needs and how the ecosystem adapts to differing water availability conditions, with 193 mm less rainfall during the second year. It also highlights the major role that ET plays in the water balance of the system.

Looking at the details of seasonal water use more closely, we see that the water consumption in the spring of 2015 was similar to that of 2014, as well as the vegetation growth (NDVI values were similar for both years), even though the rainfall was considerably lower (spring rainfall of 120 mm in 2014 and 71 mm in 2015). In this case, the vegetation water consumption in the spring-2015 may have been tapping soil water storage from an unusually high rainfall in the previous autumn (311 mm in autumn—2014 compared to 103 mm of autumn—2013), pointing out the importance of the antecedent soil moisture conditions and the recharge of the subsurface layers.

During the summers, observed ET fell to low values (around 1.1 and 1.4 mm d^{-1} on average for the months of July and August), with minima around 0.7 mm d^{-1} at the end of the dry season (in early September before the first rainfalls). In addition to this summer-time soil-moisture limited minimum, a similar shut-down of the vegetation was observed during the winter (average ET around 0.7 in January and February), due to the lack of insolation.

Despite the generally good fit of ET estimations with ECT observations during the study period, some seasonal mismatches between modeled and observed fluxes can be identified. For example, modeled ET increased at the end of the summer in 2013 (on September 3) after a light rainfall event (<2 mm), while ET observations continued to decrease. While the second peak in the observed flux series is missing due to a gap in the field data from September 28, it appears that the rise in ET outlined by Landsat and STARFM preceded the observed rise following rainfall on September 27 (36 mm event). At the end of the dry season of 2014, similar behavior was observed with a gradual increase of modeled estimates starting on September 1, although it was not until September 17 when a significant rainfall event happened (approx. 46 mm). During the wet season in mid-May of 2015, the model showed a lagged response to a drydown that caused the vegetation to dry quickly. This may be related to smoothing/gap-filling in the baseline ALEXI time series, which can tend to smooth out abrupt changes in fPET.

A comparative analysis between the different daily ET series and the observed measurements at the ECT is shown in Table 2 including the quantitative performance of the following daily ET retrievals: (a) from DisALEXI 1 km gap-filled MODIS, (b) STARFM from Landsat-MODIS, and (c) from a simple Landsat-only interpolation technique (using the potential ET as a daily scaling input, through the ratio FPET). The analysis was carried out for a period of 584 days. Results obtained with the Landsat interpolated ET series using MODIS as a daily scaling input (not shown) are similar to those obtained using FPET.

Table 2. Comparison of different daily ET retrievals modeled for the period 2013–2015.

	MODIS	MODIS-Landsat (STARFM)	Interpolated Landsat (Using FPET)
MAE (mm d^{-1})	0.589	0.539	0.596
RMSE (mm d^{-1})	0.737	0.673	0.749
MBE (mm d^{-1})	0.005	0.103	0.158

The data fusion algorithm for estimating daily 30 m ET outperformed the simpler interpolation technique based on FPET and using MODIS 1-km ET, yielding an RMSE value of 0.67 mm d^{-1}.

3.3. *Evaluation of the MOD16A2 Global ET Product*

Figure 6 shows data of the 8-day total ET layer provided by the global product MOD16A2 (500 m spatial resolution) at the tower pixel, together with the data observed at the ECT and the water consumption modeled by the STARFM approach. The MODIS ET product agreed poorly with the ECT observations (MAE = 0.79 mm d^{-1} and RMSE = 1.02 mm d^{-1}), with a significant underestimation (MBE = −0.56 mm d^{-1}). The error was unevenly distributed throughout the year, being more pronounced in the spring and summer seasons, and lessened in autumn and winter. This trend is observed in the three years and it highlights that the use of this product will underestimate ET rates over this semiarid ecosystem.

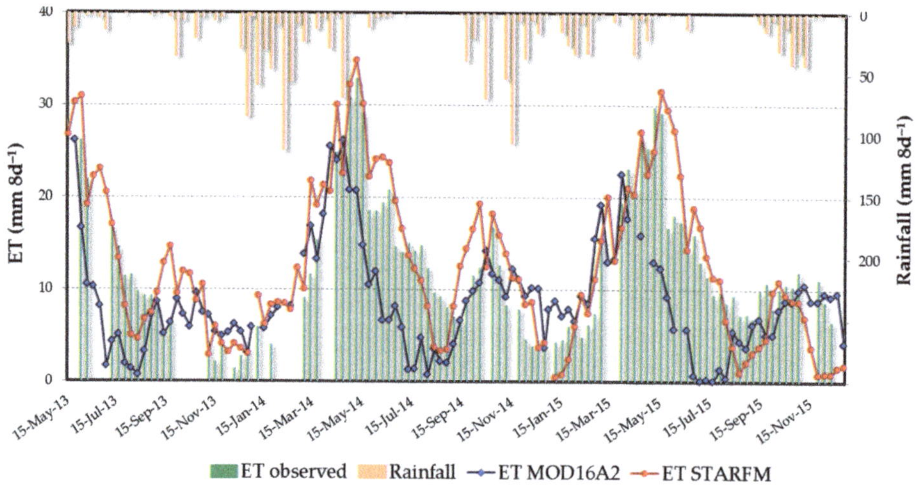

Figure 6. Comparison between observed, modeled by STARFM and MOD16A2 product ET for 2013–2015. Burnt orange bars illustrate rainfall events.

3.4. Water Resources Management at Field Scale Using High-Resolution ET Maps

At the field scale, the *dehesa* ecosystem presents variations in the distribution of vegetation strata (oak trees, grassland, crops, or scrubs) that affect the provisioning of ecosystem services and must be considered in the management of the livestock. Within this context, the evolution of daily ET at four characteristic micro-ecosystems/patches of the *dehesa* within the study area (Figure 1c), with different vegetation canopy structures and livestock feed production strategies has been analyzed at two spatial resolutions (30 m and 1 km) to evaluate the potential of Landsat-scale ET retrievals (Figure 7).

Zone A (1700 m^2) is an open grassland without trees. It is a grazing area, which exhibits high temporal and spatial variability in vegetation growth. Zone B (2400 m^2) is a pasture area where there is a land depression accumulating water through most of the year and maintaining high humidity conditions. Zone C (4300 m^2) is a mixture of tree/grass, with an average tree fractional cover of around ~25%, common in *dehesa* production systems and similar to the ECT surrounding area. Acorn production to feed Iberian pigs is usually the main economic use of these landscapes. The final area, zone D (3300 m^2), is located near a stream with a high ground fraction covered by trees and an evergreen underlayer of scrubs.

Significant differences were observed between the daily ET provided by DisALEXI/MODIS at a 1 km resolution and modeled ET with STARFM approach at a 30 m resolution, especially pronounced over zones A and D (Figure 7). Despite the relatively small area of these sites, the 30 m spatial resolution better captured the scale of heterogeneity in this landscape, whereas at 1 km, multiple vegetation covers are contributing, resulting in similar ET patterns for all the sites at this scale.

In the grassland area (Figure 7a), the STARFM 30 m ET decreases more rapidly in the spring than the MODIS-DisALEXI 1 km ET. This decrease corresponds to the drying of the herbaceous stratum, and thus the reduction in the available pasture for forage. Because there are no trees in the STARFM extraction area, transpiration flux is not appreciable during the summer months. The MODIS estimates include contributions from surrounding trees and therefore, maintain significant ET rates during the dry seasons. In each year, the pasture drying process started in May and ended in early to mid-July.

In contrast to zone A, in zone B the late spring decrease in ET is more gradual due to the high soil moisture conditions. The behavior during this season was accurately reproduced by the 30 m resolution model data. In this case, the 30 m and 1 km ET timeseries are more similar to the wet grassland (absent moisture limitations), mimics the transpiration curve of the surrounding trees. The main differences are during springtime peaks, where STARFM reaches values around 4.5–5 mm d^{-1}, and about 4 mm d^{-1} with MODIS.

In zone C, the differences between DisALEXI/MODIS and STARFM were not pronounced due to the homogeneity of the surrounding landscape. This type of landscape is representative of the most common structure of the *dehesa* ecosystem, following a behavior similar to that described in Section 3.2.

Finally, in zone D we see that the 1 km MODIS information is not able to resolve the higher ET focused in this riparian forested area. ET estimates at 30 m suggest the use of groundwater (even shallow water from the creek) by the vegetation during summer, where the ET was maintained over 1.5 mm d^{-1}, while MODIS ET incorporating surrounding grasslands drops to near 0. A more stable ET trend is seen over this river/scrub zone, where the ET reached similar maximums during the 3 years. These scrub species may have a high nutritional interest both for the domestic livestock, as well as for game species [63].

Cumulative annual water use curves, relevant for water resources management purposes, at 30 m and 1 km resolution are contrasted between relatively wet and dry hydrological years (2013/2014 and 2014/2015, respectively) over the zones A and D in Figure 8. These two areas were chosen because they showed higher scale-dependent differences (Figure 7). Total precipitation measured at the flux site was 704 mm in 2013/2014 and 511 mm in 2014/2015.

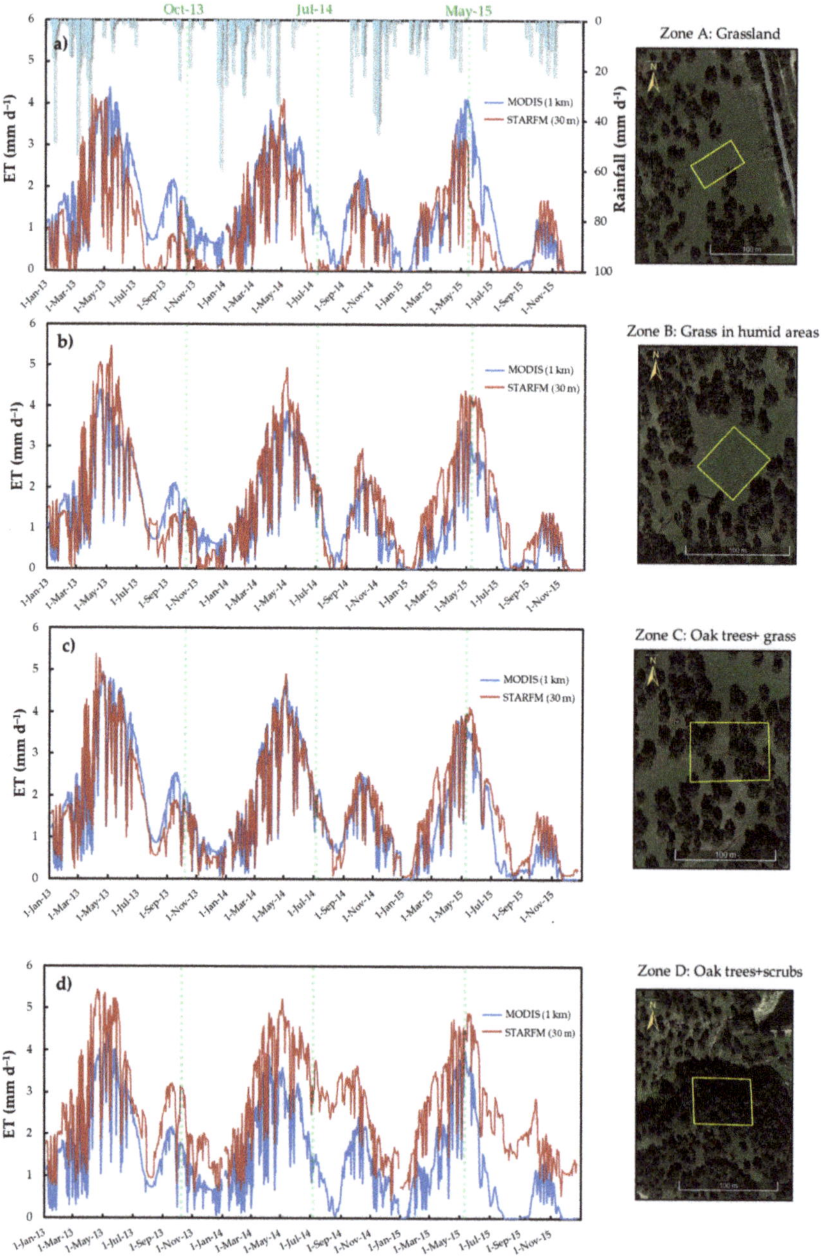

Figure 7. Evolution of daily ET generated by DisALEXI/MODIS (1 km) and STARFM (30 m) over an area (denoted by the yellow polygons) of grassland (**a**), of grass in humid areas (**b**), of combined oak tree and grassland (**c**) and of oak trees and scrub (**d**). The green vertical dashed lines correspond to dates that are analyzed in Section 3.4.

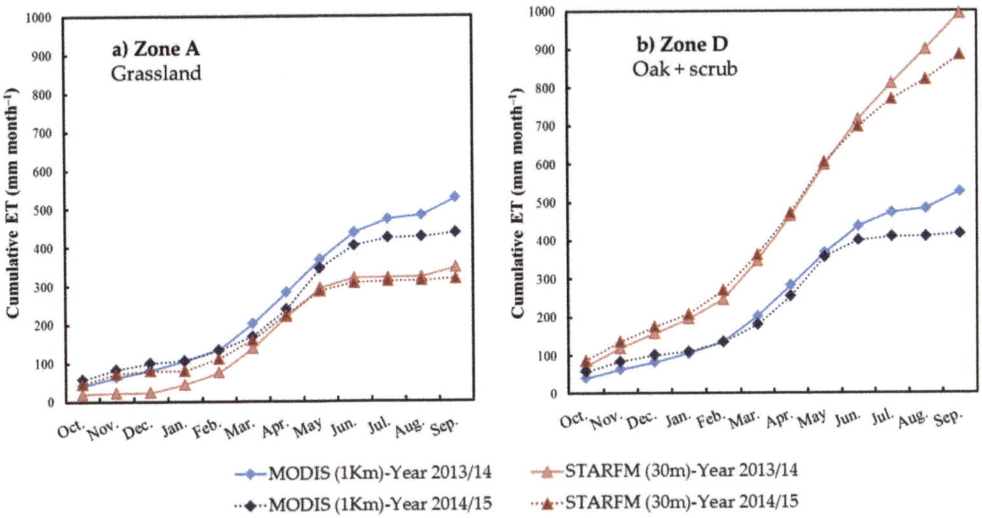

Figure 8. Evolution of cumulative monthly ET estimated by DisALEXI/MODIS and STARFM approaches for the hydrological year 2013/14 and 2014/15, over (**a**) zone A and (**b**) zone D.

At both resolutions, cumulative ET is higher in the wet year (2013/2014) as expected. While cumulative ET at 1 km resolution was similar between zones (around 450–500 mm/year), the 30 m resolution data showed substantial variations between sites, with values ranging between 350 and 900 mm/year.

In the open grassland area, zone A (Figure 8a), the DisALEXI/MODIS approach produced an overestimation of the annual ET in the order of 120–180 mm when compared with the fusion model. Differences were more pronounced during the dry season when the STARFM curve remained stable from May on due to grass senescence. In zone D (Figure 8b), with transpiring vegetation throughout the year, the cumulative ET continuously increased at 30 m resolution. Neither intra- nor inter-annual significant variations in water consumption were observed with STARFM in this area close to a stream, probably due to higher use of groundwater by trees and scrubs and a lesser coupling of water consumption with rainfall events. MODIS cumulative ET showed a slight plateau during the months of July and August, capturing the behavior of the surrounding grassland area in the analysis. An underestimation of approximately 450 mm/year was quantified in this area using 1 km resolution data.

In addition, spatially distributed maps of ET over the *Martin Gonzalo* watershed and *Santa Clotilde dehesa* farm (Figure 1c) on three days, representative of different seasons of the period 2013–2015 are presented in Figure 9, highlighting the area with a *dehesa* ecosystem. They were produced by the ALEXI model (5 km), the DisALEXI/MODIS application (1 km) and the STARFM technique (30 m).

At the watershed scale, the more detailed representation of ET at a fine resolution provides a better understanding of the variability of the *dehesa* vegetation water consumption. At 30 m the variations in topography and soils, affecting the vegetation species composition and abundance, can be observed and are largely lost at the 1 km resolution and not existent at the 5 km scale. At the farm scale, the STARFM model allowed for the observation of ET variability in the different vegetation patches previously described, highlighting the existence of micro-ecosystems and climates at the local scale.

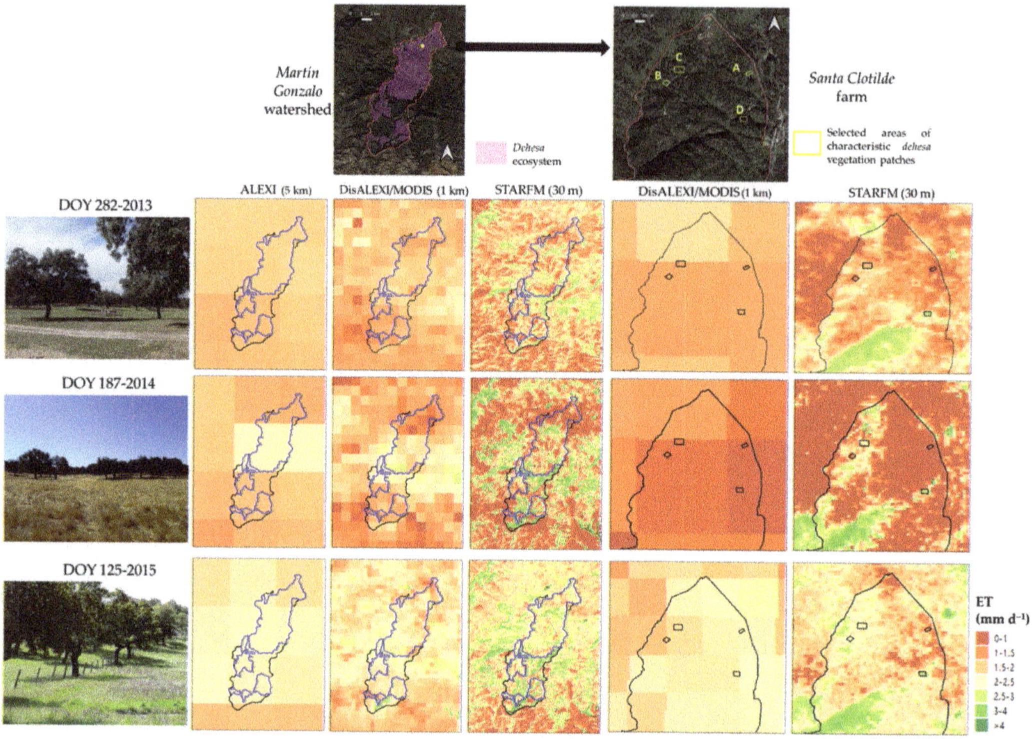

Figure 9. Distributed ET maps over the *Martin Gonzalo* watershed and *Santa Clotilde dehesa* farm derived from ALEXI (5 km), the DisALEXI/MODIS application (1 km) and the STARFM approach (30 m) on three representative dates of spring, summer, and autumn seasons. Blue lines show the *dehesa* ecosystem areas.

4. Discussion

4.1. DisALEXI Model Validation

The estimation of Rn flux presented an error (RMSE = 1.15 MJ m^{-2} d^{-1}) slightly higher compared to other applications over agricultural areas [40,41,49], but it was aligned with the errors over sparse vegetation cover crops, such as grapevines [17,42] and forested/mosaic areas [16,43]. Ideally, future filtering of insolation inputs needs to be avoided in order to build an automated application, which should include a proxy to account for the quality of the input data. An alternative could be to migrate to an MSG Land-SAF satellite insolation product. Anderson et al. [64] tested different satellite insolation products generated by geostationary satellites, which significantly improved insolation performance over CFSR, although it did not translate into comparable improvement in the ET retrieval accuracy.

Likewise, the errors in H flux (RMSE value of 1.99 MJ m^{-2} d^{-1} when using MODIS data and 2.18 MJ m^{-2} d^{-1} with Landsat) were higher than other applications over agricultural areas (in the range of 1.2–1.7 MJ m^{-2} d^{-1} observed by Cammalleri et al. [40,41] and Semmens et al. [42]), but in the same range as those obtained in more complex vegetation [17,43].

In relation to LE flux modeled by DisALEXI, a similar behavior over another *dehesa* site (fluxnet code ES-LMA) using LST data measured in the field and MODIS images, was observed by Andreu et al. [20,21]. These authors used a lower Priestley-Taylor coefficient and stated that even integrating the green fraction and reducing the coefficient, during the dry season the flux was so low as to approach zero. Furthermore, the same trend had been previously observed by Carpintero et al. [65,66] in the study area, where the scheme was applied using a different meteorological dataset source and study period. In this regard, Burchard-Levine et al. [22] implemented the TSEB model over a *dehesa* area with modifications, considering phenological dynamics of this tree-grass ecosystem and assuming a dominant vegetation structure and cover for different seasons, and different seasonal vegetation parameters. This new adaptation improved the model performance, decreasing the overestimation of LE flux and RMSD errors compared to the application of TSEB assuming a single vegetation source.

4.2. Temporal Patterns in ET Curves

The ET time series estimated for the hydrological years 2013/2014 and 2014/2015 reflected the evolution of the vegetation water consumption, depending on water availability. Both hydrological years have similar ET rates, despite the lower precipitation rates in the second one. In 2014/15 water use exceeded the rainfall, suggesting that the vegetation tapped water from other sources to sustain transpiration rates within a threshold that plants could survive. It is known that oak tree roots are able to explore a large volume of soil, with a high dependence on deep water reserves [67,68]. Moreover, oaks form both arbuscular and ectotrophic mycorrhizae, and create mycorrhizal symbioses with partners, adapting to different conditions and accessing many different resources [69].

The similarity of water consumption in spring 2015 with respect to 2014 (with a significantly lower rainfall but an unusually high rainfall in the previous autumn) highlighted the importance of significant antecedent rainfall events to recharge the subsurface layers. In this sense, Fernández [70] found a significant correlation between grassland species abundance in the *dehesa* and autumn rainfall.

The ET values observed during the summers (Figure 5; minima values of 0.7 mm d^{-1} at the end of the dry season) were comparable with David et al.'s results [71], who found that oak trees maintained transpiration rates above 0.7 mm d^{-1} during the dry season with more than 70% of the transpired water being taken from groundwater sources. In seasons with low LE values (summers and winters), the uncertainty in the data observed by the eddy covariance system [72], added to the fact that LE flux was obtained by closing the balance with the residual method must be considered in the error assessment.

The accuracy of the different modeled daily ET series, shown in Table 2, is below the target error of 0.8 mm d^{-1} suggested by Seguin et al. [73]. The errors (RMSE = 0.67 mm d^{-1} with STARFM approach) were close to those found by other authors for woody sparse semiarid crops [21,74,75], and slightly higher than the RMSE value (~0.5 mm d^{-1}) found by Campos et al. [76] and Carpintero et al. [46] in a *dehesa* ecosystem using a locally calibrated remote sensing-based soil water balance.

4.3. Performance of MOD16A2 ET

Sriwongsitanon et al. [77] showed a strong agreement of MOD16A2 values with the bulk ET computed using a water balance framework in the forested humid tropics of Thailand, on a monthly and annual scale. The estimations from that study were consistent with the soil moisture conditions and land use classes. An acceptable accuracy of this product was also found by Niyogi et al. [78] and Aguilar et al. [79] for water consumption monitoring over agricultural areas in Indiana (Midwestern United States) and Northwestern Mexico. Comparisons of this product with in-situ measurements at 15 flux tower sites with different climates and biome types ranging from croplands, grasslands, shrublands, savannas, to forests over Europe were evaluated by Hu et al. [80]. The global ET was consistent over most of Europe, with the best results over crops and meadows located in a temperate and humid climate. However, the seasonal performance of MOD16A2 was worse over vegetation under a semiarid climate, with RMSE equal to 1.17 mm d^{-1}, comparable to the error found in this work (Figure 6; RMSE = 1.02 mm d^{-1}). Global product ET was always underestimated for a *dehesa* ecosystem in Spain (similar study area and results to those of this work), demonstrating the weakness of this algorithm over limited water availability conditions [80]. The same behavior of ET underestimation was observed over a semiarid region in Iran [81], highlighting the high dependence of MOD16A2 performance with the climate type.

As shown in Figure 6, through the STARFM framework application we were capable of estimating with higher accuracy the temporal trend of ET (8 mm d^{-1}) for this semiarid ecosystem, when compared with the MOD16A2 product.

4.4. Variability of Dehesa Vegetation Water Use at Field Scale

Dehesa landscape structure features, with variations in the distribution of vegetation strata, are associated with the creation of different microclimates, directly influencing ecological processes, such as plant regeneration and growth, soil respiration, nutrient cycling, and wildlife habitats [35].

The trends observed in the grassland area (zone A; Figure 7a), where the pasture drying process started in May and ended in early to mid-July, were consistent with the pasture production cycle observed in a similar *dehesa* farm in Spain (Las Majadas). The maximum production was obtained in the spring (around 60–70%) and autumn (15–20%), while the pasture growth was at a minimum in winter (5–15%) and zero in the summer [28]. Information about the pasture drying date at Landsat-scale, for which a high temporal frequency is crucial, can benefit the assessment of the nutritional quality for livestock feed [28]. The ET behavior in zone B (Figure 7b), where high humidity conditions existed, demonstrates the greater transpiration and production capacity of these moist grasslands, important for managing grazing rotations.

Due to the great expansion of the zone C vegetation structure (Figure 7c; a mixture of grass and trees with a low coverage fraction), previous applications of models for estimating water consumption at a low spatial resolution (from 1 km to 5 km) generally have worked well over this ecosystem. These models have provided very useful information at the basin or regional scale [20,21,23]. In zone D, the high spatial resolution ET reflects the presence of dense evergreen vegetation transpiring throughout the year. These areas provide essential ecosystem services, such as shelter for livestock and wildlife (rabbits, Iberian Lynx, etc.), or the creation of specific conditions under the canopy, with different radiation interception, and soils with increased fertility and water retention [82].

The spatial variability in the ET maps is represented in Figure 9 over the *Martin Gonzalo* watershed (left) and *Santa Clotilde* farm (right). From a hydrological modeling viewpoint, the 30 m resolution ET product may be very useful to identify and delineate hydrological zones with different water storage capacities and runoff processes within the basin. Nevertheless, the coarse resolution may be enough for regional managing purposes, such as drought monitoring. At the field scale, regular availability of this type of map can lead to developing better agricultural management decisions related to grazing rotations, delineating areas containing fragile ecosystems and important ecological riparian areas with higher water holding capacity and vegetation cover that can sustain a wide diversity of plant and animal species.

5. Conclusions

The ALEXI/DisALEXI model and the STARFM data fusion technique adequately estimated the ET dynamics of a Mediterranean oak savanna at fine spatial and temporal resolution (30 m, daily) for the period 2013–2015. The energy fluxes provided by DisALEXI using MODIS images (1 km, daily) and Landsat images (60–100 m, 16 days) were compared with in-situ measurements from an eddy covariance flux tower system. The modeled fluxes compared reasonably well to field observations, with RMSE values ranging between 0.60 and 2.18 MJ m^{-2} d^{-1} depending on the sensor resolution and the flux component, a similar range to those obtained by other authors over complex landscapes. In addition, the daily 30 m ET series from STARFM model obtained a RMSE value of 0.67 mm d^{-1}, which is considered an acceptable error for *dehesa* management purposes.

The ET time series generated by both approaches, DisALEXI and STARFM with different resolutions, showed an annual bimodal behavior of the vegetation water consumption over the *dehesa* structure (grass and trees, with a tree cover fraction of 25%), with two marked peaks of different magnitudes (around 4–5 mm d^{-1} during the spring, and between 1.5 and 2.5 mm d^{-1} in the autumn). Modeled ET accurately reproduced the temporal dynamics of the water consumed by the vegetation, clearly linked to the distribution of the annual rainfall and the energy availability. Nevertheless, some difficulties of the model to reproduce certain abrupt changes in the ET behavior were identified, showing some limitations of the modeling scheme without more frequent observations and the importance of high-quality input data.

The results showed that when the data fusion model was applied, a slight improvement of RMSE, compared to using MODIS or simpler linear interpolations with Landsat data, was observed. This can be due to the combination of daily temporal frequency capturing the changes in ET and the high spatial resolution providing a better representation of the footprint of the flux tower. Modeled annual ET, close to 600 mm for both hydrological years despite their different rainfall regimes, provides an indicative value of current vegetation water needs and how the system adapts to meet these needs under the different water availability conditions. Holm oak's ability to explore a large volume of soil and access groundwater resources under higher than typical water-limited conditions can explain a lack of a difference in ET between years and the minimum modeled values of transpiration of 0.7 mm d^{-1} at the end of the dry season. However, for periods of long-term and sustained drought, the ecosystem is likely to have a different response.

The major advantage of the data fusion technique and the high spatio-temporal resolution was found in the analysis of ET dynamics over different vegetation microbiomes, characteristics of the *dehesa* landscape. The results showed that high-resolution ET maps (daily, 30 m) can provide key information to better understand the hydrological functioning of different vegetation distributions difficult to detect at a 1 km pixel resolution and assist in decision making at the farm or local level. However, 1–5 km resolution ET information may suffice for regional monitoring and water planning purposes, such as regional drought detection and impacts on water resources of different biomes.

Author Contributions: Conceptualization, E.C., M.P.G.-D., M.C.A. and A.A; methodology, M.C.A., M.P.G.-D., E.C., A.A., C.H., F.G., W.P.K.; software and software application, M.C.A., C.H., F.G., E.C.; validation, E.C. and A.A.; data curation, E.C.; field data collection and processing, E.C. and A.A.; writing, review and editing, E.C., A.A., M.P.G.-D., M.C.A.; supervision, M.P.G.-D., M.C.A., W.P.K., F.G. All authors have read and agreed to the published version of the manuscript.

Funding: This research was funded by the FPU program—University Teachers Training—from the Department of Education, Culture and Sport. Dr. Andreu's work was funded by the European Union's Horizon 2020 Research and Innovation program under the Marie Skłodowska-Curie grant agreement No 703978. Additional support was provided by the project PP.PEI.IDF2019.004, co-funded at 80% by the European Regional Development Fund (ERDF), as part of the Andalusian operational program 2014–2020, and the project PID2019-107693RR-C22 (MCIU/AEI/FEDER, UE). This research was supported in part by the U.S. Department of Agriculture, Agricultural Research Service. USDA is an equal opportunity provider and employer.

Institutional Review Board Statement: Not applicable.

Informed Consent Statement: Not applicable.

Data Availability Statement: Data available contacting the principal researcher.

Acknowledgments: Theauthors would like to thank M. Ramos, J. Alfieri, J. Prueger and L. McKee for their help with field instrumentation installation as well as the maintenance, and measurements. We also would like to thank the collaboration of Santa Clotilde managers.

Conflicts of Interest: The authors declare no conflict of interest.

References

1. Rodriguez-Iturbe, I.; Porporato, A.; Laio, F.; Ridolfi, L. Plants in water-controlled ecosystems: Active role in hydrologic processes and responce to water stress I. Scope and general outline. *Adv. Water Resour.* **2001**, *24*, 695–705. [CrossRef]
2. Rundel, P.W.; Arroyo, M.T.K.; Cowling, R.M.; Keeley, J.E.; Lamont, B.B.; Vargas, P. Mediterranean Biomes: Evolution of Their Vegetation, Floras, and Climate. *Annu. Rev. Ecol. Evol. Syst.* **2016**, *47*, 383–407. [CrossRef]
3. Gauquelin, T.; Michon, G.; Joffre, R.; Duponnois, R.; Génin, D.; Fady, B.; BouDagher-Kharrat, M.; Derridj, A.; Slimani, S.; Badri, W.; et al. Mediterranean forests, land use and climate change: A social-ecological perspective. *Reg. Environ. Chang.* **2018**, *18*, 623–636. [CrossRef]
4. Cramer, W.; Guiot, J.; Marini, K. *Climate and Environmental Change in the Mediterranean Basin–Current Situation and Risks for the Future*; First Mediterranean Assessment Report; MedECC (Mediterranean Experts on Climate and Environmental Change); Union for the Mediterranean, Plan Bleu, UNEP/MAP: Marseille, France, 2020; ISBN 978-2-9577416-0-1.
5. Milano, M.; Ruelland, D.; Fernandez, S.; Dezetter, A.; Fabre, J.; Servat, E.; Fritsch, J.M.; Ardoin-Bardin, S.; Thivet, G. Current state of Mediterranean water resources and future trends under climatic and anthropogenic changes. *Hydrol. Sci. J.* **2013**, *58*, 498–518. [CrossRef]
6. Lionello, P.; Scarascia, L. The relation between climate change in the Mediterranean region and global warming. *Reg. Environ. Chang.* **2018**, *18*, 1481–1493. [CrossRef]
7. Thiébault, S.; Moatti, J.; Ducrocq, V.; Gaume, E.; Dulac, F.; Hamonou, E.; Shin, Y.; Joel, G.; Boulet, G.; Guégan, J.; et al. *The Mediterranean Region under Climate Change: A Scientific Update*; IRD Éditions: Marseille, France, 2016; ISBN 978-2-7099-2219-7.
8. Cramer, W.; Guiot, J.; Fader, M.; Garrabou, J.; Gattuso, J.P.; Iglesias, A.; Lange, M.A.; Lionello, P.; Llasat, M.C.; Paz, S.; et al. Climate change and interconnected risks to sustainable development in the Mediterranean. *Nat. Clim. Chang.* **2018**, *8*, 972–980. [CrossRef]
9. García-Ruiz, J.M.; López-Moreno, I.I.; Vicente-Serrano, S.M.; Lasanta-Martínez, T.; Beguería, S. Mediterranean water resources in a global change scenario. *Earth-Sci. Rev.* **2011**, *105*, 121–139. [CrossRef]
10. Turco, M.; Llasat, M.C.; von Hardenberg, J.; Provenzale, A. Climate change impacts on wildfires in a Mediterranean environment. *Clim. Chang.* **2014**, *125*, 369–380. [CrossRef]
11. Bastiaanssen, W.G.; Menenti, M.; Feddes, R.A.; Holtslag, A.A.M. A remote sensing surface energy balance algorithm for land (SEBAL) 1 Formalation. *J. Hydrol.* **1998**, *212–213*, 198–212. [CrossRef]
12. Cammalleri, C.; Agnese, C.; Ciraolo, G.; Minacapilli, M.; Provenzano, G.; Rallo, G. Actual evapotranspiration assessment by means of a coupled energy/hydrologic balance model: Validation over an olive grove by means of scintillometry and measurements of soil water contents. *J. Hydrol.* **2010**, *392*, 70–82. [CrossRef]
13. Diak, G.R.; Whipple, M.S. Note on estimating surface sensible heat fluxes using surface temperatures measured from a geostationary satellite during FIFE 1989. *J. Geophys. Res.* **1995**, *100*. [CrossRef]
14. Gonzalez-Dugo, M.P.; Neale, C.M.U.; Mateos, L.; Kustas, W.P.; Prueger, J.H.; Anderson, M.C.; Li, F. A comparison of operational remote sensing-based models for estimating crop evapotranspiration. *Agric. For. Meteorol.* **2009**, *149*, 1843–1853. [CrossRef]
15. Timmermans, W.J.; Kustas, W.P.; Anderson, M.C.; French, A.N. An intercomparison of the Surface Energy Balance Algorithm for Land (SEBAL) and the Two-Source Energy Balance (TSEB) modeling schemes. *Remote Sens. Environ.* **2007**, *108*, 369–384. [CrossRef]
16. Anderson, M.; Gao, F.; Knipper, K.; Hain, C.; Dulaney, W.; Baldocchi, D.; Eichelmann, E.; Hemes, K.; Yang, Y.; Medellin-Azuara, J.; et al. Field-scale assessment of land and water use change over the California delta using remote sensing. *Remote Sens.* **2018**, *10*, 889. [CrossRef]
17. Knipper, K.R.; Kustas, W.P.; Anderson, M.C.; Alfieri, J.G.; Prueger, J.H.; Hain, C.R.; Gao, F.; Yang, Y.; McKee, L.G.; Nieto, H.; et al. Evapotranspiration estimates derived using thermal-based satellite remote sensing and data fusion for irrigation management in California vineyards. *Irrig. Sci.* **2019**, *37*, 431–449. [CrossRef]
18. Kustas, W.P.; Norman, J.M. Evaluation of soil and vegetation heat flux predictions using a simple two-source model with radiometric temperatures for partial canopy cover. *Agric. For. Meteorol.* **1999**, *94*, 13–29. [CrossRef]
19. Su, Z. The Surface Energy Balance System (SEBS) for estimation of turbulent heat fluxes. *Hydrol. Earth Syst. Sci.* **2002**, *6*, 85–100. [CrossRef]
20. Andreu, A.; Kustas, W.P.; Polo, M.J.; Carrara, A.; González-Dugo, M.P. Modeling surface energy fluxes over a dehesa (oak savanna) ecosystem using a thermal based two-source energy balance model (TSEB) I. *Remote Sens.* **2018**, *10*, 567. [CrossRef]
21. Andreu, A.; Kustas, W.P.; Polo, M.J.; Carrara, A.; González-Dugo, M.P. Modeling surface energy fluxes over a dehesa (oak savanna) ecosystem using a thermal based two source energy balance model (TSEB) II-Integration of remote sensing medium and low spatial resolution satellite images. *Remote Sens.* **2018**, *10*, 558. [CrossRef]
22. Burchard-Levine, V.; Nieto, H.; Riaño, D.; Migliavacca, M.; El-Madany, T.S.; Perez-Priego, O.; Carrara, A.; Martín, M.P. Seasonal adaptation of the thermal-based two-source energy balance model for estimating evapotranspiration in a semiarid tree-grass ecosystem. *Remote Sens.* **2020**, *12*, 904. [CrossRef]
23. González-Dugo, M.; Chen, X.; Andreu, A.; Carpintero, E.; Gómez-Giraldez, P.; Carrara, A.; Su, Z. Long-term water stress and drought monitoring of Mediterranean oak savanna vegetation using thermal remote sensing. *Hydrol. Earth Syst. Sci.* **2021**, *25*, 755–768. [CrossRef]

24. Anderson, M.C.; Norman, J.M.; Diak, G.R.; Kustas, W.P.; Mecikalski, J.R. A two-source time-integrated model for estimating surface fluxes using thermal infrared remote sensing. *Remote Sens. Environ.* **1997**, *60*, 195–216. [CrossRef]
25. Anderson, M.C.; Norman, J.M.; Mecikalski, J.R.; Otkin, J.A.; Kustas, W.P. A climatological study of evapotranspiration and moisture stress across the continental United States based on thermal remote sensing: 2. Surface moisture climatology. *J. Geophys. Res. Atmos.* **2007**, *112*, 1–13. [CrossRef]
26. Norman, J.M.; Anderson, M.C.; Kustas, W.P.; French, A.N.; Mecikalski, J.; Torn, R.; Diak, G.R.; Schmugge, T.J.; Tanner, B.C.W. Remote sensing of surface energy fluxes at 101-m pixel resolutions. *Water Resour. Res.* **2003**, *39*. [CrossRef]
27. Moreno, G.; Pulido, F.J. The Functioning, Management and Persistence of Dehesas. *Agrofor. Eur.* **2008**, *10600*, 127–160. [CrossRef]
28. Moreno, G.; Cáceres, Y. *System Report: Iberian Dehesas, Spain*; AGFORWARD; Agroforestry for Europe: Plasencia, Spain, 2016.
29. Díaz, M.; Tietje, W.D.; Barrett, R.H. Effects of Management on Biological Diversity and Endangered Species. In *Mediterranean Oak Woodland Working Landscapes*; Landscape Series, 2013; Campos, P., Huntsinger., L., Oviedo, J.L., Starrs, P.F., Díaz, M., Standiford, R.B., Montero, G., Eds.; Springer: Dordrecht, The Netherlands, 2013; Volume 16. [CrossRef]
30. Plieninger, T.; Rolo, V.; Moreno, G. Large-scale patterns of Quercus ilex, Quercus suber, and Quercus pyrenaica regeneration in central-western Spain. *Ecosystems* **2010**, *13*, 644–660. [CrossRef]
31. Coelho, C.; Ferreira, A.J.D.; Laouina, A.; Hamza, A.; Chaker, M.; Naafa, R.; Regaya, K.; Boulet, A.-K.; Keizer, J.J.; Carvalho, T.M.M. Changes in Land Use and Land Management Practices Affecting Land Degradation within Forest and Grazing Ecosystems in the Western Mediterranean. *Sustain. Agrosylvopastoral Syst. Dehesas Montados* **2004**, *37*, 137–153.
32. Baldocchi, D.D.; Xu, L.; Kiang, N. How plant functional-type, weather, seasonal drought, and soil physical properties alter water and energy fluxes of an oak-grass savanna and an annual grassland. *Agric. For. Meteorol.* **2004**, *123*, 13–39. [CrossRef]
33. Baldocchi, D.D.; Xu, L. What limits evaporation from Mediterranean oak woodlands—The supply of moisture in the soil, physiological control by plants or the demand by the atmosphere? *Adv. Water Resour.* **2007**, *30*, 2113–2122. [CrossRef]
34. Joffre, R.; Rambal, S.; Damesin, C. Functional attributes in Mediterranean-type ecosystems. In *Handbook of Functional Plant Ecology*, 2nd ed.; Pugnaire, F.I., Valladares, F., Eds.; CRC Press Books: Boca Raton, FL, USA, 2008.
35. Chen, J.; Saunders, S.C.; Crow, T.R.; Naiman, R.J.; Brosofske, K.D.; Mroz, G.D.; Brookshire, B.L.; Franklin, J.F. Microclimate in Forest Ecosystem the effects of different management regimes. *Bioscience* **1996**, *49*, 288–297. [CrossRef]
36. Johnston, M.; Andreu, A.; Verfaillie, J.; Baldocchi, D.; Gonzalez-Dugo, M.P.; Moorcroft, P. Measuring surface temperatures in a woodland savanna: Opportunities and challenges of thermal imaging in an open-canopy ecosystem. *Agric. For. Meteorol.* **2021**, *310*, 108484. [CrossRef]
37. Gao, F.; Masek, J.; Schwaller, M.; Hall, F. On the blending of the landsat and MODIS surface reflectance: Predicting daily landsat surface reflectance. *IEEE Trans. Geosci. Remote Sens.* **2006**, *44*, 2207–2218. [CrossRef]
38. Walker, J.J.; De Beurs, K.M.; Wynne, R.H.; Gao, F. Evaluation of Landsat and MODIS data fusion products for analysis of dryland forest phenology. *Remote Sens. Environ.* **2012**, *117*, 381–393. [CrossRef]
39. Zhu, X.; Chen, J.; Gao, F.; Chen, X.; Masek, J.G. An enhanced spatial and temporal adaptive reflectance fusion model for complex heterogeneous regions. *Remote Sens. Environ.* **2010**, *114*, 2610–2623. [CrossRef]
40. Cammalleri, C.; Anderson, M.C.; Gao, F.; Hain, C.R.; Kustas, W.P. A data fusion approach for mapping daily evapotranspiration at field scale. *Water Resour. Res.* **2013**, *49*, 4672–4686. [CrossRef]
41. Cammalleri, C.; Anderson, M.C.; Gao, F.; Hain, C.R.; Kustas, W.P. Mapping daily evapotranspiration at field scales over rainfed and irrigated agricultural areas using remote sensing data fusion. *Agric. For. Meteorol.* **2014**, *186*, 1–11. [CrossRef]
42. Semmens, K.A.; Anderson, M.C.; Kustas, W.P.; Gao, F.; Alfieri, J.G.; McKee, L.; Prueger, J.H.; Hain, C.R.; Cammalleri, C.; Yang, Y.; et al. Monitoring daily evapotranspiration over two California vineyards using Landsat 8 in a multi-sensor data fusion approach. *Remote Sens. Environ.* **2016**, *185*, 155–170. [CrossRef]
43. Yang, Y.; Anderson, M.C.; Gao, F.; Hain, C.R.; Semmens, K.A.; Kustas, W.P.; Noormets, A.; Wynne, R.H.; Thomas, V.A.; Sun, G. Daily Landsat-scale evapotranspiration estimation over a forested landscape in North Carolina, USA, using multi-satellite data fusion. *Hydrol. Earth Syst. Sci.* **2017**, *21*, 1017–1037. [CrossRef]
44. Campos, P.; Huntsinger, L.; Oviedo, J.; Díaz, M.; Starrs, P.; Standiford, R.; Montero, G. *Mediterranean Oak Woodland Working Landscapes: Dehesas of Spain and Ranchlands of California*; Springer: Dordrecht, The Netherlands, 2013; ISBN 978-94-007-6706-5.
45. Webb, E.K.; Pearman, G.I.; Leuning, R. Correction of flux measurements for density effects due to heat and water vapour transfer. *Q. J. R. Meteorol. Soc.* **1980**, *106*, 85–100. [CrossRef]
46. Carpintero, E.; Andreu, A.; Gómez-Giráldez, P.J.; Blázquez, Á.; González-Dugo, M.P. Remote-sensing-basedwater balance for monitoring of evapotranspiration and water stress of a mediterranean Oak-Grass Savanna. *Water* **2020**, *12*, 1418. [CrossRef]
47. Hain, C.R.; Anderson, M.C. Estimating morning changes in land surface temperature from MODIS day/night observations: Applications for surface energy balance modeling. *Geophys. Res. Lett.* **2017**, *44*, 9723–9733. [CrossRef]
48. Norman, J.M.; Kustas, W.P.; Humes, K.S. Source approach for estimating soil and vegetation energy fluxes in observations of directional radiometric surface temperature. *Agric. For. Meteorol.* **1995**, *77*, 263–293. [CrossRef]
49. Anderson, M.C.; Allen, R.G.; Morse, A.; Kustas, W.P. Use of Landsat thermal imagery in monitoring evapotranspiration and managing water resources. *Remote Sens. Environ.* **2012**, *122*, 50–65. [CrossRef]
50. Anderson, M.C.; Kustas, W.P.; Norman, J.M.; Hain, C.R.; Mecikalski, J.R.; Schultz, L.; González-Dugo, M.P.; Cammalleri, C.; d'Urso, G.; Pimstein, A.; et al. Mapping daily evapotranspiration at field to global scales using geostationary and polar orbiting satellite imagery. *Hydrol. Earth Syst. Sci. Discuss.* **2010**, *7*, 5957–5990. [CrossRef]

51. Anderson, M.C.; Norman, J.M.; Mecikalski, J.R.; Torn, R.D.; Kustas, W.P.; Basara, J.B. A multiscale remote sensing model for disaggregating regional fluxes to micrometeorological scales. *J. Hydrometeorol.* **2004**, *5*, 343–363. [CrossRef]
52. Yang, Y.; Anderson, M.C.; Gao, F.; Wood, J.D.; Gu, L.; Hain, C. Studying drought-induced forest mortality using high spatiotemporal resolution evapotranspiration data from thermal satellite imaging. *Remote. Sens. Environ.* **2021**, *265*, 112640. [CrossRef]
53. Sun, L.; Anderson, M.C.; Gao, F.; Hain, C.R.; Alfieri, J.G.; Sharifi, A.; McCarty, G.; Yang, Y. Investigating water use over the Choptank River Watershed using a multi-satellite data fusion approach. *Water Resour. Res.* **2017**, *53*, 5298–5319. [CrossRef]
54. Liang, S. Narrowband to broadband conversions of land surface albedo: I. Algorithms. *Remote Sens. Environ.* **2000**, *76*, 213–238. [CrossRef]
55. Gao, F.; Anderson, M.C.; Kustas, W.P.; Wang, Y. Simple method for retrieving leaf area index from Landsat using MODIS leaf area index products as reference. *J. Appl. Remote. Sens.* **2012**, *6*, 063554. [CrossRef]
56. Berk, A.; Bernstein, L.S.; Robertson, D.C. *MODTRAN: A Moderate Resolution Model for LOWTRAN 7*; GL-TR-89-0122; Air Force Geophysics Lab: Bedford, MA, USA, 1987; p. 38.
57. Gao, F.; Kustas, W.P.; Anderson, M.C. A data mining approach for sharpening thermal satellite imagery over land. *Remote Sens.* **2012**, *4*, 3287–3319. [CrossRef]
58. Jönsson, P.; Eklundh, L. TIMESAT—A program for analyzing time-series of satellite sensor data. *Comput. Geosci.* **2004**, *30*, 833–845. [CrossRef]
59. Saha, S.; Moorthi, S.; Pan, H.L.; Wu, X.; Wang, J.; Nadiga, S.; Tripp, P.; Kistler, R.; Woollen, J.; Behringer, D.; et al. The NCEP climate forecast system reanalysis. *Bull. Am. Meteorol. Soc.* **2010**, *91*, 1015–1057. [CrossRef]
60. Mu, Q.; Zhao, M.; Running, S.W. Improvements to a MODIS global terrestrial evapotranspiration algorithm. *Remote Sens. Environ.* **2011**, *115*, 1781–1800. [CrossRef]
61. Hsieh, C.I.; Katul, G.; Chi, T.W. An approximate analytical model for footprint estimation of scalar fluxes in thermally stratified atmospheric flows. *Adv. Water Resour.* **2000**, *23*, 765–772. [CrossRef]
62. Kustas, W.P.; Alfieri, J.G.; Evett, S.; Agam, N. Quantifying variability in field-scale evapotranspiration measurements in an irrigated agricultural region under advection. *Irrig. Sci.* **2015**, *33*, 325–338. [CrossRef]
63. Hajer, A.; López, S.; González, J.S.; Ranilla, M.J. Chemical composition and in vitro digestibility of some Spanish browse plant species. *J. Sci. Food Agric.* **2004**, *84*, 197–204.
64. Anderson, M.; Diak, G.; Gao, F.; Knipper, K.; Hain, C.; Eichelmann, E.; Hemes, K.S.; Baldocchi, D.; Kustas, W.; Yang, Y. Impact of insolation data source on remote sensing retrievals of evapotranspiration over the California delta. *Remote Sens.* **2019**, *11*, 216. [CrossRef]
65. Carpintero, E.; Semmens, K.; Anderson, M.C.; Andreu, A.; Gao, F.; Kustas, W.P.; González-Dugo, M.P. Use of remote sensing data fusion for daily evapotranspiration monitoring at watershed scale over a *dehesa* ecosystem. In Proceedings of the Fourth International Symposium on Recent Advances in Quantitative Remote Sensing, Valencia, Spain, 22–26 September 2014.
66. Carpintero, E.; González-Dugo, M.P.; Hain, C.; Nieto, H.; Gao, F.; Andreu, A.; Kustas, W.P.; Anderson, M.C. Continuous evapotranspiration monitoring and water stress at watershed scale in a Mediterranean oak savanna. In *Remote Sensing for Agriculture, Ecosystems, and Hydrology XVIII*; SPIE: Bellingham, WA, USA, 2016; ISSN 0277-786X, ISBN 9781510604001.
67. Cubera, E.; Morena, G. Effect of single Quercus ilex trees upon spatial and seasonal changes in soil water content in dehesas of central western Spain. *Ann. For. Sci.* **2007**, *64*, 355–364. [CrossRef]
68. Moreno, G.; Obrador, J.J.; Cubera, E.; Dupraz, C. Fine root distribution in Dehesas of Central-Western Spain. *Plant Soil.* **2005**, *277*, 153–162. [CrossRef]
69. Allen, M.F. How Oaks Respond to Water Limitation. In *Proceedings of the Seventh California Oak Symposium: Managing Oak Woodlands in a Dynamic World*; US Department of Agriculture, Forest Service, Pacific Southwest Research Station: Berkeley, CA, USA, 2015; pp. 13–21.
70. Fernández-Alés, R. Response of Mediterranean grassland species to changing rainfall. A reply to Figueroa and Davy. *Orsis* **1993**, *8*, 121–126.
71. David, T.S.; Henriques, M.O.; Kurz-Besson, C.; Nunes, J.; Valente, F.; Vaz, M.; Pereira, J.S.; Siegwolf, R.; Chaves, M.M.; Gazarini, L.C.; et al. Water-use strategies in two co-occurring Mediterranean evergreen oaks: Surviving the summer drought. *Tree Physiol.* **2007**, *27*, 793–803. [CrossRef] [PubMed]
72. Burba, G.; Anderson, D. Eddy Covariance Flux Measurements. *Ecol. Appl.* **2010**, *18*, 211.
73. Seguin, B.; Becker, F.; Phulpin, T.; Gu, X.F.; Guyot, G.; Kerr, Y.; King, C.; Lagouarde, J.P.; Ottlé, C.; Stoll, M.P.; et al. IRSUTE: A minisatellite project for land surface heat flux estimation from field to regional scale. *Remote Sens. Environ.* **1999**, *68*, 357–369. [CrossRef]
74. Mateos, L.; González-Dugo, M.P.; Testi, L.; Villalobos, F.J. Monitoring evapotranspiration of irrigated crops using crop coefficients derived from time series of satellite images. I. Method validation. *Agric. Water Manag.* **2013**, *125*, 81–91. [CrossRef]
75. Padilla, F.L.M.; González-Dugo, M.P.; Gavilán, P.; Domínguez, J. Integration of vegetation indices into a water balance model to estimate evapotranspiration of wheat and corn. *Hydrol. Earth Syst. Sci.* **2011**, *15*, 1213–1225. [CrossRef]
76. Campos, I.; Villodre, J.; Carrara, A.; Calera, A. Remote sensing-based soil water balance to estimate Mediterranean holm oak savanna (dehesa) evapotranspiration under water stress conditions. *J. Hydrol.* **2013**, *494*, 1–9. [CrossRef]

77. Sriwongsitanon, N.; Suwawong, T.; Thianpopirug, S.; Williams, J.; Jia, L.; Bastiaanssen, W. Validation of seven global remotely sensed ET products across Thailand using water balance measurements and land use classifications. *J. Hydrol. Reg. Stud.* **2020**, *30*, 100709. [CrossRef]
78. Niyogi, D.; Jamshidi, S.; Smith, D.; Kellner, O. Evapotranspiration Climatology of Indiana, USA Using In Situ and Remotely Sensed Products. *J. Appl. Meteorol. Climatol.* **2020**, *59*, 2093–2111. [CrossRef]
79. Aguilar, A.; Flores, H.; Crespo, G.; Marín, M.I.; Campos, I.; Calera, A. Performance Assessment of MOD16 in Evapotranspiration Evaluation in Northwestern Mexico. *Water* **2018**, *10*, 901. [CrossRef]
80. Hu, G.; Jia, L.; Menenti, M. Comparison of MOD16 and LSA-SAF MSG evapotranspiration products over Europe for 2011. *Remote Sens. Environ.* **2015**, *156*, 510–526. [CrossRef]
81. Jamshidi, S.; Zand-Parsa, S.; Pakparvar, M.; Niyogi, D. Evaluation of Evapotranspiration over a Semiarid Region Using Multiresolution Data Sources. *J. Hydrometeorol.* **2019**, *20*, 947–964. [CrossRef]
82. Joffre, R.; Rambal, S. How tree cover influences the water balance of Mediterranean rangelands. *Ecology* **1993**, *74*, 570–582. [CrossRef]

remote sensing

MDPI

Article

Analysis of the Future Land Use Land Cover Changes in the Gaborone Dam Catchment Using CA-Markov Model: Implications on Water Resources

Botlhe Matlhodi [1,*], Piet K. Kenabatho [1], Bhagabat P. Parida [2] and Joyce G. Maphanyane [1]

[1] Department of Environmental Science, Faculty of Science, University of Botswana, P/Bag UB, Gaborone 00704, Botswana; kenabatho@ub.ac.bw (P.K.K.); maphanyanej@ub.ac.bw (J.G.M.)

[2] Department of Civil and Environmental Engineering, Botswana International University of Science and Technology, Private Bag 16, Palapye, Botswana; paridab@biust.ac.bw

* Correspondence: bmatlhodi@gmail.com; Tel.: +267-355-5475

Abstract: Land use/land cover (LULC) changes have been observed in the Gaborone dam catchment since the 1980s. A comprehensive analysis of future LULC changes is therefore necessary for the purposes of future land use and water resource planning and management. Recent advances in geospatial modelling techniques and the availability of remotely sensed data have become central to the monitoring and assessment of both past and future environmental changes. This study employed the cellular automata and Markov chain (CA-Markov) model combinations to simulate future LULC changes in the Gaborone dam catchment. Classified Landsat images from 1984, 1995, 2005 and 2015 were used to simulate the likely LULCs in 2015 and 2035. Model validation compared the simulated and observed LULCs of 2015 and showed a high level of agreement with Kappa variation estimates of K_{no} (0.82), K_{loc} (0.82) and $K_{standard}$ (0.76). Simulation results indicated a projected increase of 26.09%, 65.65% and 55.78% in cropland, built-up and bare land categories between 2015 and 2035, respectively. Reductions of 16.03%, 28.76% and 21.89% in areal coverage are expected for shrubland, tree savanna and water body categories, respectively. An increase in built-up and cropland areas is anticipated in order to meet the population's demand for residential, industry and food production, which should be taken into consideration in future plans for the sustainability of the catchment. In addition, this may lead to water quality and quantity (both surface and groundwater) deterioration in the catchment. Moreover, water body reductions may contribute to water shortages and exacerbate droughts in an already water-stressed catchment. The loss of vegetal cover and an increase in built-up areas may result in increased runoff incidents, leading to flash floods. The output of the study provides useful information for land use planners and water resource managers to make better decisions in improving future land use policies and formulating catchment management strategies within the framework of sustainable land use planning and water resource management.

Keywords: CA-Markov model; land use/land cover (LULC) change; water resources; Gaborone dam catchment; Botswana

1. Introduction

Land use/land cover (LULC) changes are a crucial driver of environmental change, and are increasingly becoming a global concern due to their impact on the local, regional and global environments [1,2]. Anthropogenic activities such as increasing population, urbanisation, deforestation, etc., are known to largely drive these changes [3,4]. In addition to anthropogenic activities, biophysical factors also drive environmental changes [5], e.g., climate variability. In semi-arid areas in particular, the effects of LULC changes on water resources are of great concern, as these areas are known to be more vulnerable to both natural and human-induced changes [6]. This often results in modifications in the hydrological cycle [7–9] by disrupting hydrological processes such as the interception of water by vegetation, soil water content, runoff quantity, surface evapotranspiration, base flow, groundwater recharge, percolation and the quality of water and sediments [10–15]. This therefore changes the hydrological regime and rainfall-runoff mechanisms of an area [16]. It is, however, recognized that the documentation and modeling of LULC patterns offer a better understanding of past and future LULC patterns and their related implications [17–20] and also guide future land use policies [21]. For water resources in semi-arid areas, this ensures the sustainability of catchments through proactive planning and management [22].

Remote sensing and geographic information systems (GIS) have over the years demonstrated a great potential in understanding landscape dynamics [10] and offer valuable techniques in the wider context of environmental monitoring [23]. Additionally, a number of models to simulate future LULC changes have also been developed. The models provide suitable tools for identifying spatial patterns in LULC. One widely used model with great accuracy is the integrated cellular automata (CA) and Markov chain model (CA-Markov) [6,20,24]. The robustness of the CA-Markov model offers a convenient approach to the spatial and temporal dynamic modelling of LULC changes in complex systems [25–28]. The model's ability to integrate geospatial and remotely sensed data and biophysical and socio-economic data enables a more comprehensive simulation of LULC changes from patterns to processes [29]. Numerous studies have successfully employed the CA-Markov model in LULC simulations for different purposes. For instance, Naboureh et al. [30] and Permatasari et al. [4] applied the CA-Markov model to predict land cover changes for the purposes of development planning, informing and formulating land use policies in Jiangle County, China, and Penajam Paser Utara Regency, Indonesia, respectively. In addition, in Zimbabwe, the CA-Markov model was used to simulate future LULCs in the Shashe sub-catchment, in order to detect their impact on the wetland area [6]. Another study by Regmi et al. [31] compared the CA-Markov model with the GEOMOD model to predict future LULC changes in the Phewa Lake Watershed, Nepal, and found that the integration of the CA and Markov models was effective in projecting future LULC scenarios compared to GEOMOD. Considering the model's vast applications in the field of LULC change analysis and its ability to broaden understandings about complex spatial systems, to the best of our knowledge, no study in Botswana has employed the CA-Markov model to simulate future LULCs, particularly in the country's important catchments. Therefore, this work will be among the first to test the applicability of the CA-Markov model to simulate future LULC patterns for the Gaborone dam catchment, Botswana, and their implications on future water resource management in the catchment.

The Gaborone dam catchment is one of the most important supply areas of freshwater for the metropolitan city of Gaborone. As a result of rural–urban migration and the population increase in the city of Gaborone, the catchment has been undergoing LULC changes since the 1980s [32]. Additionally, water supply challenges arising from shortfalls in Gaborone dam water levels have also been noted [33]. Moreover, the catchment is a typical representation of an environmentally fragile area due to its semi-arid state; as a result, LULC changes might affect its water quantity and quality. It is therefore necessary to understand the spatio-temporal characteristics of future LULCs for the improved sustainability of the catchment. Based on the background described above, this study simulated future LULC changes and analyzed the expected spatio-temporal patterns for the year 2035 in the Gaborone dam catchment using the CA-Markov model. The output of this study provides useful information for land use planners and water resource managers to make better decisions in improving future land use policies and formulating catchment management strategies within the framework of sustainable land use planning and water resource management. Similarly, the results of the study will assist in lessening the potential effects of LULC modifications in land systems [34,35]. Furthermore, the results of the study can be used for the hydrological assessment of the catchment with particular emphasis on the streamflow response.

2. Materials and Methods

2.1. Description of the Study Area

The transboundary Gaborone dam catchment, with a total area of 4000 km^2 (Figure 1), is located in the southeastern part of Botswana and the northwest of South Africa. It is an upstream sub-catchment of the Notwane catchment, which forms the headwaters of the Limpopo River Basin in Southern Africa (a transboundary basin between Botswana, South Africa, Zimbabwe and Mozambique). The Notwane River, an ephemeral river [36], forms the main drainage channel, among other tributaries such as the Metsemaswaane and Taung Rivers. The Gaborone dam acts as the catchment outlet. Geographically, the catchment is bordered by latitudes $-24.98°$ to $-25.55°$ and longitudes 25.24° to 26.07°. The catchment experiences a semi-arid climate with long sunny days and dry winter conditions on most days. Summers are hot and winters are cool, with the temperature averaging 33 °C and 20 °C, respectively. The annual rainfall received in the catchment is between 475 mm and 525 mm, with the bulk of the precipitation occurring in the summer months of December to February. Annual evaporation rates are estimated at about 2000 mm [37]. The area covering the catchment is one of the most densely populated in the country [12], with an estimated population of about 474,860 in 2011 and a projected 835,625 people expected to be living within the catchment area by 2024 [38]. Water use in the catchment area is mainly domestic and the demand for water has been growing, mainly due to the increasing population and expansion in industries. The catchment is largely rural, with agriculture as a major livelihood activity for household subsistence. Built-up areas (settlements) and areas of agriculture are the predominant land uses, with agricultural activities making up the largest land use [32]. The vegetation in the catchment can be described as the hard veldt type, consisting of tree and shrub savanna.

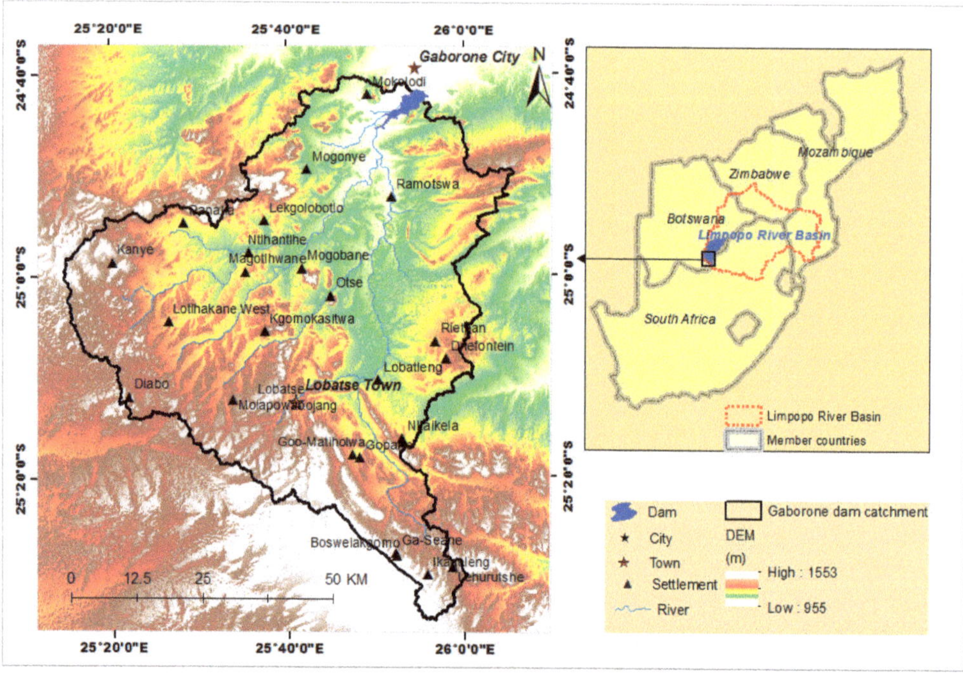

Figure 1. Locational map of the study area.

2.2. LULC Data Sources and Classification

The simulation of future LULCs in the Gaborone dam catchment was based on baseline LULC data (1984–2015) for the catchment, as described in Matlhodi et al. [32]. Therefore, this study is an extension of the previous study done in the catchment. In their previous work, Matlhodi et al. [32] acquired the dry season images of the Landsat Thematic Mapper (TM) and Operational Land Imager (OLI) for the years 1984, 1995, 2005 and 2015 (Table 1) from the United States Geological Survey (USGS) Earth Explorer (http://earthexplorer.usgs.gov, accessed on 10 March 2021). Supervised classification using the maximum likelihood classifier (MLC) algorithm in ERDAS Imagine 2015 was employed to measure the spatial and temporal patterns in LULCs in the catchment. Six major LULC categories (cropland, bare land, shrub land, built-up, tree savanna and water body) were identified based on the Food and Agricultural Organization (FAO) LULC classification system [39]. The classification results are shown in Figure 2. The overall classification accuracy ranged between 80% and 86%, with Kappa coefficients greater than 0.75 [32], implying an acceptable accuracy for LULC classification.

Table 1. Landsat image characteristics used for the study [32].

Year	Sensor	Spatial Resolution	Path/Row	No. of Bands	Date of Aquisition	Source
1984, 1995 and 2005	TM	30 m	p172 and r077/078	7	6 July 1984 5 July 1995 5 June 2005	USGS
2015	OLI	30 m	p172 and r077/078	11	12 July 2015	USGS

2.3. A Model Approach

For this study, the CA-Markov model was used to simulate future LULCs. CA-Markov is a change/time series model, integrating the standard Markov model and cellular automata (CA) for the prediction and modelling of future LULC changes. In the CA-Markov model, the Markov chain model, a discrete-temporal stochastic model, simulates LULC change probabilities using a transition probability matrix [40–42]. The transition probability between two states (t1 and t2) are then used to determine the transition trend among the different LULC states [19,43]. A disadvantage of the Markov chain model is its inability to provide a spatial distribution of the LULC occurrences, but rather an estimate of the magnitude of these changes [24,44]. Cellular automata (CA) however, have the ability to change and control complex spatially distributed processes, with a strong ability to simulate the spatio-temporal characteristics of complex systems [45,46]. The model is based on the interaction of the cellular space, grid size, cell neighborhood and transition rules [47–49] to generate rich patterns and effectively represent non-linear spatially stochastic LULC change processes [50]. The models are expressed as shown in Equations (1)–(3) (Markov chain model) [26,51,52] and Equation (4) (CA) [53].

$$P_{ij} = \begin{bmatrix} P_{11} & \cdots & P_{1n} \\ \vdots & \vdots & \vdots \\ P_{n1} & \cdots & P_{nm} \end{bmatrix} \quad (1)$$

$$(0 \leq P_{ij} < 1 \; and \; \sum_{j=1}^{n} P_{ij} = 1, \; i,j = 1,2,\ldots,n) \quad (2)$$

$$S_{t+1} = P_{ij} \times S_t \quad (3)$$

where P_{ij} is the state transition probability matrix and n is the LULC type number; S is LULC status and t; $t + 1$ is the time point.

$$S(t+1) = f(S(t), N) \quad (4)$$

where $S(t+1)$ and $S(t)$ are respectively the states of all cells at $t+1$ and t, N is the neighborhoods of cells and $f(.)$ is the transition function that defines the changes of states based on predefined rules [54]. The methodological framework used in this study is presented in Figure 3.

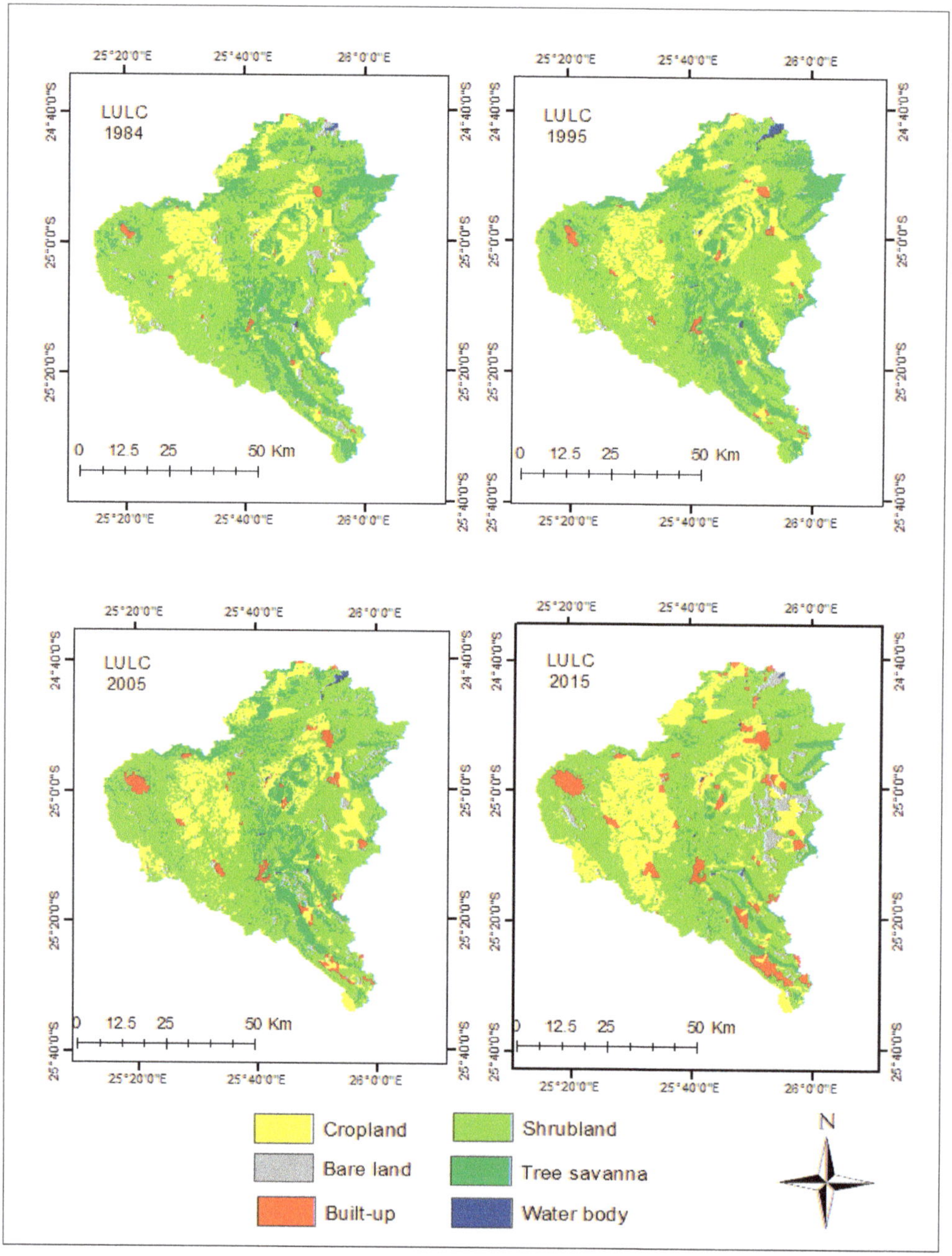

Figure 2. Classified Landsat images for 1984, 1995, 2005 and 2015 for the Gaborone dam catchment [32].

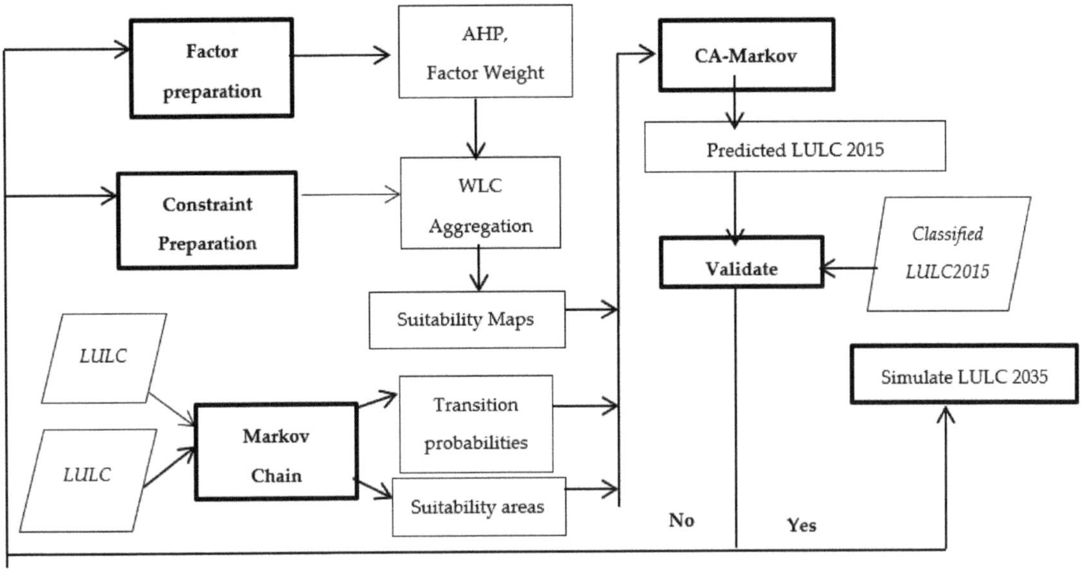

Figure 3. Modelling framework for LULC change simulation.

2.4. LULC Modelling with CA-Markov Model

The CA-Markov model in IDRISI was employed in this study to simulate future LULCs in the Gaborone dam catchment. The LULC data as derived in Matlhodi et al. [32] (see in Section 2.2) were first used to derive the transition area files and transition probabilities between 1984–1995, 1995–2005 and 2005–2015 for each LULC category using the Markov module. Transition area files carry the area of cells expected to change in the next time period, whereas the transition probabilities show the possibilities that a pixel of a given LULC category will remain the same or change to any other category in the next time period [55].

We also derived the transition suitability maps for each LULC category using a collection of factors and constraints. This was done using multi-criteria analysis (MCE) and based on expert understanding of the interactions and influence of the factors on LULC [17]. Constraints are criteria that limit LULC change and are expressed in the form of Boolean images (0 to 1), with the value of '0' representing areas restricted for suitability analysis (e.g. water bodies), whereas '1' represents areas suitable for suitability analysis. Factors, on the other hand, are mostly distance criteria and give a degree of suitability for an area to change [56] (e.g., distance to roads or rivers [26,57]). Other factors include slope (which determines the land's usefulness for humans) and elevation, which is a good predictor for areas suitable for agriculture [58]. Factors were then subjected to standardisation through fuzzy membership functions, where factors were stretched from 0–255 using different fuzzy functions and control points, e.g., as sigmoidal, J-shaped and linear [58], as shown in Table 2. The analytic hierarchy process (AHP) function was then used to derive weights for each factor based on the relative importance of the factor in influencing the suitability of a certain landscape surface for a given LULC type [59]. The factors and constraints used are shown in Figure 4.

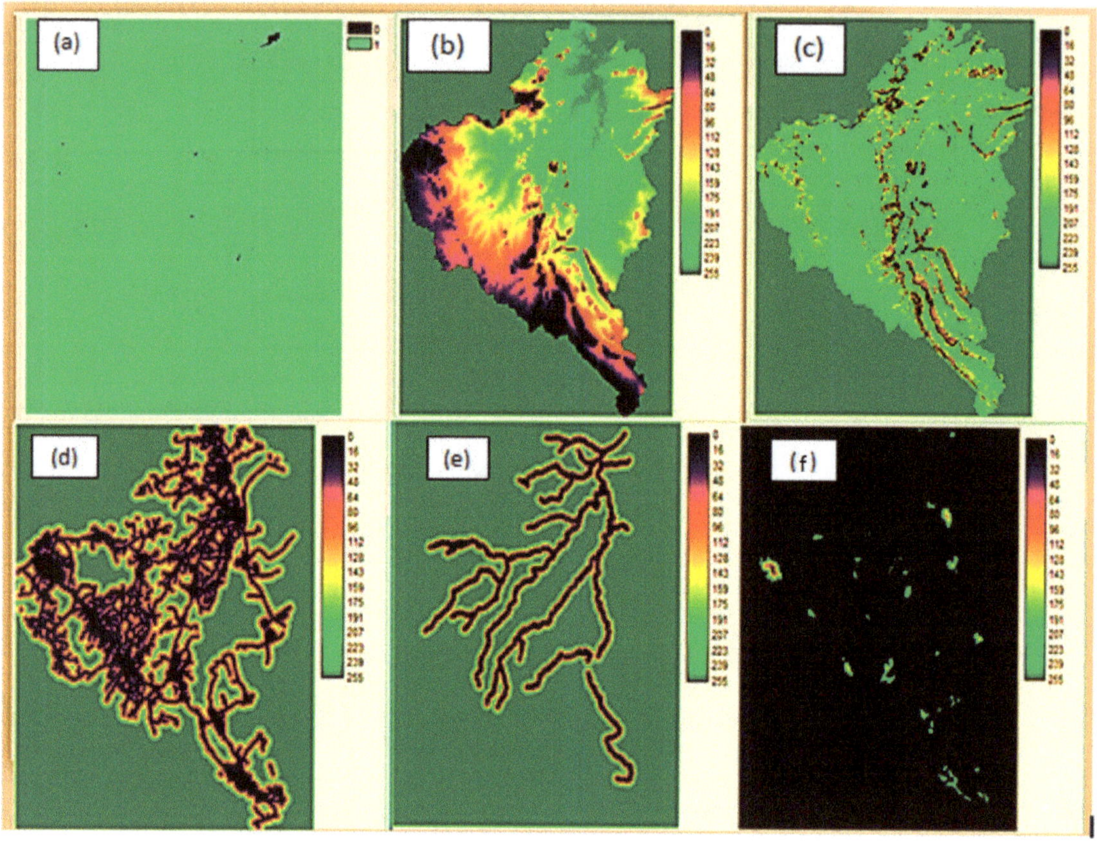

Figure 4. Factor and constraint maps—water body (**a**); elevation (**b**); slope (**c**); distance to roads (**d**); distance to rivers (**e**) distance to built-up areas (**f**).

The LULC of 2015 was then simulated using the LULC of 2005 as the base map, transition area matrix and the transition probabilities of 1995–2005 for model validation purposes. A 5 × 5 contingency filter was employed to generate a spatially explicit contiguity-weighting factor to change the state of cells based on the adjacent cellular state [60].

Table 2. Factors, constraints, membership functions and control points used for the preparation of suitability maps (modified from Gidey et al. [61] and Hyandye and Martz [25]).

Factors	Membership Function Shape	Membership Function	Control Points
Distance to road (m)	Monotonically increasing	Linear	a = 0, b = 4000
Distance to river (m)	Monotonically increasing	Sigmoidal	a = 0, b = 4000
Distance to built-up areas (m)	Symmetric	Linear	a = 1, b = 20 c = 30, d = 1700
Elevation (m)	Monotonically decreasing	Sigmoidal	c = 1000, d = 1400
Slope (°)	Monotonically decreasing	Linear	c = 0, d = 15

2.5. CA-Markov Model Validation

As stated by Eastman [55], the usefulness of any predictive change model depends on the output of the validation process. In this study, the performance of the CA-Markov model was assessed using Kappa indices, which include the Kappa for no information (K_{no}), Kappa for location ($K_{location}$) and Kappa standard ($K_{standard}$), as described in Table 3 [26,62]. Kappa indices discriminate between errors of quantity and errors of location between two qualitative maps [27,63] and are calculated using Equations (5)–(7) [64]. The result is usually between 0 and 1. The level of agreement for Kappa indices is as follows: Kappa $\leq$ 0.5 indicates rare agreement; 0.5 $\leq$ Kappa $\leq$ 0.75 indicates a medium level of agreement; 0.75 $\leq$ Kappa < 1 indicates a high level of agreement; and Kappa = 1 denotes perfect agreement [62,65].

Table 3. Definitions of Kappa indices.

Kappa Index	Definitions of the Kappa Index Agreement
K_{no}	Indicates the proportion classified correctly relative to the expected proportion classified correctly by simulation, with no ability to specify the quantity of location accurately
$K_{location}$	Measures the validation between the classified maps and the simulated map based on a specified location
$K_{standard}$	The proportion assigned correctly versus the proportion that is correct by chance

Using the "validate" module the classified LULC of 2015 and the simulated LULC of 2015 were compared. With acceptable Kappa indices, the CA-Markov model was thus adoptable for simulation of the LULC map of 2035 using the 2005–2015 transition probabilities and the 2015 classified map as the basis map.

$$K_{no} = (M(m)N(n))/(P(p) - N(n)) \quad (5)$$

$$K_{location} = (M(m)N(m))/(P(m) - N(m)) \quad (6)$$

$$K_{standard} = (M(m)N(n))/(P(p) - N(m)) \quad (7)$$

where no information is defined by $N(n)$, medium grid cell-level information by $M(m)$, $N(m)$ and $P(m)$, and perfect grid cell-level information across the landscape by $P(p)$.

3. Results

3.1. CA-Markov Model Validation

To facilitate the simulation of future LULCs in the Gaborone dam catchment using the CA-Markov model and in order to validate the CA-Markov model's ability to simulate future LULCs, the LULC for 2015 (Figure 5) was first simulated using area transition files and transition probabilities between the LULCs of 1995–2005. We then compared the simulated and classified LULCs for 2015 using Kappa variations. The simulation revealed Kappa variation estimates as follows 0.82 for K_{no}, 0.82 for $K_{location}$ and 0.76 for $K_{standard}$, revealing a high level of agreement. This shows the model's reliability and strength in simulating future LULC changes in the Gaborone dam catchment. In addition, visual analysis revealed that the LULC categories in the simulated LULC for 2015 relatively corresponded to the classified LULC categories for 2015. It also revealed a slight overestimation of the tree savanna and water body categories, whereas built-up and cropland categories were shown to be marginally under simulated. This is also cemented by the difference calculated between the classified and simulated LULCs for 2015, as presented in Table 4. The difference revealed an underestimation of about 20.07%, 23.98%, 3.64% and 14.74% in the cropland, bare land, shrubland and built-up categories, respectively. However, the model was shown to overestimate LULC categories such as tree savanna (144.16%) and water body (3.34%). This observatory analysis is also reported in Mashapa et al. [66].

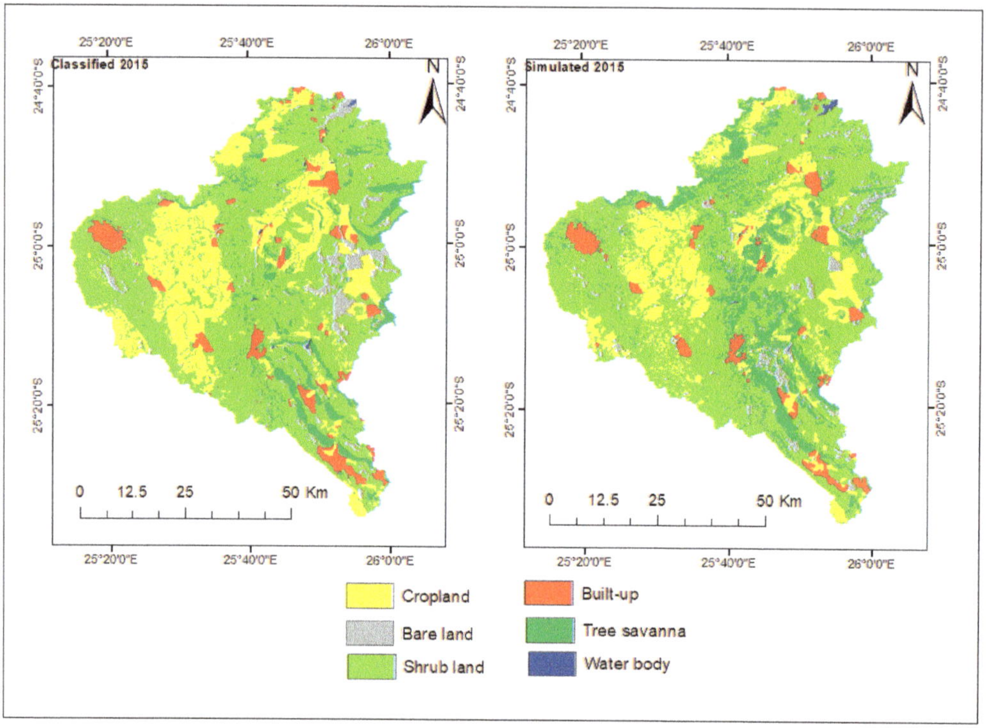

Figure 5. Classified and simulated LULC maps for 2015.

Table 4. LULC change analysis between the classified and simulated LULCs for 2015.

	Classified LULC 2015 (km^2)	Simulated LULC 2015 (km^2)	LULC Differences (km^2)	LULC Differences (%)
Cropland	1007.77	805.55	−202.21	−20.066
Bare land	179.79	136.67	−43.11	−23.981
Shrubland	2600.35	2505.80	−94.55	−3.636
Built-up	227.15	193.66	−33.49	−14.742
Tree savanna	326.85	699.99	373.14	114.164
Water body	6.77	6.99	0.23	3.344
Total	4348.66	4348.66		

3.2. Past LULC Changes and Markovian Probability Transition Matrices

Building on the first phase of the research analysis presented in Matlhodi et al. [32], six LULC classes were identified, namely, cropland, bare land, shrubland, built-up, tree savanna and water body [39] in 1984, 1995, 2005 and 2015. LULC change analysis, as shown in Figure 6, indicate a general increase in croplands, built-up and bare lands, whereas tree savanna and water body were observed to be decreasing between 1984 and 2015. Croplands, built-up and bare land categories covered about 15.02%, 2.81% and 0.61% in 1984 and 23.17%, 4.13% and 5.22% in 2015, respectively. The shrubland category, on the other hand, was shown to be consistent over the 31-year period, taking up 59.71% and 59.80% of the catchment area in 1984 and 2015, respectively. Even though tree savanna and water body categories were shown to be decreasing, an increase in these categories was observed in the period 1984–1995. LULC change analysis, as shown in Figure 7, indicated water bodies to have had the highest increase in the period 1984–1995, followed by bare land in the period 1995–2005. However, the bare land category was shown to have decreased by 74.72% in the period 1984–1995. The tree savanna category decreased by 21.52% and 56.76% in the periods 1995–2005 and 2005–2015, respectively. The increase in tree savanna and water body categories in the period 1984–1995 could be a result of the high rains received in the early 1990s. It has been reported that in semi-arid areas the health of the two categories are highly dependent on moisture availability at a particular time [32]. In contrast, the drought periods of the early 1980s and 2014–2016 experienced in Southern Africa contributed to the low areal coverages of water bodies in the Gaborone dam catchment, at about 0.13% (1984) and 0.16% (2015) [67].

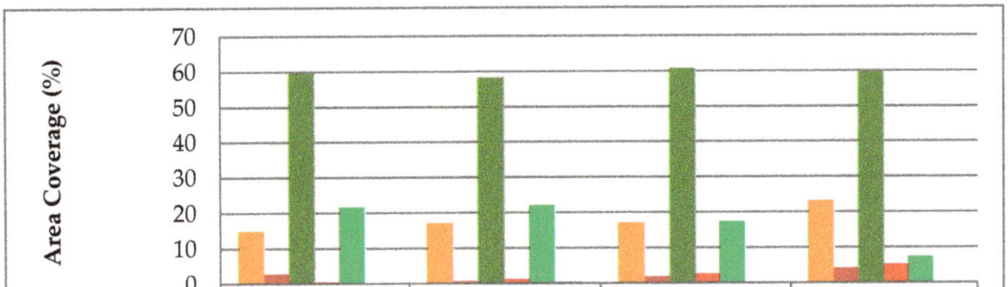

	1984	1995	2005	2015
Cropland	15.02	17.10	17.13	23.17
Bareland	2.81	0.71	1.86	4.13
Shrubland	59.71	58.33	60.72	59.80
Built-up	0.61	1.26	2.63	5.22
Tree savanna	21.72	22.15	17.38	7.52
Waterbody	0.13	0.44	0.28	0.16

Figure 6. LULC area coverages in the Gaborone dam catchment from 1984 to 2015 (modified from [32]).

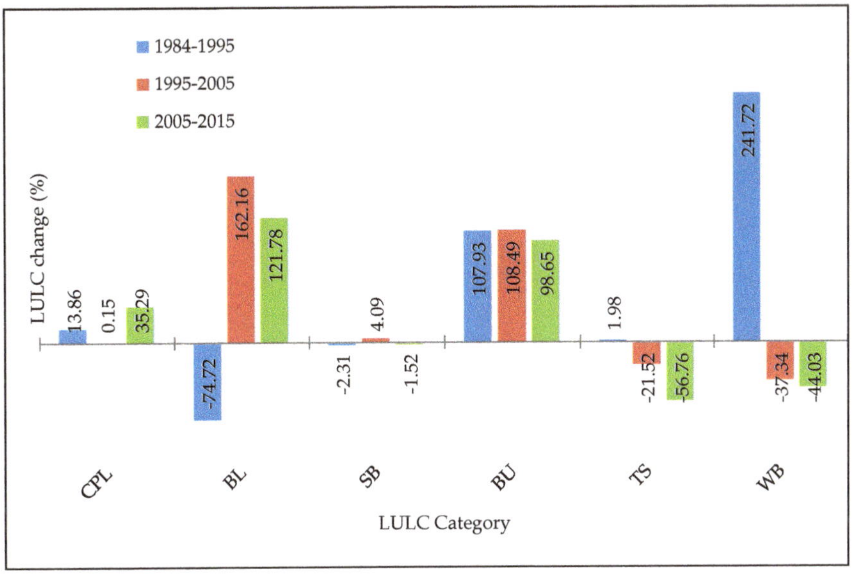

Figure 7. LULC changes in the Gaborone dam catchment from 1984 to 2015 [32]. NB: CPL—cropland; BL—bare land; SB—shrubland; BU—built-up; TS—tree savanna; WB—water body.

The Markovian transition probabilities presented in Table 5 show the probability of each class in the LULC maps remaining the same or changing to another class in the next period (i.e., 1984–1995, 1995–2005 and 2005–2015). That is, a given LULC type might change from one category of LULC to any other at any given time. This shows how the LULC change is not uni-directional in nature [61]. Diagonal values in the transition probability matrix represent the probability that each land use category remains the same from time 0 to time 1 [34]. For example, the built-up category was shown to be highly resistant to change to other categories, with its >70% likelihood of remaining the same in the periods 1984–1995 and 1995–2005. The highest percentage (>80%) of values in the built-up category to remain unchanged was recorded in the period 2005–2015. Categories such as cropland and shrubland were also shown to be persistent, with more than 60% of them likely to remain unchanged in the next period, whereas other categories were shown to be less persistent, and as a result were more likely to change to other LULC categories in the ensuing period. These include bare land and tree savanna, of which about <25% and <60% of them will remain unchanged in the next period. The water body category instead showed both high and low likelihoods of remaining unchanged during the 31-year period, with more than 79% of this category remaining unchanged in the period 1984–1995. In addition to showing persistence, the transition probability matric also shows how each LULC category changes to another. For instance, 70% or more of the tree savanna category was likely to be changed to the shrubland category in the period 2005–2015. Additionally, about 71% of the bare land category was highly likely to convert to shrublands in the period 1984–1995, indicating its less likely persistence during those periods. The high transitional potential of tree savanna to shrub lands, as explained by Mashapa et al. [66], is due to relatively close proximity of the two categories, suggesting that the CA-Markov model's simulation accuracy increases with the proportion of a given LULC category relative to others.

Table 5. Markov probability transition matrices for 1984–1995, 1995–2005 and 2005–2015 LULCs in the Gaborone dam catchment.

	Transition Years	Cropland	Bare Land	Shrubland	Built-Up	Tree Savanna	Water Body
Cropland	1984–1995	0.668	0.0053	0.2944	0.0042	0.028	0.0001
	1995–2005	0.6242	0.0097	0.3345	0.0141	0.0175	0
	2005–2015	0.7085	0.0403	0.1821	0.0644	0.0039	0.0007
Bare land	1984–1995	0.1288	0.0597	0.7106	0.0282	0.0091	0.0637
	1995–2005	0.1222	0.2309	0.5912	0.0424	0.0132	0.0001
	2005–2015	0.0988	0.2531	0.5414	0.0484	0.0534	0.0048
Shrubland	1984–1995	0.1265	0.0128	0.6795	0.0155	0.1629	0.0027
	1995–2005	0.1211	0.0406	0.6858	0.0354	0.1168	0.0003
	2005–2015	0.1976	0.0695	0.6355	0.0446	0.0523	0.0005
Built-up	1984–1995	0.039	0.0086	0.237	0.7069	0.0069	0.0015
	1995–2005	0.0695	0.0034	0.1718	0.7474	0.0079	0
	2005–2015	0.0877	0.0442	0.046	0.8149	0.0071	0.0001
Tree savanna	1984–1995	0.0074	0.0005	0.3661	0.0011	0.623	0.0019
	1995–2005	0.0082	0.0045	0.4839	0.0034	0.4997	0.0002
	2005–2015	0.0269	0.0073	0.703	0.008	0.2545	0.0002
Water body	1984–1995	0.0032	0	0.1323	0	0.0683	0.7962
	1995–2005	0.0089	0.0876	0.3678	0.0016	0.0302	0.5038
	2005–2015	0.0065	0.5933	0.037	0.0028	0.0067	0.3538

3.3. Simulation of LULC Changes for 2035

The 2035 LULC for the Gaborone dam catchment was simulated using the 2015 LULC as the base map, the 2005–2015 transition probabilities and the generated suitability images, and the results are presented in Figure 8. The respective descriptive statistics of LULC showing the areal coverage and spatio-temporal changes between 2015 and 2035 LULCs are shown in Table 6. LULC change analysis revealed that 23.17% of the total land area in the catchment was croplands in 2015 and is expected to register a 26.9% growth by 2035. Similarly, built-up areas are anticipated to double in proportion, rising from 227.15 km^2 (5.22%) in 2015 to 376.29 km^2 (8.65%) in 2035. A projected increase in the bare land category is expected from 179.79 km^2 (4.13%) in 2015 to 280.07 km^2 (6.44%) in 2035. Modelling results further indicated that shrubland, tree savanna and water body categories are expected to decrease over the next 20 years. In particular, shrublands will decrease by 416.84 km^2 (16.03%). Even the tree savanna and water body categories areal coverage were found to decrease by about 94.01 km^2 (28.76%) and 1.48 km^2 (21.89%), respectively.

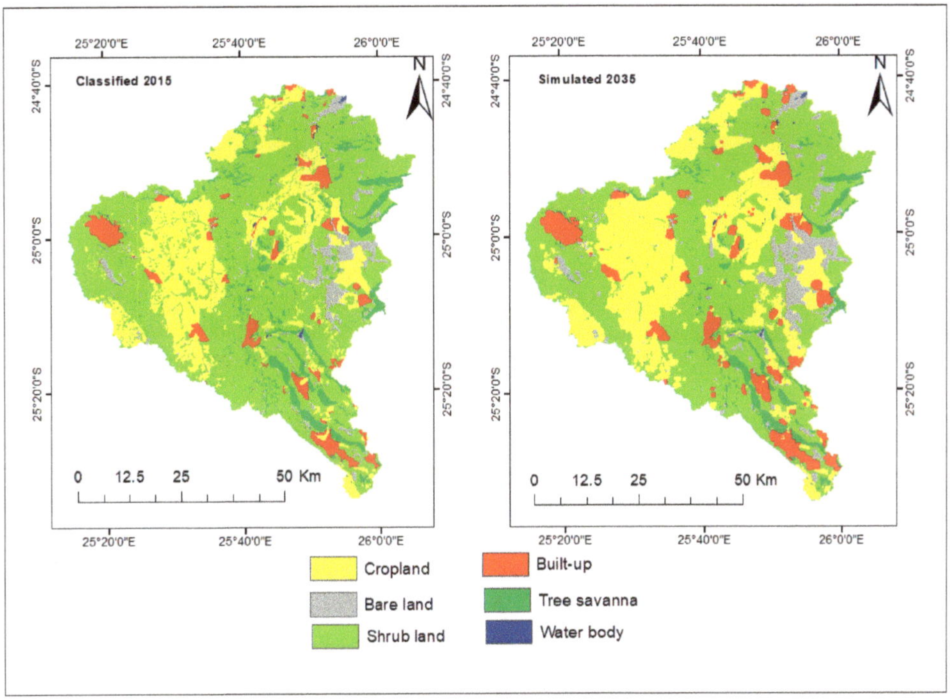

Figure 8. LULC distribution in the Gaborone dam catchment for 2015 (classified) and 2035 (simulated).

As the environment continues to change due to human pressures, LULC categories are expected to convert into other categories. Figures 9 and 10 show the projected losses and gains in LULC categories in the catchment. The results reveal that built-up, bare land and cropland categories are expected to be the main gaining LULC categories by 2035. For instance, areas of about 311.9 km^2, 108.87 km^2 and 83.79 km^2 in shrubland are projected to change into cropland, built-up and bare land categories, respectively. Moreover, tree savanna is projected to lose about 91.67 km^2 of its area coverage to the shrubland category. The tree savanna and water body categories' expected gains are insignificant compared to the other categories. The observed LULC changes in the catchment may lead to high surface runoff, low infiltration and evapotranspiration as a result of impervious surfaces due to the expansion in built-up areas and the loss of vegetal cover. Thus, the current and ongoing trends in LULCs will eventually result in significant environmental dilapidation and devastation in future. Moreover, as the area is a semi-arid catchment, this calls for appropriate management and use of land with a particular emphasis on sustainable water resource management.

Table 6. LULC change distribution from 2015–2035.

	2015		2035		LULC Changes between 2015–2035	
LULC Type	**Area (km^2)**	**Area (%)**	**Area (km^2)**	**Area (%)**	**Area (km^2)**	**Area (%)**
Cropland	1007.77	23.17	1270.67	29.22	262.91	26.09
Bare land	179.79	4.13	280.07	6.44	100.29	55.78
Shrubland	2600.35	59.80	2183.51	50.21	−416.84	−16.03
Built-up	227.15	5.22	376.29	8.65	149.13	65.65
Tree savanna	326.85	7.52	232.84	5.35	−94.01	−28.76
Water body	6.77	0.16	5.28	0.12	−1.48	−21.89

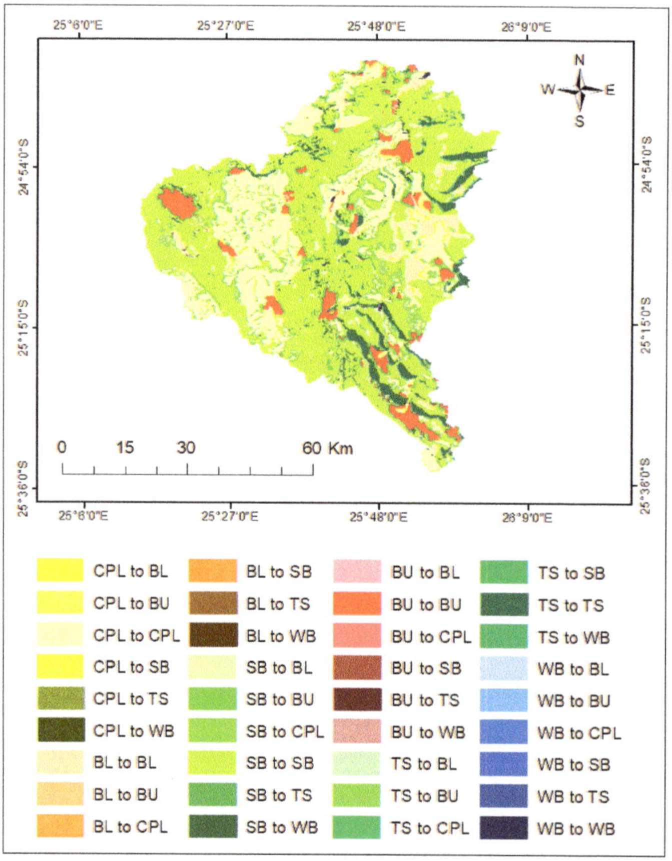

Figure 9. Projected losses and gains between 2015 and 2035. NB: CL—cropland, BL—bare land, SB;—shrubland, BU—built-up, TS—tree savanna, WB—water body.

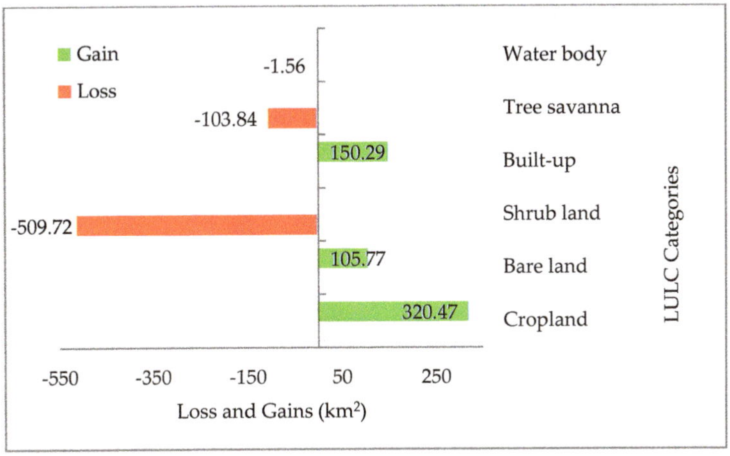

Figure 10. Transition change matrix from 2015–2035 (km^2) (losses and gains).

4. Discussion

4.1. CA-Markov Model Validation

Our simulation of future LULCs in the Gaborone dam catchment employed the CA-Markov model. To test the model's performance in simulating LULCs, model validation was done by comparing the classified and simulated LULCs of 2015. Kappa variations between 0.76 and 0.82 were recorded, indicating a high level agreement [62,65,68] and the suitability of CA-Markov to accurately simulate future LULCs in the Gaborone dam catchment. However, the model was shown to overestimate water body and tree savanna areas and underestimate cropland, built-up, bare land and shrubland LULC categories. Acceptable Kappa indices for the CA-Markov model have also been reported by other researchers (e.g., Munthali et al. [20], Hyandye and Martz [25] and Maviza and Ahmed [24], in the Dedza district, Malawi; Usangu catchment, Tanzania; and Upper Mzingwane sub-catchment, Zimbabwe, respectively). Therefore, the CA-Markov model is suitable for the accurate prediction of future spatio-temporal LULC changes in the Gaborone dam catchment.

4.2. Simulation of LULC Changes for 2035

LULC is expected to continue to change in the Gaborone dam catchment area, with expansions and contractions in the different LULC categories, which might have implications on the environment, particularly water resources. Expansions are expected in the croplands, built-up and bare land categories, whereas the tree savanna, water body and shrubland categories are projected to decrease by 2035. This will have implications on the catchment's ecological integrity and water resources [69]. Expansions in built-up areas in the catchment is an indication of population growth, largely driven by rural-urban migration to the city of Gaborone, as settlements surrounding cities and towns tend to absorb the city's overspill, as also noted in other developing countries (e.g., Gidey et al. [61], Maviza and Ahmed [24] and Munthali et al. [20]). For the Gaborone dam catchment, this means major settlements are slowly urbanizing and taking up an urban or town setting, with improved amenities thus attracting more people. Due to population growth and agricultural subsidy schemes such as the Integrated Support Program for Arable Agricultural Development (ISPAAD; 2008) [70], which is still in place, an increase in cropland is expected. The ISPAAD has so far led to an increase in cultivated acreage [71]. Vegetal cover loss, on the other hand, is mostly due to expansions in cropland and built-up areas. However, in semi-arid areas like the Gaborone dam catchment, vegetal cover loss can also be attributed to climate variability [25] as vegetation health in these areas is naturally moisture- and temperature-dependent [24,72,73], with high rainfall years signifying healthy vegetation. Continued vegetal cover loss may lead to forage reductions for those practicing pastoral farming, resulting in bare land expansions and land degradation due to animal overstocking.

The water body category was observed to be decreasing in the Gaborone dam catchment since the 1980s [32]. Fluctuations in water levels in the catchment have been observed, with the Gaborone dam in 2015 reported to have dried up, with water levels reaching below 5%, which exacerbated water supply shortages and droughts, resulting in water supply restrictions in the city of Gaborone and surrounding areas. Thus, LULC changes upstream of the catchment may disturb water components such as runoff and base flow [14,74] and could lower the dam water levels [75]. Furthermore, climate variability, as a characteristic of semi-arid areas, affects water availability and with projected mean annual reductions in rainfall over Botswana [76–78], a decline in water bodies is thus certain.

4.3. Implications of LULC Changes on Water Resources

Changes in LULC have been noted to have implications on the environment, in particular the water resources of semi-arid areas. The impact of LULC changes on the hydrological regime is, however, a complex process and depends on the land use types, the size of the affected areas and their landscape location. Among the LULC changes, deforestation is the most prevalent and is considered a major cause of changes in hydrological processes such as evapotranspiration, surface runoff, infiltration and rainfall interception [79,80]. Expansions in cropland and built-up areas have been associated with a decrease in vegetal cover in most areas of the world [79], and this has led to a decrease in the infiltration capacity of soils due to soil compaction in agricultural lands and increases in impervious surfaces in urban areas. Croplands are known to result in reductions in evapotranspiration, as crops transpire at relatively shorter intervals than forests/woodlands [81]. Nonetheless, these LULC categories show low porosity and a high release of water into the catchment outlet, leading to high water yields. Vegetal cover loss likewise leads to less infiltration, implying low groundwater recharges and baseflow [82], low evapotranspiration rates, high surface runoff and peak flows [79]. However, semi-arid areas are characterised by high temperatures and rainfall variability, so surface water evaporation is still high. For instance, surface water evaporation in Botswana is estimated at about 2000 mm per year [37], which exceeds the annual rainfall received (<600 mm), adding to the low water yield in these catchments. With high evapotranspiration, there is a decrease in baseflow and subsequently in streamflow and water yield. A decrease in water yield may result in water scarcity, distressing the water supply.

Another subject of interest is the impact of small farm dams on water yield and runoff. These structures have been found to lead to reductions in streamflow yield [83], annual runoff volumes, peak flows and increased periods of low flows [84]. Within the Gaborone dam catchment, for instance, there is evidence of an increasing number of small farm dams over the last few decades [85–87], which have been reported to lead to a 25% reduction in runoff and a 30% decline in yield at the Gaborone dam [85,86]. Additionally, small farm dams have been found to result in substantial evaporation losses [87]. Therefore, their cumulative impact on downstream yield is significant and should be considered for future water resource management. Future predictions on the effects of small farm dams warn of a significant decrease in streamflows under agricultural intensification [84].

Apart from water quantity, changes in LULC also affect the quality of water resources. Agro-chemicals, due to expansions in croplands, could leak or leach through the soil, causing pollution of nearby water sources. Settlements, especially in developing countries, lack proper sanitation and waste disposal facilities (both household and industrial) and this could result in the pollution of water sources. Groundwater pollution is found to be prevalent (e.g., in Botswana and South Africa) with high quantities of nitrates, faecal coliforms and caffeine. The village of Ramotswa (within the Gaborone dam catchment; see Figure 1) has experienced the worst cases of groundwater pollution due to nitrate and faecal coliform pollution, as stated by [67,88,89], from the use of unlined pit latrines, septic tanks and dumpsites [88]. In South Africa, the presence of nitrates, chlorides and phosphates are attributed to farming and wastewater treatment facilities [90]. With the projected LULC changes in the Gaborone dam catchment, water resources are expected to continue to be impacted; therefore, it is important that appropriate measures are taken to avert the impacts of LULC changes on water resources, especially in these water-stressed catchments.

The CA-Markov model showed great efficiency in modelling the spatio-temporal LULC changes in the Gaborone dam catchment, implying that the area is prone to change and requires extensive land use planning and water resource management by both Botswana and South Africa. Past LULC changes, used together with LULC drivers/factors, have shown their applicability in simulating future LULC changes, as observed by [34,48,91], and should be documented for future reference. However, since this study was limited in the use of driving factors, and because future LULC patterns are uncertain, factors such as population growth, climate variability, natural hazards and socio-economic developments should also be considered in order to enhance our understanding of LULC change processes.

5. Conclusions

This study attempted to simulate future LULC changes in the Gaborone dam catchment, Botswana, from 2015 to 2035 using the CA-Markov model, which assisted in giving a better understanding of possible LULC changes within the study area. The Gaborone dam catchment is expected to continue to experience LULC changes in the future, mainly expansions in the cropland, built-up areas and bare land categories. Cropland, built-up and bare land categories are expected to cover about 29.22%, 8.65% and 6.44% of the catchment area in 2035, as compared to 23.17%, 5.22% and 4.13% in 2015, respectively. On the other hand, categories such as shrubland, tree savanna and water bodies are projected to decrease. Reductions in vegetal cover will occur due to expansions in cropland and built-up areas. Urbanisation and population growth stemming from rural-urban migrations to the city of Gaborone are expected to result in built-up expansions, whereas agricultural subsidy schemes will lead to expansions in croplands in the catchment. Apart from human drivers of LULC change, as a semi-arid area, climate variability may have implications on moisture and temperature variations, which determine vegetal cover health and water body expanses. The decline of potential water sources may also increase water supply challenges in the already water-stressed catchment and exacerbate droughts. The projected changes in LULCs in the Gaborone dam catchment will have implications on water resources, such as reductions in streamflows, catchment water yield (groundwater and surface water) and evapotranspiration and water quality. Therefore, this requires land use planers and water resource managers to take cognizance of the impacts of the changing LULCs in the catchment on water resources for improved future water resource management. This study showed that the CA-Markov model has the ability to simulate LULC changes in the Gaborone dam catchment by providing realistic LULC patterns for the year 2035. However, the consideration of other LULC-forcing factors, such as climate, socio-economic and census dynamics, which was beyond the scope of this work, could further enhance our understanding of the LULC changes in rapidly urbanizing cities such as Gaborone city. The output of this study provides useful information to land use planners and water resource managers to make better decisions in improving future land use policies and formulating catchment management strategies within the framework of sustainable land use planning and integrated water resources management (IWRM). The results can also be used in hydrological assessments of the catchment for the sustainability of the Gaborone dam.

Author Contributions: Conceptualisation, methodology, validation, formal analysis, investigation, data curation, writing-original draft preparation and final report, B.M.; supervision and writing-review and editing, P.K.K., B.P.P. and J.G.M.; funding acquisition, P.K.K. All authors have read and agreed to the published version of the manuscript.

Funding: This work was supported by the Southern African Science Service Center for Climate Change and Adaptive Land Management (SASSCAL) Task 337 and the Wetland Assessment and Monitoring of transboundary basins in Southern Africa (WeMAST Project) funded under the Global Monitoring for Environment and Security (GMES) and Africa Support Programme between the African Union and the European Union, in the area of space science and technology and a key priority under the EU-Africa partnership.

Acknowledgments: This study is part of an ongoing research being undertaken for PhD work of the first author. The authors would like to acknowledge support and funding from the Southern African Science Service Center for Climate Change and Adaptive Management (SASSCAL) and the Wetland Assessment and Monitoring of transboundary basins in Southern Africa (WeMAST Project) funded under the Global Monitoring for Environment and Security (GMES) and Africa Support Programme between the African Union and the European Union.

Conflicts of Interest: The authors declare no conflict of interest regarding the publication of this manuscript.

References

1. Karki, S.; Thandar, A.M.; Uddin, K.; Tun, S.; Aye, W.M.; Aryal, K.; Kandel, P.; Chettri, N. Impact of land use land cover change on ecosystem services: A comparative analysis on observed data and people's perception in Inle Lake, Myanmar. *Environ. Syst. Res.* **2018**, *7*, 25. [CrossRef]
2. Tolessa, T.; Senbeta, F.; Kidane, M. The impact of land use/land cover change on ecosystem services in the central highlands of Ethiopia. *Ecosyst. Serv.* **2017**, *23*, 47–54. [CrossRef]
3. Benzer, N. Using the geographical information system and remote sensing techniques for soil erosion assessment. *Pol. J. Environ. Stud.* **2010**, *19*, 881–886.
4. Permatasari, R.J.; Damayanti, A.; Indra, T.L.; Dimyati, M. Prediction of Land Cover Changes in Penajam Paser Utara Regency Using Cellular Automata and Markov Model. In *Earth and Environmental Science*; IOP Conference Series; IOP Publishing: Semarang, Indonesia, 2021.
5. Hossain, F.; Moniruzzaman, M. Environmental Change Detection through Remote Sensing Technique: A Study of Rohingya Refugee Camp Area (Ukhia and Teknaf Sub-district), Cox's Bazar, Bangladesh. *Environ. Chall.* **2021**, 100024. [CrossRef]
6. Sibanda, S.; Ahmed, F. Modelling historic and future land use/land cover changes and their impact on wetland area in Shashe sub-catchment, Zimbabwe. *Model. Earth Syst. Environ.* **2021**, *7*, 57–70. [CrossRef]
7. Dibaba, W.T.; Demissie, T.A.; Miegel, K. Watershed Hydrological Response to Combined Land Use/Land Cover and Climate Change in Highland Ethiopia: Finchaa Catchment. *Water* **2020**, *12*, 1801. [CrossRef]
8. Shi, Y.; Xiao, J.; Shen, Y.; Yamaguchi, Y. Quantifying the spatial differences of landscape change in the Hai River Basin, China, in the 1990s. *Int. J. Remote. Sens.* **2011**, *33*, 4482–4501. [CrossRef]
9. Zhang, K.; Ruben, G.B.; Li, X.; Li, Z.; Yu, Z.; Xia, J.; Dong, Z. A comprehensive assessment framework for quantifying climatic and anthropogenic contributions to streamflow changes: A case study in a typical semi-arid North China basin. *Environ. Model. Softw.* **2020**, 104704. [CrossRef]
10. Armin, M.; Majidian, M.; Kheybari, V.G. Land Use/Land Cover Change Detection and Prediction in the Yasouj City Suburbs in Kohgiluyeh Va Boyerahmad Province in Iran. *Arid. Ecosyst.* **2020**, *10*, 203–210. [CrossRef]
11. Brasil, J.B.; De Andrade, E.M.; Palácio, H.A.D.Q.; Medeiros, P.H.A.; Dos Santos, J.C.N. Characteristics of precipitation and the process of interception in a seasonally dry tropical forest. *J. Hydrol. Reg. Stud.* **2018**, *19*, 307–317. [CrossRef]
12. Franchi, F.; Ahad, J.M.E.; Geris, J.; Jhowa, G.; Petros, A.K.; Comte, J.-C. Modern sediment records of hydroclimatic extremes and associated potential contaminant mobilization in semi-arid environments: Lessons learnt from recent flood-drought cycles in southern Botswana. *J. Soils Sediments* **2019**, *20*, 1632–1650. [CrossRef]
13. Mohan, C.; Western, A.W.; Wei, Y.; Saft, M. Predicting groundwater recharge for varying land cover and climate conditions—A global meta-study. *Hydrol. Earth Syst. Sci.* **2018**, *22*, 2689–2703. [CrossRef]
14. Mulungu, D.M.; Kashaigili, J.J. Dynamics of land use and land cover changes and implications on river flows in Simiyu River catchment, Lake Victoria Basin in Tanzania. *Nile Basin Water Sci. Eng. J.* **2012**, *5*, 23–35.
15. Teklay, A.; Dile, Y.T.; Setegn, S.G.; Demissie, S.S.; Asfaw, D.H. Evaluation of static and dynamic land use data for watershed hydrologic process simulation: A case study in Gummara watershed, Ethiopia. *Catena* **2019**, *172*, 65–75. [CrossRef]
16. Li, L.; Jiang, D.; Li, J.; Liang, L.; Zhang, L. Advances in hydrological response to land use/land cover change. *J. Nat. Resour.* **2007**, *22*, 211–224.

17. Behera, M.D.; Borate, S.N.; Panda, S.N.; Behera, P.R.; Roy, P.S. Modelling and analyzing the watershed dynamics using Cellular Automata (CA)–Markov model—A geo-information based approach. *J. Earth Syst. Sci.* **2012**, *121*, 1011–1024. [CrossRef]
18. Islam, M.S.; Ahmed, R. Land use change prediction in Dhaka city using GIS aided Markov chain modeling. *J. Life Earth Sci.* **2011**, *6*, 81–89. [CrossRef]
19. Mosammam, H.M.; Nia, J.T.; Khani, H.; Teymouri, A.; Kazemi, M. Monitoring land use change and measuring urban sprawl based on its spatial forms: The case of Qom city. *Egypt. J. Remote Sens. Space Sci.* **2017**, *20*, 103–116.
20. Munthali, M.; Mustak, S.; Adeola, A.; Botai, J.; Singh, S.; Davis, N. Modelling land use and land cover dynamics of Dedza district of Malawi using hybrid Cellular Automata and Markov model. *Remote. Sens. Appl. Soc. Environ.* **2020**, *17*, 100276. [CrossRef]
21. Rahman, M.T.U.; Tabassum, F.; Rasheduzzaman, M.; Saba, H.; Sarkar, L.; Ferdous, J.; Uddin, S.Z.; Islam, A.Z. Temporal dynamics of land use/land cover change and its prediction using CA-ANN model for southwestern coastal Bangladesh. *Environ. Monit. Assess.* **2017**, *189*, 565. [CrossRef]
22. Hewett, C.J.; Wilkinson, M.E.; Jonczyk, J.; Quinn, P.F. Catchment systems engineering: An holistic approach to catchment management. *Wiley Interdiscip. Rev. Water* **2020**, *7*, e1417. [CrossRef]
23. Kotaridis, I.; Lazaridou, M. Environmental Change Detection Study in the Wider Area of Lignite Mines. *Civil Eng. Archit.* **2018**, *6*, 108–114. [CrossRef]
24. Maviza, A.; Ahmed, F. Analysis of past and future multi-temporal land use and land cover changes in the semi-arid Upper-Mzingwane sub-catchment in the Matabeleland south province of Zimbabwe. *Int. J. Remote Sens.* **2020**, *41*, 5206–5227. [CrossRef]
25. Hyandye, C.; Martz, L.W. A Markovian and cellular automata land-use change predictive model of the Usangu Catchment. *Int. J. Remote. Sens.* **2017**, *38*, 64–81. [CrossRef]
26. Memarian, H.; Balasundram, S.K.; Talib, J.B.; Sung, C.T.B.; Sood, A.M.; Abbaspour, K. Validation of CA-Markov for simulation of land use and cover change in the Langat Basin, Malaysia. *Sci. Res.* **2012**, *4*, 542–554. [CrossRef]
27. Singh, S.K.; Mustak, S.; Srivastava, P.K.; Szabó, S.; Islam, T. Predicting spatial and decadal LULC changes through cellular automata Markov chain models using earth observation datasets and geo-information. *Environ. Process.* **2015**, *2*, 61–78. [CrossRef]
28. Zhang, L.; Walker, G.R.; Dawes, W.R. Water balance modelling: Concepts and applications. In *Regional Water and Soil Assessment for Managing Sustainable Agriculture in China and Australia*; ACIAR Monograph Series; ACIAR: Canberra, Australia, 2002; Volume 84, pp. 31–47.
29. Weng, Q. Land use change analysis in the Zhujiang Delta of China using satellite remote sensing, GIS and stochastic modelling. *J. Environ. Manag.* **2002**, *64*, 273–284. [CrossRef] [PubMed]
30. Naboureh, A.; Moghaddam, M.H.R.; Feizizadeh, B.; Blaschke, T. An integrated object-based image analysis and CA-Markov model approach for modeling land use/land cover trends in the Sarab plain. *Arab. J. Geosci.* **2017**, *10*, 259. [CrossRef]
31. Regmi, R.R.; Saha, S.K.; Subedi, D.S. Geospatial Analysis of Land Use Land Cover Change Modeling in Phewa Lake Watershed of Nepal by Using GEOMOD Model. *Himal. Phys.* **2017**, *6*, 65–72. [CrossRef]
32. Matlhodi, B.; Kenabatho, P.K.; Parida, B.P.; Maphanyane, J.G. Evaluating Land Use and Land Cover Change in the Gaborone Dam Catchment, Botswana, from 1984–2015 Using GIS and Remote Sensing. *Sustainability* **2019**, *11*, 5174. [CrossRef]
33. WUC. *Annual Report 2015/16*; Water Utilities Corporation: Gaborone, Botswana, 2016.
34. Halmy, M.W.A.; Gessler, P.E.; Hicke, J.A.; Salem, B.B. Land use/land cover change detection and prediction in the north-western coastal desert of Egypt using Markov-CA. *Appl. Geogr.* **2015**, *63*, 101–112. [CrossRef]
35. Verburg, P.H.; Kok, K.; Pontius, R.G.; Veldkamp, A. *Modeling Land-Use and Land-Cover Change, in Land-Use and Land-Cover Change*; Springer: Berlin/Heidelberg, Germany, 2006; pp. 117–135.
36. Alemaw, B.; Keaitse, E.; Chaoka, T. Management of water supply reservoirs under uncertainties in arid and urbanized environments. *J. Water Resour. Prot.* **2016**, *8*, 990. [CrossRef]
37. Altchenko, Y.; Lefore, N.; Villholth, K.G.; Ebrahim, G.; Genco, A.; Pierce, K.; Woolf, R.; Mosetlhi, B.B.; Moyo, T.; Kenabatho, P. *Resilience in the Limpopo Basin: The Potential Role of the Transboundary Ramotswa Aquifer-Baseline Report*; USAID: Pretoria, Southern Africa, 2016.
38. Statistics-Botswana. *Botswana Population Projections 2011–2026*; Government Printers: Gaborone, Botswana, 2015.
39. FAO. Africover Land cover Classification. In *Environment and Natural Resources Service*; Organisation, F.a.A., Ed.; Food and Agriculture Organisation: Rome, Italy, 1997.
40. Eastman, J.R. *IDRISI Andes Guide to GIS and Image Processing*; Clark University: Worcester, MA, USA, 2006; p. 328.
41. Mishra, M.; Mishra, K.; Subudhi, A.; Phil, M. Urban sprawl mapping and land use change analysis using remote sensing and GIS: Case study of bhubaneswar city, orissa. In Proceedings of the Geo-Spatial World Forum, Hyderabad, India, 18–21 January 2011.
42. Tavangar, S.; Moradi, H.; Bavani, A.M.; Gholamalifard, M. A futuristic survey of the effects of LU/LC change on stream flow by CA–Markov model: A case of the Nekarood watershed, Iran. *Geocarto Int.* **2019**, 1–17. [CrossRef]
43. Wu, L.; Shi, P.; Gao, H. State Estimation and Sliding-Mode Control of Markovian Jump Singular Systems. *IEEE Trans. Autom. Control.* **2010**, *55*, 1213–1219. [CrossRef]
44. Yang, X.; Zheng, X.-Q.; Lv, L.-N. A spatiotemporal model of land use change based on ant colony optimization, Markov chain and cellular automata. *Ecol. Model.* **2012**, *233*, 11–19. [CrossRef]
45. Guan, D.; Li, H.; Inohae, T.; Su, W.; Nagaie, T.; Hokao, K. Modeling urban land use change by the integration of cellular automaton and Markov model. *Ecol. Model.* **2011**, *222*, 3761–3772. [CrossRef]

46. Yang, X.; Zheng, X.-Q.; Chen, R. A land use change model: Integrating landscape pattern indexes and Markov-CA. *Ecol. Model.* **2014**, *283*, 1–7. [CrossRef]
47. Al-Sharif, A.A.A.; Pradhan, B. Monitoring and predicting land use change in Tripoli Metropolitan City using an integrated Markov chain and cellular automata models in GIS. *Arab. J. Geosci.* **2014**, *7*, 4291–4301. [CrossRef]
48. Keshtkar, H.; Voigt, W. A spatiotemporal analysis of landscape change using an integrated Markov chain and cellular automata models. *Model. Earth Syst. Environ.* **2016**, *2*, 1–13. [CrossRef]
49. Liang, X.; Liu, X.; Li, D.; Zhao, H.; Chen, G. Urban growth simulation by incorporating planning policies into a CA-based future land-use simulation model. *Int. J. Geogr. Inf. Sci.* **2018**, *32*, 2294–2316. [CrossRef]
50. Liu, Z.; Verburg, P.H.; Wu, J.; He, C. Understanding Land System Change Through Scenario-Based Simulations: A Case Study from the Drylands in Northern China. *Environ. Manag.* **2016**, *59*, 440–454. [CrossRef]
51. Ma, C.; Zhang, G.; Zhang, X.; Zhao, Y.; Li, H. Application of Markov model in wetland change dynamics in Tianjin Coastal Area, China. *Procedia Environ. Sci.* **2012**, *13*, 252–262. [CrossRef]
52. Wang, Y.; Yu, X.; He, K.; Li, Q.; Zhang, Y.; Song, S. Dynamic simulation of land use change in Jihe watershed based on CA-Markov model. *Trans. Chin. Soc. Agric. Eng.* **2011**, *27*, 330–336.
53. Sang, L.; Zhang, C.; Yang, J.; Zhu, D.; Yun, W. Simulation of land use spatial pattern of towns and villages based on CA–Markov model. *Math. Comput. Model.* **2011**, *54*, 938–943. [CrossRef]
54. Berberoğlu, S.; Akın, A.; Clarke, K.C. Cellular automata modeling approaches to forecast urban growth for adana, Turkey: A comparative approach. *Landsc. Urban Plan.* **2016**, *153*, 11–27. [CrossRef]
55. Eastman, J.R. *IDRISI Taiga Guide to GIS and Image Processing*; Clark Labs Clark University: Worcester, MA, USA, 2009.
56. Mishra, V.; Rai, P.; Mohan, K. Prediction of land use changes based on land change modeler (LCM) using remote sensing: A case study of Muzaffarpur (Bihar), India. *J. Geogr. Inst. Jovan Cvijic SASA* **2014**, *64*, 111–127. [CrossRef]
57. Gashaw, T.; Tulu, T.; Argaw, M.; Worqlul, A.W. Evaluation and prediction of land use/land cover changes in the Andassa watershed, Blue Nile Basin, Ethiopia. *Environ. Syst. Res.* **2017**, *6*, 17. [CrossRef]
58. Ruben, G.B.; Zhang, K.; Dong, Z.; Xia, J. Analysis and Projection of Land-Use/Land-Cover Dynamics through Scenario-Based Simulations Using the CA-Markov Model: A Case Study in Guanting Reservoir Basin, China. *Sustainability* **2020**, *12*, 3747. [CrossRef]
59. Malczewski, J. GIS-based multicriteria decision analysis: A survey of the literature. *Int. J. Geogr. Inf. Sci.* **2006**, *20*, 703–726. [CrossRef]
60. Li, Z.; Deng, X.; Yin, F.; Yang, C. Analysis of Climate and Land Use Changes Impacts on Land Degradation in the North China Plain. *Adv. Meteorol.* **2015**, *2015*, 1–11. [CrossRef]
61. Gidey, E.; Dikinya, O.; Sebego, R.; Segosebe, E.; Zenebe, A. Cellular automata and Markov Chain (CA_Markov) model-based predictions of future land use and land cover scenarios (2015–2033) in Raya, northern Ethiopia. *Modeling Earth Syst. Environ.* **2017**, *3*, 1245–1262. [CrossRef]
62. Singh, S.K.; Laari, P.B.; Mustak, S.; Srivastava, P.K.; Szabó, S. Modelling of land use land cover change using earth observation data-sets of Tons River Basin, Madhya Pradesh, India. *Geocarto Int.* **2018**, *33*, 1202–1222. [CrossRef]
63. Subedi, P.; Subedi, K.; Thapa, B. Application of a Hybrid Cellular Automaton—Markov (CA-Markov) Model in Land-Use Change Prediction: A Case Study of Saddle Creek Drainage Basin, Florida. *Appl. Ecol. Environ. Sci.* **2013**, *1*, 126–132. [CrossRef]
64. Omar, N.Q.; Ahamad, M.S.S.; Hussin, W.M.A.W.; Samat, N.; Ahmad, S.Z.B. Markov CA, Multi Regression, and Multiple Decision Making for Modeling Historical Changes in Kirkuk City, Iraq. *J. Indian Soc. Remote. Sens.* **2013**, *42*, 165–178. [CrossRef]
65. Pontius, G.J.; Schneider, L.C. Land-cover change model validation by an ROC method for the Ipswich watershed, Massachusetts, USA. *Agric. Ecosyst. Environ.* **2001**, *85*, 239–248. [CrossRef]
66. Mashapa, C.; Gandiwa, E.; Muboko, N.; Mhuriro-Mashapa, P. Land Use and Land Cover Changes in a Human-Wildlife Mediated Landscape of Save Valley Conservancy, South-Eastern Lowveld of Zimbabwe. *J. Anim. Plant Sci.* **2021**, *31*. [CrossRef]
67. McGill, B.M.; Altchenko, Y.; Hamilton, S.K.; Kenabatho, P.K.; Sylvester, S.R.; Villholth, K.G. Complex interactions between climate change, sanitation, and groundwater quality: A case study from Ramotswa, Botswana. *Hydrogeol. J.* **2019**, *27*, 997–1015. [CrossRef]
68. Pontius, R.; Millones, M. Death to Kappa: Birth of quantity disagreement and allocation disagreement for accuracy assessment. *Int. J. Remote Sens.* **2011**, *32*, 4407–4429. [CrossRef]
69. Yang, C.; Wu, G.; Chen, J.; Li, Q.; Ding, K.; Wang, G.; Zhang, C. Simulating and forecasting spatio-temporal characteristic of land-use/cover change with numerical model and remote sensing: A case study in Fuxian Lake Basin, China. *Eur. J. Remote. Sens.* **2019**, *52*, 374–384. [CrossRef]
70. Motlhwa, G.; James, M.Y.; Yunxian, Y. Effects of Integrated Support Programme for Arable Agricultural Development (ISPAAD) on the Income of Sorghum Farmers: Evidence from Kweneng District, Botswana. *Asian J. Agric. Ext. Econ. Sociol.* **2019**, 1–10. [CrossRef]
71. Seleka, T.B.; Mmopelwa, D. Effects of Input Subsidies on Subsistence Crop Acreage Diversity in Botswana. In *BIDPA Working Paper*; Botswana Institute for Development and Policy Analysis (BIDPA): Gaborone, Botswana, 2018.
72. Chagumaira, C.; Rurinda, J.; Nezomba, H.; Mtambanengwe, F.; Mapfumo, P. Use patterns of natural resources supporting livelihoods of smallholder communities and implications for climate change adaptation in Zimbabwe. *Environ. Dev. Sustain.* **2015**, *18*, 237–255. [CrossRef]

73. Gandhi, G.M.; Parthiban, S.; Thummalu, N.; Christy, A. Ndvi: Vegetation Change Detection Using Remote Sensing and Gis—A Case Study of Vellore District. *Procedia Comput. Sci.* **2015**, *57*, 1199–1210. [CrossRef]
74. Monteiro, J.A.; Kamali, B.; Srinivasan, R.; Abbaspour, K.; Gücker, B. Modelling the effect of riparian vegetation restoration on sediment transport in a human-impacted Brazilian catchment. *Ecohydrology* **2016**, *9*, 1289–1303. [CrossRef]
75. Yang, Y.; Tian, F. Abrupt change of runoff and its major driving factors in Haihe River Catchment, China. *J. Hydrol.* **2009**, *374*, 373–383. [CrossRef]
76. James, R.; Washington, R. Changes in African temperature and precipitation associated with degrees of global warming. *Clim. Chang.* **2012**, *117*, 859–872. [CrossRef]
77. Kenabatho, P.; Parida, B.; Moalafhi, D. The value of large-scale climate variables in climate change assessment: The case of Botswana's rainfall. *Phys. Chem. Earth Parts A/B/C* **2012**, *50*, 64–71. [CrossRef]
78. Moise, A.F.; Hudson, D.A. Probabilistic predictions of climate change for Australia and southern Africa using the reliability ensemble average of IPCC CMIP3 model simulations. *J. Geophys. Res. Space Phys.* **2008**, *113*, 15. [CrossRef]
79. Hassaballah, K.; Mohamed, Y.; Uhlenbrook, S.; Biro, K. Analysis of streamflow response to land use and land cover changes using satellite data and hydrological modelling: Case study of Dinder and Rahad tributaries of the Blue Nile (Ethiopia–Sudan). *Hydrol. Earth Syst. Sci.* **2017**, *21*, 5217–5242. [CrossRef]
80. Obahoundje, S.; Ofosu, E.A.; Akpoti, K.; Kabo-Bah, A.T. Land Use and Land Cover Changes under Climate Uncertainty: Modelling the Impacts on Hydropower Production in Western Africa. *Hydrology* **2017**, *4*, 2. [CrossRef]
81. Yan, B.; Fang, N.; Zhang, P.; Shi, Z. Impacts of land use change on watershed streamflow and sediment yield: An assessment using hydrologic modelling and partial least squares regression. *J. Hydrol.* **2013**, *484*, 26–37. [CrossRef]
82. Huang, J.; Wu, P.; Zhao, X. Effects of rainfall intensity, underlying surface and slope gradient on soil infiltration under simulated rainfall experiments. *Catena* **2013**, *104*, 93–102. [CrossRef]
83. Neal, B.; Nathan, R.J.; Schreider, S.; Jakeman, A.J. Identifying the Separate Impact of Farm Dams and Land Use Changes on Catchment Yield. *Australas. J. Water Resour.* **2002**, *5*, 165–176. [CrossRef]
84. Thompson, J.C. *Impact and Management of Small Farm Dams in Hawke's Bay, New Zealand, in School of Geography, Environment and Earth Sciences 2012*; Victoria University of Wellington: Wellington, New Zealand, 2012.
85. DWA. *A Study on the Impact of Small Dam Construction on Downstream Water Resources*; Department of Water Affairs: Gaborone, Botswana, 1992.
86. DWA. *A Study on the Impact of Small Dam Construction on Downstream Water Resources (Updated)*; Department of Water Affairs: Gaborone, Botswana, 2014.
87. Meigh, J. The impact of small farm reservoirs on urban water supplies in Botswana. *Nat. Resour. Forum* **1995**, *19*, 71–83. [CrossRef]
88. Ranganai, R.T.; Gotlop-Bogatsu, Y.; Maphanyane, J.; Tladi, B. Hydrochemical and Geophysical evaluation of groundwater pollution in the Ramotswa wellfield, SE Botswana. *BIE2001 Tech. Pap.* **2001**, 193–200.
89. Vogel, H.; Keipeile, K.; Kgomanyane, J.; Zwikula, T.; Pontsho, M.; Mafa, B.; Matthes, L.; Staudt, M.; Beger, K.; Güth, T. *Groundwater Quality Case Studies in Botswana*; Groundwater Pollution in Africa; Taylor & Francis: London, UK, 2006; pp. 253–264.
90. Adelana, S.; Xu, Y. Contamination and Protection of the Cape Flats Aquifer, South Africa. In *Groundwater Pollution in Africa*; Xu, U., Ed.; Taylor and Francis/Balkema: Leiden, The Netherlands, 2006; Volume 13, p. 978.
91. Luo, G.; Amuti, T.; Zhu, L.; Mambetov, B.T.; Maisupova, B.; Zhang, C. Dynamics of landscape patterns in an inland river delta of Central Asia based on a cellular automata-Markov model. *Reg. Environ. Chang.* **2015**, *15*, 277–289. [CrossRef]

remote sensing

MDPI

Technical Note

Application of the Random Forest Classifier to Map Irrigated Areas Using Google Earth Engine

James Magidi [1], Luxon Nhamo [2,3,*], Sylvester Mpandeli [2,4] and Tafadzwanashe Mabhaudhi [4]

1 Geomatics Department, Tshwane University of Technology, Pretoria 0001, South Africa; MagidiJT@tut.ac.za
2 Water Research Commission of South Africa (WRC), Lynnwood Manor, Pretoria 0081, South Africa; sylvesterm@wrc.org.za
3 Centre for Transformative Agricultural and Food Systems (CTAFS), School of Agricultural, Earth and Environmental Sciences, University of KwaZulu-Natal, Scottsville, Pietermaritzburg 3209, South Africa
4 School of Environmental Sciences, University of Venda, Thohoyandou 0950, South Africa; Mabhaudhi@ukzn.ac.za
* Correspondence: luxonn@wrc.org.za

Abstract: Improvements in irrigated areas' classification accuracy are critical to enhance agricultural water management and inform policy and decision-making on irrigation expansion and land use planning. This is particularly relevant in water-scarce regions where there are plans to increase the land under irrigation to enhance food security, yet the actual spatial extent of current irrigation areas is unknown. This study applied a non-parametric machine learning algorithm, the random forest, to process and classify irrigated areas using images acquired by the Landsat and Sentinel satellites, for Mpumalanga Province in Africa. The classification process was automated on a big-data management platform, the Google Earth Engine (GEE), and the *R*-programming was used for post-processing. The normalised difference vegetation index (NDVI) was subsequently used to distinguish between irrigated and rainfed areas during 2018/19 and 2019/20 winter growing seasons. High NDVI values on cultivated land during the dry season are an indication of irrigation. The classification of cultivated areas was for 2020, but 2019 irrigated areas were also classified to assess the impact of the Covid-19 pandemic on agriculture. The comparison in irrigated areas between 2019 and 2020 facilitated an assessment of changes in irrigated areas in smallholder farming areas. The approach enhanced the classification accuracy of irrigated areas using ground-based training samples and very high-resolution images (VHRI) and fusion with existing datasets and the use of expert and local knowledge of the study area. The overall classification accuracy was 88%.

Keywords: spatial analysis; water management; change detection; vegetation indices; image classification

1. Introduction

Existing global food and water insecurity challenges are being compounded by increasing temperatures and drought recurrences and increasing demand for water and food due to population growth [1,2]. The need to meet the growing food requirements is intensifying pressure on freshwater resources, which is a cause for concern under climate change [3–6]. The warming climate is exacerbating the challenge of water scarcity, resulting in increased aridity and shifts in agro-ecological zones, affecting crop yields [7]. Irrigation already uses more than 70% of available freshwater resources, on only 18% of cultivated areas globally [8]. Thus, climate change and the increasing demand for food drive the huge increases in irrigated agriculture, changes that require innovative approaches that improve the productive use of water [3,7]. However, strategies on irrigation expansion require accurate information on the spatial extent of irrigated agriculture, which is currently scant [6,9].

Accurate irrigated area maps are critical for informing policy and supporting decision-makers to formulate coherent and transformative strategies with irrigation as an indispensable climate change adaptation strategy [6]. Besides, accurate information on the extent of irrigated areas goes beyond water resources management; its importance cascades into water and food security and assessing the impact of climate change on the agriculture sector [6,10]. However, knowledge on the actual extent and spatial distribution of irrigated areas remains scant, despite the many attempts that have been made to map irrigated areas worldwide [11–13]. Existing irrigated area datasets developed at varying spatial scales and resolutions, including the FAO (Food and Agriculture Organization of the United Nations) database [14], the Global Map of Irrigated Areas version 5 (GMIA 5.0) [15], the MIRCA2000 product [16], and IWMI's (International Water Management Institute) irrigated area map [17]. In 2016, IWMI further developed an improved irrigated area map for Asia and Africa using canonical correlation analysis and time-lagged regression at 250 m spatial resolution for 2000 and 2010 [13]. Although these datasets are important for indicating irrigated areas, they generally over-estimate the areas due to the low spatial resolution data used [18]. A recent study in South Africa has shown how low spatial resolution data can result in over-estimates of irrigated areas, particularly in smallholder fields that are about 2 ha in size, very small to be detected by low spatial resolution satellites [18].

Advances in remote sensing technologies in conjunction with the emergence of big data and cloud-based processing platforms such as Google Earth Engine (GEE) are facilitating the classification of irrigated areas within improved accuracies in a time and cost-effective manner, enhancing the monitoring of these at both local and global scales [19,20]. This is aided by freely available high-resolution remotely sensed products and novel non-parametric machine learning algorithms for land use classification [18,21,22]. Supervised image classification machine learning algorithms include Support Vector Machine (SVM), Random Forest (RF), decision tree algorithms, and Extreme Gradient Boosting (XGboost) [23]. Managing the increasingly large volumes of remotely sensed data and products is challenging by traditional processing techniques [9]. These cumbersome and slow pre-processing, processing, and post-processing traditional approaches are being replaced by novel machine learning algorithms embedded in cloud computing big data platforms such as GEE [9,20]. The GEE provides access to many petabytes of remotely sensed spatial datasets, enabling their timely geo-processing using machine learning algorithms and batch processing [9].

Methods that improve the accuracy of irrigated areas are critical for supporting decision-making on irrigation expansion. Advances in machine learning and big data management and processing platforms are improving the accuracy and speed of generating cultivated areas. This study used a non-parametric machine learning algorithm, the random forest, to classify near accurate irrigated areas using high-resolution satellite images of Mpumalanga Province, South Africa. The objective was to develop a spatial model that assists policy and decision-makers to make informed assessments on crop water requirements, water allocation, agricultural land planning, crop evapotranspiration patterns, basin hydrology, and the impact of different types of irrigation at any spatial scale on an annual basis.

2. Methods

2.1. Description of the Study Area

Mpumalanga Province (Figure 1) lies in the east of South Africa, sharing international borders with Eswatini and Mozambique. In South Africa, it shares its borders with the following provinces: Limpopo to the north, Gauteng to the west, the Free State to the southwest, and KwaZulu-Natal to the south. Mpumalanga Province has a total land area of about 76,495 km^2, representing 6.3% of the total land area of South Africa, making it the second-smallest province after Gauteng. It is divided into three district municipalities, namely Ehlanzeni, Nkangala, and Gert Sibande. The topography of the province is characterised by a Great Escarpment, dividing its land area into two major sections: (a) the Plateau area with an elevation of more than 2000 m to the west, and (b) the Lowveld area in the east [24]. The two topographic zones determine the province's climate where a temperate climate dominates the Highveld, whilst a sub-tropical climate dominates the Lowveld areas [25].

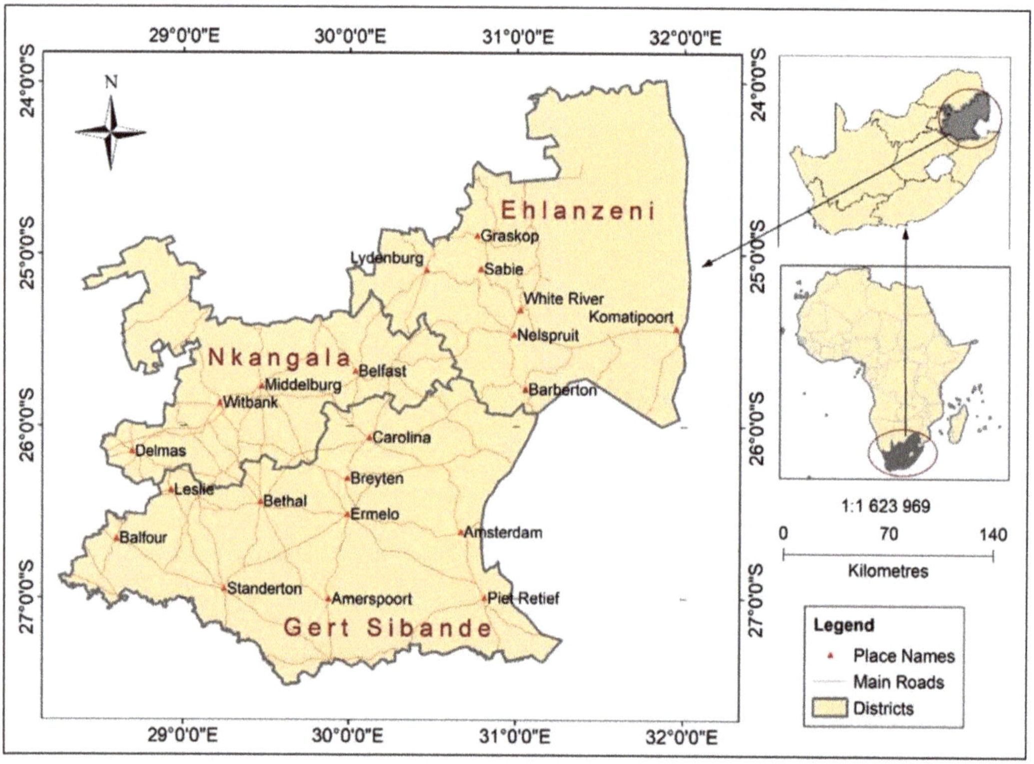

Figure 1. Map of Mpumalanga Province and is location in in South Africa and Africa.

Vegetation is predominantly grassland. Agriculture is mostly irrigated in both commercial and smallholder farming areas. Smallholder irrigation is predominant in former-homelands, also called Bantustans, areas allocated to indigenous black people during the apartheid era and are generally poorly resourced. A small section in the east lies within the Kruger National Park. The rivers within the province hold some of the most valued ecologically rich systems in the whole of South Africa [26], apart from its rich mineral base that includes huge coal reserves, whose mining is, however, threatening the ecological infrastructure [24].

Rainfall is seasonal, occurring during the summer season (October to February), as indicated in Figure 2. Like the rest of South Africa, the Mpumalanga Province is water-scarce with almost all its available freshwater resources now allocated, leaving little room for further development [25]. The winter season (April to August) is generally dry and cold (Figure 2), with occasional light snow in the southwestern divide. The average annual temperature is about 20 °C, and the mean annual rainfall ranges from 400 mm to 1500 mm [24].

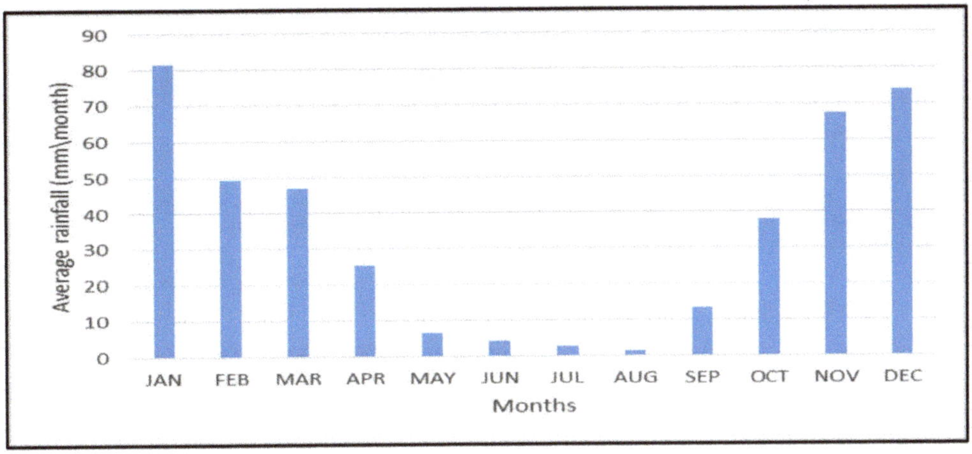

Figure 2. Mean monthly average rainfall of Mpumalanga Province (1972–2018).

2.2. *Methodological Framework*

The methodological framework (Figure 3) describes the processes followed in classifying irrigated and rainfed areas in Mpumalanga Province. The study was informed by the increasing warming climate and the intensity of drought, which are compounding water and food insecurity, increased aridity, declining yields, and land degradation [3]. The challenges require informed land use planning and irrigated area development by accurately mapping land use. The Normalised Difference Vegetation Index (NDVI) was used to distinguish irrigated from rainfed areas and compare changes in irrigated areas during 2018/19 and 2019/20 winter growing seasons. The procedure to classify irrigated areas followed four main stages that include (i) data collection, image correction, and pre-processing, (ii) feature extraction, (iii) data integration, and (iv) use of vegetation indices to separate irrigated areas from the rest of cultivated areas (Figure 3). The step-by-step procedure included the mosaicking and compositing of images, signature development and evaluation of samples, use of the random classifier to map cultivated areas, accuracy assessment, and lastly, the use of vegetation indices to separate irrigated areas from other cultivated areas, respectively. The premise was to classify both irrigated and rainfed areas separately, improve the mapping accuracy as an initial step to enhance irrigation and agricultural land use planning, improve agricultural water management and ensure water and food security (Figure 3).

The classification of cultivated areas and the seasonal comparisons in NDVI variations was done using the GEE (see Supplementary Information 1), a big data and cloud computing platform, on cloud-free Sentinel-2 and Landsat-8 images [27]. Satellite images from the Sentinel from July to August for 2020 were mosaicked using GEE. Training sites that were digitised included water, natural vegetation, cultivated land, and built-up areas. The classification was realised through a non-parametric machine learning algorithm, the random forest classifier. The random forest was preferred due to its flexibility and capability to produce, even without hyper-parameter tuning, near accurate land use classifications and is applicable for both classification and regression tasks [28,29]. The classification accuracy was enhanced by integrating existing land use datasets. Average monthly NDVI values were calculated from June to October, a dry season in the study area for 2019 and 2020.

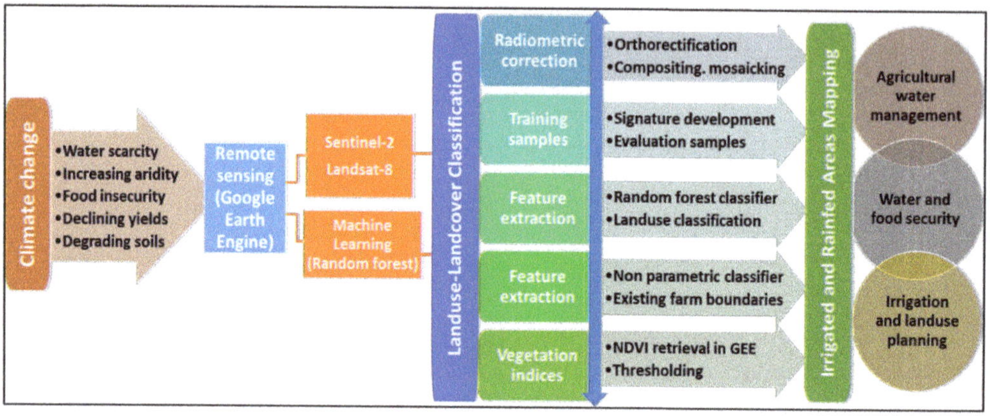

Figure 3. Methodological framework to classify irrigated areas using a non-parametric machine learning algorithm, random forest.

2.3. Data Sources

Various time-series and static input data were used for the random forest (RF) classification model on the GEE computing platform. The static input variables refer to existing datasets, terrain, geographic location, and dynamic inputs represent data derived from remote sensing such as NDVI and satellite images and climate data. The premise was to classify cultivated lands at a provincial scale and separate irrigated areas during the dry season that extends from June to October. Data were either obtained from or uploaded to the GEE cloud computing platform for the classification procedure [27].

High-resolution satellite images from the Sentinel 2 sensor (20 m spatial resolution) and the 30 m resolution Landsat images for 2019 and 2020 were accessed from the GEE cloud computing platform (https://developers.google.com/earth-engine/datasets/catalog/). Monthly NDVI data were accessed from Landsat images, also in GEE. Existing agricultural fields and land use datasets were obtained from the Department of Agriculture, Land Reform, and Rural Development's (DALRRD) Crop Estimates Consortium (CEC) [30] and GeoTerraImage dataset [31], respectively.

2.4. The Random Forest Algorithm

The random forest classifier is an ensemble of classification methods consisting of several decision tree models and is expressed as [32]:

$$\{DT(x,\theta_k)\} \begin{matrix} T \\ k=1 \end{matrix} \tag{1}$$

where x is the input vector, and θ_k denotes a random vector, which is sampled independently but with the same distribution as the previous $\theta_{k,}$ ... , θ_{k-1}. T bootstrap samples are initially derived from the training data. A no-pruned classification and regression tree (CART) is drawn from each bootstrap sample β where only one of M randomly selected features is chosen for the split at each node of CART [32].

The random forest is more robust after minor changes in the input data, and it enhances land use classification accuracy a achieving classifier stability. A k number of samples extracted from the training sample set using bootstrap sampling, and the sample size of each sample, are the same as that of the original training set. A k number of trees were created for k samples, and a k number of classification results were obtained. According to the classification results, each record was classified to determine its final land use category. The random forest classifier improves the classification accuracy through its object-based processing algorithms, and its use in remote sensing provides the following advantages:

- It enhances the accuracy of land use/cover mapping as compared to other popular similar algorithms [33]. It maintains the classification error balance when the class size distribution is unbalanced [34].
- The random forest classifier requires little or no manual intervention as it determines data characteristics by itself, simplifying its design process [35].
- Although the random forest algorithm provides various data characterisations, it generally has a quick processing speed [36].

2.4.1. Landcover Classes and Reference Data

Landcover categories were grouped into four types to separate cultivated areas from the rest of the land uses. These included cultivated land, forest, grassland, water, built-up area, and bare land (Table 1). A random sampling method was used to balance the number of training points. Therefore, the number of training points for each landcover class was proportional to the total pixels using the random forest classifier. Existing datasets were used to train and validate the landcover classifications. The reference training and testing datasets were gathered from a time series of both Sentinel and Landsat images and visual interpretation using Google Earth images. The training samples were validated using the Department of Agriculture, Land Reform, and Rural Development's (DALRRD) cropped area dataset [30] and the GeoTerraImage dataset [31]. The classification resulted in the derivation of the cultivated area dataset of the study area.

Table 1. Land-cover classification types that were used as training samples.

Land Classification	Description
Cultivated land	Land planted with crops, newly opened cropped areas, fallow land
Natural vegetation	Shrublands, forested lands, grasslands, natural or planted forests
Water	All water bodies, including rivers, wetlands, reservoirs, etc
Built-up area	All settlements, including industrial areas

2.4.2. Crop Phenology Derivation from NDVI

Time series data from NDVI was essential for obtaining croplands' phenological data and for distinguishing irrigated from rainfed areas. Unlike other vegetation types such as grassland and forests, croplands have unique characteristics in the stages of emergence, vegetative growth, senescence, and harvest [37]. These features enable the separation of croplands from other vegetation types using a time series of satellite images (see Supplementary Information 1). NDVI time-series data derived from Landsat images, compiled using Google Earth Engine, were then used to classify and distinguish irrigated from rainfed areas. This was achieved by applying NDVI thresholds ranging between 0.19 and 0.25 for June, July, August, September, and October. The NDVI thresholds were derived using the histogram equalisation and the sigmoid contrast stretch method to distinguish low and high NDVI vigour (Figure 4). Between June and October, it does not rain in the study area, and rainfed agriculture is entirely absent. Therefore, the presence of healthy and green crops signifies the presence of irrigation [38]. Thus, vegetation indices' seasonal pattern provided the foundation for separating irrigated from rainfed areas through time-series NDVI data.

Histogram equalisation (Figure 4a) and sigmoid contrast stretch (Figure 4b) were used to establish the NDVI thresholds [9,11] used to distinguish vegetated (cropped) and non-vegetated (non-cropped) areas (Figure 4c) from where cultivated lands were derived (Figure 4d). NDVI thresholds ranged between 0.19 and 0.25, and they also confirmed the results of previous studies [38,39]. Cell statistics was applied to combine the reclassified NDVI values (vegetated and non-vegetated) from where the cultivated areas were derived and, ultimately, the separation of irrigated from rainfed areas.

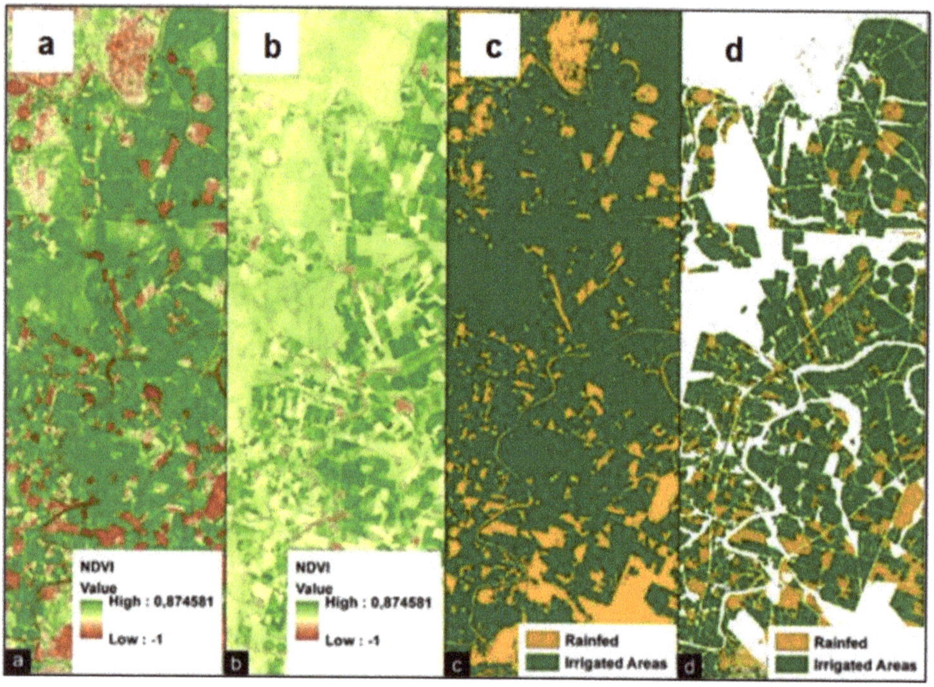

Figure 4. (**a**) Histogram equalisation of the NDVI average of July 2019; (**b**) sigmoid stretch of the NDVI average of July 2019; (**c**) The reclassified NDVI of July 2019 using a threshold; and (**d**) the classified cultivated land.

2.5. NDVI Trend Analysis over Cultivated Land

Of note is that outliers and seasonality typically characterise a time series of NDVI data in regions characterised by dry and wet season [40]. The single modal annual rainfall pattern of the study area shows that rainfall (or agricultural season) starts in October and ends in April, indicating a time lag between rainfall and crop phenophases [41] (Figure 5).

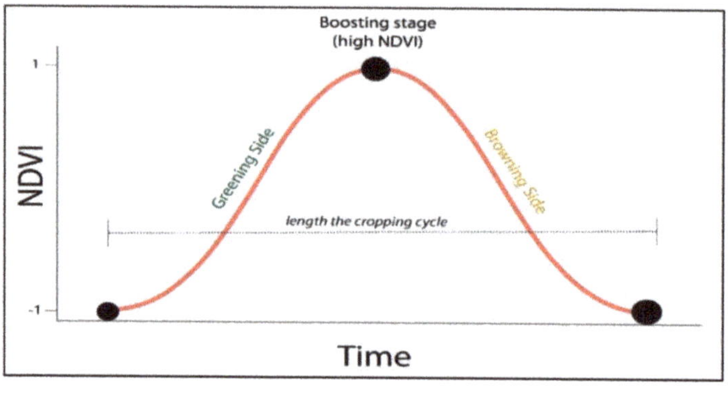

Figure 5. An example of single growing season and related phenological measures analysed through NDVI.

As rainfall is a crucial factor in the seasonal crop cycle, there is a positive correlation between rainfall and the seasonal crop growth pattern [42]. Thus, vegetation growth and greening begin in October/November, and browning begins in May/June (Figure 5). This process also applies to rainfed crops in areas with a single modal crop growth cycle. In irrigated areas, greening starts at any time of the dry period from June to October. Croplands could be at a greening stage, yet others are at the boosting stage, while some are at the browning stage (Figure 5).

In most cases, and due to farming relying on rainfall during the wet season, the rainfed crop will start greening in November and browning in May. The greening of irrigated crops begins around June, and the browning in October (Figure 5). Thus, the classification of irrigated areas was assessed from June to October, which coincides with the dry season.

2.6. *Distinguishing Irrigated from Rainfed Areas*

Healthy vegetation reflects more infrared and absorbs more red and blue portions of the electromagnetic spectrum. The red and near-infrared bands are important for calculating NDVI and other vegetation indices as the blue portion is affected by atmospheric scattering [43]. In areas with a single modal annual rainfall pattern, there is less absorption of the visible light and low reflection of the infrared light during the dry season (Figure 5), resulting in low NDVI values. High NDVI values on croplands during the dry season signify irrigation as there is generally no rainfall. NDVI values drop around May to June but picks-up between July and September (Figure 5). Thus, the NDVI threshold ranging between 0.19 and 0.25 was used to isolate irrigated from non-irrigated areas between July and August during the dry season.

3. Results and Discussion

3.1. *Cultivated Areas of Mpumalanga Province*

The initial product developed from the remote sensing procedure was a dataset of cultivated areas (Figure 6) incorporating both irrigated and rainfed areas for 2020. Twenty-seven percent of the total land areas of Mpumalanga Province is used for agriculture (Figure 6), highlighting the importance of agriculture in the province.

The distribution of cultivated land is uneven as determined by topography, distribution of the river network and soil types, and the location of former homelands. The highest concentration of cultivated land is mainly in the Lowveld area in former homelands, mainly in Nkangala and Gert Sibande districts municipalities. The third district, Ehlanzeni, has limited agriculture activity as most of its area lies within the Kruger National Park and other conservancies.

3.2. *Accuracy Assessment*

Accuracy assessment of the mapped cultivated areas was verified by comparing the developed dataset with 130 ground truth points derived from high-resolution Google Earth images (Figure 7). This was enhanced by combining the ground-truth points generated from Google Earth with those from existing datasets using the accuracy assessment tool in ArcMap. This visual interpretation, coupled with the expert and prior knowledge of the study area's features, contributed to the accuracy assessment using the kappa assessment coefficient algorithm. The ground-truth points were randomly chosen. The accuracy assessment was essential for determining the quality of the classified cultivated area derived through the random forest classifier in remote sensing. The classified cultivated areas' accuracy was 88%, with a kappa coefficient of 80%, which indicates a very good mapping accuracy. Although relatively high, accuracy can still be improved through post-classification enhancement methods using drones [22].

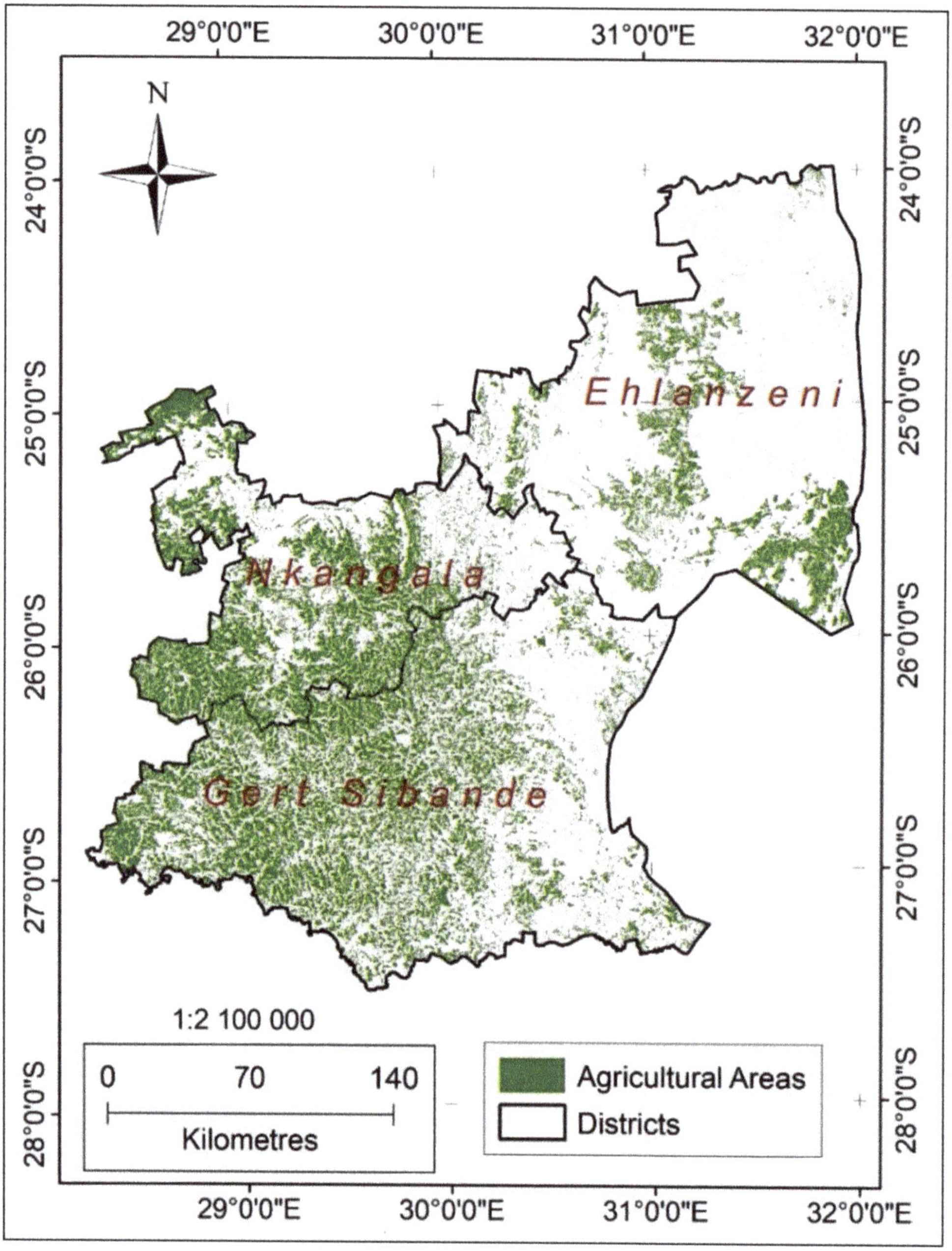

Figure 6. Cultivated lands of Mpumalanga Province derived using the random forest classifier.

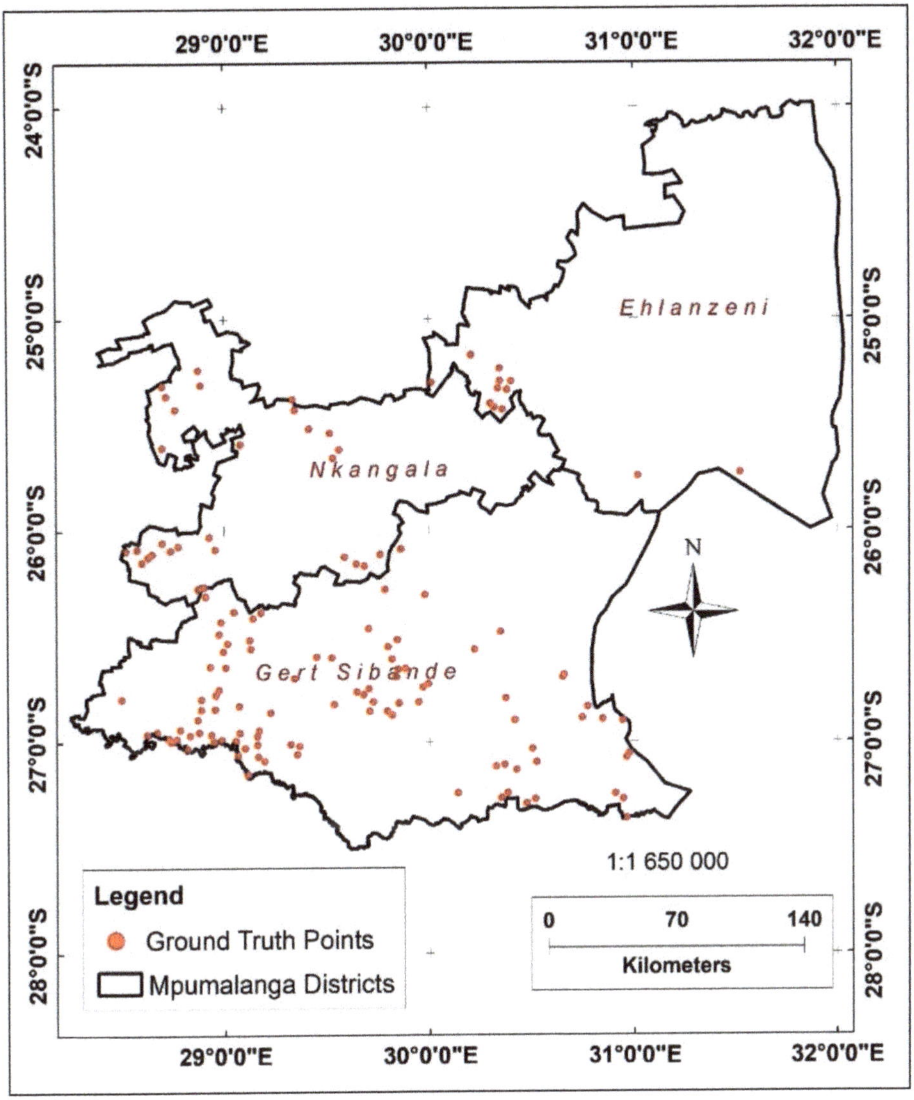

Figure 7. Distribution of accuracy assessment validation points.

3.3. Separating Irrigated from Rainfed Areas

The map presented in Figure 8 shows the irrigated and rainfed areas in Mpumalanga, giving the proportion of the two cultivated systems per district municipality. At 75% of the total cultivated area, irrigated land is predominant in the province, where the rainfed area accounts for less than 25% of the total cultivated area. The pie-charts (Figure 8) confirm that the land under irrigation is predominant in the water management area (with an overall proportion of over 75% of the cultivated area), a scenario also given in Table 2. The irrigated area's predominance also confirms the allocation of most of the available freshwater resources towards irrigation. The situation also confirms the observations from other studies, which also found out that the province's available freshwater resources are almost fully allocated [25,26].

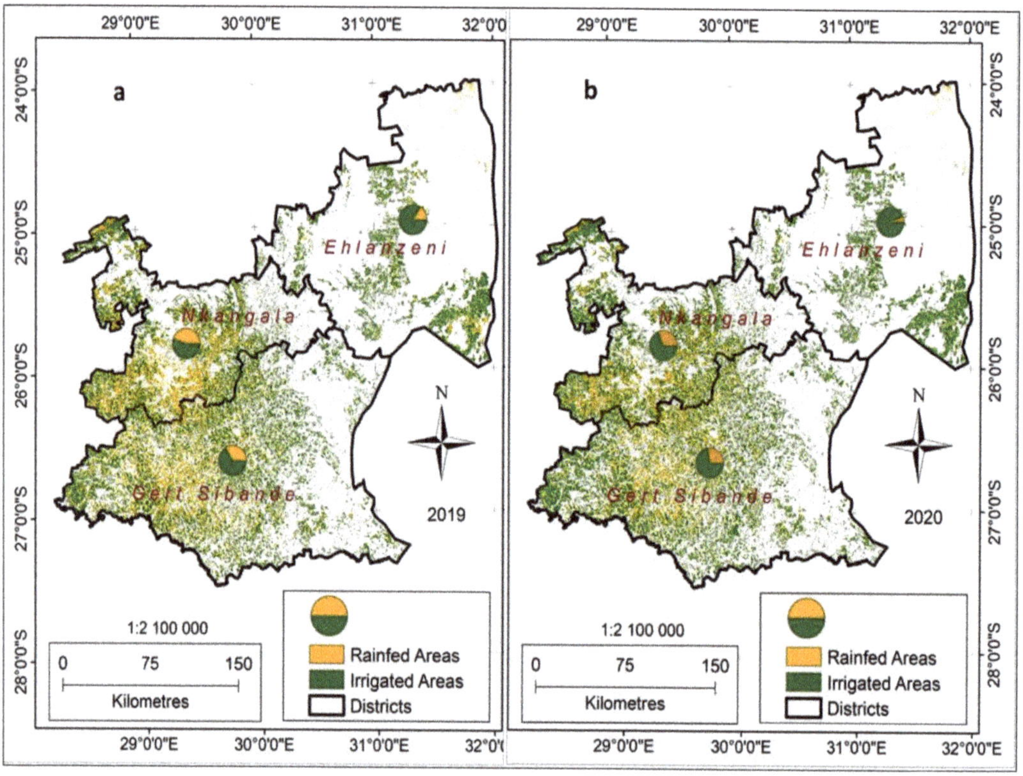

Figure 8. Distribution of cultivated lands in Mpumalanga Province and the proportion between irrigated and rainfed areas per district as well as the changes that took place between 2019 (map **a**) and 2020 (map **b**).

Table 2. Disaggregated proportions of cultivated areas per cultivation type between 2019 and 2020.

District	Area (ha)	Rainfed Area (ha)	Irrigated Area (ha)	Cultivated Area (ha)	Rainfed Area (%)	Irrigated Areas (%)	Cultivated Area (%)
				2020			
Gert Sibande	3,189,239.14	284,963.35	756,614.12	1,041,577.47	27.36	72.64	32.66
Nkangala	1,679,938.90	201,492.75	445,151.50	646,644.25	31.16	68.84	38.49
Ehlanzeni	2,789,686.33	22,592.30	326,176.48	348,768.78	6.48	93.52	12.50
Mpumalanga	7,658,864.36	509,048.40	1,527,942.10	2,036,990.50	24.99	75.01	26.60
				2019			
Gert Sibande	3,189,239.14	352,240.02	689,337.45	1,041,577.47	33.82	66.18	32.66
Nkangala	1,679,938.90	288,002.67	358,641.59	646,644.25	44.54	55.46	38.49
Ehlanzeni	2,789,686.33	54,785.44	293,983.34	348,768.78	15.71	84.29	12.50
Mpumalanga	7,658,864.36	695,028.12	1,341,962.38	2,036,990.50	34.12	65.88	26.60

Over time, irrigation trends in Mpumalanga Province indicate that the irrigated area is bigger than rainfed areas, an indicator of high water use in the agriculture sector (Figure 9). Therefore, most of the cultivated land in the province is irrigated as most agricultural land is commercial [44]. The rainfed area has been continuously declining since 2016, except in 2018 when it increased as there was limited irrigation due to drought [3]. However, it significantly declined in 2019 at a time when irrigated further increased (Figure 9). These differences between agricultural types between 2019 and 2020 are noticeable in the maps shown in Figure 8, where it is noticeable that in 2020 most of the cultivated land was irrigated, as indicated by the predominantly green colour.

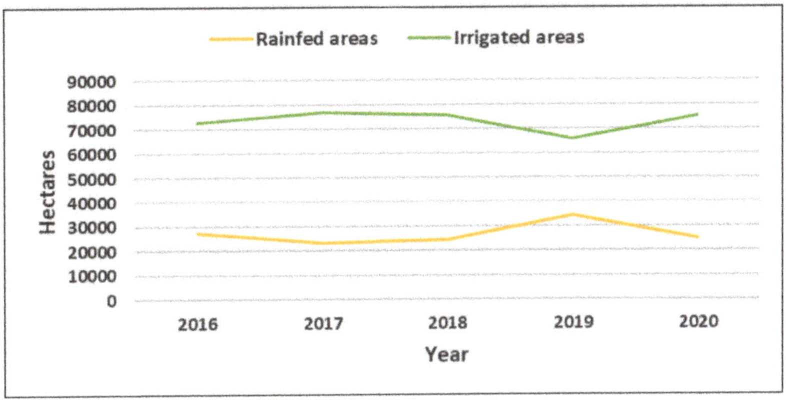

Figure 9. Trends in the irrigated area versus rainfed area in in Mpumalanga Province from 2016 to 2020.

Irrigation expansion in the province is likely due to ongoing government efforts to equip such areas with irrigation potential and target smallholder farming areas to enhance food security and achieve sustainability by 2030 [45]. However, of concern is that irrigation in smallholder farming areas is usually exempt from water use licences as the water withdrawals are assumed to be negligible [46]. However, this study's results highlight the need to monitor irrigation expansion at all scales regardless of users, and plan in a more sustainable manner.

3.4. Changes in the Irrigated Area between 2019 and 2020

Significant increases are notable in irrigated areas in Mpumalanga Province between the 2019 and 2020 winter growing seasons, a characteristic evident in all districts of the province (Figure 8). The increase only shows that some cultivated areas that were not irrigated in 2019 were irrigated in 2020, not necessarily a change in land area under irrigation. As the rainfed area is less than irrigated in the province, still, between 2019 and 2020, the area that continued as rainfed further decreased by 36.53%. During the same period, the land under irrigation increased by over 12.17% (Table 3), putting a further strain on already scarce water resources [25]. The increase in irrigated area conforms with the National Development Plan (NDP), which promotes the doubling of irrigated areas by 2030 [45], but without considering the availability of other important sectors linked to irrigation like water and energy availability. The land area under irrigation increased during the 2020 Covid-19 pandemic lockdowns when it was expected to decrease. The main contributing factor to the gains in the irrigated area in 2020 is mainly due to agriculture being considered as an essential sector that was allowed to continue operating during the height of the Covid-19 pandemic lockdowns. With many people having lost their other income sources during the lockdown, agriculture became an alternative source of income. In South Africa, there is already irrigation infrastructure that is not being fully utilised [46,47].

Table 3. Percentage change in the cultivated areas between 2019 and 2020. Negative changes represent a decrease and positive increases.

	2019	2020	% Change
Rainfed areas (ha)	695,028.12	509,048.40	−36.53
Irrigated areas (ha)	1,341,962.38	1,527,942.10	12.176
Total cultivated areas (ha)	2,036,990.50	2,036,990.50	

Whilst increasing the land under irrigation is necessary to ensure food and water security, there is a need for holistic approaches to ensure that the scarce water resources are equitably distributed with other sectors. Like the rest of South Africa, Mpumalanga Province is water-scarce, and its available freshwater resources are almost fully allocated, leaving little room for future development [25]. Therefore, there is a need for transformative approaches to address resource management in an integrated manner.

The continued decrease in rainfed agricultural area is problematic for policy and decision-makers as increasing the land under irrigation is not always the best-fit solution. Given the worsening water scarcity in South Africa, the reality is that there is a limit to expand the land under irrigation [3,12,48]. A significant part of the solution must be addressed by improving crop water productivity and optimising rainfed systems [49]. Thus, the challenge requires integrated and cross-sectoral approaches in irrigation expansion strategies, translating irrigation into climate change adaptation and mitigation. Previous studies have highlighted the importance of addressing multi-sector and intertwined challenges holistically and avoid the challenges of transferring problems to other sectors [10,50–52]. This calls for transformative approaches such as nexus planning, circular economy, sustainable agricultural systems, and scenario planning in agricultural development [53,54]. For example, this noble initiative to increase irrigation area should also consider water and energy availability. Uninformed irrigation development will only compound the existing water, energy, and energy insecurity [55]. To inform policy and decision-making, there is always a need to have accurate knowledge of the irrigated area's spatial extent from where water productivity and crop water requirements can be modelled with improved accuracy.

3.5. Pros and Cons of Using the Random Forest in Land Use Classification

There are more advantages than disadvantages to using the random forests algorithm in land use classification, although there is still room for further improvements. This non-parametric land use classification method offers a more superior method for working with missing data. It substitutes missing values with a variable appearing the most in a particular node [56]. However, in our case, the most important advantage of using the random forest classifier has been the improvement in the mapping accuracy. Furthermore, integrating the random forest technique with big data management platforms facilitated the processing of large data sets with numerous variables within a short space of time. Thus, the technique automatically balances data sets when a class is more infrequent than other classes in the data. It can handle variables fast, making it suitable for complicated tasks [36]. In addition to these advantages, some of the benefits for using the random forest algorithm include (i) the ease with which to interpret the rules using a tree algorithm, (ii) as a non-parametric model, it is easy to add a range of numeric or categorized data layers, (iii) the unimodal training data distribution type is not a prerequisite, and (iv) the classification process is time and cost-effective as long as the programming rules are well-set [29,57].

Although the algorithm has been shown to improve the mapping accuracy and to handle large datasets by integrating with big data management platforms like the GEE, its major disadvantage is that it tends to over-fit the training data, which was the main reason for not providing even better accuracy. This is generally evident when the decision-tree grows too large, with the terminal nodes representing very small subsets of the training data [57]. However, this drawback is overcome by a technique called "pruning", which simplifies the tree by removing terminal nodes that could be linked to noise in the data [57].

4. Conclusions

Advances in remote sensing technologies, alongside the development of machine learning classification algorithms, have facilitated cost and time effective delineation of more accurate irrigated areas. The choice of the most appropriate image classification algorithm is determined by the objectives that need to be achieved; however, the random forest machine learning algorithm used in this study is applicable at any scale and is time and cost-efficiency. This was enhanced using the big-data management platform of the Google Earth Engine (GEE). The high spatial resolution (20 m resolution) cultivated land dataset was developed using a two-year analysis (2019 and 2020), 10-day Sentinel, and 16-day Landsat data. The random forest algorithm handled large volumes of remotely sensed products and reference training and validation datasets from various sources courtesy of the big-data management and processing capabilities of the GEE cloud-computing platform. The combination of methods and approaches in GEE facilitated the rapid classification of more accurate irrigated areas with petabyte volume big-data. The developed, cultivated areas dataset has an overall accuracy of over 88%. The enhanced outputs of the irrigated area mapping are essential for policy and decision-makers to assess vast and complex irrigation systems' performance in detail. They are critical for accurate monitoring of irrigation activities from the field to transboundary or national scales.

Supplementary Materials: The following are available online at https://www.mdpi.com/2072-4292/13/5/876/s1, Code S1: Classification of land use using Sentinel, and Extracting monthly average NDVI.

Author Contributions: Conceptualization, J.M., L.N. and T.M.; methodology, J.M. and L.N.; validation, L.N., S.M. and T.M.; formal analysis, J.M.; investigation, L.N.; resources, J.M. and T.M.; data curation, J.M. and L.N.; writing—original draft preparation, L.N.; writing—review and editing, S.M. and T.M..; visualization, J.M. and L.N.; supervision, S.M.; project administration, S.M.; funding acquisition, J.M. All authors have read and agreed to the published version of the manuscript.

Funding: The Tshwane University of Technology funded this research and the APC. The Water Research Commission's (WRC) Research, Development and Innovation Branch is acknowledged for funding part of the study. This research was funded, in part, by the Wellcome Trust [Grant number 205200/Z/16/Z]. For the purpose of open access, the author has applied a CC BY public copyright licence to any Author Accepted Manuscript version arising from this submission.

Institutional Review Board Statement: Not applicable.

Informed Consent Statement: Not applicable.

Data Availability Statement: Publicly available datasets were analysed in this study. This data can be found here: [https://developers.google.com/earth-engine/datasets/catalog/].

Acknowledgments: The authors would like to acknowledge the support received from Department of Agriuclture, Land Reform and Rural Development (DALRRD for providing background maps used in accuracy assessmnet.

Conflicts of Interest: The authors declare no conflict of interest.

References

1. Kuil, L.; Carr, G.; Prskawetz, A.; Salinas, J.L.; Viglione, A.; Blöschl, G. Learning from the Ancient Maya: Exploring the Impact of Drought on Population Dynamics. *Ecol. Econ.* **2019**, *157*, 1–16. [CrossRef]
2. Tian, J.; Bryksa, B.C.; Yada, R.Y. Feeding the world into the future–food and nutrition security: The role of food sci-ence and technology. *Front. Life Sci.* **2016**, *9*, 155–166. [CrossRef]
3. Nhamo, L.; Mathcaya, G.; Mabhaudhi, T.; Nhlengethwa, S.; Nhemachena, C.; Mpandeli, S. Cereal Production Trends under Climate Change: Impacts and Adaptation Strategies in Southern Africa. *Agriculture* **2019**, *9*, 30. [CrossRef]

4. Cosgrove, W.J.; Loucks, D.P. Water management: Current and future challenges and research directions. *Water Resour. Res.* **2015**, *51*, 4823–4839. [CrossRef]
5. Mo, X.-G.; Hu, S.; Lin, Z.-H.; Liu, S.-X.; Xia, J. Impacts of climate change on agricultural water resources and adaptation on the North China Plain. *Adv. Clim. Chang. Res.* **2017**, *8*, 93–98. [CrossRef]
6. Cai, X.; Magidi, J.; Nhamo, L.; van Koppen, B. *Mapping Irrigated Areas in the Limpopo Province, South Africa*; 9290908513; International Water Management Institute (IWMI): Colombo, Sri Lanka, 2017; p. 42.
7. Nhemachena, C.; Nhamo, L.; Matchaya, G.; Nhemachena, C.R.; Muchara, B.; Karuaihe, S.T.; Mpandeli, S. Climate Change Impacts on Water and Agriculture Sectors in Southern Africa: Threats and Opportunities for Sustainable Development. *Water* **2020**, *12*, 2673. [CrossRef]
8. Mancosu, N.; Snyder, R.L.; Kyriakakis, G.; Spano, D. Water Scarcity and Future Challenges for Food Production. *Water* **2015**, *7*, 975–992. [CrossRef]
9. Teluguntla, P.; Thenkabail, P.; Oliphant, A.; Xiong, J.; Gumma, M.K.; Congalton, R.G.; Yadav, K.; Huete, A. A 30-m landsat-derived cropland extent product of Australia and China using random forest machine learning algorithm on Google Earth Engine cloud computing platform. *ISPRS J. Photogramm. Remote. Sens.* **2018**, *144*, 325–340. [CrossRef]
10. Mpandeli, S.; Nhamo, L.; Moeletsi, M.; Masupha, T.; Magidi, J.; Tshikolomo, K.; Liphadzi, S.; Naidoo, D.; Mabhaudhi, T. Assessing climate change and adaptive capacity at local scale using observed and remotely sensed data. *Weather. Clim. Extrem.* **2019**, *26*, 100240. [CrossRef]
11. Meier, J.; Zabel, F.; Mauser, W. A global approach to estimate irrigated areas—A comparison between different data and statistics. *Hydrol. Earth Syst. Sci.* **2018**, *22*, 1119–1133. [CrossRef]
12. Angelakιs, A.N.; Zaccaria, D.; Krasilnikoff, J.; Salgot, M.; Bazza, M.; Roccaro, P.; Jimenez, B.; Kumar, A.; Yinghua, W.; Baba, A.; et al. Irrigation of World Agricultural Lands: Evolution through the Millennia. *Water* **2020**, *12*, 1285. [CrossRef]
13. Siddiqui, S.; Cai, X.; Chandrasekharan, K. *Irrigated Area Map Asia and Africa (International Water Management Institute*; International Water Management Institute (IWMI): Colombo, Sri Lanka, 2016.
14. Siebert, S.; Henrich, V.; Frenken, K.; Burke, J. *Update of the Digital Global Map of Irrigation Areas to Version 5*; Food and Agriculture Organization of the United Nations (FAO): Rome, Italy, 2013; p. 117.
15. Siebert, S.; Kummu, M.; Porkka, M.; Döll, P.; Ramankutty, N.; Scanlon, B.R. A global data set of the extent of irri-gated land from 1900 to 2005. *Hydrol. Earth Syst. Sci.* **2015**, *19*, 1521–1545. [CrossRef]
16. Portmann, F.T.; Siebert, S.; Döll, P. MIRCA2000-Global monthly irrigated and rainfed crop areas around the year 2000: A new high-resolution data set for agricultural and hydrological modeling. *Glob. Biogeochem. Cycles* **2010**, *24*, 24. [CrossRef]
17. Thenkabail, P.; GangadharaRao, P.; Biggs, T.; Krishna, M.; Turral, H. Spectral matching techniques to determine historical land-use/land-cover (LULC) and irrigated areas using time-series 0.1-degree AVHRR pathfinder datasets. *Photogramm. Eng. Remote. Sens.* **2007**, *73*, 1029–1040.
18. Nhamo, L.; Magidi, J.; Nyamugama, A.; Clulow, A.D.; Sibanda, M.; Chimonyo, V.G.P.; Mabhaudhi, T. Prospects of Improving Agricultural and Water Productivity through Unmanned Aerial Vehicles. *Agriculture* **2020**, *10*, 256. [CrossRef]
19. Amani, M.; Ghorbanian, A.; Ahmadi, S.A.; Kakooei, M.; Moghimi, A.; MirMazloumi, S.M.; Moghaddam, S.H.A.; Mahdavi, S.; Ghahremanloo, M.; Parsian, S.; et al. Google Earth Engine Cloud Computing Platform for Remote Sensing Big Data Applications: A Comprehensive Review. *IEEE J. Sel. Top. Appl. Earth Obs. Remote. Sens.* **2020**, *13*, 5326–5350. [CrossRef]
20. Xiong, J.; Thenkabail, P.S.; Tilton, J.C.; Gumma, M.K.; Teluguntla, P.; Oliphant, A.; Congalton, R.G.; Yadav, K.; Gorelick, N. Nominal 30-m Cropland Extent Map of Continental Africa by Integrating Pixel-Based and Object-Based Algorithms Using Sentinel-2 and Landsat-8 Data on Google Earth Engine. *Remote. Sens.* **2017**, *9*, 1065. [CrossRef]
21. Basukala, A.K.; Oldenburg, C.; Schellberg, J.; Sultanov, M.; Dubovyk, O. Towards improved land use mapping of irrigated croplands: Performance assessment of different image classification algorithms and approaches. *Eur. J. Remote. Sens.* **2017**, *50*, 187–201. [CrossRef]
22. Nhamo, L.; Van Dijk, R.; Magidi, J.; Wiberg, D.; Tshikolomo, K. Improving the accuracy of remotely sensed irrigated areas using post-classification enhancement through UAV capability. *Remote. Sens.* **2018**, *10*, 712. [CrossRef]
23. Abdi, A.M. Land cover and land use classification performance of machine learning algorithms in a boreal land-scape using Sentinel-2 data. *GISci. Remote. Sens.* **2020**, *57*, 1–20. [CrossRef]
24. Du Plessis, A. Current Water Quality Risk Areas for Limpopo, Olifants and the Inkomati-Usuthu WMAs. In *Water as an Inescapable Risk*; Springer: Berlin/Heidelberg, Germany, 2019; pp. 183–212.
25. Denby, K.; Movik, S.; Mehta, L.; van Koppen, B. The'trickle down'of IWRM: A case study of local-level realities in the Inkomati Water Management Area, South Africa. *Water Alternatives* **2016**, *9*, 473–492.
26. Riddell, E.S.; Govender, D.; Botha, J.; Sithole, H.; Petersen, R.M.; Shikwambana, P. Pollution impacts on the aquatic ecosystems of the Kruger National Park, South Africa. *Sci. Afr.* **2019**, *6*, 00195. [CrossRef]
27. Gorelick, N.; Hancher, M.; Dixon, M.; Ilyushchenko, S.; Thau, D.; Moore, R. Google Earth Engine: Planetary-scale geospatial analysis for everyone. *Remote. Sens. Environ.* **2017**, *202*, 18–27. [CrossRef]
28. Sheykhmousa, M.; Mahdianpari, M.; Ghanbari, H.; Mohammadimanesh, F.; Ghamisi, P.; Homayouni, S. Support Vector Machine vs. Random Forest for Remote Sensing Image Classification: A Meta-analysis and systematic review. *IEEE J. Sel. Top. Appl. Earth Obs. Remote. Sens.* **2020**, *13*, 18. [CrossRef]

9. Berhane, T.M.; Lane, C.R.; Wu, Q.; Autrey, B.C.; Anenkhonov, O.A.; Chepinoga, V.V.; Liu, H. Decision-Tree, Rule-Based, and Random Forest Classification of High-Resolution Multispectral Imagery for Wetland Mapping and Inventory. *Remote. Sens.* **2018**, *10*, 580. [CrossRef]
30. CEC. *Field Crop Boundary Data Layer 2019*; Crop Estimates Consortium (CEC), Department of Agriculture, Land Reform and Rural Development (DALRRD): Pretoria, South Africa, 2019; Volume 2019.
31. GTI. *2013–2014 South African National Land-Cover Dataset: Data User Report and Metadata*; GeoTerraImage (GTI), Department of Environmental Affairs (DEA): Pretoria, South Africa, 2015.
32. Breiman, L. Random forests. *Mach. Learn.* **2001**, *45*, 5–32. [CrossRef]
33. Zeferino, L.B.; De Souza, L.F.T.; Amaral, C.H.D.; Filho, E.I.F.; De Oliveira, T.S. Does environmental data increase the accuracy of land use and land cover classification? *Int. J. Appl. Earth Obs. Geoinf.* **2020**, *91*, 102128. [CrossRef]
34. Toosi, N.B.; Soffianian, A.R.; Fakheran, S.; Pourmanafi, S.; Ginzler, C.; Waser, L.T. Comparing different classifica-tion algorithms for monitoring mangrove cover changes in southern Iran. *Glob. Ecol. Conserv.* **2019**, *19*, e00662. [CrossRef]
35. Hatwell, J.; Gaber, M.M.; Azad, R.M.A. CHIRPS: Explaining random forest classification. *Artif. Intell. Rev.* **2020**, *53*, 5747–5788. [CrossRef]
36. Schmidt, J.; Marques, M.R.; Botti, S.; Marques, M.A. Recent advances and applications of machine learning in solid-state materials science. *NPJ Comput. Mater.* **2019**, *5*, 1–36. [CrossRef]
37. Da Silva, V.S.; Salami, G.; Da Silva, M.I.O.; Silva, E.A.; Junior, J.J.M.; Alba, E. Methodological evaluation of vegetation indexes in land use and land cover (LULC) classification. *Geol. Ecol. Landsc.* **2019**, *4*, 159–169. [CrossRef]
38. Nhamo, L.; Ebrahim, G.Y.; Mabhaudhi, T.; Mpandeli, S.; Magombeyi, M.; Chitakira, M.; Magidi, J.; Sibanda, M. An assessment of groundwater use in irrigated agriculture using multi-spectral remote sensing. *Phys. Chem. Earth, Parts A/B/C* **2020**, *115*, 102810. [CrossRef]
39. Hoare, D.; Frost, P. Phenological description of natural vegetation in southern Africa using remotely-sensed vegetation data. *Appl. Veg. Sci.* **2004**, *7*, 19–28. [CrossRef]
40. Ambika, A.K.; Wardlow, B.; Mishra, V. Remotely sensed high resolution irrigated area mapping in India for 2000 to 2015. *Sci. Data* **2016**, *3*, 160118. [CrossRef]
41. Sujariya, S.; Jongrungklang, N.; Jongdee, B.; Inthavong, T.; Budhaboon, C.; Fukai, S. Rainfall variability and its ef-fects on growing period and grain yield for rainfed lowland rice under transplanting system in Northeast Thailand. *Plant Prod. Sci.* **2020**, *23*, 48–59. [CrossRef]
42. Adole, T.; Dash, J.; Atkinson, P.M. Large-scale prerain vegetation green-up across Africa. *Glob. Chang. Biol.* **2018**, *24*, 4054–4068. [CrossRef] [PubMed]
43. Lee, K.; Kim, K.; Lee, S.-G.; Kim, Y. Determination of the Normalized Difference Vegetation Index (NDVI) with Top-of-Canopy (TOC) Reflectance from a KOMPSAT-3A Image Using Orfeo ToolBox (OTB) Extension. *ISPRS Int. J. Geo-Inf.* **2020**, *9*, 257. [CrossRef]
44. StatsSA. *Statistics South Africa Releases Census of Commercial Agriculture 2017 Report*; Statistics South Africa (StatsSA): Pretoria, South Africa, 2020.
45. NDP. *National Development Plan: Vision 2030*; National Planning Commission (NDP): Pretoria, South Africa, 2013; p. 489.
46. Van Averbeke, W.; Denison, J.; Mnkeni, P. Smallholder irrigation schemes in South Africa: A review of knowledge generated by the Water Research Commission. *Water SA* **2011**, *37*, 797–808. [CrossRef]
47. Fanadzo, M.; Ncube, B. Challenges and opportunities for revitalising smallholder irrigation schemes in South Africa. *Water SA* **2018**, *44*, 436–447. [CrossRef]
48. Vanschoenwinkel, J.; Van Passel, S. Climate response of rainfed versus irrigated farms: The bias of farm heteroge-neity in irrigation. *Clim. Chang.* **2018**, *147*, 225–234. [CrossRef]
49. Nhamo, L.; Mabhaudhi, T.; Magombeyi, M. Improving Water Sustainability and Food Security through Increased Crop Water Productivity in Malawi. *Water* **2016**, *8*, 411. [CrossRef]
50. Mabhaudhi, T.; Mpandeli, S.; Nhamo, L.; Chimonyo, V.G.P.; Senzanje, A.; Naidoo, D.; Liphadzi, S.; Modi, A.T. Emerging Water-Energy-Food Nexus Lessons, Experiences, and Opportunities in Southern Africa. In *Environmental Management of Air, Water, Agriculture, and Energy*; CRC Press: Boca Raton, FL, USA, 2020; pp. 141–158.
51. Mabhaudhi, T.; Mpandeli, S.; Nhamo, L.; Chimonyo, V.G.P.; Nhemachena, C.; Senzanje, A.; Naidoo, D.; Modi, A.T. Prospects for Improving Irrigated Agriculture in Southern Africa: Linking Water, Energy and Food. *Water* **2018**, *10*, 1881. [CrossRef]
52. Mpandeli, S.; Naidoo, D.; Mabhaudhi, T.; Nhemachena, C.; Nhamo, L.; Liphadzi, S.; Hlahla, S.; Modi, A.T. Climate Change Adaptation through the Water-Energy-Food Nexus in Southern Africa. *Int. J. Environ. Res. Public Health* **2018**, *15*, 2306. [CrossRef] [PubMed]
53. Nhamo, L.; Ndlela, B. Nexus planning as a pathway towards sustainable environmental and human health post Covid-19. *Environ. Res.* **2021**, *192*, 110376. [CrossRef] [PubMed]
54. Nhamo, L.; Ndlela, B.; Nhemachena, C.; Mabhaudhi, T.; Mpandeli, S.; Matchaya, G. The Water-Energy-Food Nexus: Climate Risks and Opportunities in Southern Africa. *Water* **2018**, *10*, 567. [CrossRef]
55. Nhamo, L.; Mabhaudhi, T.; Mpandeli, S.; Dickens, C.; Nhemachena, C.; Senzanje, A.; Naidoo, D.; Liphadzi, S.; Modi, A.T. An integrative analytical model for the water-energy-food nexus: South Africa case study. *Environ. Sci. Policy* **2020**, *109*, 15–24. [CrossRef]

56. Gudmann, A.; Csikós, N.; Szilassi, P.; Mucsi, L. Improvement in Satellite Image-Based Land Cover Classification with Landscape Metrics. *Remote. Sens.* **2020**, *12*, 3580. [CrossRef]
57. Jin, Y.; Liu, X.; Chen, Y.; Liang, X. Land-cover mapping using Random Forest classification and incorporating NDVI time-series and texture: A case study of central Shandong. *Int. J. Remote. Sens.* **2018**, *39*, 8703–8723. [CrossRef]

MDPI AG
Grosspeteranlage 5
4052 Basel
Switzerland
Tel.: +41 61 683 77 34

Remote Sensing Editorial Office
E-mail: remotesensing@mdpi.com
www.mdpi.com/journal/remotesensing

www.ingramcontent.com/pod-product-compliance
Lightning Source LLC
Chambersburg PA
CBHW061821210926
51850CB00006B/770
* 9 7 8 3 7 2 5 8 5 3 2 9 8 *